MECHANISMS & MECHANICAL DEVICES SOURCEBOOK

OTHER McGRAW-HILL BOOKS OF INTEREST

Baumeister & Marks   *Marks' Standard Handbook for Mechanical Engineers*

Brady & Clauser   *Materials Handbook*

Ettlie & Stoll   *Design for Manufacturing and Assembly*

Ganic & Hicks   *The McGraw-Hill Handbook of Essential Engineering Information and Data*

Harris   *Handbook of Noise Control*

Harris & Crede   *Shock and Vibration Handbook*

Hicks   *Standard Handbook of Engineering Calculations*

Juran & Gryna   *Juran's Quality Control Handbook*

Kamm   *Designing Cost-Efficient Mechanisms*

Karassik et al.   *Pump Handbook*

Parmley   *Standard Handbook of Fastening and Joining*

Rohsenow, Hartnett,
  & Ganic   *Handbook of Heat Transfer Fundamentals*

Rohsenow, Hartnett,
  & Ganic   *Handbook of Heat Transfer Applications*

Shigley & Mischke   *Standard Handbook of Machine Design*

Tuma   *Engineering Mathematics Handbook*

Tuma   *Handbook of Numerical Calculations in Engineering*

Wadsworth   *Handbook of Statistical Methods for Engineering and Science*

Young   *Roark's Formulas for Stress and Strain*

NICHOLAS P. CHIRONIS

MECHANISMS & MECHANICAL DEVICES SOURCEBOOK

McGraw-Hill, Inc.
New York   St. Louis   San Francisco   Auckland   Bogotá
Caracas   Hamburg   Lisbon   London   Madrid
Mexico   Milan   Montreal   New Delhi   Paris
San Juan   São Paulo   Singapore
Sydney   Tokyo   Toronto

1 2 3 4 5 6 7 8 9 0   HAL/HAL   9 7 6 5 4 3 2 1

ISBN 0-07-010918-4

*The sponsoring editor for this book was Harold B. Crawford, the
editing supervisor was Fred Bernardi, the designer was Ron Lane,
and the production supervisor was Thomas G. Kowalczyk.*

Printed and bound by Arcata Graphics/Halliday.

CONTENTS

Chapter 8. TORQUE-LIMITING, TENSIONING AND GOVERNING DEVICES 335

Chapter 9. NONMECHANICAL METHODS OF MACHINE AND MECHANISM CONTROL 365

Chapter 10. FASTENING, LATCHING, CLAMPING AND CHUCKING DEVICES   399

Chapter 11. KEY EQUATIONS AND CHARTS FOR DESIGNING MECHANISMS   415

445

PREFACE

The classical view of an inventor is that of a bushy-haired individual who conceives a solution out of thin air, relying on a sort of divine revelation. But in reality, when a designer is confronted with a problem, he or she generally draws upon past experiences with such problems and on collections of books and published descriptions and illustrations of various devices and concepts. In other words, a knowledge of what has already been conceived is an excellent starting point. "Let's not reinvent the wheel" is a commonly heard criticism in the design fields, and it is a good one. Find out first what solutions have already been conceived, and then put your mind to work in improving or modifying them.

This book, therefore, brings together extensive compilations of modern mechanisms, classical linkages, and machinery devices that provide a wide variety of motions and functions. Drawn largely from product-design and design-engineering publications, and containing over 2,000 illustrations, the book also displays how mechanical components — such as mechanisms, linkages, cams, variable-speed drives, gears, clutches, brakes, belts, flexures, chucks, clamps and springs — can be combined successfully with electrical, hydraulic, pneumatic, optical, thermal, and photoelectric devices to perform complex tasks. Such integrated designs are being used in automatic production machines, aerospace technology, vehicles and instrumentation, as well as in numerous consumer and commercial products; in fact, whenever some motion or action is to be worked into a design or product.

Simply stated, there is no other book that can match the sheer scope of this extensive work, both as a first-step, idea-initiator, as well as for design. The book, therefore, should prove invaluable to design engineers, production engineers, machine and product designers, draftspersons and inventors, as well as to instructors and students of mechanical design.

Nicholas P. Chironis

MECHANISMS & MECHANICAL DEVICES SOURCEBOOK

CHAPTER 1.

PARTS-HANDLING MECHANISMS

Orienting parts for assembly

**In mechanical assembly, you first have to capture
the part, then perhaps rotate it to a specific position.
Here are some ideas on how to do the job mechanically**

Hopper and bowl feed tracks have been developed to feed parts of almost any possible shape and size from a bulk container and down a track. But often the part must be reoriented, or at least checked, which means it must be freed from a confining track. At the same time, the mechanism must prevent the part from becoming lost.

The devices that assembly engineers dream up to handle this kind of manipulation are often clever, and are frequently as simple as possible (hopefully) so that they will maintain high reliability.

Usually, a small part has some minor irregularity that marks one end from the other, and almost always the orienting mechanism has to work with this difference, no matter how slight. Sometimes one end weighs more than the other, or one end is cupped while the other is rounded, or perhaps the difference is simply geometric shape. The following examples indicate some of the possible approaches.

Weight or drag

The first example really depends on gravity, but it is the geometry of the part that makes the orienting device work.

The component being oriented (Fig. 1) has a head diameter which is only 0.050 mm to 0.066 mm greater in diameter than the body of the component and difficulty was experienced in feeding components with the required orientation of the head leading.

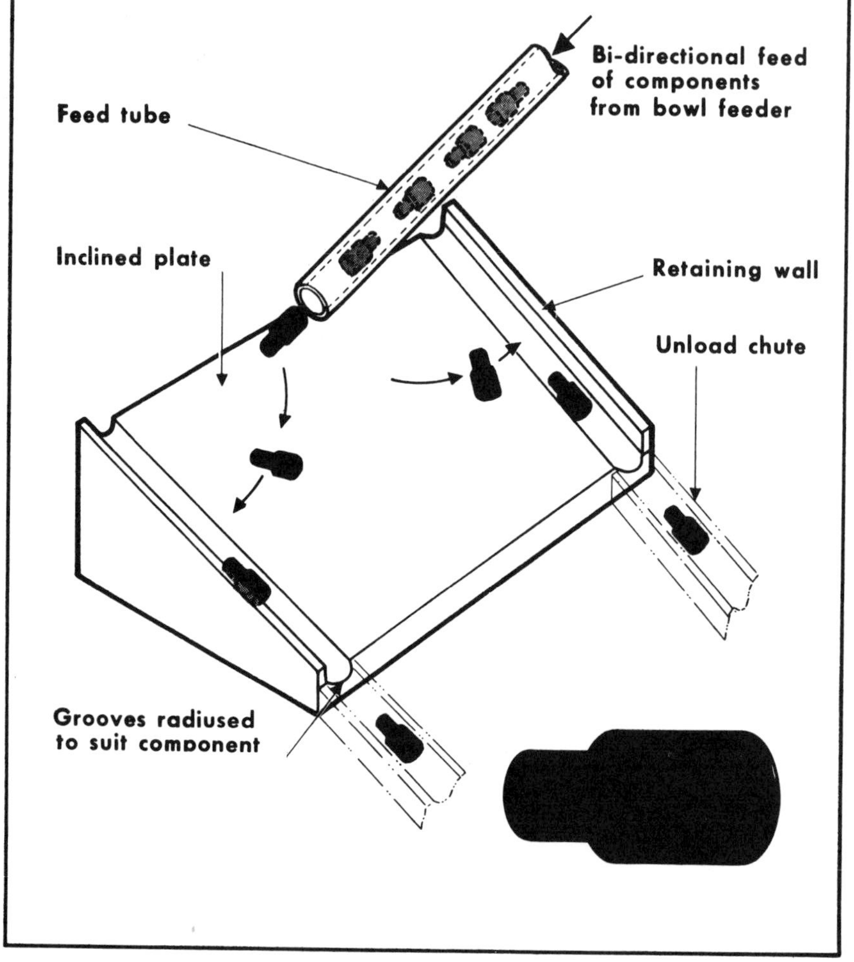

The components, fed axially from a conventional bowl feeder, are randomly disposed end to end, and are discharged onto an inclined plate whose gradient is normal to the axis of feeding. Along each sloping edge of the plate a groove is formed, and as the components roll by gravity down the inclined plate, they tend to take an arcuate path, terminating with their heads downwards. They locate into

Excerpted by American Machinist from material gathered by The Institution of Production Engineers, London UK.

Fig. 1

Feed tube

Bi-directional feed of components from bowl feeder

Inclined plate

Retaining wall

Unload chute

Grooves radiused to suit component

one or other of the two grooves in the edges of the plate, so they then feed down in parallel lines.

With this simple method, it is essential that the center of gravity of the stepped component be located within the smaller diameter, so that the component in rolling down the incline has its direction monitored by differing diameters. Taper pins can be handled with the same tooling, as well as a broad variety of flanged components.

Components which are cylindrical with a short flat at one end (Fig. 2) are fed through a feed chute with their major axes in a horizontal plane. They

fall into a tank containing a suitable liquid.

As the components fall through the liquid, the flat section end has a higher viscous drag, thus causing that end to lag.

A funnel-shaped chute is positioned beneath the surface of the liquid so that the components fall into it correctly oriented. The sides of the mouth of this chute are tapered into a diameter which prevents any change of the orientation.

The chute bends radially so that the components are discharged horizontally, to locate on a toothed conveyor

Orienting parts for assembly

belt running across the discharge orifice of the chute.

A convenient method of obtaining the toothed belt is to turn a standard timing belt of appropriate pitch inside out, giving a conveyor with 'built in' driving dogs.

Second jobs for liquids

The liquid used in the tank may be selected to perform a secondary function on the component. For example, if the component is required to be degreased, a solvent such as carbon tetraehloride may be used, whereas a rust inhibitor or light oil can be used in applications for a component where the subsequent assembly will benefit

from such a treatment. More viscous liquids, of course, will slow the action.

Parts with a high length-to-diameter, ratio, having a small hole in one end only, must be probed. The device can be arranged to feed the components with the hole trailing or leading.

The components (Fig. 3) are fed by rolling singly along a feed chute and come to rest in the 'vee' formed by a pair of movable jaws. A sensing device is positioned at one end of the vee and, depending upon the sensor signal determined by the position of the hole, one or the other of the moving jaws is caused to retract, thus allowing the component to drop into a split exit chute.

As the component falls into the chute, it strikes one or two orienting blocks positioned on either side of the center plate. The blocks can be positioned in one of two ways. Either each block is positioned so as to contact the component at the end containing the hole, thus causing the component to pivot and fall with the hole in a trailing position, or each block is positioned to contact the component at the end without the hole.

The two sections of the exit chute may be combined into one below the center plate. A suitable escapement mechanism may be incorporated in the feed chute near the exit to insure that the components feed to the orienting mechanism singly. This type of chute must be kept full to prevent the possibility of the component feeding 'end on.'

In some cases, the components (Fig. 4) must be presented at the assembly station with the blind hole trailing. The orienting sequence commences when the component carried in the work plate is indexed to the vertical position directly beneath the aperture in the face of the orienting plate.

A probe assembly consisting of an outer sleeve, a tipped probe, and a spring-loaded clamping jaw, descends through the aperture in the orienting plate and probes the end face of the component.

If the hole in the component is presented, the probe tip enters and allows the spring-loaded jaw to grip on the outside walls. The probe assembly then retracts, carrying the component with it until a position is reached

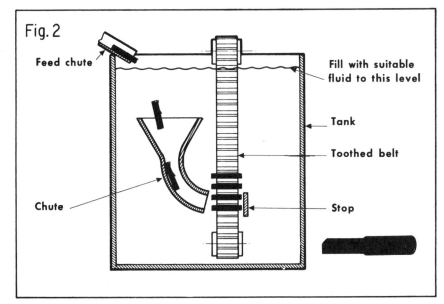

Fig. 2

Feed chute

Fill with suitable fluid to this level

Tank

Toothed belt

Chute

Stop

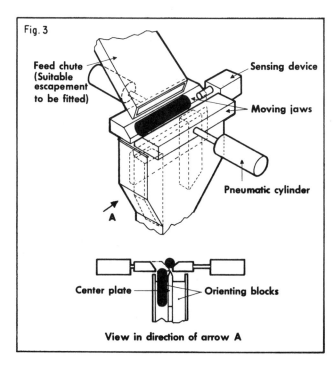

Fig. 3

Feed chute (Suitable escapement to be fitted)

Sensing device

Moving jaws

Pneumatic cylinder

A

Center plate — Orienting blocks

View in direction of arrow A

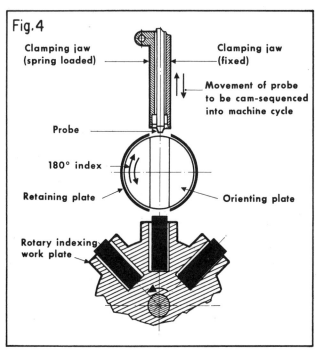

Fig. 4

Clamping jaw (spring loaded)

Clamping jaw (fixed)

Movement of probe to be cam-sequenced into machine cycle

Probe

180° index

Retaining plate

Orienting plate

Rotary indexing work plate

where the probe tip is just clear of the top of the orienting plate.

A cam arrangement (not shown on the drawing) pries the spring loaded jaw open, thus releasing the component, which drops clear of the probe and withdraws into the orienting plate.

The indexing of the orienting plate is sequenced by camming to the probe assembly movement. The component thus reenters the machine work plate location in a reverse position to that previously assumed. The orienting plate subsequently reverses through 180°, returning to its initial position.

If the hole in the component is not presented to the probe, the probe terminates its descent when contact is made with the closed end of the component. The spring-loaded jaw operates ineffectually and the probe assembly withdraws, leaving the already correctly oriented component undisturbed. Feedrate achieved with this component is 240 parts an hour.

Peelers and catchers

A different mechanism might be designed to feed cylindrical shapes having a large hole in one end (Fig. 5). In this case, the components are fed continuously down a vertically mounted feed chute. At the base of the chute one component at a time is released by the reciprocating action of a probe. The probe height and diameter are such that it will enter freely into the hole in the lower component.

If the closed end of the component faces the probe, the component is pushed forward beyond the edge of the funnel-shaped chute until it falls naturally with the open end leading. It will be guided in falling in the desired orientation by contacting the wedge surface of a spring-mounted pawl mounted on the opposite side of the chute.

The primary action of the pawl, however, is to act as a stripper. If the open end of the component faces the probe, the probe enters the hole and in its traverse carries the component across the entrance of the exit chute until it rests with its closed end on the lip located on the further side.

As the component is carried forward on the probe it deflects the pawl, which springs back eventually into position behind the component. When the probe retracts, the component is stripped by the retaining action of the pawl. The lip of the exit chute supports the closed end of the component, thus guiding the component to fall with its open end leading. Feedrate achieved with this component is 2100 per hour.

Another variation on this basic idea (Fig. 6) is shown with a shorter,

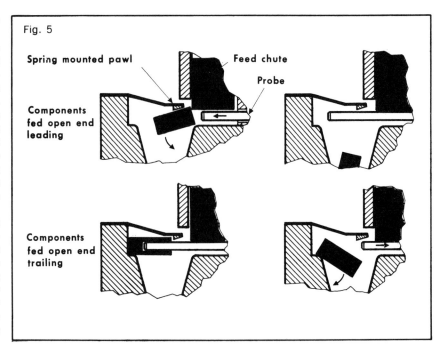

Fig. 5

Spring mounted pawl   Feed chute
Probe

Components fed open end leading

Components fed open end trailing

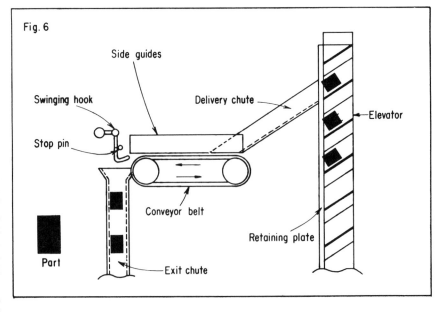

Fig. 6

Side guides
Swinging hook   Delivery chute
Elevator
Stop pin
Conveyor belt
Retaining plate
Part   Exit chute

chunkier part. The components feed horizontally through a chute, but are moved by a short conveyor belt. At the end of the belt is a counterbalanced hook. The chute retains the components in line until they either fall free or are picked up by the hook.

If the hole end is forward, the part slides onto the hook, conveyor motion carries the part forward, and the hook swings back. The hook end must be so designed that the part (as soon as it is free from the conveyor and chute) slides off and drops down the exit chute with the hole end trailing. This may require some trials to get the shape just right.

If the part comes along the conveyor with the hole end trailing, of course, it merely drops down off the conveyor into the exit chute with the hole end still trailing.

Self-keying

If the components are relatively simple, or if the difference between ends is quite obvious, it may be possible to design a mechanism that lets the part select its own action (Fig. 7). In this case the cylindrical part, with one end open and the other end rounded, is fed down a track into a reversing mechanism.

If the part drops down open-end leading, the end of the part will key into the plug at the bottom of the yoke. When the yoke rotates, it rotates the inner fixture and the part, which drops down with the open end trailing after the yoke has rotated 180°.

Orienting parts for assembly

A second gate below the orienting gate does simple reversing; this is a common way to turn a part around once it has been oriented. In many cases, as in this one, it is simpler to orient parts into a particular direction. If what's actually needed is the parts traveling in the opposite direction, then a reversing gate of some kind is required.

Another self-keying mechanism is illustrated in Fig. 8. This mechanism is based on the fact that the part is simply a cylinder with a small stepped diameter on one end to make the part choose its own orientation.

As the parts drop down through a chute, they fall one at a time into a nest in a reciprocating slide. If the part has the small end down, this end projects through the central portion of the slide and engages a slot in the outer sleeve of the slide. As the slide moves to the right, the outer sleeve is automatically rotated 180° by a pin and helical track. Since the inner slide is keyed by the part to the outer sleeve, the part must be rotated.

As it comes over the escape chute, it drops through small end trailing. If the part originally came down small end trailing, the small end would lie in a clearance space in the inner slide, not keyed to the outer sleeve, which would then rotate as before, but without inverting the part. As the part

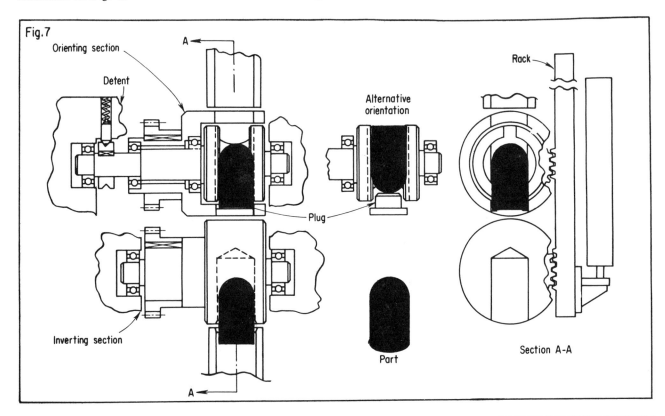

Fig.7

Orienting section

Detent

A

Alternative orientation

Rack

Plug

Inverting section

Part

Section A-A

A

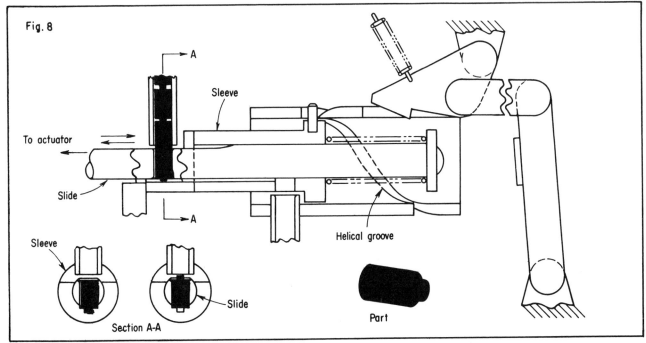

Fig. 8

A

Sleeve

To actuator

Slide

A

Helical groove

Sleeve

Slide

Section A-A

Part

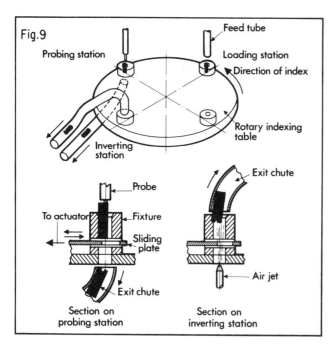

Fig.9

Probing station

Feed tube

Loading station

Direction of index

Inverting station

Rotary indexing table

Probe

To actuator

Fixture

Sliding plate

Exit chute

Exit chute

Air jet

Section on probing station

Section on inverting station

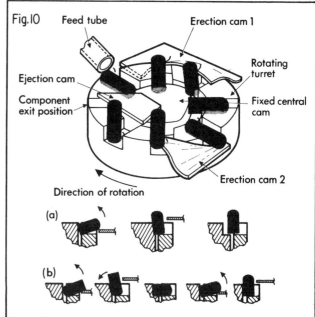

Fig.10

Feed tube

Erection cam 1

Rotating turret

Fixed central cam

Ejection cam

Component exit position

Direction of rotation

Erection cam 2

(a)

(b)

comes over the escape chute, it drops through as before.

Detectors

Something different is necessary for components such as slotted or socket setscrews (Fig. 9). Here, allowance must be made for the slot or socket to be fed in the leading or trailing end of the component.

Equi-spaced around a rotary indexing table are a series of fixtures (four in this instance), and a single component is deposited into each fixture as the table indexes beneath the feed tube.

At the second station, the end face of the component is probed to ascertain whether or not the slot or socket is uppermost. In this case it is necessary to feed the component with this feature trailing.

If the travel of the probe indicates that the slot or socket is uppermost, a sliding plate in the fixture is moved so that a clearance hole in this plate comes into line with the component location in the fixture. This allows the component to fall through the fixture into an exit chute.

If the travel of the probe indicates that the feature in the component face is not uppermost, then the component passes to the next indexed position, where a jet of air from beneath the table blows the component with the slot or socket trailing upwards out of the fixture into an overhead subsidiary exit chute. This subsidiary exit chute is combined later with the other exit chute, or it may be led to another assembly station.

The sliding plate may be operated by an actuator mounted on the ma-chine frame and should be spring-loaded to return to the original position. Feedrate achieved with this component is 3600 per hour.

A special mechanism is necessary to handle fragile components having a general cylindrical configuration terminating in a spherical shape. In this example (Fig. 10), the components are required to be fed in a vertical attitude with the radiused end located uppermost.

The components feed randomly from a hopper feeder through a feed tube. A suitable escapement is incorporated in the feed tube so that one component at a time is released into a constantly rotating turret. This turret head has a series of suitably shaped radial slots equispaced around its upper face as illustrated on the drawing.

On the inside of the rotating turret is a fixed cam which serves to locate the component inside its respective slot in the turret.

As the turret rotates, the ends of the components protrude sufficiently beyond the outside diameter of the turret to enable the stationary erecting cams situated around the periphery of the turret to effect erection of the component as the turret revolves.

The orienting process is illustrated. If the component is fed with the domed end leading, both erection cams are brought into use. If the component is fed with the domed end trailing, only the first erection cam is needed to achieve orientation.

The fixed control cam also serves to station the correctly oriented component on the platform of the radial slot so that the ejection mechanism functions correctly.

Air jets can trigger actuators to reverse parts, and they are especially useful when the parts are fragile. The illustrated flat components with notched ends also had a small hole pierced in one corner, and this corner had to be trailing when the parts were pushed along a feed track. To orient the part, three air jets were employed (Fig. 11).

Two jets were located at a rotary gate. One made sure that the part was oriented with the notched corners at upper left and lower right. A second jet detected the hole in the part, if present.

If the hole was present, the rotary gate did nothing, and parts slid along the track to the right angle corner, and were pushed on down. If the air jet met resistance, the hole was therefore on the wrong side, and air back pressure tripped the rotary gate, indexing 180°. This put the part right, so it moved along with the others, in proper sequence.

A third air jet was required at the chute corner, to make sure there was a new part present before the push mechanism engaged to move parts down the final chute to subsequent assembly operations.

A similar device can be made with a photoelectric cell acting as the prime decision-making element. The illustrated long strap, with a pierced hole in one end (Fig. 12) had to be fed with the holes all at the trailing end of the part, though they come down the track randomly oriented.

The photocell is lined up at one side of the rotary gate to detect light from a lamp under the gate (through symmetrical holes in the gate). If the part

Orienting parts for assembly

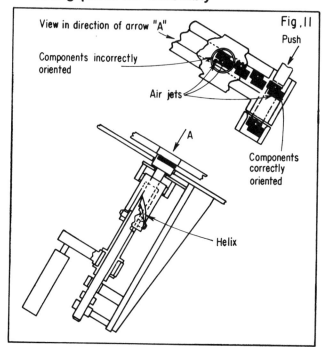

View in direction of arrow "A"

Components incorrectly oriented

Push

Air jets

Components correctly oriented

A

Helix

Fig. 11

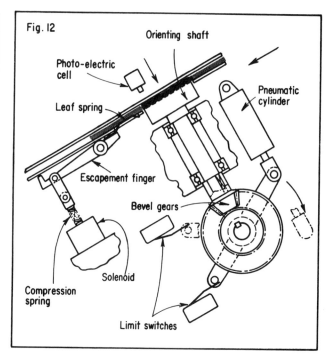

Fig. 12

Orienting shaft

Photo-electric cell

Leaf spring

Escapement finger

Bevel gears

Pneumatic cylinder

Solenoid

Compression spring

Limit switches

comes through with hole trailing, the part blocks the light, the photocell does not respond, and the part goes on through.

Note the escapement mechanism just to the left of the gate. This insures that a part is stopped in the gate in the correct position.

If a part comes through with the hole leading, light fires the photocell and the air cylinder operates through the gearing to index the gate 180° and reverse the part. Limit switches return the gate to original position after the part has escaped.

Indexing

A device for feeding plain cylindrical components or cylindrical shapes, and for changing them from vertical to horizontal is shown in Fig. 13.

The components are fed along a feed chute into one of a series of equi-pitched locations positioned at the periphery of an indexing turret. The turret is then indexed step by step so that each component in turn becomes aligned with the exit chute. A pusher bar then moves the components out.

A static block shaped similarly to a bevel wheel blank is positioned inside the profile of the indexing turret. The bevel surface of this block provides a base on which the components locate and slide as they are retained in the indexing turret before the exit chute position is reached and the actuator operates.

In the example illustrated, a bevel gear is cut on the outside of the indexing turret and this meshes with a bevel

pinion mounted on a drive shaft, the shaft being coupled to a suitable intermittent drive. A spring detent may be necessary to locate the indexing turret in the required positions for feed and discharge.

This method of orientation is not limited to 90°, as illustrated, and the direction of discharge can be varied widely by adjusting the angular position of the static block.

It may be necessary to fit a retaining plate on the outer face of the indexing turret between the inlet and outlet positions to restrain the components in their locations.

Obviously the part-orienting devices described above are not the only means for achieving the desired ends. These are merely examples, but they are examples already proved by usage. ■

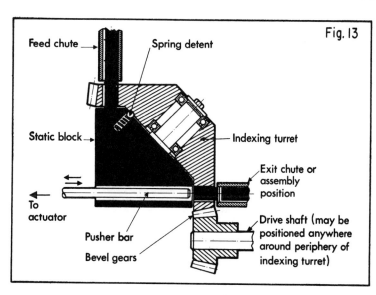

Feed chute

Spring detent

Fig. 13

Static block

Indexing turret

Exit chute or assembly position

To actuator

Pusher bar

Bevel gears

Drive shaft (may be positioned anywhere around periphery of indexing turret)

Your latest group of mechanisms to
Sort, feed, or weigh

Sooner or later, you may be faced with the need to design such devices for your plant. These 19 selections are easily modified to suit your product

NICHOLAS P. CHIRONIS

ORIENTING DEVICES

Orienting short, tubular parts

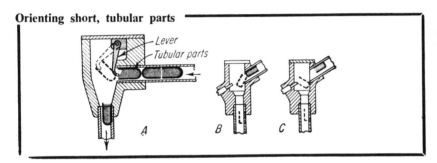

Here's a common problem: Parts come in either open-end or closed-end first; you need a device which will orient all the parts so they feed out facing the same way. In (A), when a part comes in open-end first, it is pivoted by the swinging lever so that the open end is up. When it comes in closed-end first, the part brushes away the lever to keel over head-first. Fig B and C show a simpler arrangement with pin in place of lever.

Orienting dish-like parts

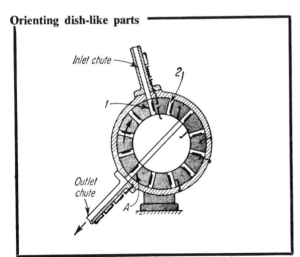

Part with open-end facing to the right (part 1) falls on to a matching projection as the indexing wheel begins to rotate clockwise. The projection retains the part for 230 deg to point *A* where it falls away from the projection to slide down the outlet chute, open-end up. An incoming part facing the other way (2) is not retained by the projection, hence slides *through* the indexing wheel so that it, too, passes through the outlet with the open-end up.

Orienting pointed-end parts

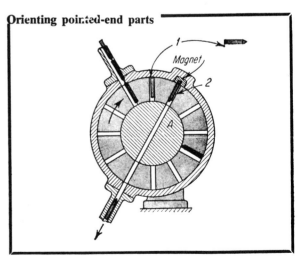

Main principle here is that the built-in magnet cannot hold on to a part as it passes by if the part has its pointed end facing the magnet. Such a correctly oriented part (part 1) will fall through the chute as the wheel indexes to a stop. An incorrectly oriented part (part 2) is briefly held by the magnet until the indexing wheel continues on past the magnet position. The wheel and the core with the slot must be made of nonmagnetic material.

Orienting U-shaped parts

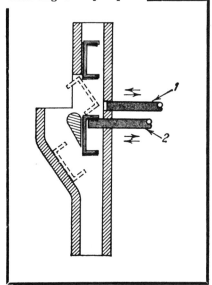

Key to this device is two pins which reciprocate one after another in the horizontal direction. The parts come down the chute with the bottom of the U facing either to the right or left. All pieces first strike and rest on pin 2. Pin 1 now moves into the passage way, and if the bottom of the U is facing to the right, the pin would kick over the part as shown by the dotted lines. If on the other hand the bottom of the U had been to the left, motion of pin 1 would have no effect, and as pin 2 withdrew to the right, the part would be allowed to pass down through the main chute.

Orienting cone-shaped parts

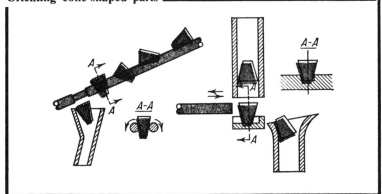

Regardless of which end of the cone faces forward as the cones slide down the cylindrical rods, the fact that both rods rotate in opposite directions causes the cones to assume the position shown in section A-A (above). When the cones reach the thinned-down section of the rods, they fall down into the chute as illustrated.

In the second method of orienting cone-shaped parts (left), if the part comes down small end first, it will fit into the recess. The reciprocating rod, moving to the right, will then kick the cone over into the exit chute. But if the cone comes down with its large end first, it sits on top of the plate (instead of inside the recess), and the rod merely pushes it into the chute without turning it over.

Orienting stepped-disk parts

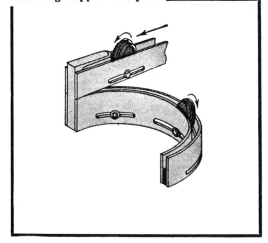

Parts rolling down the top rail to the left drop to the next rail which has a circular segment. The parts, therefore, continue to roll on in the original direction but their faces have now been rotated 180 deg. The idea of dropping one level may seem over simplified, but it avoids the use of camming devices which is the more common way of accomplishing this job.

SIMPLE FEEDING DEVICES

Feeding a fixed number of parts

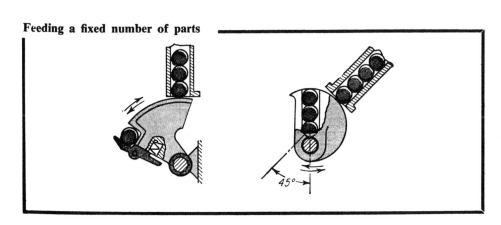

The oscillating sector picks up the desired number of parts, left diagram, and feeds them by pivoting the required number of degrees. The device for oscillating the sector must be able to produce dwells at both ends of the stroke to allow sufficient time for the parts to fall in and out of the sector.

The circular parts feed down the chute by gravity, and are separated by the reciprocating rod. The parts first roll to station 3 during the downward stroke of the reciprocator, then to station 1 during the upward stroke; hence the time span between parts is almost equivalent to the time it takes for the reciprocator to make one complete oscillation.

The device in (B) is similar to the one in (A), except that the reciprocator is replaced by an oscillating member.

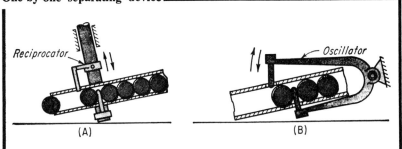

(A)   (B)

Mixing different parts together

Two counter rotating wheels form a simple device for alternating the feed of two different workpieces.

Pausing until actuated

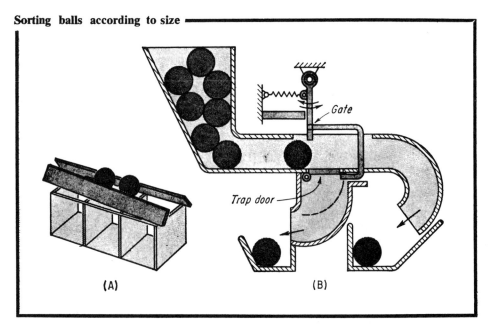

Each gear in this device is held up by a pivotable cam sector until the gear ahead of it moves forward. Thus, gear 3, rolling down the chute, kicks down its sector cam but is held up by the previous cam. When gear 1 is picked off (either manually, or mechanically), its sector cam pivots clockwise because of its own weight. This permits gear 2 to move into the place of gear 1—and frees cam 2 to pivot clockwise. Thus, all gears in the row move forward one station.

SORTING DEVICES

Sorting balls according to size

In the simple device (A) the balls run down two inclined and slightly divergent rails. The smallest balls, therefore, will fall into the left chamber, the medium-size ones into the middle-size chamber, and the largest ones into the right chamber.

In the more complicated arrangement (B) the balls come down the hopper and must pass a gate which also acts as a latch for the trapdoor. The proper-size balls pass through without touching (actuating) the gate. Larger balls, however, brush against the gate which releases the catch on the bottom of the trapdoor, and fall through into the special trough for the rejects.

(A)   (B)

Sorting according to height

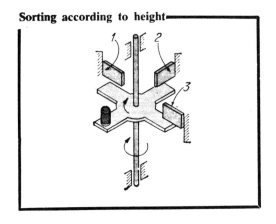

This is a simple device in which an assembly worker can place the workpiece on a slowly rotating cross-platform. Bars 1, 2, and 3 have been set at decreasing heights beginning with the highest bar (bar 1), and the lowest bar (bar 3). The workpiece is therefore knocked off the platform at either station 1, 2, or 3 depending on its height.

WEIGHT-REGULATING ARRANGEMENTS

By varying the vibration amplitudes

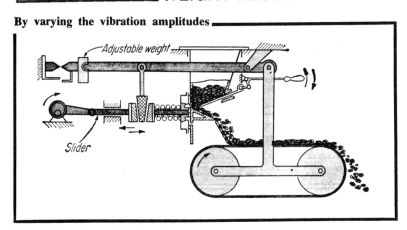

The material in the hopper is fed to a conveyor by means of vibration actuated by the reciprocating slider. The pulsating force of the slider passes through the rubber wedge and on to the actuating rod. The amplitude of this force can be varied by moving the wedge up or down. This is done automatically by making the conveyor pivot around a central point. As the conveyor becomes overloaded, it pivots clockwise to raise the wedge which reduces the amplitudes—and the feed rate of the material.

Further adjustments in feed rate can be made by shifting the adjustable weight or by changing the speed of the conveyor belt.

By linkage arrangement

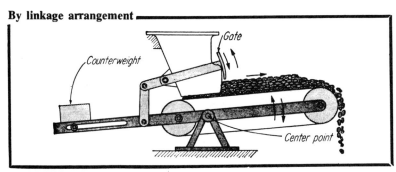

The loose material comes down the hopper and is fed to the right by the conveyor system which can pivot about the center point. The frame of the conveyor system also actuates the hopper gate, so that if the material on the belt exceeds the required amount the conveyor pivots clockwise and closes the gate. The position of the counterweight on a frame determines the feed rate of the system.

By electric-eye and balancer

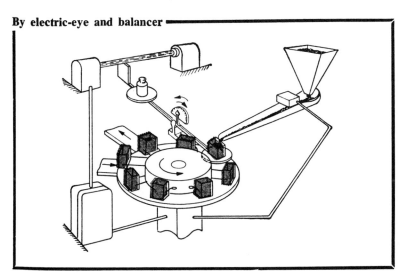

The indexing table automatically comes to a stop at the feed station. As the material drops into the container, its weight pivots the screen upward to cut off the operation of the photocell relay. This in turn shuts the feed gate. Reactuation of the indexing table can be automatic after a time interval, or by the cutoff phase of the electric eye.

More selections of
Machinery mechanisms

Clamping and cutting device

Pressing the foot pedal of this ingenious device down causes the top knife and the clamp to move downward. However, when the clamp presses on to the material it (and link *EDO*) will not be able to move any further, Link *AC* will now begin to pivot around point *B*. This draws up the lower knife to begin the cutting action.

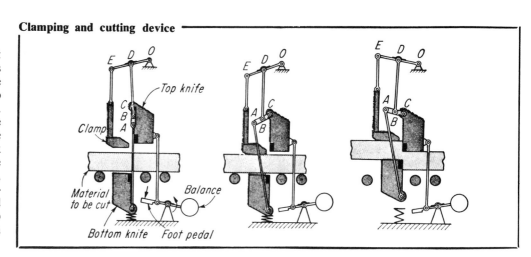

Four-bar cutter devices

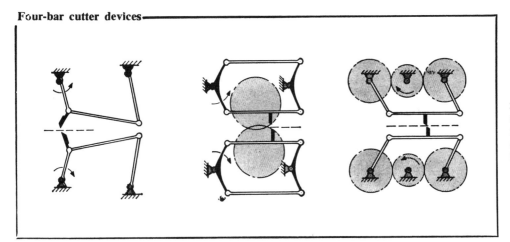

These three arrangements provide a very stable, strong cutting action by coupling two sets of links to obtain a four-bar arrangement.

Parallel cutter mechanisms

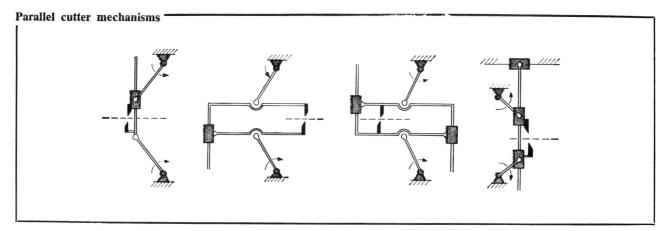

The cutting edges of the knives in the four arrangements move parallel to each other, and also remain vertical at all times to cut the material while the material is in motion. The two cranks are rotated with constant velocity by means of a 1:1 gear system (not shown) which also feeds the material through the mechanism.

Curved-motion cutter

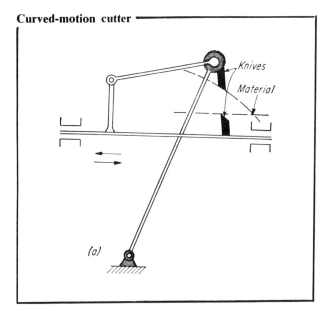

(a)

The material is cut while in motion by the reciprocating action of the horizontal bar. As the bar with the bottom knife moves to the right, the top knife will arc downward to perform the cutting operation.

Vertical cutter motion

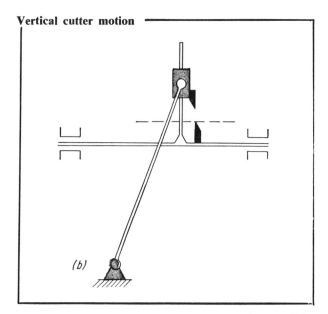

(b)

The top knife in this arrangement remains parallel to the bottom knife at all times during cutting to provide a true scissor-like action, but friction in the sliding member can limit the cutting force.

Slicing mechanism

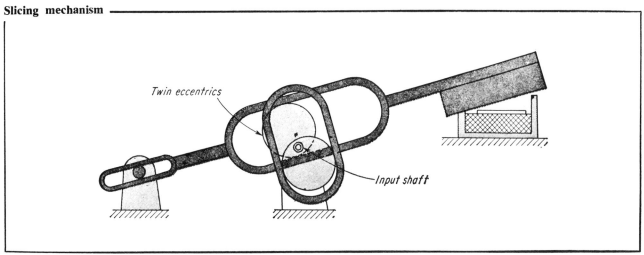

Slicing motion is obtained from the synchronized effort of two eccentric disks. The two looped rings actuated by the disks are welded together. In the position shown, the bottom eccentric disk provides the horizontal cutting movement and the top disk the up-and-down force necessary for the cutting.

Web-cutting mechanism

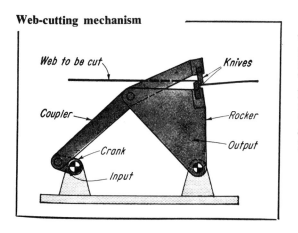

This four-bar linkage with an extended coupler can cut a web on the run at high speeds. The four-bar linkage is proportioned to provide a knife velocity during the cutting operation equal to the linear velocity of the web.

The mechanism can turn over a flat piece by driving two four-bar linkages from one double crank. The two flippers are actually extensions of the fourth members of the four-bar linkages. Link proportions are selected so that both flippers come up the same time to meet at a line slightly off the vertical to transfer the piece from one flipper to the other by the momentum of the piece.

Turn over device

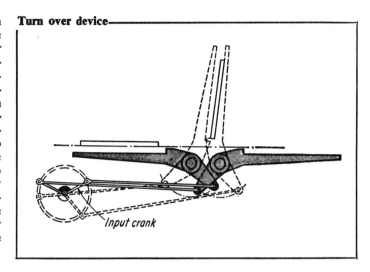

Input crank

Upside-down flipper

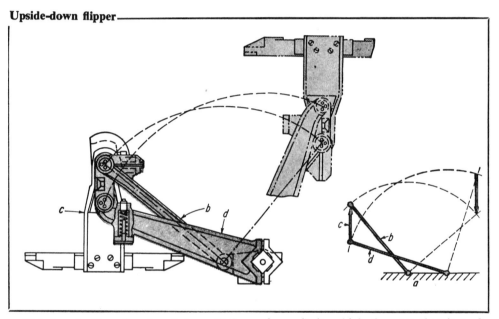

This is actually a four-bar linkage (links *a, b, c, d*) in which the part that is to be turned over is the coupler *c* of the linkage. For the proportions shown, the 180-deg rotation of link *c* is accomplished during 90-deg rotation of the input link.

Vibrating mechanism

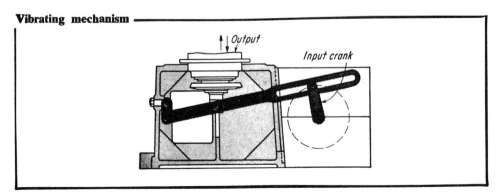

Output

Input crank

As the input crank rotates, the slotted link, which is fastened to the frame with the aid of an intermediate link, oscillates to vibrate the output table up and down.

FEEDER MECHANISMS—types

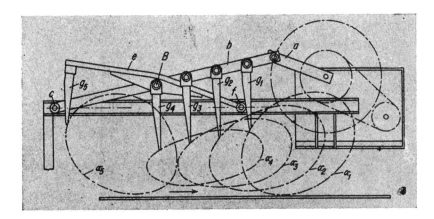

H SCHAEDLER and G MARX
Institute for Fundamental Research
Brunswick, Germany

How to perform systematic analysis of individual motions and translate them into mechanisms—the basic steps for precise design of feeder mechanisms.

Feeding parts into a continuous or piece-processing machine involves many complex motions. First step in a systematic analysis is to catalog the motions, determine the minimum actual time duration, and coordinate complementary motions. Second step is to translate these motions into hardware.

Feeder mechanisms can be classified according to type and stage of the process, Fig 1. This diagram is divided vertically, between continuous and piece-processing; and horizontally into four process stages. Continuous-flow material can be granular, bulk, fluid, or continuous strip. Furthermore, the production process can change the form of

the material from continuous-flow to piece-processing or vice versa as shown by dotted lines in the chart.

Generally, mechanisms are more complex in the piece-processing operation, although some continuous-flow operations require auxiliary mechanisms for feeding or removal. In sheetmetal processing, auxiliary mechanisms may be needed on fixed or adjustable universal strip-metal presses, special-purpose self-actuating presses, or shearing, bending and forming machines. Mechanisms for continuously advancing strip and sheetmetal combine feeding, advancing and removal; and employ cams, tongs or grippers, or rolling clamps.

By contrast, a piece-processing mechanism performs these functions:

 • Hopper loading, which requires supplying and introducing the material.
 • Arranging, separating, turning, shaking, and directing.
 • Holding with means for charging or removing from the machine.
 • Transferring product to the next machine.

Small parts can be advanced by gripping jaw or turntable carriers with cam, geneva or gear drives; by compressed-air-actuated feed mechanism; by sliding track or plunger; or by a pushing mechanism.

Preliminary handling functions

In a continuous-flow process, the handling functions that precede actual processing largely consist of continuously withdrawing material from a steady source of supply, and feeding it into the process at a predetermined rate. Mechanisms for such functions are relatively simple, as shown in (A) of Fig 2.

1 FEEDER-MECHANISM REQUIREMENTS here are classified according to type of material and processing stage.

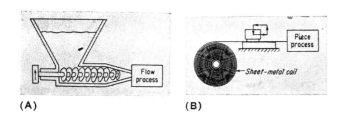

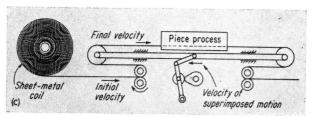

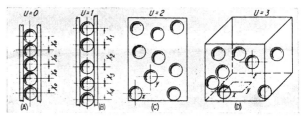

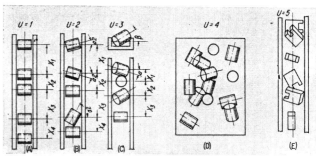

2 CONTINUOUS FLOW. (A) Screw feeders in many forms are used to feed continuously flowing bulk material at a predetermined rate.
(B) Continuous sheetmetal is fed into a piece process by an intermittent motion—which can be furnished by a crank and lever coupled to a ratchet mechanism.
(C) Intermittent motion can be applied to a sheetmetal coil by superimposing a second motion of a cam brake.

3 DEGREES OF DISORDER of spherical parts depend on number of dimensions required to specify the position.

4 FOR CYLINDRICAL PARTS, degrees of disorder depend on number of dimensions and angles required to specify the position. Similarly, in (E) degrees of disorder of odd-shape parts depends on number of dimensions and angles, and their relation to a reference plane.

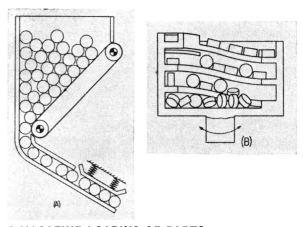

5 MAGAZINE LOADING OF PARTS can be done a number of ways. In (A) gravity-fed balls are slowed down by a spring-actuated device. In (B) a vibrating or oscillating drum elevates parts that fall on grooves specially designed to fit the parts.

A continuous-flow process feeding into a piece process, Fig 2(B), sometimes requires intermittent motion, as in stamping presses, sewing machines and film projectors. Another example of continuous material feeding into a piece process is shown in Fig 2(C) where a secondary motion by a cam is superimposed on the uniform initial velocity of the sheetmetal coil in such a way that the final velocity of the sheet is zero.

In the piece process, degree of difficulty required to produce some orderly arrangement of the pieces can be defined in terms of parameter U, or the "degree of disorder" of the individual pieces. For example, if uniform spherical pieces fit a grooved guide, then the distance between them is nominally zero. This is sufficient to determine their relative positions; thus $U = 0$ as at (A) in Fig 3. However, if the balls are to be spaced at different intervals, as in (B), one dimension becomes necessary to determine their positions, and $U = 1$. To establish positions of the balls in contact with a given plane, two coordinates are needed as in (C), and $U = 2$. But to establish a systematic order of the balls within the space of a container, three dimensions are required as in (D); here $U = 3$.

Handling cylindrical pieces is somewhat more complicated than handling spherical pieces. Fig 4(A) shows that $U = 1$ when only one dimensional factor is required to establish the positions. When one dimensional factor and one angular factor are required, $U = 2$, as in (B). When one dimensional factor and two angular factors are required, $U = 3$, as in (C). To establish positions of the cylinders in contact with a given plane, two coordinates and two angular factors are required, which makes $U = 4$ as in

(D). To establish a position for the cylinders in space, the value of U increases to 5. The degree of disorder of odd-shape objects, as in (E), is also 5 since their position is determined by two geometric axes and by the surface planes of the object.

Mechanisms for preliminary handling

Three general methods of sorting and arranging, independent of the degree of disorder are: continuous sorting, sorting into groups, arranging by individual separation.

Continuous sorting sets in order a group of disordered parts steadily yet not in any exactly predictable time, Fig 5. When sorting into groups a few disordered parts are

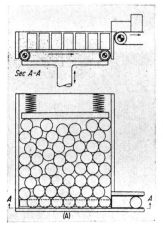

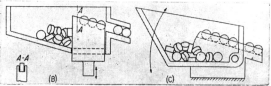

6 SORTING INTO GROUPS by a spring-actuated device can be applied either horizontally or vertically as in (A); can also be done by vertical oscillation (B), or by angular motion (C).

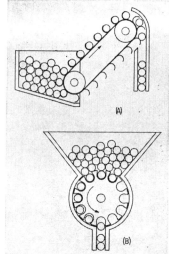

7 ARRANGING BY SEPARATING: (A) by a belt conveyor with specially fitted cogs; (B) by rotating grooves.

separated and forced into a desired orderly position, Fig 6. To arrange by individual separation, a single piece is selected from a disordered quantity, and taken off in a constrained manner as shown by the several varieties in Fig 7.

Mechanisms for input and removal

Mechanisms for control and positioning the part involve two principal factors:

- Space limitations surrounding the machine entrance.
- Accessibility of the mechanism to the central working position of the machine.

Also, the number of free or forced movements must be considered. Free or gravity movements require mechanisms to control their direction and velocity, but forced movements require mechanisms that actuate as well as guide the movements, and more complexity adds to cost. However, mechanisms that actuate forced movements are positive, and therefore more reliable.

Removal of parts from a processing position requires fewer movements, and consequently less-complex mechanisms. Primarily, this is a transfer operation, in which the parts are forced out, or fall out of the machine. Fig 8 shows an arrangement for simultaneous input and removal from a machine. The tubelike workpiece b rolls out of the feed magazine a into a recess in the slider c. It is then advanced into the work station by the forward stroke of the slide, and carried into the machine by a ramrod (not shown). At the same time a previously finished part rolls down the inclined surface of the slider, and on the next stroke is pushed into the transfer channel.

An arrangement that uses complex forced movements for feeding and gravity movements for removal of parts is shown in Fig 9. Three power cylinders a, b, c connected to an adjustable bell crank are used to generate the forced movements required to feed sheetmetal into a machine. Removal by gravity occurs after the part has been directed by forced movements into the desired direction.

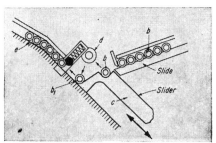

8 INPUT AND REMOVAL OF PARTS can sometimes be done with a single mechanism. In this unit, slider c simultaneously pushes incoming part b and outgoing part b_1.

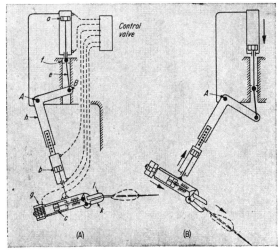

9 ADJUSTABLE BELL CRANK pivoted at A provides the angular motion while 3 air cylinders with a central control valve provide the linear motion. Counterbalancing is provided by adjustable weights g.

Knowing the basic hoppers, feeders, and conveyors
will be helpful in developing new feeder combinations.

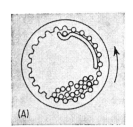

1 . . . EIGHT SPIRAL-SHELF HOPPERS

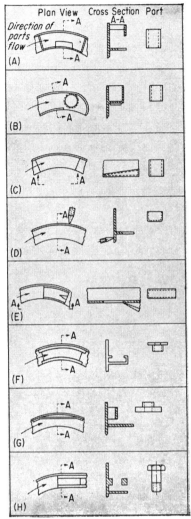

RAILED SHELF with a cutout carries only those cylindrical parts that remain upright. Those that fall horizontal are ejected through the cutout.

HOLE IN SHELF ejects those cylindrical parts that are upright, lets horizontal ones pass.

INCLINED SECTION on shelf turns cylinders into upright position.

AIR JET will pass cup-like pieces in the upright position, and eject them in any other position.

RECTANGULAR HOLE in shelf with spear will catch long U-shape cylinders with open end in front, and let fall those with open end in rear.

FLANGED SHELF carries T-shape cylinders in upright position only.

CHORD SECTION on the shelf carries inverted T-shape cylinders only.

NECK IN SHELF passes bolts with heads up or down, ejects bolts aligned otherwise.

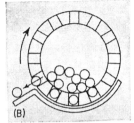

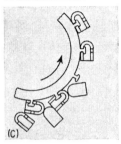

2 . DRUM HOPPERS. Cylindrical parts are spaced for any suitable interval (A) and injected to the machine by a ramrod. Spherical parts are spaced in a drum hopper with holes (B) and gravity-fed into machine. U-shape parts are handled by hooks on the outside of the drum (C).

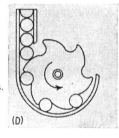

STAR FEEDERS for spherical or cylindrical parts (D) can be made with various dwell periods.

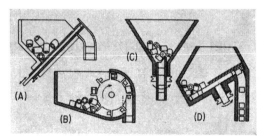

VARIETY OF HOPPERS for the same part: Oscillating piston and tube (A) elevate and arrange the disordered parts. Rotary pickup (B) performs the same task. Rotary vertical tube (C), or inclined rotary plate (D) use gravity feed.

3 . . . SIX WAYS TO FEED ALIGNED PARTS

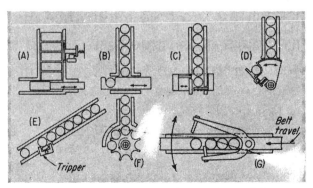

RECIPROCATING LINEAR MOTION is used for feeding parts in (A), (B) and (C). Angular motion combines with a gravity slide to deliver balls in (D) and (E). Star feeder (F) with or without dwell periods, feeds parts in round-table fashion. Belt conveyor with an angular motion device (G) separates individual cylinders.

7 BASIC SELECTORS for PARTS

PETER C. NOY

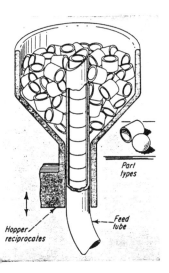

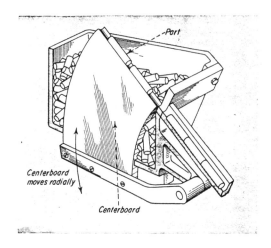

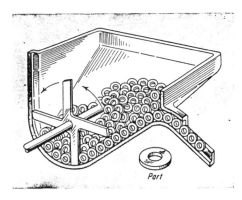

Paddle wheel . . .

is effective for disk-shaped parts if they are stable enough. Thin, weak parts would bend and jam. Such designs must be avoided if possible—especially if automatic assembly methods will be employed.

...eciprocating feed . . .

..r spheres or short cylinders is perhaps the ...mplest feed mechanism. Either the hopper ...the tube reciprocates. The hopper must be ... kept topped-up with parts unless the tube ...n be adjusted to the parts level.

Centerboard selector . . .

is similar to reciprocating feed. The centerboard top can be milled to various section shapes to pick up moderately complex parts. It works best, however, with cylinders too long to be fed with the reciprocating hopper. Feed can be continuous or as required.

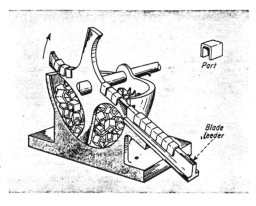

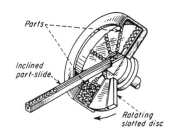

...ry centerblades . . .

...small U-shaped parts effectively if their legs are not ...ng. Parts must also be resilient enough to resist perma-...set from displacement forces as blades cut through ...f parts. Feed is usually continuous.

Rotary screw-feed . . .

handles screws, headed pins, shouldered shafts and similar parts. In most hopper feeds, random selection of chance-orientated parts necessitates further machinery if parts must be fed in only one specific position. Here, however, all screws are fed in the same orientation (except for slot position) without separate machinery.

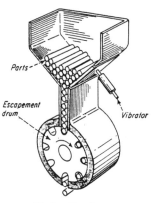

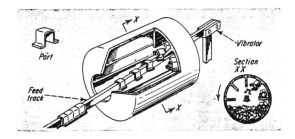

...ong-cylinder feeder . . .

...s a variation of the first two hoppers. If the ...ylinders have similar ends, the part can be ...ed without pre-positioning, thus assisting ...utomatic assembly.
...t cylinder with differently shaped ends re-...uires extra machinery to orientate the part ...efore it can be assembled.

Barrel hopper . . .

is most useful if parts tend to tangle. The parts drop free of the rotating-barrel sides. By chance selection some of them fall onto the vibrating rack and are fed out of the barrel. Parts should be stiff enough to resist excessive bending because the tumbling action can subject parts to relatively severe loads. The tumbling sometimes helps to remove sharp burrs.

12 Ideas

WILLIAM SCHWARTZ

SMALL PARTS, FASTENINGS, AND PACKAGE COMPONENTS are usually brought to processing and assembly operations in random heaps. Where these operations are to be mechanical, the parts must each be separated from their pile, oriented into the correct position, and fed in proper numbers and timing. Several firms manufacture ingeniously contrived hoppers applicable to many of these feeding, sorting, and counting operations, but except in the case of standard screws, nuts, washers, caps, and parts approximating them in shape, modification or a special design is usually necessary. Ten designs by the author are

Materials and Shapes	1	2	3	4	5	6	7	8	9	10	11	12
Round and Flat	X				X	X	X		X	X	X	X
Long Cylinders			X	X		X	X	X		X		
Cubes				X				X	X	X		X
Irregular Shapes			X		X		X					X
Symmetrical Parts having Projections			X	X		X	X	X	X			X
Spheres	X				X				X			
Fragile or Brittle	X		X				X			X		X
Soft	X			X	X	X	X	X	X	X		X
Nesting or Tangling		X		X		X		X				
Delivery Rate over 100 parts per minute	X	X		X	X	X	X	X				X
Delivery Rate under 100 parts per minute		X					X		X	X	X	
Multiple Stream Delivery	X				X			X				X

Best Applications of Designs Illustrated: The selections noted above can only be a rough guide to designers. Where several mechanism could handle a given type the simplest designs have been noted; special considerations must be weighed in all cases.

1. Pierced Drum Feeding, Sorting, and Counting Hopper

A drum driven by a ring gear and supported on idler rolls has stationary troughs fitted into the ends which hold the supply of parts. One line of holes around the drum, closely spaced, will remove a stream of parts when they drop onto a slide at the exit point. A stationary external guard keep parts from falling through except at the discharge point, and an internal guard lining the upper half keeps parts not discharged from falling back. The drum is slightly thicker than the height of the part, a bearing cap in the illustration. The drum revolves slowly enough so the supply of parts does not tumble, but rides in the bottom. A pair of slides projecting far enough into the chute to pick up an upturned rim but clearing those turned down, will remove improperly oriented pieces. These may be dumped into a bucket at the end of the slides, or into a hopper equipped with a conveyor to return them to the feeding troughs. A counting roll fitted with spokes projecting up through slots in the delivery slide, will hold parts on the slide until needed. The slide will fill back to the drum, and

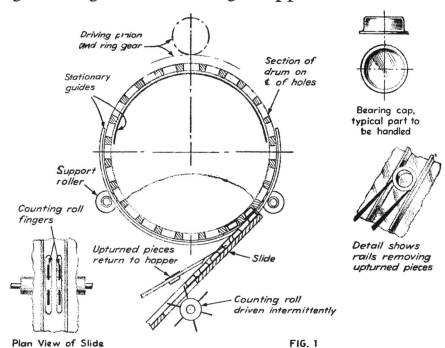

Plan View of Slide

FIG. 1

successive parts carried to this point by the drum will pass over the last piece on the slide, remaining in their hole in the drum until they pass the drum delivery point again and find it unoccupied. Dividing the counting roll into the proper number of segments, and rotating it through part of one revolution will count pieces as required by different applications.

for design of feeding, sorting and counting mechanisms

illustrated and are presented as a guide to the designer faced with this type of problem. Two other designs, interesting examples of devices now being made and sold as standard equipment by well known companies, have been included to broaden the scope and utility of the article. These 12 designs should solve a big percentage of industry's feeding, sorting, and counting problems. However, they are not presented as the definitive work on this subject. Variations are of course infinite; each part must be handled as a special case, but the development work involved is made increasingly worthwhile by rising labor costs and the distaste, even of unskilled labor, for monotony. The

recommendations checked in the accompanying table will furnish a rough guide to designers. Where several selections can handle a given type of material or component the simplest design should be chosen, and must be modified to meet the special requirements of the particular job. Quantities of pieces to be sorted, space requirements, speeds required, and other such factors will also enter into the selection of these mechanisms. Full-scale models must be built and thoroughly proven before these mechanisms can be placed in production, but their construction is not too expensive because wood, plastics, and sheet metal can be used for many parts.

2. Reciprocating Wiper Feeding and Sorting Device

A clip angle with unequal legs is to be chute-fed. The parallel wiper bars K-K move between the end rails B and C, positioned by the rods L-L and driven by the crank N. As they reach the right end of travel, projecting bar G strikes the pivoting gate H, which swings to pass a few parts down the chute from the hopper. The timing is such that they fall between K-K. The spring J keeps the gate closed unless struck, and feed can be regulated by lowering or raising G. As the wiper bars move to the left, the mass of parts is piled against the right hand one, and as it pushes them over the slot in the table between belts E and F, some will drop one flange into the slot. These are carried outwards by the belts, and restrained from falling through by the guides M-M. Clearance between the bottom of the wiper blades and the table is such that the wipers pass over parts falling into the slot. If only one line of clips is needed, belt E dumps its load into a bucket below the delivery end pulley. Belt F is also carrying parts, but half will have their long leg lying across its top surface. A

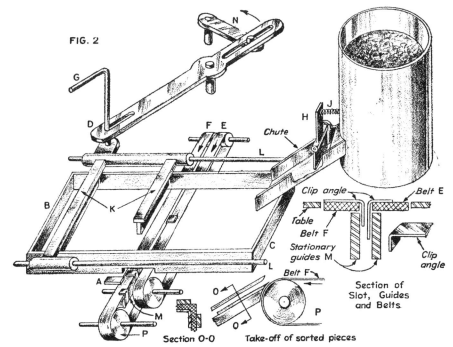

FIG. 2

Section of Slot, Guides and Belts

Section O-O

Take-off of sorted pieces

block A positioned just above belt F and projecting inwards just far enough to be struck by the long leg of the clip but not by the short one, will knock off any clips which are im-

properly oriented. Takeoff from the belt to a chute is shown in Section O-O, where a knifedge angle will pick clips off the belt.

(continued on next page)

3. Sorting Mechanism for Tube-Fed Long Parts

This device, shown end-sorting studs so that all studs enter the funnel with the stud end down, has the advantage that all parts continue in one stream; synchronization of the shaft drive with other equipment can time or count the discharge. The spider has at the end of each arm a cup with an OD equal to that of the part and an ID slightly larger than the small end. While the shaft is making a quarter turn, the edge of the rubber cam projects into the tube, holding the next piece from falling until one of the cutouts appears. Timing is adjusted so that a piece hits the cup just after it is aligned with the tube centerline. If the stub end of the part is down, it will rest across the rim of the cup. After the dwell period of the intermittent-drive the spider moves again, and the cup jerks out from under the stud, the stud falling into the

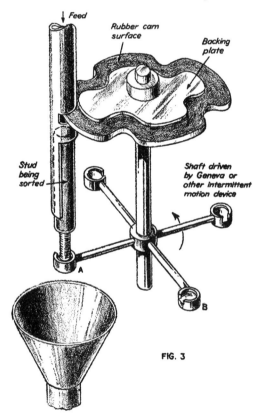

FIG. 3

funnel below, the stub end still down. When a part drops small end down, the small end will drop into the cup. The arm starting to move, the end is held by the cup, and it moves away from the tube, which is split here so that the upper end slides along it. When the cup reaches some position between A and B the top end is no longer supported, and it drops. When the stub end is below the level of the cup the small end slides off the rim, and the piece drops stub end down.

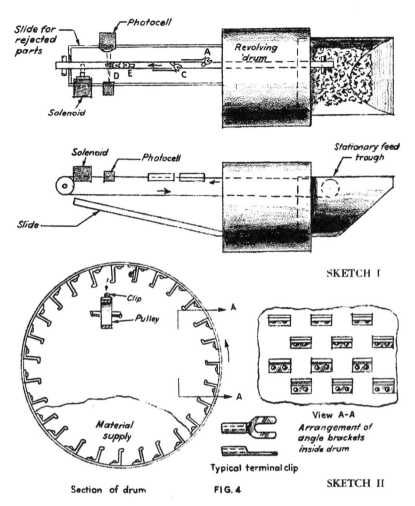

SKETCH I

Section of drum   FIG. 4

View A-A
Arrangement of angle brackets inside drum

Typical terminal clip

SKETCH II

4. Drum-and-Belt Feeding, Sorting Hopper

The drum is end-fed from a fixed trough, and is lined with angle brackets staggered in arrangement and having small lips. The brackets are just large enough to hold one piece, though in operation others will cling. As the terminal clips illustrated ride into the upper right quadrant of sketch II, most parts not securely resting on the brackets will fall. The belt is so placed that parts tipping out of the brackets above will fall onto it; many will of course bounce off if the drop is too great or the belt too resilient. The belt carries the clips past the fixed guides A and B, which either line them up with the belt or push them off. The photocell light is interrupted by the high part of the clip, (the wire ferrule), which, with a suitable time delay circuit corresponding to the time of belt travel from the cell to the solenoid, energizes the solenoid. The solenoid plunger extends and retracts quickly; if the clip is turned like D on sketch I, the plunger hits nothing, but if it is turned opposite like E, the forked part of the clip is in the way and the solenoid plunger knocks the clip off the belt. A trough below the belt slopes down, carrying rejected clips back into the drum. If this feature is unnecessary, the lips on the angle brackets inside the drum may be omitted, and the belt located just above the center and near the right side of the drum. In the case of the clips illustrated, the proportions and speed of the drum and belt may be adjusted so that only pieces falling to the belt with their flat sides down will remain on it. This type of hopper is adaptable to a very wide variety of parts which tangle, nest, or have wires attached. When the drum and belt are properly proportioned a rain of parts strikes the belt in such numbers that even though a small percentage are properly oriented the output is large.

5. Turntable Hopper

Any solid part which will easily orient itself into a round hole and which is not too thin may be handled by this design. The square hopper has a rubber boot around its outlet, which ends just above the turntable. Radial lines of holes in the turntable pick up pieces as they pass under the boot, and a few others besides will lock into the edge of a hole and be pulled out of the boot. These latter are removed by the stationary fence. After passing under the fence, a filled line of holes passes to the discharge sector opposite the hopper, where a dropping gate occupies a slot in the stationary plate under the turntable. Obviously all the holes in a line may not be filled, and it would not be desirable to dump these. A line of microswitches, supported slightly above the turntable, have rollers with "feel" for the piece, which must project slightly above the turntable surface. These switches are normally open circuit, and are wired in series. If all of an entire line of holes is filled, they will all close simultaneously, energizing the solenoid which pulls out the dropping gate. A belt is illustrated carrying off the discharge, but a tube or chute might be located under each hole at the discharge station. Another electrical device can energize the solenoid only when discharge is desired in synchronization with another piece of equipment; the holes will remain filled until the dropping gate empties them.

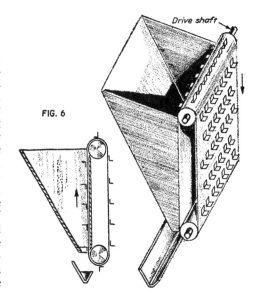

FIG. 6

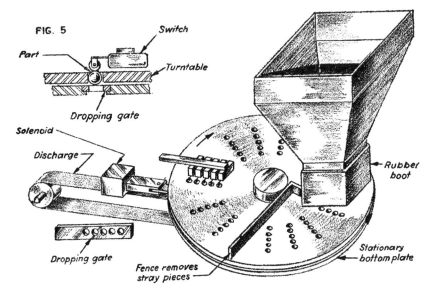

FIG. 5

Switch

Part

Turntable

Dropping gate

Solenoid

Discharge

Dropping gate

Fence removes stray pieces

Rubber boot

Stationary bottom plate

6. Non-clogging Stationary Hopper

Stationary hoppers handling granular materials can be equipped with electrical or mechanical vibrators to prevent clogging, but objects having a major dimension more than twice their minor dimension will clog unless the discharge opening is made several times their length. One positive anti-clogging device is the belt illustrated. The cleats are staggered in arrangement, wide enough so that the entire length of the discharge opening is frequently cleared, but too small and narrow to support a piece lifted out of the top of the material heap. The hopper may drop parts to a belt below, or be tilted so that the discharge opening is parallel to a sloping chute. When almost empty this type of hopper increases its rate of discharge, but the flow is quite constant from full to about one quarter full.

7. Coin Changer Feeding Mechanism

In this device, applicable only to round, flat objects, the tube is stationary. The funnel reciprocates, driving the tube end through the material heap; at each downward stroke one or more parts will be passed through the slot in the top of the tube. Lubricant on the OD of the tube will dirty and contaminate the pieces handled. If this is objectionable the bearing bushing must be dry, but lengthening it will extend its useful life. Parts which are too heavy will batter the end of the tube, and burrs will reduce the number fed into it. The clearance between the OD of the piece and the ID of the tube may be sufficient so the design can be applied to parts not too uniform in size.

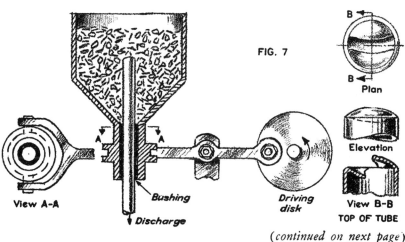

FIG. 7

B

B

Plan

Elevation

View B-B
TOP OF TUBE

View A-A

Bushing

Driving disk

Discharge

(continued on next page)

8. Bar-sided Drum Hopper

This design can be used in the random discharge version (shown in sketch I) for parts which tangle, since pieces not discharged are thrown back into the heap at the bottom. Sketch II shows the arrangement of the bars; the outer guide ends at the discharge slide or belt, with the bar at the discharge point sloping at the same angle as the slide.

Sketch III shows how the design can be adapted to discharge more than one stream of material by dividing the openings with separators, but parts which tangle would clog the openings. Cylindrical objects can be discharged at a timed rate or synchronized with other mechanisms according to the operation of the gate solenoid.

Looking at sketch II upside down, the bottom pair of bars at the centerline will be seen to form a funnel shape between surfaces A and B, which facilitates the parts dropping onto surface C, on which they ride in the lower half of the drum. An inner guide may be fitted into the upper half of the drum, in which case the spaces between the bars will remain filled until each slot passes discharge point and finds the gate open.

9. Power Screwdriver Hopper

One of the oldest and most popular feeding and sorting devices, this design has been adapted to a variety of symmetrically shaped parts. The wedge-shaped blocks fastened to the rotating ring are spaced so as to permit the part to fall between them when they pass under the supply stored inside the cover. The baffle plate inside the rotating ring is stationary, being attached to the fixed central shaft which also supports the discharge track. This baffle plate keeps the 50 caliber bullet cores illustrated from falling inwards until they are carried up to the end of the track. The selector guard attached

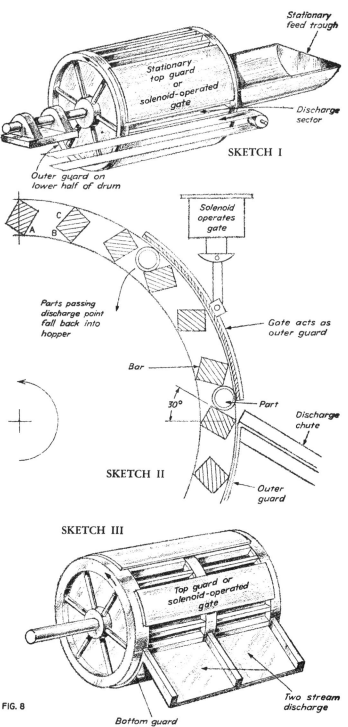

SKETCH I

SKETCH II

SKETCH III

FIG. 8

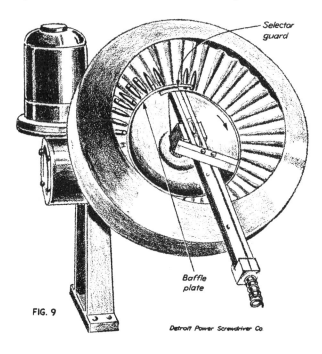

FIG. 9

Detroit Power Screwdriver Co.

to the baffle plate allows bullet cores oriented point first to pass into the track, but rejects those having their blunt end down. For parts which are bulky and which must be fed at a high rate this hopper must be equipped with an auxiliary device to dump batches of parts inside the cover every few minutes, since the cover cannot hold a very large supply. This hopper is usually driven by a constant speed motor, a feed limiting device or escapement being used to regulate the discharge. Feeds as high as 300 parts per minute can be obtained, but light plastic or sheet metal objects must be discharged at a much lower rate.

10. Squirrel Cage Feeding and Sorting Hopper

Rubber parts are difficult to sort because of their high coefficient of friction and resilience. These rubber casters, pressed into the bottom of small electrical appliances, could not be handled in any device dropping them even a short distance, and their center of gravity is such that orientation by movement would be difficult. However, lying in the outer cover of this hopper, they are tumbled until each piece happens to orient itself with the shank down. A slot passes under them, and they drop into it when other pieces permit. As the hopper turns clockwise, the caster lying with its shank in the slot between two wedges is carried into the upper left quadrant, where the inner guard keeps it from falling out. The inner guard ends at top center, where the chute picks up whatever pieces are in each slot and removes them. The weight of the piece is taken at

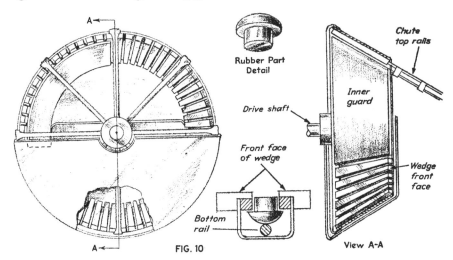

FIG. 10

point contact with the bottom rail, the top rails being stabilizers. This type of chute is best for rubber and other high-friction materials because it is open for clearing jams, has little fric-

tional resistance to movement, and is easy to fabricate. It will handle cubes, long cylinders and other shapes, and is especially good for material that have a high coefficient of friction.

11. Piston Hopper

Materials too fragile to be tumbled or agitated can be fed from this design. The chute angle should be approximately equal to the angle of repose of the material. The drive pinion may be driven by a variable speed motor and gear train, or the screw feed replaced by a hydraulic cylinder. A surprisingly constant flow of material can be maintained, but additional regulation by adjustment of the pivoting gate is difficult unless accompanied by control of the piston feed.

12. Vibratory Bowl Feeder

In this design a bowl is placed above a vibrator and shaken at 60 cycles by an AC solenoid. The center of the dish remains relatively motionless, since the direction of vibration is normal to a radius. Vibration sets up forces which drive parts to the periphery of the bowl, where a spiral track guides them up to the rim of the bowl. The base containing the vibrator is a standard unit, but the shape of the bowl and track are determined by the objects to be handled. Rate of feed is controlled by a rheostat which varies the amplitude of vibration; it is not normally necessary to vary the frequency. Multiple streams of parts can be discharged by making the helical track of multiple pitch like a screw thread. Gates, fences, air blows, or variations in the shape of the track at one point can discharge pieces which are improperly oriented. With proper design and regulation the discharge can be blocked awaiting the needs of the associated equipment, or the vibrator can be turned on and off by counting devices.

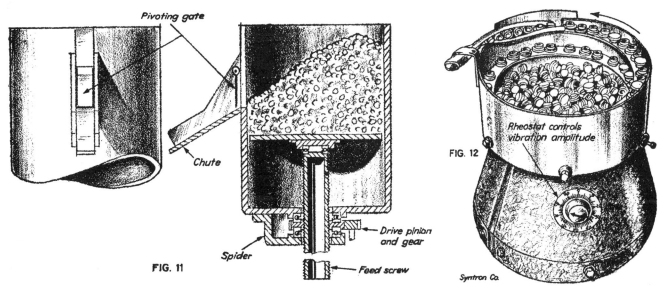

FIG. 11

FIG. 12

From Japan: Parts handling mechanisms

SHIGENOBU ARICHIKA

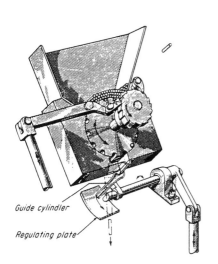

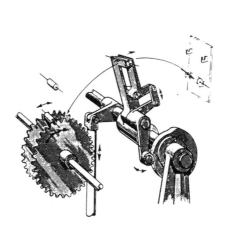

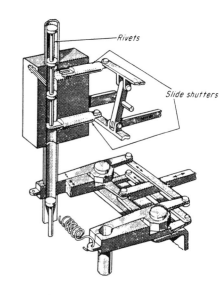

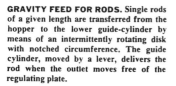

GRAVITY FEED FOR RODS. Single rods of a given length are transferred from the hopper to the lower guide-cylinder by means of an intermittently rotating disk with notched circumference. The guide cylinder, moved by a lever, delivers the rod when the outlet moves free of the regulating plate.

FEEDING ELECTRONIC COMPONENTS. Condensers, for example, are delivered by a pair of intermittently rotating disks with notched circumferences. Then a pick-up arm lifts the condenser and carries it to the required position by the action of a cam and follower.

FEEDING HEADED RIVETS. Headed rivets, correctly oriented, are supplied from a parts-feeder in a given direction. They are dropped, one by one, by the relative movement of a pair of slide shutters. Then the rivet falls through a guide cylinder to a clamp. Clamp pairs drop two rivets into corresponding holes.

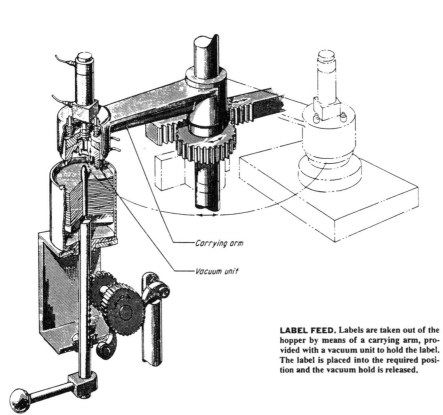

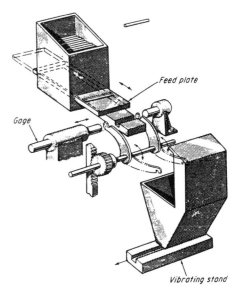

LABEL FEED. Labels are taken out of the hopper by means of a carrying arm, provided with a vacuum unit to hold the label. The label is placed into the required position and the vacuum hold is released.

HORIZONTAL FEED FOR FIXED-LENGTH RODS. Single rods of a given length are brought from the hopper to the slot of a fixed plate by means of a moving plate. After gaging in the notched portion of the fixed plate, the rod is moved to the shoot by means of a lever, and is taken out from the shoot by a vibrating table.

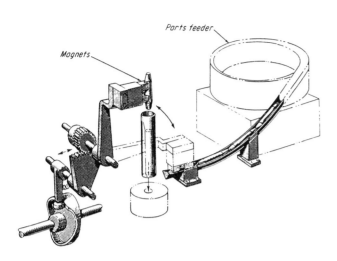

PIN INSERTER. Pins, supplied from the parts-feeder, are raised to the vertical position by means of a magnet arm. The pin drops through a guide cylinder when the magnetic hold is turned off.

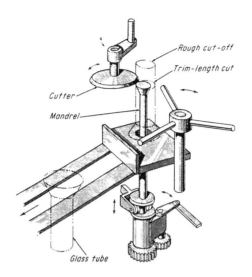

CUTOFF AND TRANSFER DEVICE FOR GLASS TUBES. Rotating glass tube has its upper portion held by a chuck (not shown). The moment the cutter sections it to a given length, the mandrel comes down and a spring member (not shown) drops the tube on the chute.

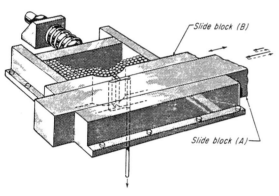

VERTICAL FEED FOR WIRES. Wires of fixed length are stacked vertically as illustrated. They are taken out, one by one, as blocks A and B are slid by means of cam and lever (not shown) while the wires are pressed into the hopper by a spring.

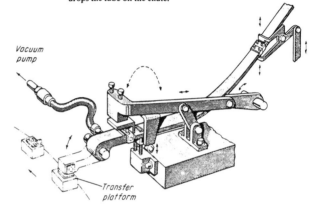

FEEDING SPECIAL-SHAPED PARTS. Parts of such special shapes as shown are taken out, one by one, in a given direction, and then moved into the corresponding indents on transfer platforms.

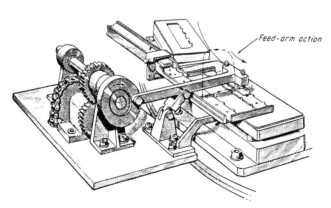

LATERAL FEED FOR PLAIN STRIPS. Strips supplied from the parts-feeder are put into the required position, one by one, by an arm that is part of a D-drive linkage.

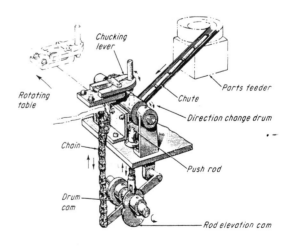

VERTICAL FEED FOR RODS. Rods supplied from the parts-feeder are fed vertically by means of a directing drum and a pushing bar. The rod is then drawn away by a chucking lever.

INNOVATIVE AUTOMATIC FEED MECHANISMS

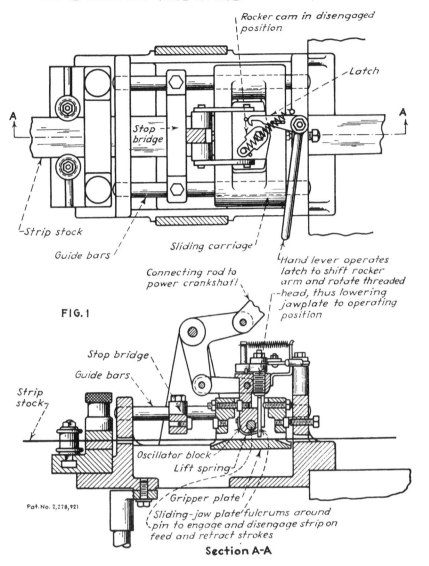

Rocker cam in disengaged position

Latch

Stop bridge

A — — A

Strip stock

Guide bars

Sliding carriage

Hand lever operates latch to shift rocker arm and rotate threaded head, thus lowering jawplate to operating position

FIG. 1

Connecting rod to power crankshaft

Stop bridge

Guide bars

Strip stock

Oscillator block

Lift spring

Gripper plate

Sliding-jaw plate fulcrums around pin to engage and disengage strip on feed and retract strokes

Pat. No. 2,278,921

Section A-A

Design of feed mechanisms for automatic or semi-automatic machines depends largely upon such factors as size, shape, and character of materials or parts being fed into a machine, and upon the type of operations to be performed. Feed mechanisms may be merely conveyors, may give positive guidance in many instances, or may include tight holding devices if the parts are subjected to processing operations while being fed through a machine. One of the functions of feed mechanism is to extract single pieces from a stack or unassorted supply of stock or, if the stock is a continuous strip of steel, roll of paper, long bar, and the like, to maintain intermittent motion between processing operations. All of these conditions are illustrated in the accompanying feed mechanisms.

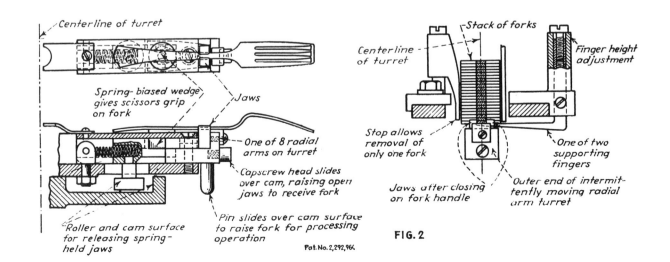

Centerline of turret

Stack of forks

Centerline of turret

Finger height adjustment

Spring-biased wedge gives scissors grip on fork

Jaws

One of 8 radial arms on turret

Capscrew head slides over cam, raising open jaws to receive fork

Stop allows removal of only one fork

One of two supporting fingers

Outer end of intermittently moving radial arm turret

Jaws after closing on fork handle

Roller and cam surface for releasing spring-held jaws

Pin slides over cam surface to raise fork for processing operation

Pat. No. 2,292,961

FIG. 2

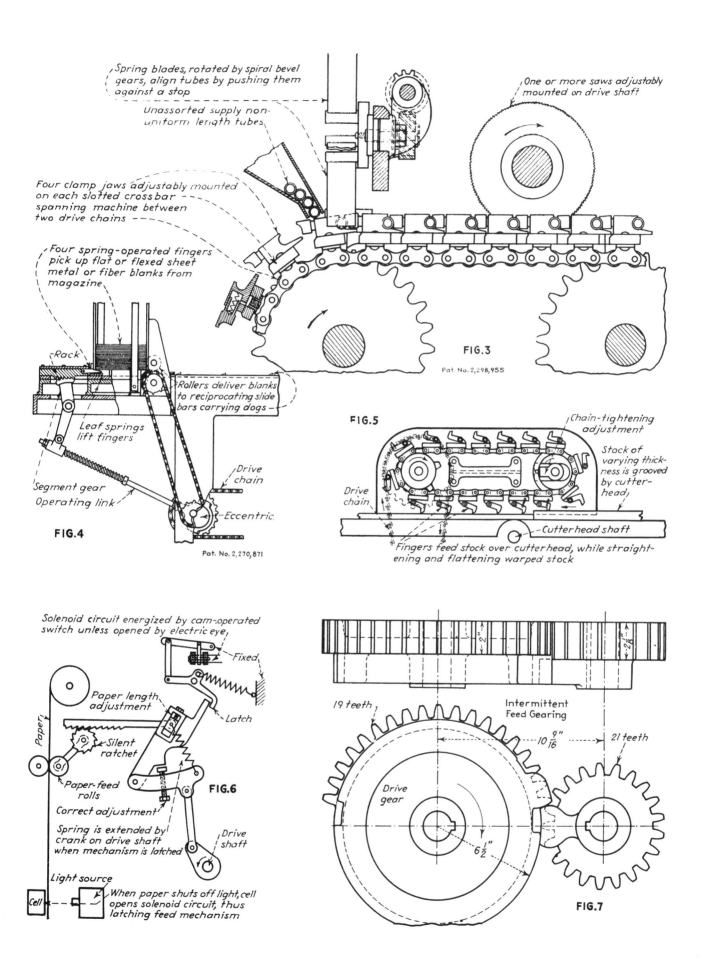

Spring blades, rotated by spiral bevel gears, align tubes by pushing them against a stop

Unassorted supply non-uniform length tubes

One or more saws adjustably mounted on drive shaft

Four clamp jaws adjustably mounted on each slotted crossbar spanning machine between two drive chains

Four spring-operated fingers pick up flat or flexed sheet metal or fiber blanks from magazine

FIG. 3

Pat. No. 2,298,955

Rack

Rollers deliver blanks to reciprocating slide bars carrying dogs

Leaf springs lift fingers

Drive chain

Segment gear
Operating link

Eccentric

FIG. 4

Pat. No. 2,270,871

FIG. 5

Chain-tightening adjustment

Stock of varying thickness is grooved by cutterhead

Drive chain

Cutterhead shaft

Fingers feed stock over cutterhead, while straightening and flattening warped stock

Solenoid circuit energized by cam-operated switch unless opened by electric eye

Fixed

Paper length adjustment

Latch

Silent ratchet

Paper-feed rolls

Correct adjustment

Spring is extended by crank on drive shaft when mechanism is latched

Drive shaft

Light source

Cell

When paper shuts off light, cell opens solenoid circuit, thus latching feed mechanism

FIG. 6

Paper

19 teeth

Intermittent Feed Gearing

$10\frac{9}{16}$"

21 teeth

Drive gear

$6\frac{1}{2}$"

2"

$2\frac{6}{8}$"

FIG. 7

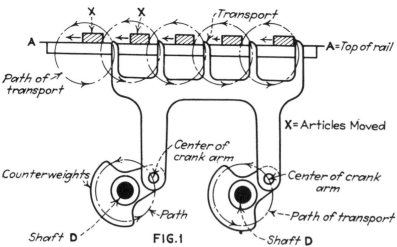

Fig. 1—In this design a rotary action is used. The shafts *D* rotate in unison and also support the main moving member. The shafts are carried in the frame of the machine and may be connected by either a link motion, a chain and sprocket, or by an intermediate idler gear between two equal gears keyed on the shafts. The rail *A-A* is fixed rigidly on the machine. A pressure or friction plate may be used to hold the material against the top of the rail and prevent any movement during the period of rest.

TRANSPORT mechanisms are generally used for moving material. The motion, although unindirectional, gives an intermittent advancement of the material being conveyed. The essential characteristic of such a motion is that all points in the main moving members follow similar and equal paths. This is necessary in order that the members may be subdivided into sections with projecting portions. The purpose of the projections is to push the articles during the forward motion of the material being transported. The transport returns by a different path from that which it follows in its advancement, and the material is left undisturbed until the next cycle begins. During this period of rest while the transport is returning to

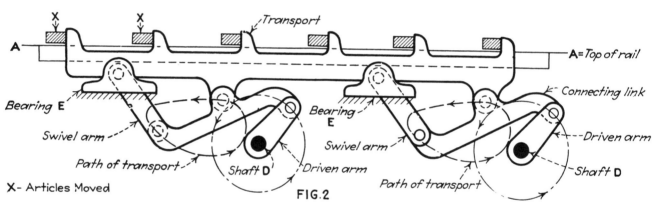

Fig. 2—Here is shown a simple form of link motion which imparts a somewhat "egg-shaped" motion to the transport. The forward stroke is almost a straight

line. The transport is carried on the connecting links. As in design in Fig. 1, the shafts *D* are driven in unison and are supported in the frame of the machine.

Bearings *E* are also supported by the frame of the machine and the rail *A-A* is fixed. The details of operation can be understood readily from the figure.

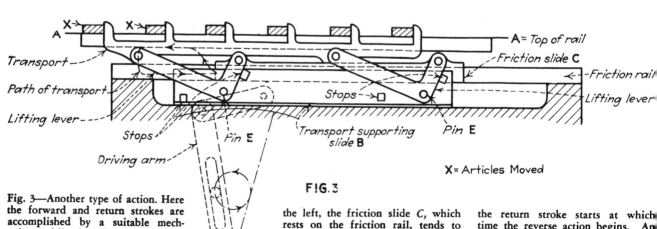

Fig. 3—Another type of action. Here the forward and return strokes are accomplished by a suitable mechanism, while the raising and lowering is imparted by a friction slide. Thus it can be seen from a study of the figure that as the transport supporting slide *B* starts to move to

the left, the friction slide *C*, which rests on the friction rail, tends to remain at rest. As a result, the lifting lever starts to turn in a clockwise direction. This motion raises the transport which remains in its raised position against stops until

the return stroke starts at which time the reverse action begins. An adjustment should be provided for the amount of friction between the slide and its rail. It can readily be seen that this motion imparts a long straight path to the transport.

Transport Mechanisms

F. R. ZIMMERMAN

The designs shown here represent a summary of typical constructions for obtaining intermittent advancement. Although they are specifically for conveying material, these same mechanisms might be adopted for other purposes

its starting position, various operations may be progressively performed.

Selection of the particular type of transport mechanism best suited to any case depends to some degree on the arrangement which may be obtained for the driving means and also the path desired. A slight amount of over-travel is always required in order that the projection on the transport can clear the material when going into position for the advancing stroke.

The designs illustrated here have been selected from numerous sources and are typical of the simplest solutions of such problems. The paths, as indicated in these illustrations, can be varied by changes in the cams, levers and associated parts. Usually the customary cut-and-try method should be used to obtain the best solution.

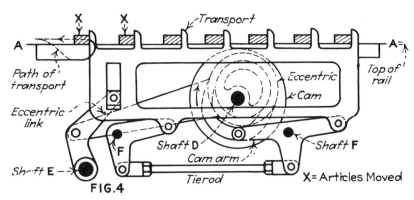

Fig. 4—Here is illustrated an action such that the forward motion is imparted by an eccentric while the raising and lowering of the transport is accomplished by means of a cam. The shafts, F, E and D are located by the frame of the machine. Special bell cranks support the transport and are interconnected by means of a tierod.

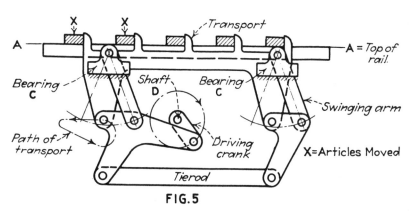

Fig. 5—This is another form of transport mechanism wherein a link motion is used. The bearings C are supported by the frame as is the driving shaft D.

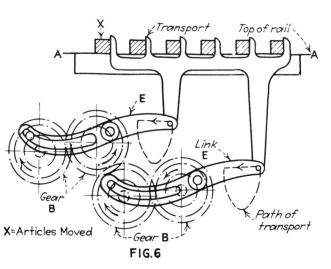

Fig. 6—An arrangement of interconnected gears of equal diameters which will impart a transport motion to a mechanism, the gear and link mechanism imparting both the forward motion and the raising and lowering. The gear shafts are supported in the frame of the machine.

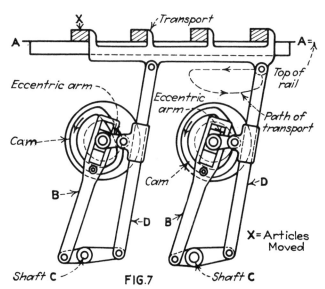

Fig. 7—In this transport mechanism the forward and return strokes are accomplished by the eccentric arms while the vertical motion is performed by the cams.

CONVEYORS may be divided into two classes: those that are a part of a machine used in processing a product and those that are used to move products, which are in various stages of fabrication, from one worker to another or from one part of a plant to another. For the most part the accompanying group of conveyors are of the first class and are elements of machines taking part in processing various articles. Both continuous and intermittently moving equipment are illustrated.

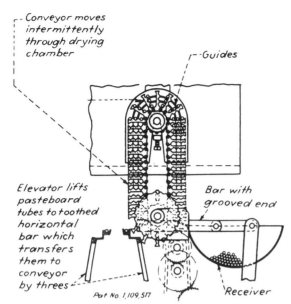

FIG. 1 – *Intermittently moving grooved bar links convey pasteboard tubes through drying chamber*

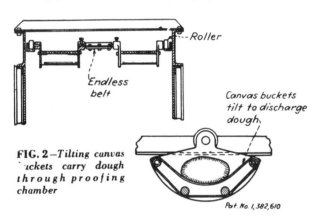

FIG. 2 – *Tilting canvas buckets carry dough through proofing chamber*

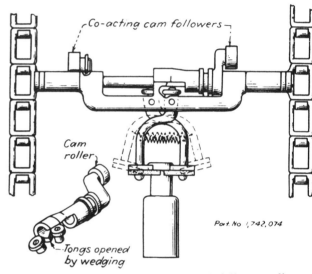

FIG. 3 – *Co-acting cams in paths of follower rollers open and close tongs over bottle necks by wedging action*

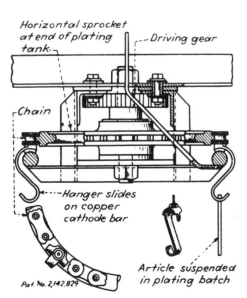

FIG. 4 – *Chain-driven conveyor hooks move articles through plating bath*

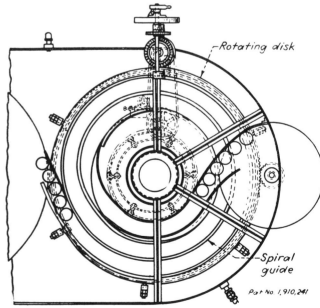

FIG. 5 – *Rotating disk carries food cans in spiral path between stationary guides for pre-sealing heat-treatment*

PRODUCTION MACHINES

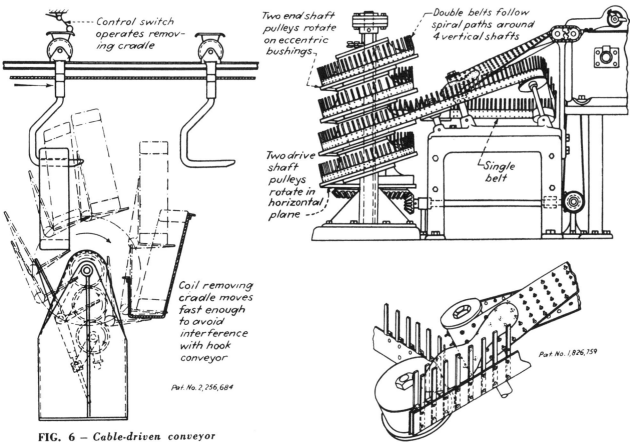

Control switch operates removing cradle

Coil removing cradle moves fast enough to avoid interference with hook conveyor

Pat. No. 2,256,684

Two end shaft pulleys rotate on eccentric bushings

Double belts follow spiral paths around 4 vertical shafts

Two drive shaft pulleys rotate in horizontal plane

Single belt

Pat. No. 1,826,759

FIG. 6 – *Cable-driven conveyor hooks and automatic removing cradle*

FIG. 7 – *Double belt sandwiches shoe soles during cycle around spiral system and separates to discharge soles*

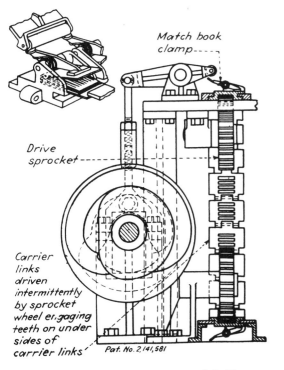

Match book clamp

Drive sprocket

Carrier links driven intermittently by sprocket wheel engaging teeth on under sides of carrier links

Pat. No. 2,141,581

FIG. 8 – *Matchbook carrier links with holding clips are moved intermittently by sprockets*

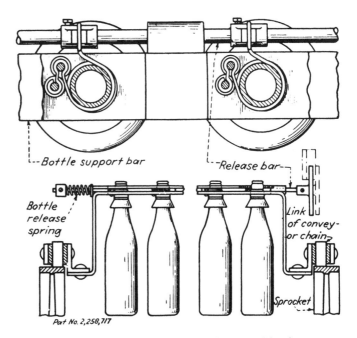

Bottle support bar

Release bar

Bottle release spring

Link of conveyor chain

Sprocket

Pat. No. 2,258,717

FIG. 9 – *One of several possible types of bottle clips with release bars for automatic operation*

Conveyor systems, continued

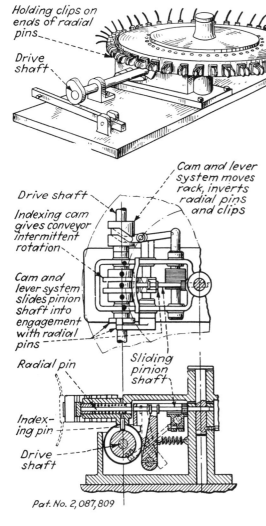

This group of conveyor mechanisms includes principles varying considerably among themselves. Each has been applied in many ways, which are dependent upon the characteristics of the article handled. In addition to the methods shown in this and the previous group, other conveyors have employed vibration, reciprocated jerking motion, suction and magnetic holders, forming carriers, and other principles and mechanisms requiring more space than is here available for adequate explanation.

Pat. No. 2,087,809

Fig. 1—*Intermittent rotary conveyor inverts electrical condensers, to be sealed at both ends, by engaging radial pins to which holding clips are attached*

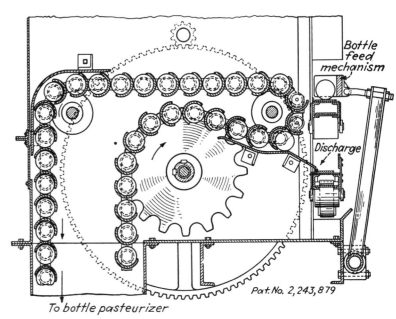

Pat. No. 2,243,879

To bottle pasteurizer

Conveyor Detail

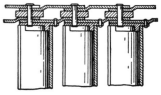

Fig. 2 — *Pasteurizer carrier links lock bottles in place on straightways*

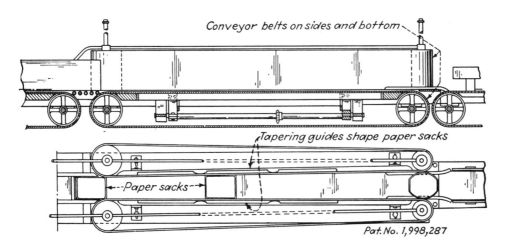

Pat. No. 1,998,287

Fig. 3 — *Wedging action of side belts shapes paper sacks for wrapping and packing*

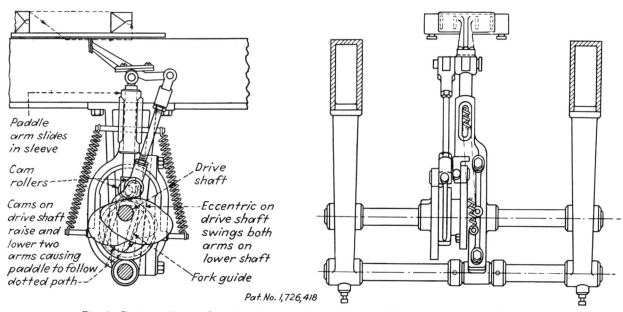

Paddle arm slides in sleeve

Cam rollers

Cams on drive shaft raise and lower two arms causing paddle to follow dotted path

Drive shaft

Eccentric on drive shaft swings both arms on lower shaft

Fork guide

Pat. No. 1,726,418

Fig. 4 Reciprocating pusher plate is activated by eccentric disk and two cams on drive shaft

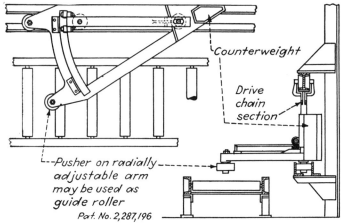

Counterweight

Drive chain section

Pusher on radially adjustable arm may be used as guide roller

Pat. No. 2,287,196

Fig. 5—Pusher type conveyor can be used with drive on either side

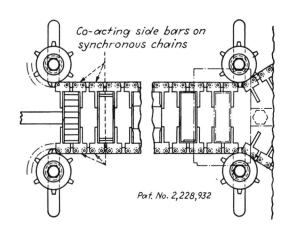

Co-acting side bars on synchronous chains

Pat. No. 2,228,932

Fig. 6—Synchronous chains with side arms grasp and move packages

Fig. 7—Rotary conveyor transfers articles from one belt conveyor to another without disturbing their relative positions

Hanger rotates on sleeve

Fig. 8—Cable conveyor with clip hangers

Woven wire cable

Clips

Pat. No. 2,156,353

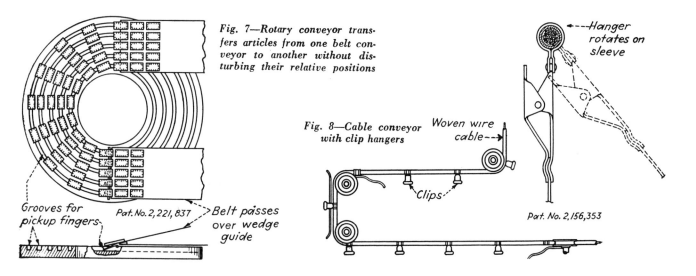

Grooves for pickup fingers

Pat. No. 2,221,837

Belt passes over wedge guide

Traversing Mechanisms Used

The seven mechanisms shown below are used on different types of yarn and coil winding machines. Their fundamentals, however, may be applicable to other machines which require

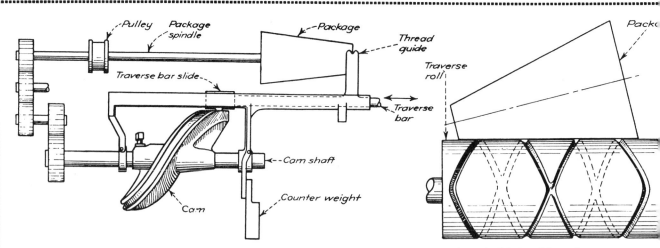

FIG. 1. Package is mounted on belt driven shaft on this precision type winding mechanism. Cam shaft imparts reciprocating motion to traverse bar by means of cam roll that runs in cam groove. Gears determine speed ratio between cam and package. Thread guide is attached to traverse bar. Counterweight keeps thread guide against package.

FIG. 2. Package is friction-driven from traverse roll. Yarn is drawn from the supply source by traverse roll and is transferred to package from the continuous groove in the roll. Different winds are obtained by varying the grooved path.

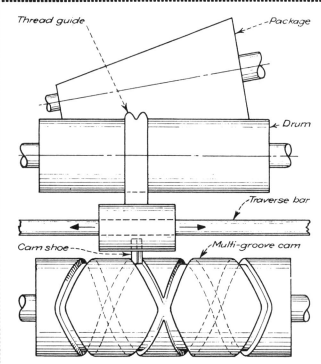

FIG. 4. Drum drives package by friction. Pointed cam shoe which pivots in the bottom side of the thread guide assembly rides in cam grooves and produces reciprocating motion of the thread guide assembly on the traverse bar. Plastic cams have proved quite satisfactory even with fast traverse speeds. Interchangeable cams permit a wide variety of winds.

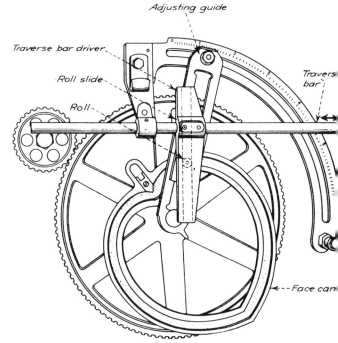

FIG. 5. Roll that rides in heart-shaped cam groove engages slot in traverse bar driver which is attached to the traverse bar. Maximum traverse is obtained when adjusting guide is perpendicular to the driver. As angle between guide and driver is decreased, traverse decreases proportionately. Inertia effects limit this type mechanism to slow speeds.

on Winding Machines

E. R. SWANSON

similar changes of motion. Except for the lead screw as used, for example on lathes, these seven represent the operating principles of all well-known, mechanical types of traversing devices.

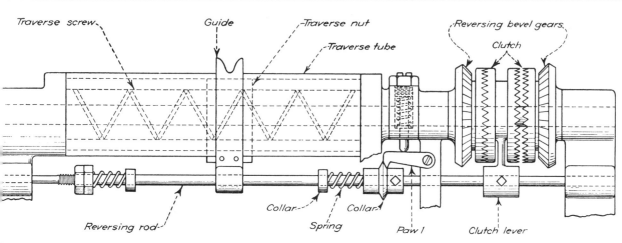

FIG. 3. Reversing bevel gears which are driven by a common bevel gear, drive the shaft carrying the traverse screw. Traverse nut mates with this screw and is connected to the yarn guide. Guide slides along the reversing rod. When nut reaches end of its travel, the thread guide compresses the spring that actuates the pawl and the reversing lever. This engages the clutch that rotates the traverse screw in the opposite direction. As indicated by the large pitch on the screw, this mechanism is limited to low speeds, but permits longer lengths of traverse than most of the others shown.

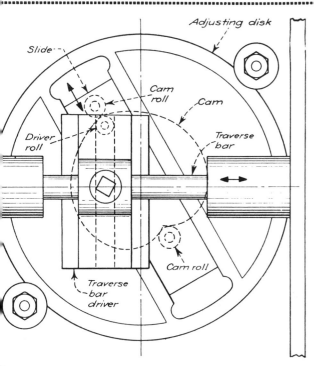

FIG. 6. Two cam rolls that engage heart shaped cam are attached to the slide. Slide has a driver roll that engages a slot in the traverse bar driver. Maximum traverse (to capacity of cam) occurs when adjusting disk is set so slide is parallel to traverse bar. As angle between traverse bar and slide increases, traverse decreases. At 90 deg traverse is zero.

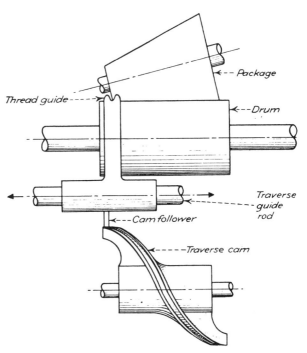

FIG. 7. Traverse cam imparts reciprocating motion to cam follower which drives thread guides on traverse guide rods. Package is friction driven from drum. Yarn is drawn from the supply source through thread guide and transferred to the drum-driven package. Speed of this type of mechanism is determined by the weight of the reciprocating parts.

How to collect and stack die stamping

FEDERICO STRASSER

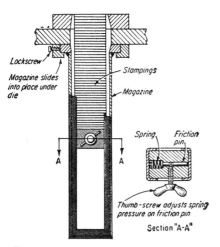

Lockscrew

Magazine slides into place under die

Stampings

Magazine

Spring

Friction pin

Thumb-screw adjusts spring pressure on friction pin

Section "A-A"

1. SLIDING MAGAZINE

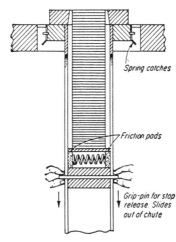

Spring catches

Friction pads

Grip-pin for stop release. Slides out of chute

2. HANGING MAGAZINE

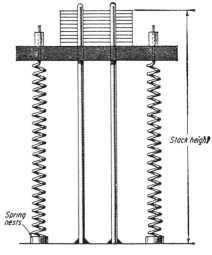

Stack height

Spring nests

3. SPRINGS — stack height is constant when stamping weight divided by spring rate equals stamping thickness

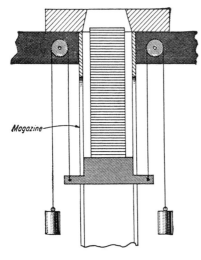

Magazine

4. COUNTERBALANCE METHOD does not need high stacking friction as in first two methods shown above

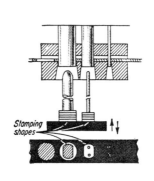

Stamping shapes

5. PEGGED MAGAZINE lets stampings be stacked by gravity, guided by hole in stamping

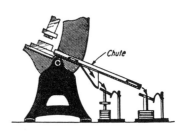

Chute

6. CHUTED STACKING WITH PEGGED MAGAZINE "sieves" stampings from two-stage die

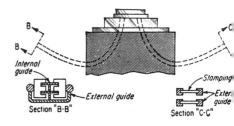

B

B

Internal guide

Section "B-B"

External guide

C

Stamping

External guide

Section "C-C"

9. SPECIAL SHAPES ARE CHUTED IN CORRECT ORIENTATION

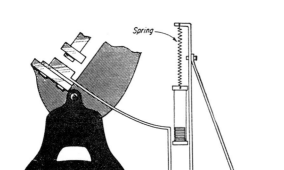

Spring

7. CHUTED STACKING WITH SPRING-SUSPENDED MAGAZINE

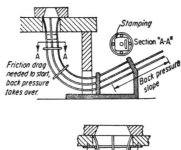

Stamping

Section "A-A"

Friction drag needed to start, back pressure takes over.

Back pressure slope

Flange

8. RODS FORM CHUTE

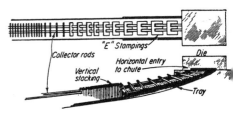

"E" Stampings

Collector rods

Vertical stacking

Horizontal entry to chute

Die

Tray

10. E-STAMPINGS ARE CHUTED FROM HORIZONTAL TO VERTICAL STACKING ON RODS

VACUUM PICKUP POSITIONS PILLS

Carrying tablet cores to moving dies, placing cores accurately in coating granulation, and preventing formation of tablets without cores

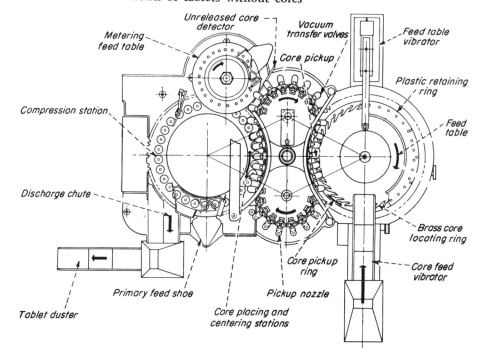

CORES ARE HOPPER FED to a rotating feeder disk through a tablet duster. This disk is vibrated clockwise under a slotted pickup ring which rotates counterclockwise. Each slot in pickup ring holds two cores and lets broken tablets pass through to an area under feeder table. Cores are picked from ring slots, carried to tablet press dies and deposited in dies by vacuum nozzles fastened to a chain driven by press die table. This chain also drives the pickup ring to synchronize motion of ring slots and pickup nozzles. Coating granulation is fed into dies ahead of and after station where vacuum pickup deposits a core in each die. Compressing rolls are at left side of machine. Principal design objectives were to evolve a machine to apply dry coatings at speeds which lowered costs below those of liquid coating techniques and was adapted to positive brand identification.

UNIT APPLIES LABELS FROM STACKS OR ROLLERS

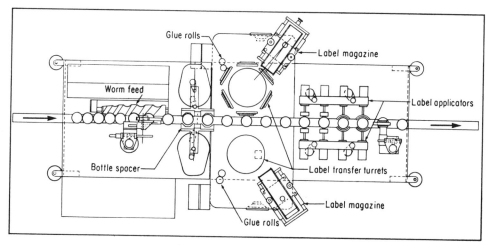

Flow of containers through versatile new labeler is indicated by top-view drawing of machine. Bottle spacers insure that containers remain 7½ in. apart on conveyor. Dual label-transfer turrets allow for simultaneous application of front and back labels.

Pony Pacer Labeler—available from New Jersey Machine—is a versatile unit that can use either conventional glue-label application or heat-seal labels in cut or roll form. The device labels front and back of round or odd-shaped containers at speeds of 60 to 160 containers per minute. Containers handled range from 1 in. diameter or thickness to 4¼ in. diameter by 5½ in. wide. Container height can vary from 2 to 14 inches. The unit handles labels ranging from ⅞ to 5½ in. wide and ⅞ to 6½ in. high. The label hopper is designed for labels substantially rectangular in shape, although it can be modified to handle irregular shapes. Provision has been made in design of the unit, according to the manufacturer, to allow labels to be placed at varying heights on the containers

Unit's cut-and-stacked label capacity is 4,500. An electric eye is provided for cutting labels in web-roll form. Model number of the unit is 202-RL-2. *New Jersey Machine Corp., Hoboken, N.J.*

From Modern Packaging

DEVICES FOR AUTOMATIC DIE OPERATION

EMIL LOEFFEL

First step toward automated press operation is development of proper die-protection devices. Many die designers fail to provide such items in the mistaken belief that they are complicated and costly.

Actually, the devices to be described are simple and relatively inexpensive. Some of them have been used exclusively in our plant with satisfying results. Among the most important safety devices is the:

From *American Machinist*

Misfeed Detector

Progressive dies utilize piercing and piloting stations to establish the amount of stock feed per press stroke. Shown in Fig. 1 is a device that is used to protect this kind of die. The misfeed detector is nothing more than a simple push button. The metal cup making contact with the punch holder completes a circuit, Fig. 2, which holds a relay closed, allowing the press to run. Should the stock misfeed, the pilot will contact the stock as the press ram moves down, causing the metal cup to leave the punch holder, breaking the circuit and stopping the press.

This simple protective switch offers reliability, instantaneous action and simplicity of installation. It requires one hole above the probe pilot, slightly larger than the probe pilot head, and an intersecting hole or groove for the insulated wire. The ease of installation and the small space requirements make this switch popular for building protection into existing dies.

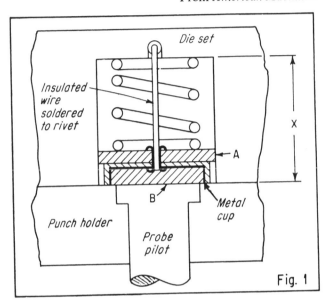

Fig. 1

Design Details

Fiber washer *A* insulates metal cup from spring and guides unit in hole. Fiber washer *B* insulates metal cup from pilot head. Distance *X* should be at least two times the distance the pilot will travel back during a misfeed and hence allow for reasonable spring compression. The spring should be stiff enough to prevent the metal cup from bouncing while running, but not stiff enough to become "solid" during misfeed.

Basic Circuit

An understanding of the basic circuit will enable the designer to design protective devices for many types of dies. The double-pole relay shown at left in Fig. 2 is the heart of the protective system. One set of contacts, held closed by the solenoid, supply current to the solenoid itself. Any break, even a momentary one, in the solenoid circuit—either through operation of the misfeed detector, the stock-buckling relay, or the end-of-stock switch—will cause the solenoid to de-energize, thus releasing both sets of contacts and immediately stopping the press. Once the solenoid contacts open, the solenoid cannot be re-energized by re-establishing the current path through the detectors. The solenoid can be re-energized only by pushing the reset button, and then only if the detectors are in their normal positions. However, the press should not start if the reset button is pushed. Accidental starting can be hazardous. Avoid this condition by wiring the protective circuit into the press stop button, if available. In this way the start button must be pushed to restart the press.

The addition of a normal-operation indicator shows when a misfeed has occurred by going out. This feature is helpful when running a bank of presses and it is difficult to hear if a press is actually running.

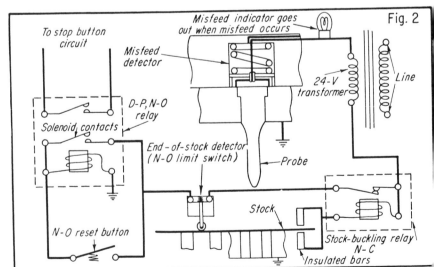

Fig. 2

End-of-Stock Detector

When one man is servicing many presses, an end-of-stock detector is a valuable asset. The device is easy to install, and is independent of the type of die, and can thus be used for all jobs run on the press. As shown in Fig. 2, the detector is a normally open snap-action switch held closed by the stock. The switch is sometimes mounted so as to utilize a small probe which lies on the stock and activates the switch at the end of the coil, as shown in Fig. 3. The normal-operation indicator will go out when the stock has run out and the press has stopped.

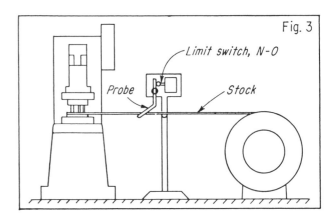

Fig. 3

Limit switch, N-O

Probe

Stock

Stock-Buckling Detector

When stock is fed automatically, it is usually better to pull rather than to push the material through a die. In the event that a punch is pulled out or a slug is wedged in the die—or any circumstance prevents the stock from feeding—there is a good chance that the probe pilot will re-enter the same hole, not detect any malfunction, and permit the press to continue to run. Should this occur on a press set up to pull stock through the die, the worst that can happen is that the feeding equipment will slip or a short piece of stock can be ripped from the die. However, if the stock catches in a die equipped with a push feed, the stock can buckle and cause serious damage.

Many dies are fed by pushing because they leave little or no stock to pull on. In order to protect these dies a stock-buckling detector should be used. One design is shown in Fig. 2. If the stock buckles

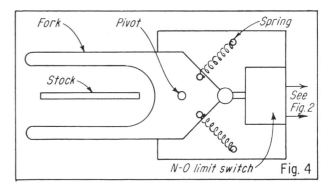

Fork   Pivot   Spring

Stock

See Fig. 2

N-O limit switch   Fig. 4

enough to touch either one of the insulated bars, the press will stop. Shown in Fig. 4 is a method sometimes preferred because it eliminates the use of a relay. The N-O (normally open) limit switch is held in the closed position by the fork.

(continued)

Rules for Designing and Using Protective Devices

Now that we have seen how protective devices can be used, it is reasonable to set down the rules for their design and use:

1. Protective circuits should always be designed as live current-carrying circuits. Thus, if a break in the wiring or a power failure occurs, the press will stop and not continue to run unprotected.

2. Design protective devices to detect malfunctions and to stop the press at the earliest possible moment. Hence, keep the feed pilot and the probe-pilot detector as long as possible without interfering with feeding action.

3. When using protective circuits which have a grounded connection, never use more than 24 volts in the ground circuit.

Avoid using 110 volts in the protective circuit; a 24-v transformer and a relay cost very little as compared to an accident.

4. Protective circuits should be wired into the stop button if available, so that pushing the reset button can never trip the press. The starting button must always be used to initiate starting action.

5. Always provide an air-oil hole for the probe pilot bushing. This hole will prevent unintentional shut-off due to build-up of air or oil.

6. Isolate electronic components of the system from vibration. Use shock mounts or —better yet—mount the components off the press.

7. Make periodic checks to see that the protective system is functioning in the proper manner.

Feed Pilot Used As a Probe

A feed pilot can be used as a misfeed detector, Fig. 5, when stamping stock up to 1/16 in. thick. Scored pilots and pulled pilot bushings are prevented, because the pilot can retract during misfeeds. An added advantage of the method is that on certain dies the stock can be placed right into the feeding equipment and the press started up without the need for hand feeding the stock across the die. This use saves press time and allows a lower-skilled press operator to start new coils. The first few pieces made will be incomplete, but these can be removed without stopping the press, as they come out of the die.

This type of misfeed detector should not be used on stock thicker than 1/16 in., because the pilot cannot position the stock and an unintended shutoff will occur.

Existing progressive dies can be easily protected by giving the feed pilot the dual role of both locating the stock and acting as a probe. For best results the pilot should have a long gradually curved tip.

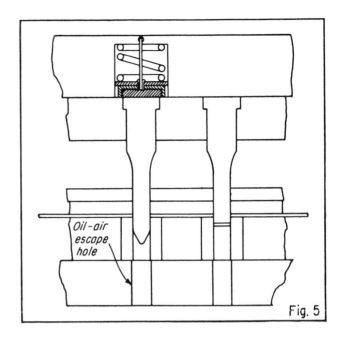

Oil-air escape hole

Fig. 5

Misfeeds Detected by Probe

When the stock is thicker than 1/16 in. and a spring-loaded pilot would not be satisfactory, use the separate probe shown in Fig. 6. The feed pilot, in this case, should be made as long as possible without interfering with feeding of stock. Length A of the probe pilot should be made as long as the feed pilot minus the curved locating tip. With this setup, the feed pilot, not the probe pilot, locates the stock, and a misfeed is detected at the earliest possible moment.

The separate probe pilot method, in most cases, cannot be used to add protection to existing dies. Since experience has proved this method superior to using the feed pilot as a probe, it is recommended that separate probe pilots be used for new tooling.

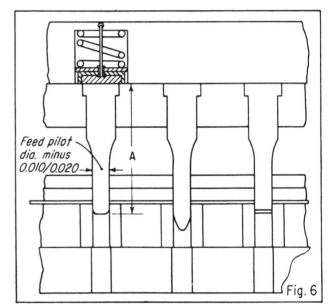

Feed pilot dia. minus 0.010/0.020

A

Fig. 6

Separate Probe Pilot Used in Scrap

Misfeed detection can be extended to probing a hole in the scrap strip, Fig. 7. This method is frequently used to add protection to existing dies. Many dies are made with a solid die block as shown. Application of a separate pilot-misfeed detector into a hardened die block is expensive. However, adding a small die block section is easy and inexpensive. Progressive dies that produce several different items by changing inside diameters are conveniently protected by this method. The added punch plate and die block do not have to be changed to suit the size of hole being punched, because the outside diameter of the blank remains constant.

Dies that produce stampings of irregular outline can be protected by using a round probe which just fits into the opening of the scrap strip. The probe is placed so that it enters the scrap strip after the pilot has registered the stock. Where a round probe would not be practical, a sleeve of the proper shape can be attached to the bottom of the probe.

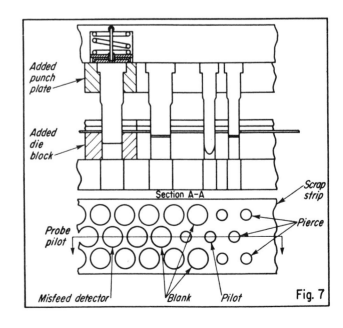

Added punch plate

Added die block

Section A-A

Scrap strip

Pierce

Probe pilot

Misfeed detector   Blank   Pilot   Fig. 7

Insulated Probe Pilot

Fast misfeed detection is provided by the insulated probe, Fig. 8. If the insulated probe comes in contact with the strip, as from a misfeed, the circuit is closed, a solenoid is energized and the normally closed contacts in the latching relay are opened, thus stopping the press. To avoid unnecessary shutdowns the probe diameter A should be made 10 to 20% smaller than the hole in the strip and pilot bushing. Stray metal chips and cocking of the strip might otherwise act to trigger the device.

Under normal operation the wire attached to the probe does not carry any current. However, current is carried during a misfeed when the probe comes in contact with the stock. Should the probe wire break and not come into contact with a grounded surface, the press would continue to run unprotected. But this method offers the advantage that as many probes as desired can be used on the same die with the same relay, each probe affording individual protection.

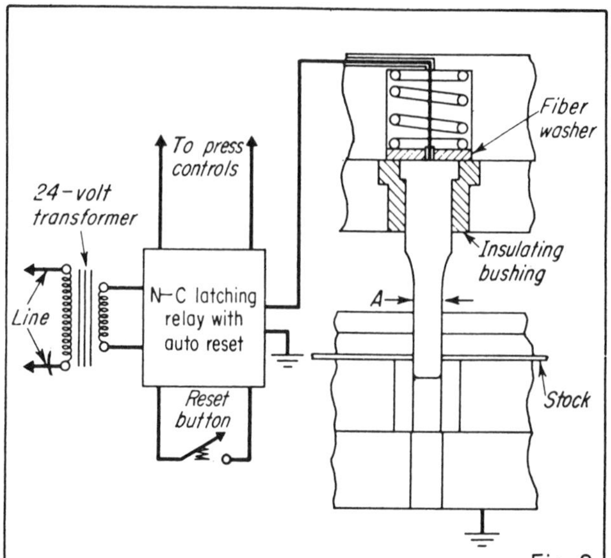

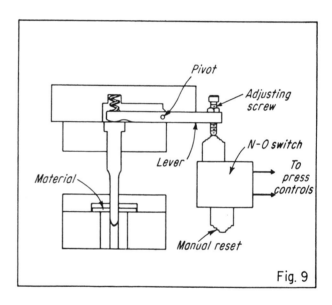

Fig. 8

Snap-Action Misfeed Detector

Here is a popular misfeed detector that uses a manual reset switch, Fig. 9. More space and machine work are required for the device than other methods, but the switch eliminates the relay. This simplifies the electrical part of the system. However, the use of a relay is recommended if the switch is to carry a high voltage. Pushing the manual reset button should not start the press if the switch is properly wired into the stop button.

Adjusting screw should be positioned to trip the switch upon slight movement of the probe pilot, so that a misfeed will be detected as early as possible.

The switch is attached to the bottom die shoe. Sufficient over-travel is built in, to avoid excessive strain on the switch plunger during a misfeed. The probe pilot can be used either as a separate probe pilot or as a feed pilot.

Fig. 9

Cam-Actuated Snap-Action Switch

A cam, instead of a lever, can be used to actuate a snap-action switch, as shown in Fig. 10. A small flat is ground in the pilot to prevent rotation and insure that the cam will always re-enter the cam groove after a misfeed. Make certain that the switch has enough over-travel to prevent damage during misfeeds.

This misfeed detector can also be used as a feed pilot. However, you should remember that the device and its connecting wires move with the press ram and are subject to jolts and possible breakage. If more than 24 volts pass through the switch, a hazardous condition will exist.

When ordinary wires are subjected to constant flexing, they invariably work harden and break. Braided wire has performed satisfactorily and is recommended when flexing is unavoidable.

(continued)

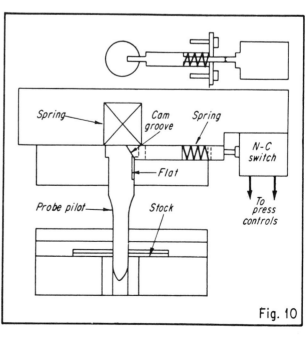

Fig. 10

43

Missing-Part Detector

Detection of parts missing from the carrier strip is the function of the detector shown in Fig. 11. When progressive dies run at high speeds, many scrap parts can be produced in a very short time, wasting stock, and worse, mixing bad parts with good ones. Difficulty of separation can result in rejection of a large bin of parts because of a small percentage of rejects. However, more important than detecting rejects is the prevention of smash-ups due to the pile-up of missing parts. This switch will detect such a condition and stop the press before damage can occur.

NORMAL OPERATION—As the die closes and the part is in its proper position, the probe will be forced upward, preventing it from activating the switch. The press is allowed to cycle.

ABSENCE OF PART—If a part is missing or not in correct position, the probe will drop, allowing the adjusting screw to activate the switch, stopping the

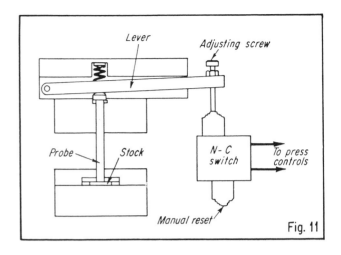

Fig. 11

press. After the malfunction is corrected, the press is made operative by resetting the switch in a normally closed position.

Switch Protects Die

Should the pressure pad of the die shown in Fig. 12 stick or fail to function properly, the press will stop instantly and prevent serious die damage.

NORMAL OPERATION—At bottom stroke the pressure pad is pushed up. Now the cam moves right, where it would activate the switch if the roller were opposite the cam. However, the cam is below the switch roller and the circuit is not opened. As the ram moves upward the spring pin pushes the pressure pad down, and the cam is pushed left before it can strike the switch. Therefore, the press can finish its cycle and begin the next one.

DETECTING A MALFUNCTION—In normal operation the cam is continually moving in and out, both motions taking place beneath the switch. If the pad fails to return to its normal position, the cam will

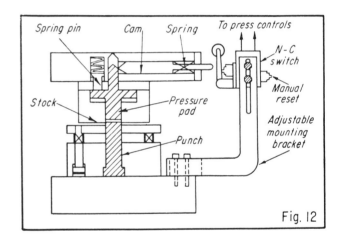

Fig. 12

be held out and will trip the switch as the ram continues upward.

Auxiliary Circuit Protects Dies

Many cut-off dies for heavy stock are built without feed pilots. Because a pilot-protection method cannot be used, another protective method must be employed. Shown in Fig. 13 is a method for protecting a die which cuts off against a stop block.

NORMAL OPERATION—The stock feeds left to right against the insulated stop block, completing the solenoid circuit through the ground connection. The cam then comes into contact with the snap-action switch, opening an alternate supply circuit to the solenoid. As the ram continues down, the cam leaves the snap-action switch, closing the alternate solenoid circuit. The stamping is pushed past the stop-block, opening the solenoid circuit through the ground connection.

MISFEED—The stock does not reach the stop block and cannot complete the solenoid circuit through the ground connection. The cam will then contact the snap-action switch, stopping the press. In essence the cam exposes the press to a shutting-off

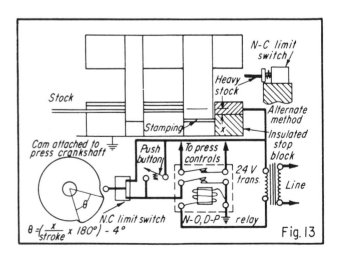

$\theta = \left(\dfrac{x}{stroke} \times 180°\right) - 4°$

Fig. 13

action during the time that the stock should be against the stop block. Should the stock feed up against the stop block and then bounce away the press will stop.

Two Switches in Parallel Protect Dies With Pads and Strippers

Any die that uses a pad or stripper plate can be protected by a circuit that uses two snap-action switches in parallel, Fig. 14. The circuit will stop the press should two parts be stamped or if the pad should stick in its lower position.

NORMAL OPERATION—As press descends, switch No. 1 is closed and current to the solenoid is maintained. Further movement of the ram causes the pad to drop, which then opens switch No. 2. As the ram rises, the pad contacts switch No. 2, closing it. Further movement of the ram causes the spring-loaded finger to leave switch No. 1, opening it. Hence, the switches are alternately opening and closing with a small overlap, so that one switch is always supplying current to solenoid.

DETECTING A MALFUNCTION — Should the pad contain two stampings, switch No. 2 will open before switch No. 1 can close, and the press will stop. During the upstroke if the pad sticks, switch No. 2 will not be closed, switch No. 1 will open, and the press will stop.

The trick in this method is to adjust the overlap to a value which is less than a thickness of stock.

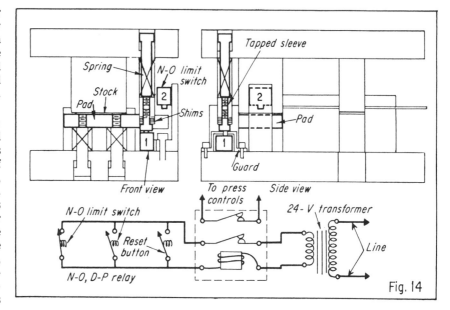

Fig. 14

This adjustment is easily made by placing a piece of stock between the punch and the pad and lowering the ram by hand until the pad just closes switch No. 2. A meter placed across the switch contacts, with the switch wires removed, will indicate when it has been actuated. Enough shim washers are removed from the spring-loaded finger to leave a small gap between the screw and the switch plunger.

A thickness gage is then used to determine the shim washers required to just actuate switch No. 1. The determined thickness of washers plus about 0.010 in. (overlap amount) is added to the spring-loaded finger. The die is now ready to run automatically. The tool should be tested by adding a 0.010 in. plus shim to the pad along with the stock, and tripping the press to see if shut-off occurs.

Cup Switch Protects Compound Dies

A cup switch (see Fig. 1) can be used to protect compound dies. A typical layout is shown in Fig. 15. Because of ejection problems, these dies are particularly vulnerable to breakage. A failure of the knock-out system and a pile-up of stampings can ruin an expensive die and destroy customer good will because of failure to deliver parts on time. This trouble can be avoided by installing the cup switch. Should the knock-out system fail and the stampings pack-up in the die block, the knock-out will push the pin against the insulated cup, breaking the ground connection to the relay circuit (see Fig. 2), immediately stopping the press.

When designing this kind of protection into your compound dies, always make dimension A a minimum of two stock thicknesses. Thus, if the press should make another stroke after shut-off, the die will not hit solid. However, if dimension A is made too large, it will be necessary to shorten the knock-out as the die is ground, in order not to trip the protective device during normal operation.

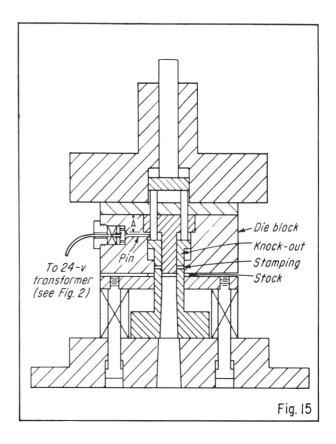

Fig. 15

Stepping-Off Tube Lengths

A pair of electromagnets are now stepping off tube lengths in an extremely ingenious arrangement. The magnets are placed on 1-ft centers about $\frac{3}{32}$ in. above the tube OD and aligned with it (sketch). One magnet is the *write* head, the other the *read* head.

As the tube moves past the heads, a magnetic spot is imprinted on the tube surface by a short electrical pulse through the write head. When the spot reaches the read head an electric current is generated in the head windings. This triggers the next pulse of the write head to imprint a second magnetic spot, which is picked up by the read head to repeat the sequence. Once started, the system is self-sustaining.

The magnetic "imprinter" has been added to a conveyor at the Tube Investment Ltd plant in Walsall, Staffordshire, England. The box-shaped carrier arm is hinged to the conveyor structure. The arm is raised and lowered by an air cylinder which operates against two damping springs on the down stroke. This ensures gentle contact between the *fractional measuring contact wheel*, which holds the read and write heads, and the tube. The measuring wheel has a 1-ft circumference to give accurate fractional measurement.

System was developed by the Steel Tube Development Engineering Dept of the company, which reports that the spots are counted when imprinted to record tube length in feet.

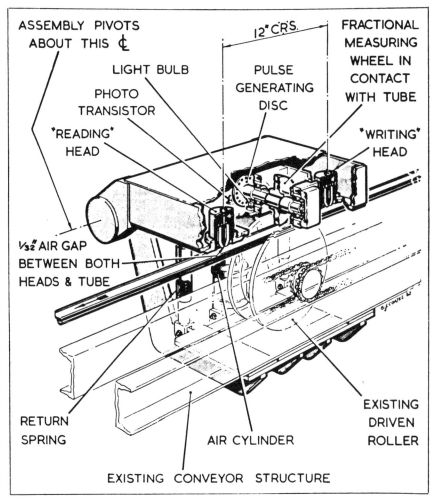

Projector Threads Itself

Film inserted into the mechanism of an 8-mm movie projector is threaded automatically through the shuttle and the drive sprockets. Loop-forming guides determine film path until take-up reel exerts tension.

Designed by Bell & Howell to simplify setting up for home movies, the Auto Load projector threads itself in less than 3 seconds.

Film is inserted . . .

at A after loop-forming guides have been closed. Guides are linked (not shown) so pressure at B will move both of them to the threading position. Top sprocket drives film under roller, through upper loop-forming guide, between aperture and pressure plates, and past shuttle to lower loop-forming guide. Then film is engaged by bottom sprocket and discharged at C. Operator leads film under roller and back to take-up reel. Linkage between roller and guides causes them to move out of threading position when reel absorbs slack and applies tension to film.

HOW LABELING MACHINES WORK

Fingers, suction and glue-pads are some of the ways these machines perform high-speed sleight of hand with stacks of cut paper labels.

J A CUCKSON

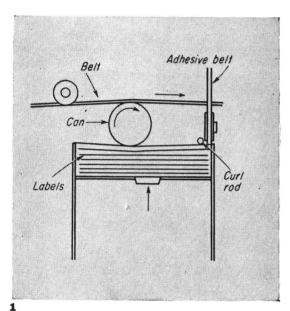

1

There are two ways to put a label on a package. One way is to coat the package with glue so that it picks up the label as flypaper does a fly. The other way is to coat the label with glue and stick it on like a postage stamp.

Flypaper feed

Can-labeling machines label 600 to 750 cans per hour this way. Fig 1 shows how a belt rolls a can streaked with glue over the pile of labels. As it rolls, the can wraps itself in a label. The adhesive belt smears glue on the end of the label to keep it from lifting up.

Labeling boxes is more complex, because boxes won't roll. The machine in Fig 2 uses three stations for the job. A roller puts glue on a rubber pad, which transfers the glue to the package at station I. At station II, a cross-slide presses a stack of labels against the package. When the slide pulls back, it leaves a label behind. Another rubber pad at station III presses the label more firmly to the package.

(continued next page)

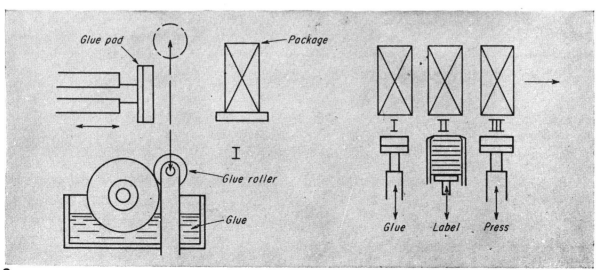

2

LABELING MACHINES — continued

Postage-stamp method

Labels may, indeed, be like postage stamps—preglued so they only have to be moistened. Other labels have pressure-sensitive adhesives, or thermoplastic backing that sticks when it is heated.

However, the machine in Fig 3 puts wet glue on the label just before it goes on the package. A roller spreads glue on each face of the octagonal drum. When the glued face reaches the bottom, it picks a label off the pile that pops up to meet it. At the top of rotation, fingers snatch the label off the drum.

One disadvantage of feeding labels off the top of the pile is the need for a mechanism like a counterweighted piston to keep the stack rising. A stack that deals from the bottom slides down by its own weight. Another advantage of bottom feed is that an operator can replenish the pile without stopping the machine.

Instead of a drum, some machines use a picker (Fig 4). The picker is a plate with a slot big enough for the package to fit through. After picking a label, the picker lowers down over the package and lays on the label.

Suction pickup and transfer

Suction heads can pick up labels dry. In Fig 5 the suction head pulls down one corner of the label, and the

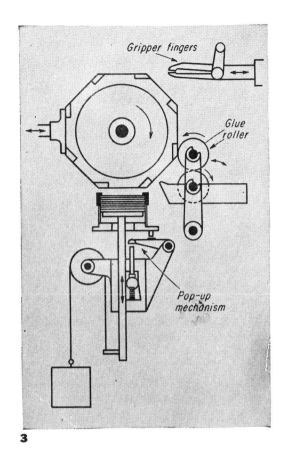

3

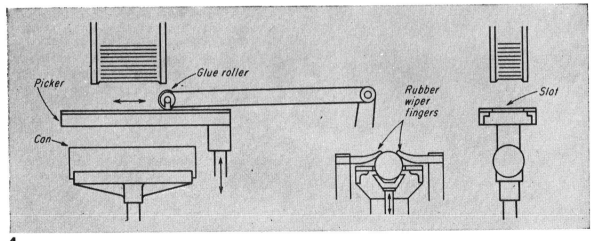

4

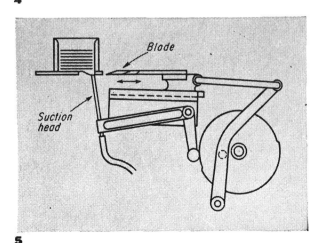

5

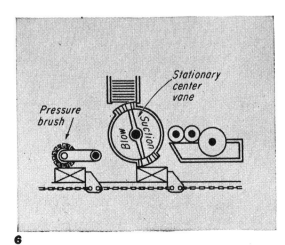

6

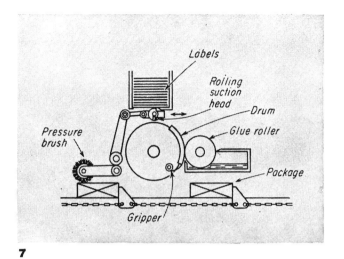

7

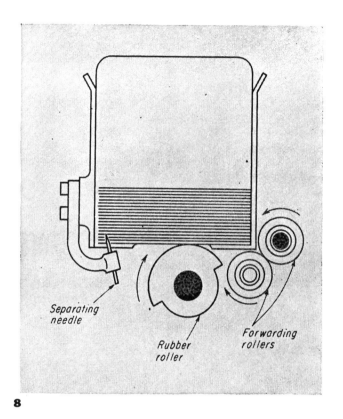

8

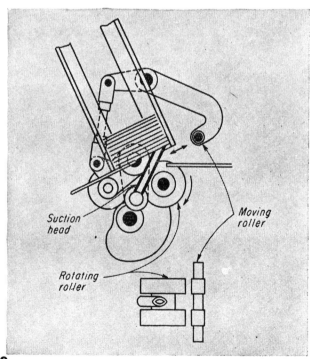

9

blade slides across and peels it off the bottom of the pile, at the same time pushing up the hopper. With suction holes in the bottom, the blade can carry away the label on the return stroke.

The cylinder in Fig 6 both inhales and exhales. Suction picks up the label and holds it for gluing. When the holes rotate to the other side of the center vane, air blows out of them to apply the label.

Friction feed

Between two sheets of smooth, dry paper there is less friction than between the paper and a rubber roller. Needles (Fig 8), which pierce the first five or six sheets on the bottom of the pile, ensure only one sheet leaves at a time. The rubber feed roller pushes the labels into the forwarding rollers.

The machine in Fig 9 uses a combination of suction nozzle and friction. In sequence, the suction nozzle pulls down the end of the label, and then a moving roller moves in against a fixed rotating roller, which pushes the label on its way.

Making labels stick

The last step is to press on the label after it is in place to make sure it sticks. The machine in Fig 2 uses direct pressure at station III. Because each package must stop momentarily under the rubber pad, the machine can run only so fast. Moving the pad in a D-shape path and never stopping the package beats the speed limit that stop-start motion imposes. During the straight-line part of the cycle the pad follows the moving package and presses on the label.

Direct-pressure pads can cover only 140° of a cylinder. Substitute methods are wiper fingers (Fig 4) or rollers arranged to cradle the cylinder.

Flat packages ride on belts under roller brushes, (Fig 6) or, for higher pressures, between roller.

Troubles

Three things commonly cause labels to stick together in the magazine: (1) Edges locked together during cutting. (2) Embossing made by print. (3) Ink that dried after the

labels were stacked. Bending, twisting or riffling will loosen the sheets.

With needles, the grain of the paper must be in the direction of feed so that the tear is clean.

REFERENCE:

"Devices for Feeding Small Cut Sheets," by J. A. Cuckson. Published in *Mechanical World*

ADHESIVE APPLICATORS FOR

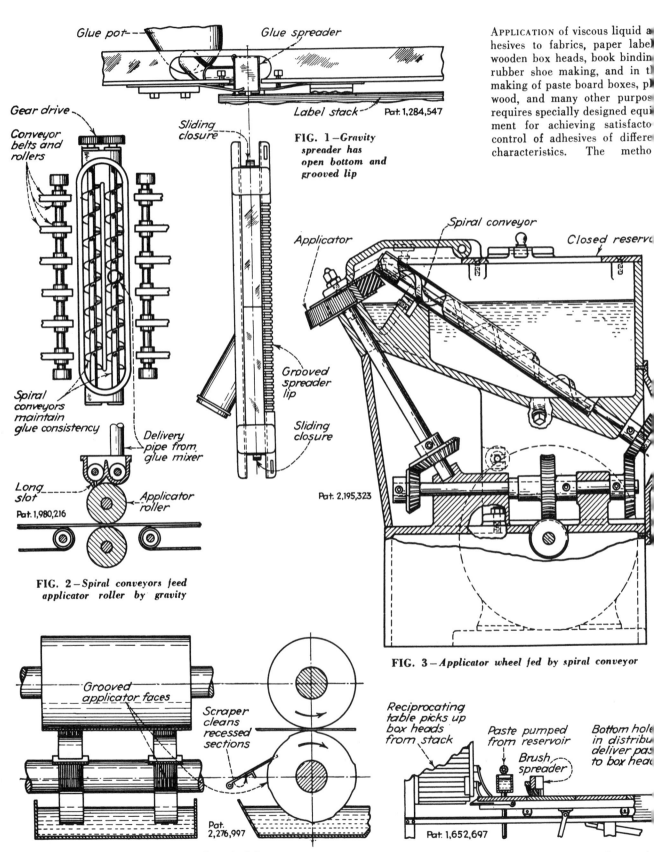

Glue pot

Glue spreader

Sliding closure

Label stack   Pat. 1,284,547

FIG. 1 – *Gravity spreader has open bottom and grooved lip*

Gear drive

Conveyor belts and rollers

Spiral conveyors maintain glue consistency

Delivery pipe from glue mixer

Long slot

Pat. 1,980,216

Applicator roller

Applicator

Grooved spreader lip

Sliding closure

Pat. 2,195,323

FIG. 2 – *Spiral conveyors feed applicator roller by gravity*

Spiral conveyor

Closed reservo

FIG. 3 – *Applicator wheel fed by spiral conveyor*

Grooved applicator faces

Scraper cleans recessed sections

Pat. 2,276,997

FIG. 4 – *Adhesive pattern produced by raised faces on applicator roll*

Reciprocating table picks up box heads from stack

Paste pumped from reservoir

Brush spreader

Bottom hole in distribu deliver pas to box hea

Pat. 1,652,697

FIG. 5 – *Gravity spreader with flow from bottom h*

APPLICATION of viscous liquid a
hesives to fabrics, paper label
wooden box heads, book bindin
rubber shoe making, and in th
making of paste board boxes, pl
wood, and many other purpos
requires specially designed equi
ment for achieving satisfacto
control of adhesives of differe
characteristics. The metho

HIGH-SPEED MACHINES

hown here have been devised to ncorporate the application of adesives in production machines; may be applicable for applying iquid finishes such as lacquers nd paints also.

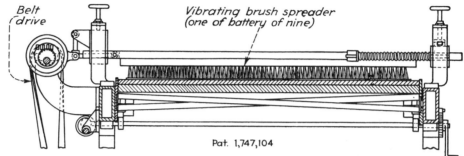

Belt drive

Vibrating brush spreader (one of battery of nine)

Pat. 1,747,104

FIG. 6—*Vibrating brushes spread coating after application by cylindrical brush*

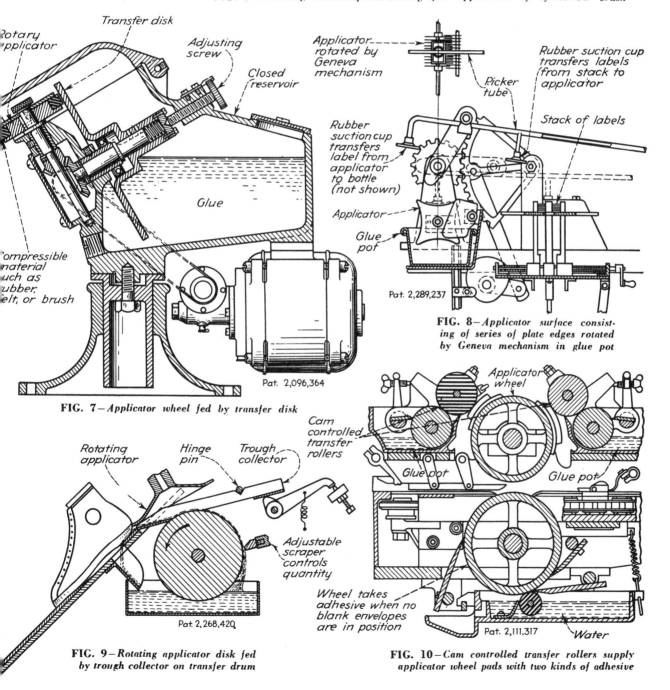

Rotary pplicator

Transfer disk

Adjusting screw

Closed reservoir

Glue

ompressible naterial uch as ubber, elt, or brush

Pat. 2,096,364

FIG. 7—*Applicator wheel fed by transfer disk*

Applicator rotated by Geneva mechanism

Picker tube

Rubber suction cup transfers labels from stack to applicator

Rubber suction cup transfers label from applicator to bottle (not shown)

Stack of labels

Applicator

Glue pot

Pat. 2,289,237

FIG. 8—*Applicator surface consisting of series of plate edges rotated by Geneva mechanism in glue pot*

Rotating applicator

Hinge pin

Trough collector

Adjustable scraper controls quantity

Pat. 2,268,420

FIG. 9—*Rotating applicator disk fed by trough collector on transfer drum*

Applicator wheel

Cam controlled transfer rollers

Glue pot

Glue pot

Wheel takes adhesive when no blank envelopes are in position

Pat. 2,111,317

Water

FIG. 10—*Cam controlled transfer rollers supply applicator wheel pads with two kinds of adhesive*

Adhesive applicators, continued

METHODS OF APPLYING LIQUID ADHESIVES include rotary applicators on movable axes and otherwise movable between adhesive pick-up position and applying position, endless belt applicators, applicators in the form of moving daubers, plates, and the like, reciprocating dies exuding measured quantities of cement, and spray nozzles. All of these mechanisms are used or are applicable on production machines such as for making pasteboard boxes or cartons, pasting labels or envelopes, and making shoes or other products involving the use of liquid adhesives.

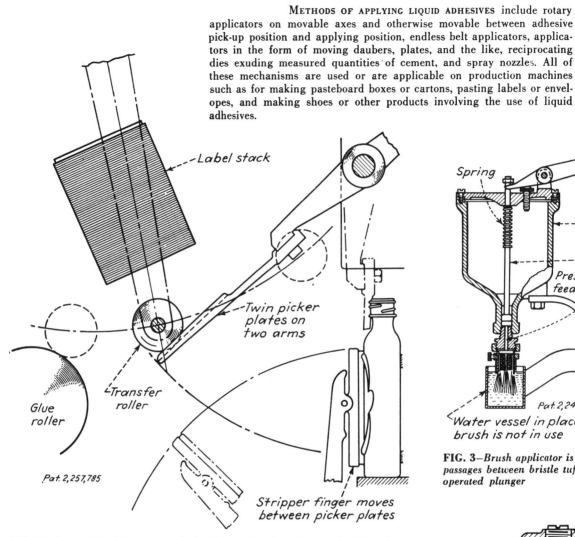

FIG. 1 – Bottom label is spread with glue by two abutting glue-coated picker plates, which separate during contact with label stack, then carry label to bottle

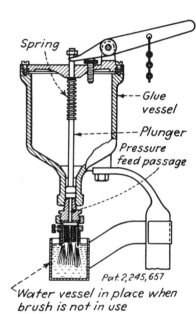

FIG. 3–Brush applicator is fed through passages between bristle tufts by spring operated plunger

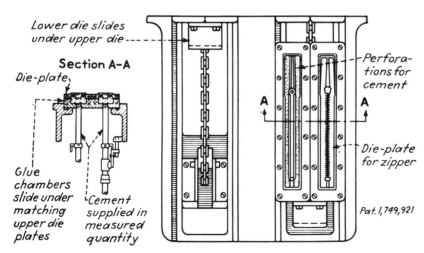

FIG. 2 – Measured quantities of cement are forced through perforations in specially designed upper and lower die plates, which are closed hydraulically over zippers. Lower die only is shown

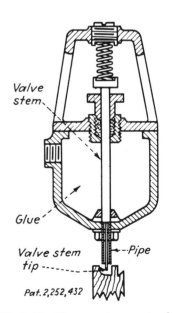

FIG. 4–Shoulder on valve stem in glue chamber retains glue until pressure on tip opens bottom valve

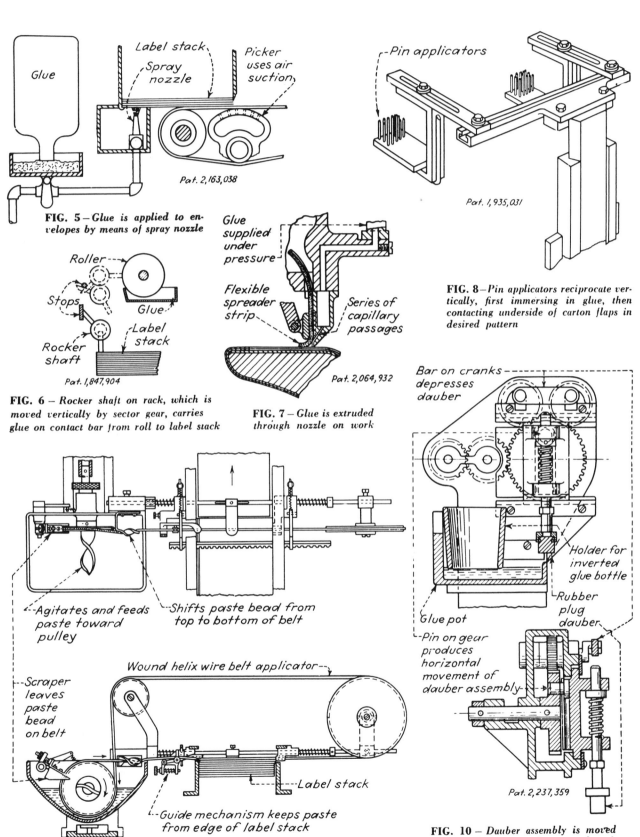

Glue

Label stack

Spray nozzle

Picker uses air suction

Pat. 2,163,038

FIG. 5 – *Glue is applied to envelopes by means of spray nozzle*

Roller

Stops

Glue

Rocker shaft

Label stack

Pat. 1,847,904

FIG. 6 – *Rocker shaft on rack, which is moved vertically by sector gear, carries glue on contact bar from roll to label stack*

Glue supplied under pressure

Flexible spreader strip

Series of capillary passages

Pat. 2,064,932

FIG. 7 – *Glue is extruded through nozzle on work*

Pin applicators

Pat. 1,935,031

FIG. 8 – *Pin applicators reciprocate vertically, first immersing in glue, then contacting underside of carton flaps in desired pattern*

Agitates and feeds paste toward pulley

Shifts paste bead from top to bottom of belt

Scraper leaves paste bead on belt

Wound helix wire belt applicator

Label stack

Guide mechanism keeps paste from edge of label stack

Pat. 2,206,964

FIG. 9 – *Paste belt applicator passes around pulley in pastepot and slides over label stack*

Bar on cranks depresses dauber

Holder for inverted glue bottle

Glue pot

Rubber plug dauber

Pin on gear produces horizontal movement of dauber assembly

Pat. 2,237,359

FIG. 10 – *Dauber assembly is moved horizontally between glue pot and work by eccentric pin on gear. Vertical movements are produced by crank operated bar over dauber shaft*

AUTOMATIC STOPPING MECHANISMS

Many machines, particularly automatically operated production machines, may damage themselves or parts being processed unless they are equipped with devices that stop the machine or cause it to skip an operation when something goes wrong. The accompanying patented mechanisms show principles that can be employed to interrupt normal machine operations: Mechanical, electrical or electronic, hydraulic, pneumatic, or combinations of these means. Endless varieties of each method are in use.

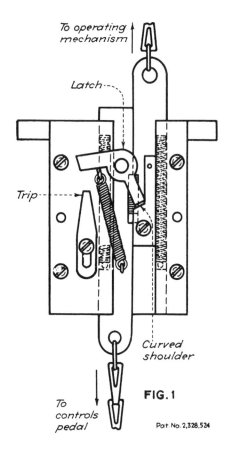

FIG. 1

Pat. No. 2,328,524

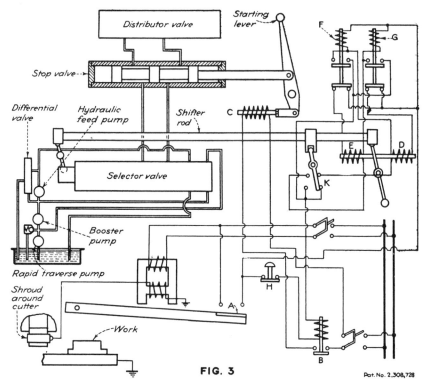

FIG. 3

Pat. No. 2,308,728

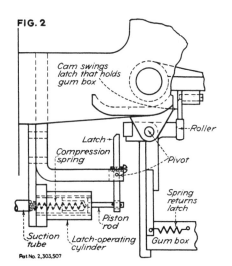

FIG. 2

Pat. No. 2,303,507

Fig. 1—Repetition of machine cycle is prevented if pedal remains depressed. Latch carried by left slide pushes right slide downward by means of curved shoulder until latch is disengaged by trip member.

Fig. 2—Gumming of suction picker and label carrier when label is not picked up by the suction, is prevented by insufficient suction on latch-operating cylinder, caused by open suction holes on picker. When latch-operating cylinder does not operate, gum box holding latch returns to holding position after cyclic removal by cam and roller, thus preventing gum box and rolls from rocking to make contact with picker face.

Fig. 3—Damage to milling cutter, work or fixtures is prevented by shroud around cutter, which upon contact closes electric circuit through relay, thus closing contact A. This causes contact B to close, thus energizing relay C to operate stop valve, and closes circuit through relay D, thus reversing selector valve by means of shifter rod so that bed travel will reverse on starting. Simultaneously, relay F opens circuit of relay E and closes a holding circuit that was broken by the shifter lever at K. Relay G also closes a holding circuit and opens circuit through relay D. Starting lever, released by push button H, releases contact A and returns circuit to normal. If contact is made with shroud when bed travel is reversed, interchange D and E, and F and G in above sequence of operations.

FOR FAULTY MACHINE OPERATION

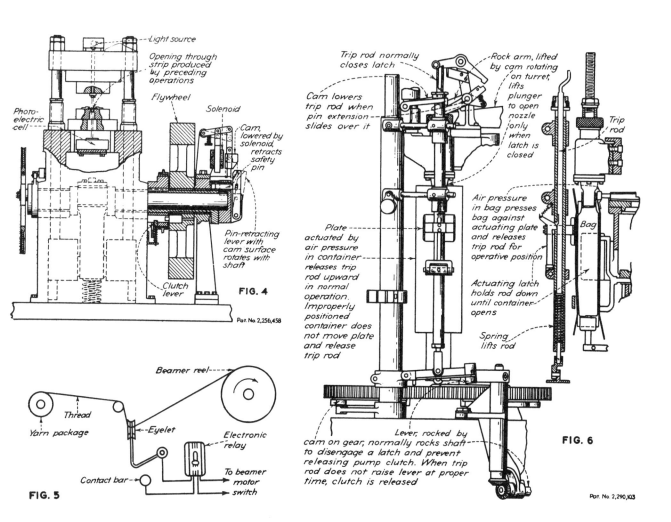

Fig. 4—High-speed press is stopped when metal strip advances improperly so that hole punched in strip fails to match with opening in die block to permit passage of light beam. Intercepted light beam to photo-electric cell results in energizing solenoid and withdrawal of clutch pin.

Fig. 5—Broken thread permits contact bar to drop, thereby closing electronic relay circuit, which operates to stop beamer reeling equipment.

Fig. 6—Nozzle on packaging machine does not open when container is not in proper position.

Fig. 7—Obstruction under explorer foot of wire-stitching machine prevents damage to machine by raising a vertical plunger, which releases a latch lever so that rotary cam raises lever that retains clutch operating plunger.

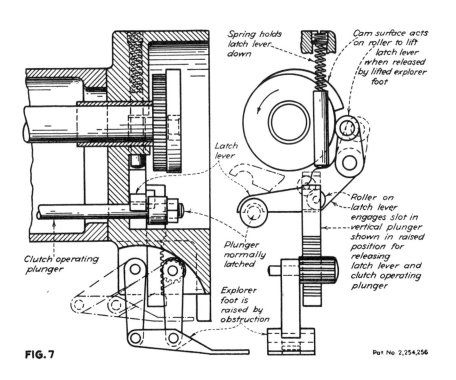

Such devices, which prevent automatic machines from damaging themselves or the work passing through them, make use of mechanical, electrical, hydraulic and pneumatic principles. Typical mechanisms illustrated prevent excess speed, misweaving, jamming of toggle press and food canning machines, operation of printing press when the paper web breaks, improper feeding of wrapping paper, and uncoordinated operation of a glass making machine.

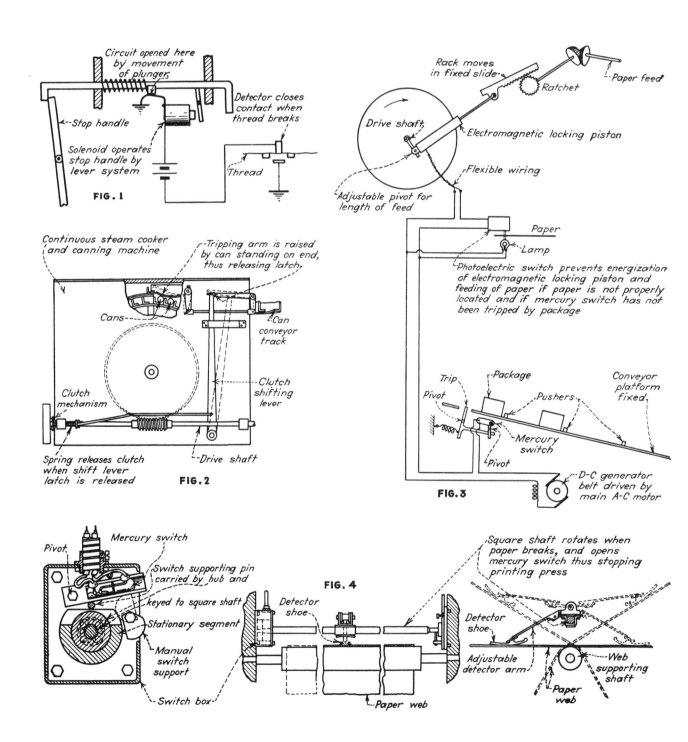

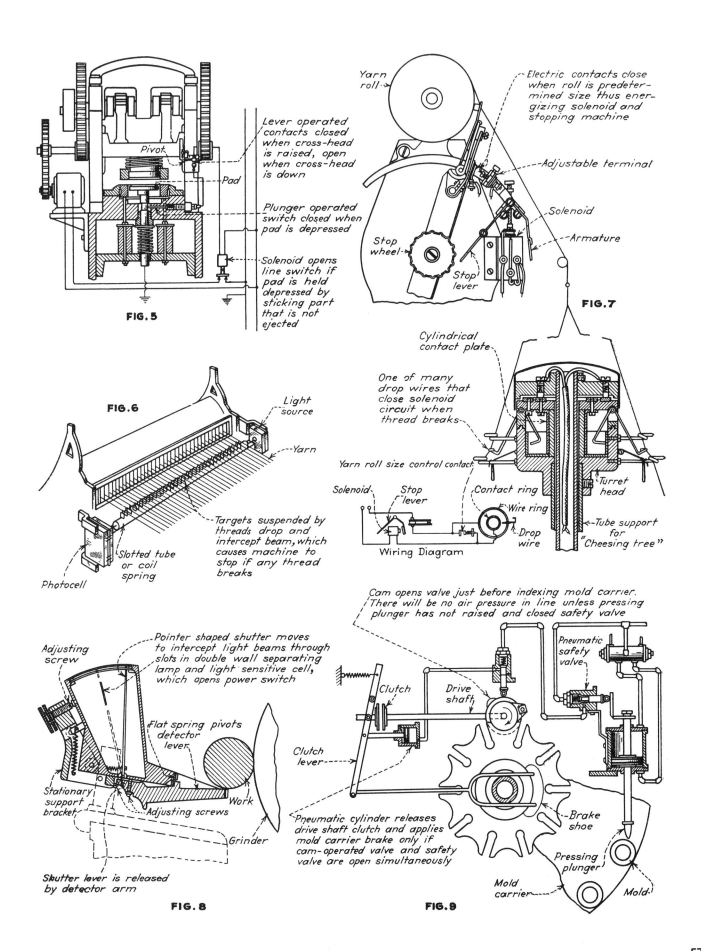

FIG. 5

Lever operated contacts closed when cross-head is raised, open when cross-head is down

Pivot

Pad

Plunger operated switch closed when pad is depressed

Solenoid opens line switch if pad is held depressed by sticking part that is not ejected

Yarn roll

Electric contacts close when roll is predetermined size thus energizing solenoid and stopping machine

Adjustable terminal

Solenoid

Armature

Stop wheel

Stop lever

FIG. 7

FIG. 6

Light source

Yarn

Targets suspended by threads drop and intercept beam, which causes machine to stop if any thread breaks

Slotted tube or coil spring

Photocell

Cylindrical contact plate

One of many drop wires that close solenoid circuit when thread breaks

Yarn roll size control contact

Solenoid

Stop lever

Contact ring

Wire ring

Drop wire

Turret head

Tube support for "Cheesing tree"

Wiring Diagram

Adjusting screw

Pointer shaped shutter moves to intercept light beams through slots in double wall separating lamp and light sensitive cell, which opens power switch

Flat spring pivots detector lever

Stationary support bracket

Adjusting screws

Work

Grinder

Shutter lever is released by detector arm

FIG. 8

Cam opens valve just before indexing mold carrier. There will be no air pressure in line unless pressing plunger has not raised and closed safety valve

Pneumatic safety valve

Clutch

Drive shaft

Clutch lever

Pneumatic cylinder releases drive shaft clutch and applies mold carrier brake only if cam-operated valve and safety valve are open simultaneously

Brake shoe

Pressing plunger

Mold carrier

Mold

FIG. 9

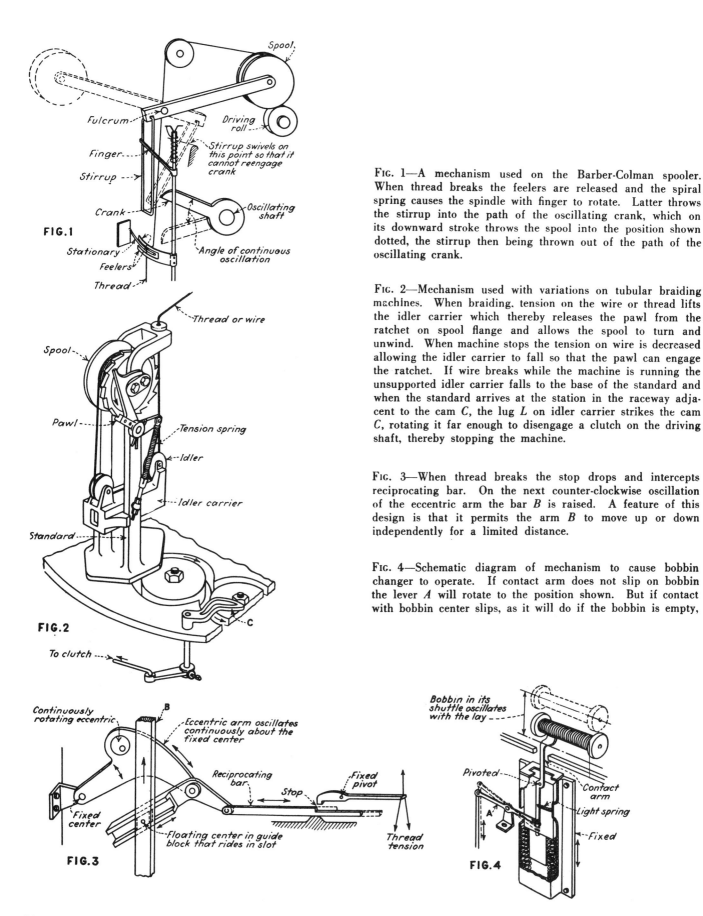

FIG.1

FIG.2

FIG.3

FIG.4

Fig. 1—A mechanism used on the Barber-Colman spooler. When thread breaks the feelers are released and the spiral spring causes the spindle with finger to rotate. Latter throws the stirrup into the path of the oscillating crank, which on its downward stroke throws the spool into the position shown dotted, the stirrup then being thrown out of the path of the oscillating crank.

Fig. 2—Mechanism used with variations on tubular braiding machines. When braiding, tension on the wire or thread lifts the idler carrier which thereby releases the pawl from the ratchet on spool flange and allows the spool to turn and unwind. When machine stops the tension on wire is decreased allowing the idler carrier to fall so that the pawl can engage the ratchet. If wire breaks while the machine is running the unsupported idler carrier falls to the base of the standard and when the standard arrives at the station in the raceway adjacent to the cam C, the lug L on idler carrier strikes the cam C, rotating it far enough to disengage a clutch on the driving shaft, thereby stopping the machine.

Fig. 3—When thread breaks the stop drops and intercepts reciprocating bar. On the next counter-clockwise oscillation of the eccentric arm the bar B is raised. A feature of this design is that it permits the arm B to move up or down independently for a limited distance.

Fig. 4—Schematic diagram of mechanism to cause bobbin changer to operate. If contact arm does not slip on bobbin the lever A will rotate to the position shown. But if contact with bobbin center slips, as it will do if the bobbin is empty,

Designs shown here diagrammatically were taken from textile machines, braiding machines and packaging machines. Possible modifications of them to suit other applications will be apparent

lever *A* will not rotate to position indicated by dashed line, thereby causing bobbin changer to come into action

F\ɪɢ. 5—Simple type of stop mechanism for limiting the stroke of a reciprocating machine member. Arrows indicate the direction of movements.

F\ɪɢ. 6—When the predetermined weight of material has been poured on the pan, the movement of the scale beam pushes the latch out of engagement, allowing the paddle wheel to rotate and thus dump the load. The scale beam drops, thereby returning the latch to the holding position and stopping the wheel when the next vane hits the latch.

F\ɪɢ. 7—In this textile machine any movement that will rotate the stop lever counter-clockwise will bring it in the path of the continuously reciprocating shaft. This will cause the catch lever to be pushed counter-clockwise and the hardened steel stop on the clutch control shaft will be freed. A spiral spring then impels the clutch control shaft to rotate clockwise, which movement throws out the clutch and applies the brake. Initial movement of the stop lever may be caused by the breaking of a thread, a moving dog, or any other means.

F\ɪɢ. 8—Arrangement used on some package loading machines to stop machine if a package should pass loading station without receiving an insert. Pawl finger *F* has a rocking motion obtained from crankshaft, timed so that it enters the unsealed packages and is stopped against the contents. If the box is not filled the finger enters a considerable distance and the pawl end at bottom engages and holds a ratchet wheel on driving clutch which disengages the machine driving shaft.

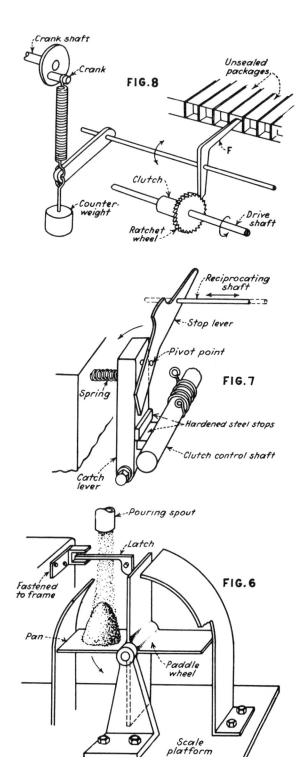

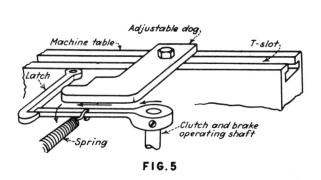

FIG. 5

ELECTRICAL AUTOMATIC-STOP MECHANISMS

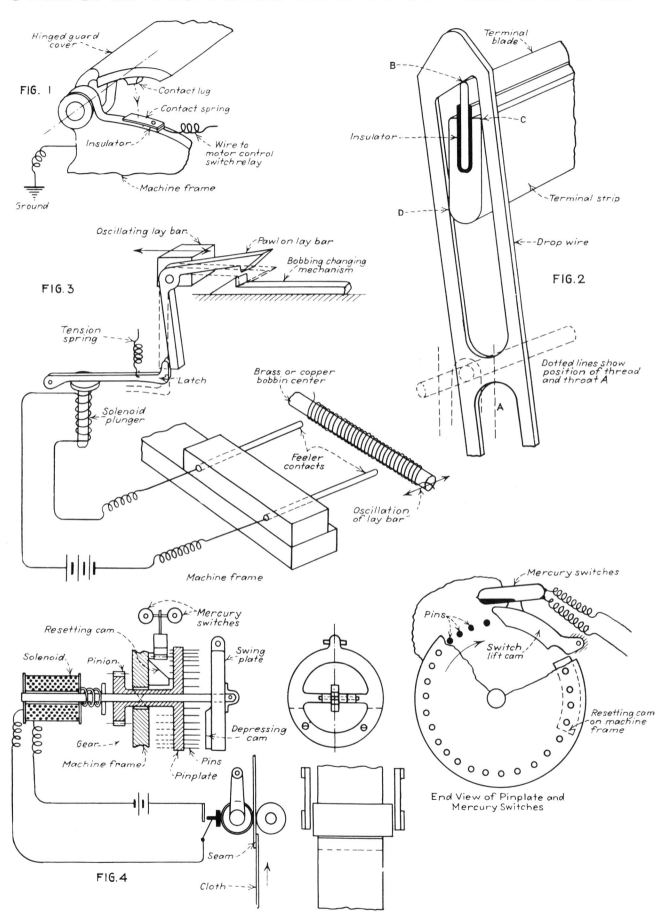

FIG. 1

Hinged guard cover

Contact lug

Contact spring

Insulator

Wire to motor control switch relay

Machine frame

Ground

FIG. 2

Terminal blade

B

Insulator

C

D

Terminal strip

Drop wire

Dotted lines show position of thread and throat A

A

FIG. 3

Oscillating lay bar

Pawl on lay bar

Bobbing changing mechanism

Tension spring

Latch

Solenoid plunger

Brass or copper bobbin center

Feeler contacts

Oscillation of lay bar

Machine frame

Mercury switches

Resetting cam

Solenoid

Pinion

Swing plate

Depressing cam

Gear

Pins

Machine frame

Pinplate

Seam

FIG. 4

Cloth

Mercury switches

Pins

Switch lift cam

Resetting cam on machine frame

End View of Pinplate and Mercury Switches

Arrangements and designs diagrammatically illustrated here were taken from packaging machines and textile machines. Modifications of them to perform other operations can be easily devised

Fig. 1—Safety arrangement used on some machines to stop motor when guard cover is lifted. Circuit is complete only when cover is down, in which position contact lug has metal to metal connection with contact spring, completing circuit to relay.

Fig. 2—Electrical 3-point wedging type "warp stop" shown after thread has broken and drop wire has fallen and tilted to close circuit. Dotted lines indicate normal position of drop wire when riding on thread. When thread breaks the drop wire falls and strikes the top of terminal blade at B, the inclined top of the slot causing a wedging effect which tilts drop wire against the terminal strip at C and D intensifying the circuit closing action.

Fig. 3—Bobbin changer. When bobbin is empty the feeders contact the metal bobbin center, completing the circuit through a solenoid which pulls a latch that causes bobbin changing mechanism to operate and put a new bobbin in the shuttle. As long as the solenoid remains deenergized, the pawl on the lay bar will be raised clear of the hook on the bobbin changing mechanism.

Fig. 4—Control for automatic shear. When a seam of two thicknesses of cloth passes between the rolls the swing roller is moved outward and closes a sensitive switch which energizes a solenoid. Action of solenoid pulls in an armature the outer end of which is attached to a hinged ring to which a cam plate is fastened. The cam plate depresses a number of pins in a rotating plate. As the plate rotates the depressed pins lift a hinged cam arm on which are mounted two mercury switches which, when tilted, complete circuits in two motor controls. Fastened on the frame of the machine is a resetting cam for pushing the depressed pins back to their original position. In this arrangement two motors are stopped and reversed until seam has passed through rollers, then stopped and reversed again.

Fig. 5—Electric stop for loom. When thread breaks or slackens the drop wire falls and contact A rides on contact C. The drop wire being supported off center swings so that contact B is pulled against inner terminal strip D completing solenoid circuit.

Fig. 6—Automatic stop for folder or yarder to stop machine always in the same position when a seam in the cloth passes between the rolls. A seam passing between the rolls causes swivel mounted roll to lift slightly. This motion closes contacts in sensitive switch which throws relay in control box, so that the next time the cam closes the limit switch the power of motor with integral magnetic brake is shut off. The brake stops the machine always in the same place.

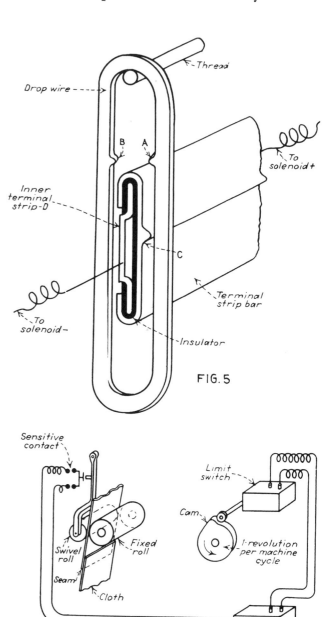

FIG. 5

FIG. 6

AUTOMATIC SAFETY MECHANISMS

THE most satisfactory automatic guard mechanisms for preventing injury to machine operators are those that have been designed with the machine. When properly designed they (1) do not reduce visibility, (2) do not impede the operator, (3) do not cause painful blows on the operator's hand in avoiding serious injury, (4) are safe with respect to wear in the safety mechanism, (5) are sensitive and instantaneous in operation, and (6) render the machine inoperative if tampered with or removed.

Safety devices range from those that keep both hands occupied with controls away from the work area to guards that completely inclose the work during operation of the machine and prevent operation of the machine unless so protected. The latter might include the "electric eye," which is the activating means of one of the mechanisms illustrated.

Perspective of Slide

Pat. No. 2,301,817

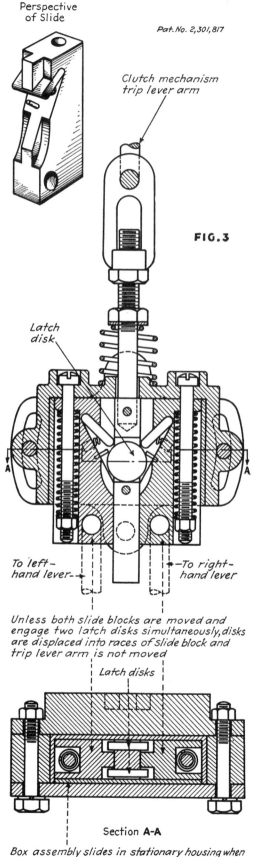

Clutch mechanism trip lever arm

FIG.3

Latch disk

To left-hand lever →

← To right-hand lever

Unless both slide blocks are moved and engage two latch disks simultaneously, disks are displaced into races of slide block and trip lever arm is not moved

Latch disks

Section A-A

Box assembly slides in stationary housing when slide blocks move together

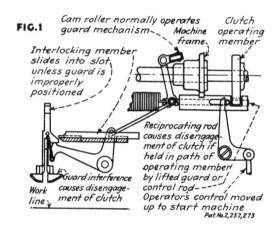

FIG.1

Cam roller normally operates guard mechanism

Machine frame

Clutch operating member

Interlocking member slides into slot unless guard is improperly positioned

Reciprocating rod causes disengagement of clutch if held in path of operating member by lifted guard or control rod. Operator's control moved up to start machine

Guard interference causes disengagement of clutch

Work line

Pat. No. 2,257,273

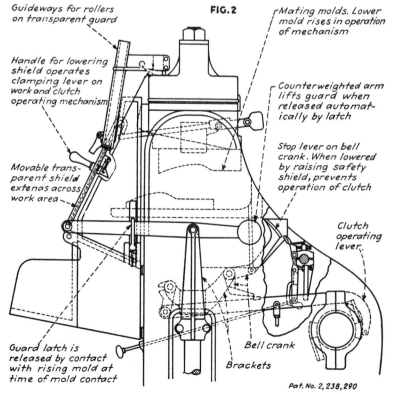

Guideways for rollers on transparent guard

FIG.2

Mating molds. Lower mold rises in operation of mechanism

Handle for lowering shield operates clamping lever on work and clutch operating mechanism

Counterweighted arm lifts guard when released automatically by latch

Stop lever on bell crank. When lowered by raising safety shield, prevents operation of clutch

Movable transparent shield extends across work area

Clutch operating lever

Guard latch is released by contact with rising mold at time of mold contact

Bell crank

Brackets

Pat. No. 2,238,290

FOR OPERATING MACHINES

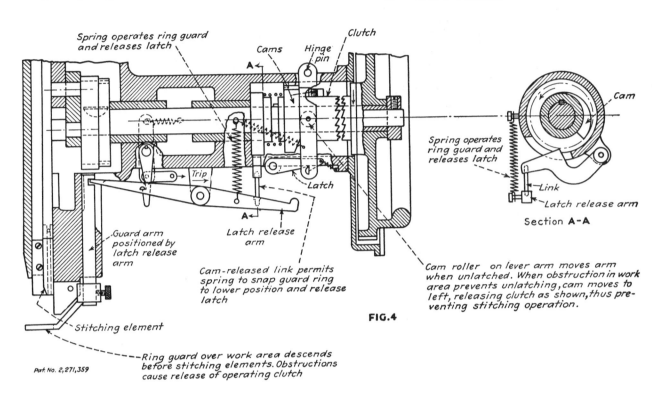

Spring operates ring guard and releases latch

Cams

Hinge pin

Clutch

Trip

Latch

Guard arm positioned by latch release arm

Latch release arm

Cam-released link permits spring to snap guard ring to lower position and release latch

Stitching element

Pat. No. 2,271,359

Ring guard over work area descends before stitching elements. Obstructions cause release of operating clutch

Spring operates ring guard and releases latch

Cam

Link

Latch release arm

Section A-A

Cam roller on lever arm moves arm when unlatched. When obstruction in work area prevents unlatching, cam moves to left, releasing clutch as shown, thus preventing stitching operation.

FIG.4

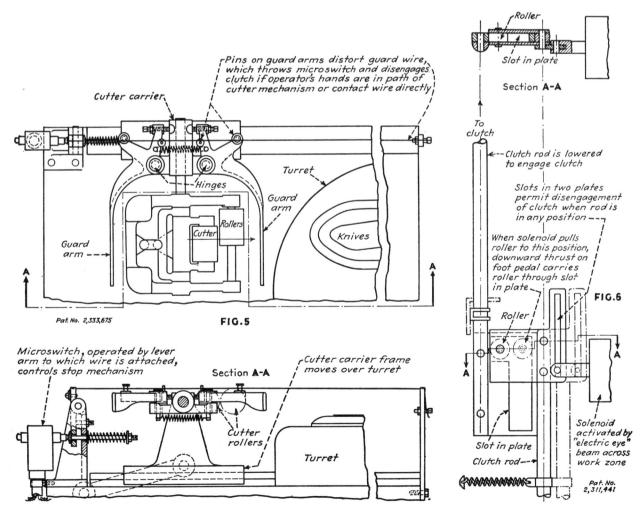

Pins on guard arms distort guard wire, which throws microswitch and disengages clutch if operator's hands are in path of cutter mechanism or contact wire directly

Cutter carrier

Hinges

Turret

Guard arm

Cutter

Rollers

Knives

Guard arm

Pat. No. 2,333,675

FIG.5

Microswitch, operated by lever arm to which wire is attached, controls stop mechanism

Section A-A

Cutter carrier frame moves over turret

Cutter rollers

Turret

Roller

Slot in plate

Section A-A

To clutch

Clutch rod is lowered to engage clutch

Slots in two plates permit disengagement of clutch when rod is in any position

When solenoid pulls roller to this position, downward thrust on foot pedal carries roller through slot in plate

Roller

FIG.6

Slot in plate

Clutch rod

Solenoid activated by "electric eye" beam across work zone

Pat. No. 2,311,441

NEW MECHANISM STOPS SPIN

MERL D. CREECH
RICHARD K. FERGI...

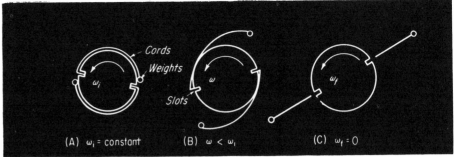

1 . . THREE STAGES OF DESPIN: (A) Spin velocity, ω_i, constant; two weights held in place. (B) Weights have been released and have begun to unwrap; spin velocity has slowed. (C) Spin velocity is zero; weights and cords fly out of slots. Photo shows test model.

How can you instantly and inexpensively stop the spin of a high-speed rotating mass—such as a gas turbine rotor—in an emergency? A problem faced in stopping the spin of a space vehicle in flight has brought a solution to this general problem. This answer is a device unofficially dubbed the "yo-yo despinner."

The device does its job by releasing two small diametrically opposed weights that are connected by cords (actually, Hugo will use flexible metal tapes) to the spinning body, Fig 1. This transfers the total angular momentum of the spinning mass to the two weights—which are automatically jettisoned at the end of the unwinding cycle.

• Once the proper weights and cord length have been selected for the spinning mass, release of the weights will stop the spinning no matter what the initial speed —1 rpm or 100,000 rpm.

• The faster the spin, the quicker will be the stop.

• Braking is smooth—the spinning mass is not jerked to a stop. There is no rebound.

• The device can be designed to cause the spinning mass to reverse its direction of rotation—to a predetermined reverse-spin velocity.

• Tolerances on cord length and weights are fairly liberal to obtain desired results.

Here's how the yo-yo stops spin: The weights are released and allowed to unwind from the spinning cylindrical section, Fig. 1 (B). Path of the weight relative to the cylinder is a circular involute, from point A to point B. At point B, the cord is tangential to the point of attachment, point P. From point B to point B′, the weight follows a circular path, pivoting about point P.

The system can be designed to stop the spin when the weights are at the end of the involute path, B. But the weights must be dropped off when the spin stops, and this is complicated at point B. However, the system can be designed so that spin doesn't stop until the weights are at B′ on the circular path; if the cords are retained only by a slot, they will fly free at the right instant— when the weight is in line with OP.

If the weights are not released at the proper time, the tension in the cord would cause the central body to have an angular acceleration in the direction of its initial spin. This acceleration would continue until the cords were fully wound in the opposite direction, at which time

the angular velocity of the system would be the same as the initial rate. Following this instant, the small weights would unwind in straight lines tangential to the central cylinder and start a recycling process.

Two sets of equations have been developed—the first set is for the case where despinning occurs at end of the involute path (tangential release of the weights); the second set is for the case where despinning occurs at end of the circular path (radial release).

DESIGNING FOR TANGENTIAL RELEASE

Maximum cord tension:

$$T_{max} = \frac{3 I \omega^2}{4}\left[\frac{3m}{2(I + 2ma^2)}\right]^{1/2} \quad (1)$$

Length of unwound cord:

$$R = a\omega t \quad (2)$$

Elapsed time to stop the central body:

$$t_{\theta=0} = \left[\frac{I + 2ma^2}{2ma^2\omega^2}\right]^{1/2} \quad (3)$$

Required length of cord to unwind tangentially from the cylinder to cause the central body to stop its spin:

$$R_{\theta=0} = \left[\frac{I + 2ma^2}{2m}\right]^{1/2} \quad (4)$$

Final spin velocity for other cord lengths

$$\theta = \left[\frac{I + 2ma^2 - 2mR^2}{I + 2ma^2 + 2mR^2}\right]\omega \quad (5)$$

DESIGNING FOR RADIAL RELEASE

Despin-mechanism design is simplified by allowing the weights to continue their motion until the cords are extended radially outward from the central body. This also allows looser tolerances of the component parts. The equation for the required length of cord to produce zero spin velocity at the end of the circular path has been derived as:

$$R_{\theta=0} = \left[\frac{I + 2ma^2}{2m}\right]^{1/2} - a \quad (6)$$

This equation shows the same independence of initial spin velocity as does Eq (4). A comparison shows that for radial release, the cord length is reduced by an amount equal to the radius of the central body.

SYMBOLS

a = Radius of central-body cylindrical section, ft

m = Point mass and mass of point mass, slugs

R = Length of unwound cord or metal tape, ft

t = Elapsed time from initial start, sec

T = Tension in cord, lb

I = Centroidal mass moment of inertia of central body about spin axis, slug-ft²

θ = Spin velocity, radians/sec

ψ = $(\beta + \theta - \phi)$, additional angle used for solution

ω = θ at $t = 0$ (initial spin velocity), radians/sec

NUMERICAL PROBLEM—INVOLUTE PATH

$m = 2.07$ lb$_m$ = 0.0643 slug (where lb$_m$ is lb-mass); M = mass of central body = 40 lb$_m$; K = radius of gyration of central body = 0.219 ft.; a = 0.279 ft.; $\omega = 20 \pi$ rad/sec

from which $I = MK^2/g = 0.06$ slug-ft²

and $2\,ma^2 = 0.01$ slug-ft²

CALCULATIONS

Using Eq (1) for maximum cord tension:

$$T_{max} = \frac{(3)\,(0.06\ \text{slug-ft}^2)\,(20\pi)^2}{(4)\,(\text{sec}^2)} \times$$

$$\left[\frac{(3)\,(0.0643\ \text{slug})}{(2)\,(0.06 + 0.01)\,(\text{slug-ft}^2)}\right]^{1/2}$$

$T_{max} = 208.5$ lb force

Using Eq (3) for the elapsed time t for the central body to stop spinning:

$$t_{\theta\,\to\,0} = \left[\frac{(0.06 + 0.01)\,(\text{slug-ft}^2\text{-sec}^2)}{(0.01)\,(\text{slug-ft}^2)\,(20\pi)^2}\right]^{1/2}$$

$$= 0.042\ \text{sec}$$

Using Eq (4) for the required length of cord to be unwound tangentially from the cylinder to cause the central-body spin to decrease to zero:

$$R_{\theta\,-} = \left[\frac{0.06 + 0.01}{2\,(0.0643)}\right]^{1/2} = 0.739\ \text{ft}$$

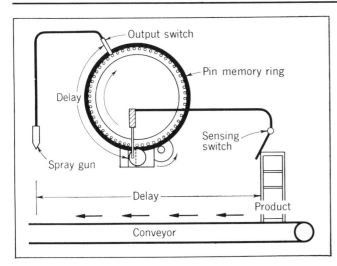

Electromechanical memory ring

Memory ring receives an electrical signal, records it, and delays the signal mechanically. Then, it transmits the signal electrically to perform a function.

Unaffected by timing and speed variations, it can be synchronized with a conveyor or other moving equipment and stops, starts, increases, or decreases speed without affecting the signal output.

Used in electrical circuits that require a memory or delay function related to mechanical motion, it contains 120 sliding pins in a rotating ring. These sliding pins represent the memory storage units. Binks Mfg. Co., 3114 W. Carroll Ave., Chicago, Ill. 60612.

Mechanical time-delay device controls luggage conveyers

Additional microswitches can be added, depending on control function.

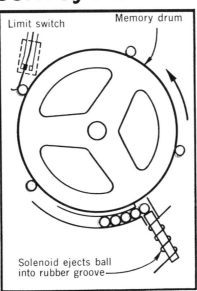

Solenoid ejects a steel ball into grooved rubber track fitted to rim of drum.

A simple mechanical time delay controls the luggage conveyor system for one of the major airlines at New York's Kennedy International Airport.

To handle the growing traffic volume at the main airports in the U.S., conveyor systems must be able to sort out 600 pieces of luggage every hour and ensure that each piece gets to the right aircraft.

This mechanical timer was designed by Robert J. Ebbert of Ebbert Engineering, Troy, Mich. It is made up of several drums mounted on a common shaft that is driven directly from the conveyer by a fixed gearing system. Each drum is fitted with a ball-inserting device, a limit switch, and a ball-storage tube that also removes the balls from a grooved rubber track that's fitted to the rim of the drum.

continued, next page

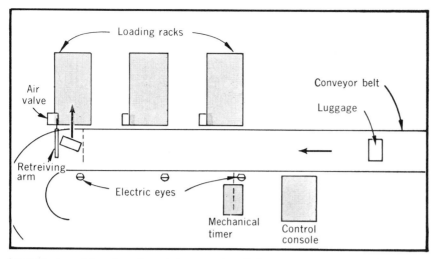

Luggage is retrieved and sent to correct exit by an arm controlled by an air valve. Air valve is activated by signals from drum microswitch and electric eye.

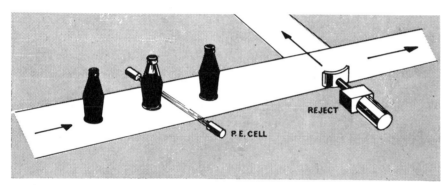

In bottle-filling, an electric eye detects deficient level of the liquid in filled bottles and signals the ball memory. The bottle is then transferred to a refill station.

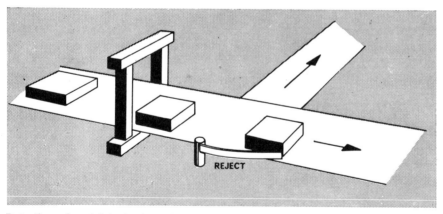

Detection of metal in food products actuates ball memory. When the ball passes under the limit switch, a deflection arm removes the product from the conveyor.

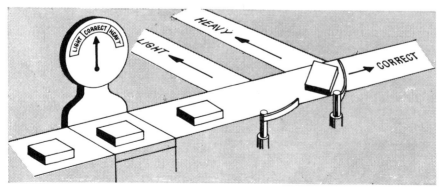

In sorting by weight, the scale automatically starts the memory sequence in proper weight classification. The wheel actuates the discharge fence as item arrives.

On the rim of each wheel is a grooved Neoprene track in which balls are wedged to "follow" the travel of the item that's being monitored.

How the timer works. When a piece of luggage is put on the conveyer belt at the check-in counter, it is tagged so the dispatcher can direct it to the required loading rack. When the luggage nears the dispatcher, he identifies the flight number and presses a button on an exit-control console, which activates a solenoid on the mechanical timer.

The energized solenoid then ejects a steel ball into the rubber grooved track, and the rotating drum carries the ball around until it hits a microswitch near the rim of the drum (diagram, p.65). When the switch is activated by the ball, it transmits a signal to an air valve that eventually operates an arm that retrieves the luggage from the conveyer belt and directs it onto a trolley for the correct flight.

Microswitch and electric eye. After the ball passes by the microswitch, it continues in the rim until it is picked up by the return tube. As soon as the microswitch is closed, it again transmits a signal, this time to an electric eye mounted near the correct exit on the conveyer belt.

When the baggage crosses the electric eye, a signal is transmitted to the solenoid-operated air valve. The air valve solenoid must receive two signals to operate the retrieving arm; one from the drum microswitch, and the other from the electric eye.

The mechanical time-delay system at Kennedy Airport has eleven drums, and each drum actuates a different luggage exit. Should more exits be needed as air traffic increases, then more drums can be added to the timer.

Ebbert said that, in other applications, more microswitches added to each drum could increase the scope of the timer. One advantage of this memory system is that unlike a cam, which is limited to a particular action, the mechanical-timer drum forgets after one cycle.

When operated, the solenoids can release a ball into the drum at any time interval, thereby varying the number of times the switches are operated. On the other hand, a cam is limited to a fixed number.

CHAPTER 2.

RECIPROCATING AND GENERAL-PURPOSE MECHANISMS

Gears and eccentric disk combine in quick indexing

This versatile indexing mechanism . . .

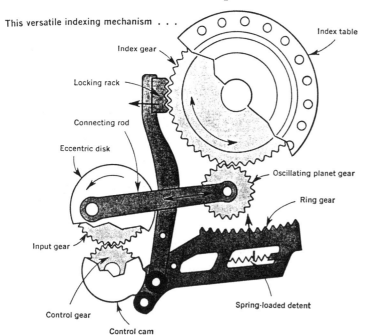

Index table

Index gear

Locking rack

Connecting rod

Eccentric disk

Oscillating planet gear

Ring gear

Input gear

Control gear

Control cam

Spring-loaded detent

Indexing mechanism is installed at base of rotary table

. . provides choice of indexing modes

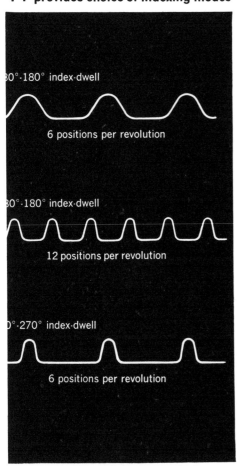

0°-180° index-dwell

6 positions per revolution

0°-180° index-dwell

12 positions per revolution

0°-270° index-dwell

6 positions per revolution

Both stops and dwell are adjustable.

An ingenious intermittent mechanism that looks Rube Goldbergish with its multiple gears, gear racks, and levers provides new smoothness and flexibility in converting constant rotary motion into a start-and-stop type of indexing.

It works equally well for high-speed operations, as fast as 2 sec per cycle, including index and dwell, or for slow-speed assembly functions. Its builder, Gilman Engineering & Mfg Co, Janesville, Wis, is confident enough to adopt it as the mechanical brain of its new Indexomatic rotary table.

The new device minimizes shock loads and offers more versatility than the indexing cams and genevas usually employed to convert rotary motion into start-stop indexing. The number of stations (stops) per revolution of the table can easily be changed, and so can the period of dwell during each stop.

Advantages. This flexibility broadens the scope of the automatic machine operations—feeding, sorting, packaging, weighing, etc—that the rotary table can perform. But the Gilman design offers other advantages, too:

• Use of gears instead of cams makes the device cheaper to manufacture, because gears are simpler to machine.

• The all-mechanical interlocked system achieves an absolute time relationship between motions.

• Gearing is arranged so that the machine automatically goes into a dwell when it is overloaded, preventing damage during jam-ups.

• Its built-in anti-backlash gear system averts rebound effects, play, and lost motion during stops.

How it works. Input from a single motor drives an eccentric disk and connecting rod. In the position shown in the drawing, the indexing gear and table are locked by the rack—the planet gear rides freely across the index gear without imparting any motion to it. Indexing of the table to its next position begins when the control cam simultaneously releases the locking rack from the index gear and causes the spring control ring gear to pivot into mesh with the planet.

We now have a planetary gear system containing a stationary ring gear, a driving planet gear, and a "sun" index gear. As the crank keeps moving to the right, it begins to accelerate the index gear with harmonic motion—a desirable type of motion because of its low acceleration-deceleration characteristics while imparting high-speed transfer to the table.

At the end of 180-deg rotation of the crank, the control cam pivots the ring-gear segment out of mesh and, simultaneously, engages the locking rack. As the connecting rod is drawn back, the planet gear rotates freely over the index gear, which is locked in place.

The cam control is so synchronized that all toothed elements are in full engagement for a few brief instants when the crank arm is in full toggle at both the beginning and end of index. The device can be operated just as easily in the other direction.

Overload protection. The ring gear segment includes a spring-load detent mechanism, (simplified in the illustration) that will hold the gearing in full engagement under normal indexing forces. If rotation of the table is blocked at any point in index, the detent spring force is overcome and the ring gear pops out of engagement with the planet gear.

A detent roller (not shown) will then snap into a second detent position, which will keep the ring gear free during the remainder of the index portion of cycle. After that, the detent will automatically reset itself.

Incomplete indexing is detected by an electrical system that stops the machine at the end of the index cycle.

Easy change of settings. To change indexes for a new job setup, the eccentric is simply replaced with one having a different crank radius, which gives the proper drive stroke for 6, 8, 12, 16, 24, 32, or 96 positions per table rotation.

Because indexing occurs during one-half revolution of the eccentric disk, the input gear must rotate at two or three times per cycle to accomplish indexing of ½, ¼, or $\frac{1}{16}$ of the total cycle time (which is equivalent to index-to-dwell cycles of 180/180, 90/270 or 60/300 deg). To change the cycle time, it is only necessary to mount a different set of change gears between input gear and control cam gear.☐

Timing belts, four-bar linkage team up for smooth indexing

A new class of intermittent mechanisms that uses timing belts, pulleys, and linkages (drawing below) instead of the usual genevas or cams is proving capable of cyclic start-and-stop motions with smooth acceleration and deceleration.

Developed by Eric S. Buhayar and Eugene E. Brown of the Engineering Research Div., Scott Paper Co. (Philadelphia), the new mechanisms are slated to be employed in automatic assembly lines.

These new mechanisms, moreover, can be used as phase adjusters in which the rotational position of the input shaft can be shifted as desired in relation to the output shaft. Such phase adjusters are frequently used in the textile and printing industries to change the "register" of one roll with that of another, when both rolls are driven by the same input.

Outgrowth from chains. Intermittent-motion mechanisms frequently take on ingenious shapes and configurations. They are all old-timers, having been used in watches and in production machines for many years. Recently, there has been interest in the chain type of intermittent mechanism (drawing, page 70), which ingeniously routes a chain around four sprockets to produce a dwell-and-index type of output.

The input shaft of such a device has a sprocket eccentrically fixed to it. The input also drives another shaft through a one-to-one gearing. This second shaft mounts a similar eccentric sprocket that is, however,

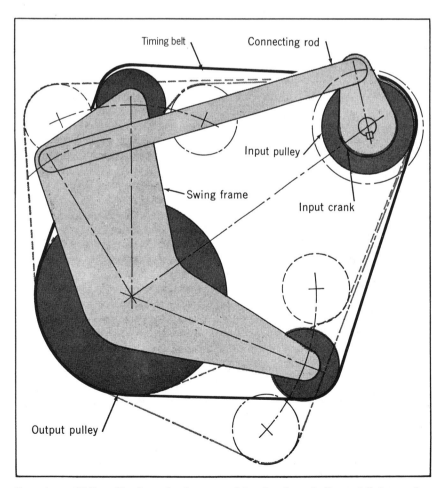

New class of intermittent mechanisms consists of pulleys, belts, and linkages. Input pulley drives belt, while input crank oscillates swing frame to modulate output.

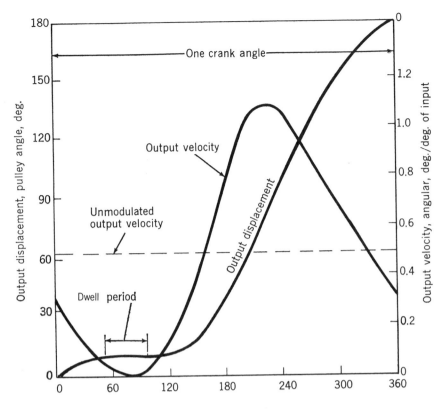

One crank angle

Output velocity

Output displacement

Unmodulated output velocity

Dwell period

Input crank angle, deg.

full circle of rotation, it continues at a slower rate and begins to repeat its slow-down—dwell—speed-up cycle. □

Modified ratchet drive

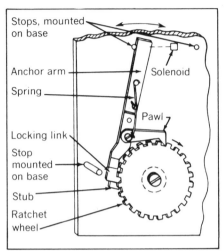

Stops, mounted on base

Anchor arm

Solenoid

Spring

Pawl

Locking link

Stop mounted on base

Stub

Ratchet wheel

free to rotate. The chain passes first around an idler pulley and then a second pulley, which is the output.

As the input gear rotates, it also pulls the chain around with it, producing a modulated output rotation. Two spring-loaded shoes, however, must be employed because the perimeter of the pulleys is not a constant figure, so the drive has a varying slack built into it.

Commercial type. A chain also

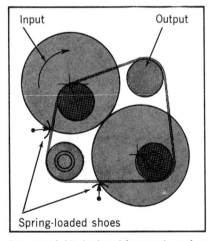

Input

Output

Spring-loaded shoes

Commercial indexing drive needs spring-loaded shoes to take up chain slack.

forms the basis for a commercial phase-adjuster drive. A handle is moved to change the phase between the input and output shafts. The theoretical chain length is constant.

In trying to improve this chain device, Scott engineers decided to keep the input and output pulleys at fixed positions and maintain the two idlers on a swing frame. The variation in wrap-around length turned out to be surprisingly little, enabling them to employ a timing belt without spring-loaded tensioners instead of a chain.

If the swing frame is maintained in one position, the device produces a constant-speed output. Shifting the swing frame to a new position automatically shifts the phase relationship between input and output.

Computer consulted. To obtain intermittent motion, a four-bar linkage is superimposed on the device by adding a crank to the input shaft and a connecting rod to the swing frame. The developers chose an iterative program on a computer to optimize certain variables of the four-bar version.

In the design of one two-stop drive, a dwell period of approximately 50 deg. is obtained. The output displacement moves slowly at first, coming to a "pseudo dwell," in which it is virtually stationary. The output then picks up speed smoothly until almost two-thirds of the input rotation has elasped (240 deg.). After the input crank completes a

A new ratchet drive designed for the Atomic Energy Commission is now available for royalty-free licensing and commercial use.

It's designed to assure movement, one tooth at a time, without overriding and only in one direction. Key element is a small stub that moves along from the bottom of one tooth well, across the top of the tooth, and into an adjacent tooth well, while the pawl remains at the bottom of another tooth well.

The locking link, which carries the stub, along with the spring, comprise a system that tends to hold the link and pawl against the outside circumference of the wheel and to push the stub and pawl point toward each other and into differently spaced wells between the teeth. A biasing element, which may be another linkage or solenoid, is provided to move the anchor arm from one side to the other, between the stops, as shown by the double arrow. The pawl will move from one tooth well to the next only when the stub is at the bottom of a tooth well and in position to prevent counter-rotation.

Odd shapes in planetary give smooth stop and go

New type of intermittent-motion mechanism, useful in automatic processing machinery, combines gears with lobes in which some pitch curves are circular, some noncircular

A new type of intermittent-motion mechanism combining circular gears with noncircular gears in a planetary arrangement (drawing) is expected to find wide application in automatic production and assembly machines.

The mechanism, developed by Ferdinand Freudenstein, noted kinematician and professor of mechanical engineering at Columbia Univ., is of the type in which continuous rotation applied to the input shaft produces a smooth, stop-and-go unidirectional rotation in the output shaft, even at high speeds.

Such jar-free intermittent motion is sought in machines designed for packaging, production, automatic transfer, and processing.

Varying differential. The basis for Freudenstein's invention—he received a patent on the device this year—is the varying differential motion obtained between two sets of gears, one set with lobular pitch circles in which curves are partly circular and partly noncircular.

The circular portions of the pitch curves cooperate with the remainder of the mechanism to provide a dwell time or stationary phase, or phases, for the output member. The noncircular portions act with the remainder of the mechanism to provide a motion phase, or phases, for the output member.

The manufacture of such noncircular gears, Freudenstein finds, is a straightforward process with modern fabrication techniques.

Competing genevas. The main competitors to his "pulsating planetary" mechanism, Freudenstein believes, are the external genevas and starwheels. These devices have a number of limitations. Freudenstein finds, including:

• Need for a means, separate from the driving pin, for locking the output member during the dwell

phase of the motion. Moreover, accurate manufacture and careful design are required to effect a smooth transition from rest to motion and vice versa.

• Kinematic characteristics in the geneva that are not favorable for high-speed operation, except when the number of stations (i.e. the number of slots in the output member) is large. For example, there is a sudden change of acceleration of the output member at the beginning and end of each indexing operation.

• Relatively little flexibility in the design of the geneva mechanism. One factor alone (the number of slots in the output member) determines the characteristics of the motion. As a result, the ratio of the

time of motion to the time of dwell cannot exceed one-half, the output motion cannot be uniform for any finite portion of the indexing cycle, and it is always opposite in sense to the sense of input rotation. The output shaft, moreover, must always be offset from the input shaft.

"Many modifications of the standard external geneva have been proposed, including multiple and unequally spaced driving pins, double rollers, and separate entrance and exit slots. These proposals have, however, been only partly successful in overcoming these limitations," says Freudenstein.

Differential motion. In deriving the operating principle of his device, Freudenstein first considered a conventional epicyclic (planetary) drive in which the input to the cage or arm causes a planet set with gears *2* and *3* to rotate the output "sun," gear *4*, while another sun, gear *1*, is kept fixed (drawing below).

Letting r_1, r_2, r_3, r_4 equal the pitch radii of the circular *1*, *2*, *3*, *4*, then the output ratio, defined as:

$$R = \frac{\text{angular velocity of output gear}}{\text{angular velocity of arm}}$$

is equal to: $R = 1 - \dfrac{r_1 r_3}{r_2 r_4}$

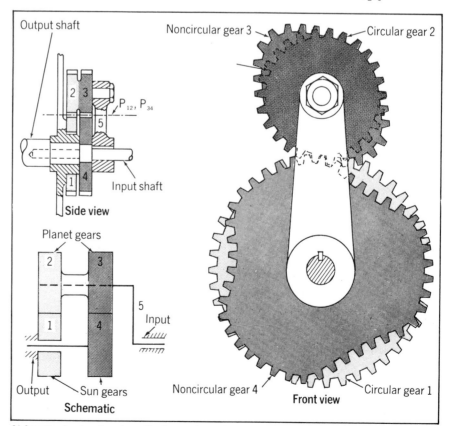

At heart of new planetary (in front view, circular set stacked behind noncircular set), two sets of gears when assembled (side view) resemble conventional unit (schematic).

Now, if $r_1 = r_4$ and $r_2 = r_3$, there is no "differential motion" and the output remains stationary. Thus if one gear pair, say 3 and 4, is made partly circular and partly noncircular, then where $r_2 = r_3$ and $r_1 = r_4$ for the circular portion, gear 4 dwells. Where $r_2 \neq r_3$ and $r_1 \neq r_4$ for the noncircular portion, gear 4 has motion. The magnitude of this motion depends on the difference in radii, in accordance with the previous equation. In this manner, gear 4 undergoes an intermittent motion (graph below, right).

Advantages. The pulsating planetary comes off with some highly useful characteristics for intermittent-motion machines:

• The gear teeth serve to lock the output member during the dwell as well as to drive that member during motion.

• Superior high-speed characteristics are obtainable. The profiles of the pitch curves of the noncircular gears can be tailored to a wide variety of desired kinematic and dynamic characteristics. There need be no sudden terminal acceleration change of the driven member, so the transition from dwell to motion, and vice versa, will be smooth, with no jarring of machine or payload.

• The ratio of motion to dwell time is adjustable within wide limits. It can even exceed unity, if desired. The number of indexing operations per revolution of the input member also can exceed unity.

• The direction of rotation of the output member can be in the same or opposite sense relative to that of the input member, according to whether the pitch axis P_{34} for the noncircular portions of gears 3 and 4 lies wholly outside or wholly inside the pitch surface of the planetary sun gear 1.

• Rotation of the output member is coaxial with the rotation of the input member.

• The velocity variation during motion is adjustable within wide limits. Uniform output velocity for part of the indexing cycle is obtainable; by varying the number and shape of the lobes, a variety of other desirable motion characteristics can be obtained.

• The mechanism is compact and has relatively few moving parts, which can be readily dynamically balanced.

Design hints. The design techniques work out surprisingly simply, says Freudenstein. First the designer must select the number of lobes L_3 and L_4 on the gears 3 and 4. In the drawings, $L_3 = 2$ and $L_4 = 3$. Any two lobes on the two gears (i.e. any two lobes of which one is on one gear and the other on the other gear) that are to mesh together must have the same arc length. Thus, every lobe on gear 3 must mesh with every lobe on gear 4, and $T_3/T_4 = L_3/L_4 = 2/3$, where T_3 and T_4 are the numbers of teeth on gears 3 and 4. T_1 and T_2 will denote the numbers of teeth on gears 1 and 2.

Next, select the ratio S of the time of motion of gear 4 to its dwell time, assuming a uniform rotation of the arm 5. For the gears shown, $S = 1$. From the geometry,

$$(\theta_{30} + \Delta\theta_3)L_3 = 360°$$

and

$$S = \Delta\theta_3/\theta_{30}$$

Hence

$$\theta_{30}(1 + S)L_3 = 360°$$
For $S = 1$ and $L_3 + 2$,
$$\theta_{30} = 90°$$
and

$$\Delta\theta_3 = 90°$$

Now select a convenient profile for the noncircular portion of gear 3. One such profile (drawing, p 116) that Freudenstein found to have favorable high-speed characteristics for stop-and-go mechanisms is

$$r_3 = R_3$$
$$\left[1 + \frac{\lambda}{2}\left(1 - \cos\frac{2\pi(\theta_3 - \theta_{30})}{\Delta\theta_3}\right)\right]$$

The profile defined by this equation has, among other properties, the characteristic that, at the transition from rest to motion and vice versa, gear 4 has zero acceleration for uniform rotation of arm 5.

In the above equation, λ is the quantity which, when multiplied by R^3, gives the maximum or peak value

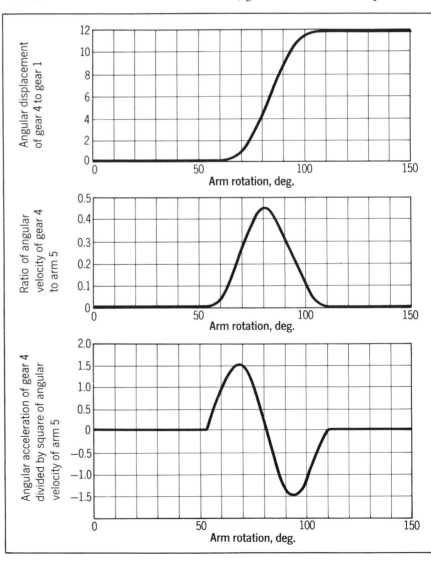

Output motion (upper curve) has long dwell periods; velocity curve (center) has smooth transition from zero to peak; acceleration at transition is zero (bottom).

of $r_3 - R_3$, differing by an amount h' from the radius R^3 of the circular portions of the gear. The noncircular portions of each lobe are, moreover, symmetrical about their midpoints, the midpoints of these portions being indicated by m.

To evaluate the quantity λ, Freudenstein worked out the equation:

$$\lambda = \frac{1-\mu}{\mu} \times$$

$$\frac{[S+\alpha-(1+\alpha)\mu][\alpha-S-(1+\alpha)\mu]}{[\alpha-(1+\alpha)\mu]^2}$$

where $R_3\lambda =$ height of lobe

$$\mu = \frac{R_3}{A} = R_3/(R_3 + R_4)$$

$$\alpha = S + (1+S)L_3/L_4$$

To evaluate the above equation,

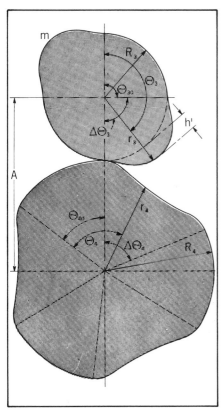

Profiles for noncircular gears are circular arcs blended to special cam curves.

select a suitable value for μ that is a reasonably simple rational fraction, i.e. a fraction such as $\frac{3}{8}$ whose numerator and denominator are reasonably small integral numbers.

Thus, without having to resort to a computer or to lengthy trial-and-error procedures, the designer can select the configuration that will achieve his purpose of smooth intermittent motion.

—Nicholas P. Chironis

Gear system taps kinetic force to control oil: gasoline ratio

Engineers at North American Aviation, Rocketdyne Div., in California, are making use of a liquid-fluid propellant's kinetic energy to control the fuel-mixture ratio. By tapping this energy to operate a planetary gear system, which is coupled to a servo system, they save power. This is critical in long-range missions.

According to James D. McGroarty, who designed the gear system, the idea could be used to control the oil:gasoline ratio in two-stroke engines. In these engines, auto makers, such as Saab and Auto-Union, have found it necessary to have independent supply lines for the oil and gasoline because of the trouble involved in refueling where it is necessary to be careful in calculating the correct mixture.

Corrects mixture ratio. In the North American design, the planetary gears are attached to shafts that are linked to a flow-measuring and driving device in the main fuel lines. Knowing the ratio of the mixture, by volume, McGroarty designs the gears so that the high-volume flow rotates the sun gear, and the low-volume flow rotates the ring gear.

For instance, if the mixture ratio

were nearly 12:1, by correcting for slippage in the flow-measuring device, the gear ratio needed would be 12:1. When the mixture is flowing at the correct ratio, an arm, carrying a planet gear, has a null reaction on the servo control linkage, and the planet gear merely acts as an idler.

However, should only one of the flows change, then the planet gear will move (the direction depending upon which gear changes speed) causing a servo valve (sketch below) to operate. Again, depending on whether the flow has to be increased or decreased, a servo valve will open or close, so a throttle valve in the fuel lines will adjust the fuel flow to the correct valve.

While this is taking place, the planet gear and the planetary arm will be moving back to the null position. When the correct ratio is attained, the servo system stops sending signals to the throttle valve.

Damper prevents overshoot. To prevent drastic flow adjustments during throttling operations, a bias piston assembly attached to the gear linkage acts in the opposite direction from the servo valve so there is a gradual return to normal flow. Thus, the bias control signal acts as a damper on the servo signal. □

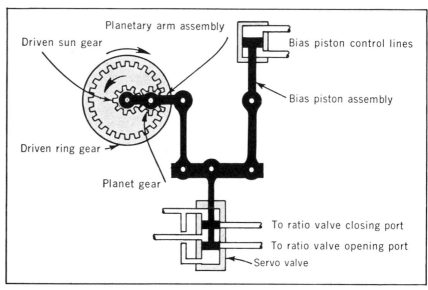

When liquid-fluid ratio is correct the planet gear acts as an idler. However, if ratio changes the planet gear moves, causing servo to correct the ratio.

Cams and gears team up in programmed motion

Pawls and ratchets are eliminated in this design, which is adaptable to the smallest or largest requirements; it provides a multitude of outputs to choose from at low cost

A new and extremely versatile mechanism provides a programmed rotary output motion simply and inexpensively. It has been sought widely for filling, weighing, cutting, and drilling in automatic and vending machines.

The mechanism, which uses overlapping gears and cams (drawing below), is the brainchild of mechanical designer Theodore Simpson of Nashua, N. H.

Based on a patented concept that could be transformed into a number of configurations (photo right), PRIM (Programmed Rotary Intermittent Motion), as the mechanism is called, satisfies the need for smaller devices for instrumentation without using spring pawls or ratchets.

It can be made small enough for a wristwatch or as large as required.

Versatile output. Simpson reports the following major advantages:
- Input and output motions are on a concentric axis.
- Any number of output motions of varied degrees of motion or dwell time per input revolution can be provided.
- Output motions and dwells are variable during several consecutive input revolutions.
- Multiple units can be assembled on a single shaft to provide an almost limitless series of output motions and dwells.
- The output can dwell, then snap around.

How it works. The basic model (drawing, below left) repeats the output pattern, which can be made complex, during every revolution of the input.

Cutouts around the periphery of the cam give the number of motions, degrees of motion, and dwell times desired. Tooth sectors in the program gear match the cam cutouts.

Simpson designed the locking lever so one edge follows the cam and the other edge engages or disengages, locking or unlocking the idler gear and output. Both program gear and cam are lined up, tooth segments to cam cutouts, and fixed to the input shaft. The output gear rotates freely on the same shaft, and an idler gear meshes with both output gear and segments of the program gear.

As the input shaft rotates, the teeth of the program gear engage the idler. Simultaneously, the cam releases the locking lever and allows the idler to rotate freely, thus driving the output gear.

Reaching a dwell portion, the teeth of the program gear disengage from the idler, the cam kicks in the lever to lock the idler, and the output gear stops until the next program-gear segment engages the idler.

Dwell time is determined by the

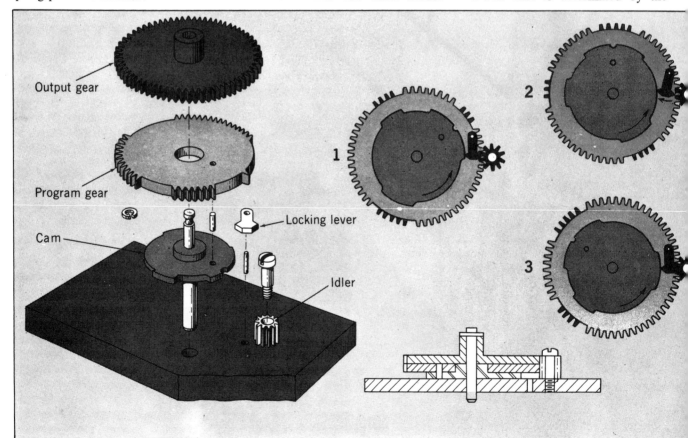

Output gear

Program gear

Cam

Locking lever

Idler

1

2

3

Basic intermittent-motion mechanism, at left in drawings, goes through the rotation sequence as numbered above.

space between the gear segments. The number of output revolutions does not have to be the same as the number of input revolutions. An idler of a different size would not affect the output, but a cluster idler with a matching output gear can increase or decrease the degrees of motion to meet design needs.

For example, a step-down cluster with output gear to match could reduce motions to fractions of a degree, or a step-up cluster with matching output gear could increase motions to several complete output revolutions.

Snap action. A second cam and a spring are used in the snap-action version (drawing below). Here, the cams have identical cutouts.

One cam is fixed to the input and the other is lined up with and fixed to the program gear. Each cam has a pin in the proper position to retain a spring; the pin of the input cam extends through a slot in the program gear cam that serves the function of a stop pin.

Both cams rotate with the input shaft until a tooth of the program gear engages the idler, which is locked and stops the gear. At this point, the program cam is in position to release the lock, but misalignment

Designer Theodore Simpson shows plastic models of some of the many variations of his mechanism. Most are unique, but unit in foreground resembles the geneva.

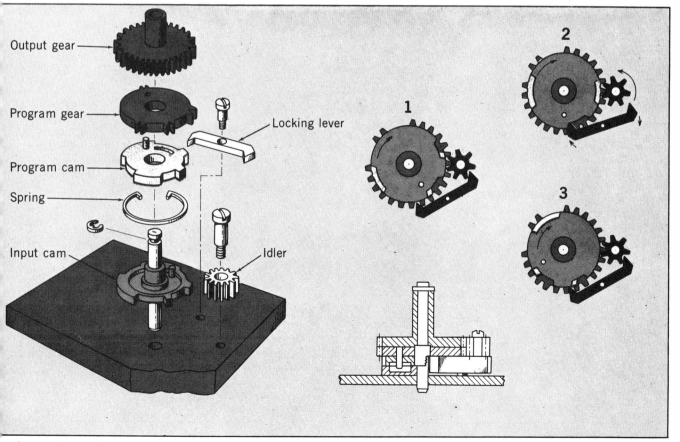

Output gear

Program gear

Locking lever

Program cam

Spring

Input cam

Idler

1

2

3

Snap-action version, with a spring and with a second cam fixed to the program gear, works as shown in numbered sequence.

of the peripheral cutouts prevents it from doing so.

As the input cam continues to rotate, it increases the torque on the spring until both cam cutouts line up. This positioning unlocks the idler and output, and the built-up spring torque is suddenly released. It spins the program gear with a snap as far as the stop pin allows; this action spins the output.

Although both cams are required to release the locking lever and output, the program cam alone will relock the output—a feature of convenience and efficient use.

After snap action is complete and the output is relocked, the program gear and cam continue to rotate with the input cam and shaft until they are stopped again when a succeeding tooth of the segmented program gear engages the idler and starts the cycle over again.

Gear with straight teeth rotates table in precise steps

A stepping drive has been designed for precise incremental angular positioning of rotary tables.

The drive (drawing below) was developed by Walter Kaspareck of NASA's Marshall Space Flight Center, Huntsville, Ala., to position scale

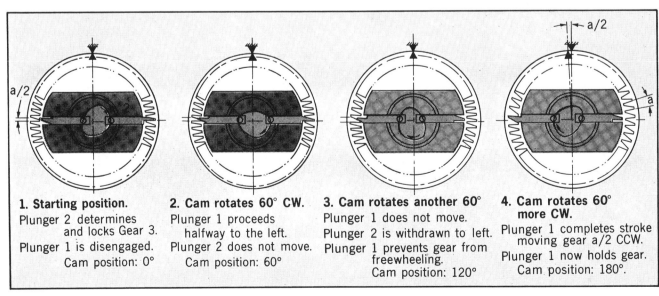

Stepping drive employs cam with groove (in color) to reciprocate dual plungers.

models of spacecraft about a horizontal axis so their radar-receiving and transmitting characteristics can be measured accurately for application to full-scale hardware.

The invention, owned by NASA, is available on a royalty-free basis. Inquiries about license rights should be addressed to NASA, Code GP, Washington, D.C. 20546.

Application. The drive replaces a conventional gearhead drive that has proved inadequate in precision and torque output. NASA expects the new system, which lends itself to a straightforward design, to find wide use in rotating and positioning indexing tables used in production, and in instruments of various types.

A motor drives a groove cam that, during 180 deg. of rotation, rotates the output gear one "step," or an angle of $a/2$, where a is the angle between two adjacent teeth (drawing below).

Details of operation. The stepping drive functions as follows:

At the starting position, the output gear is locked and held in precise position by plunger 2. The output gear has internal teeth of a right-triangle shape; one side is radially parallel with a diametral line, and the other is side-beveled at an angle of 24 deg. The gear in this stepping-drive design has 90 teeth.

The cam then begins a 180-deg. clockwise rotation. During the first 60 deg., the cam shifts plunger 1 to the left, halfway into the space between two teeth. Plunger 2 is still in the locked position.

During the second 60 deg., the cam withdraws plunger 2 completely from the gear. The final 60-deg. rotation moves plunger 1 all the way to the left. This movement, by wedge action, rotates the output gear an angle of:

$$\tfrac{1}{2}a = \tfrac{1}{2}\,\frac{360}{90} = 2 \text{ deg. motion}$$

Upon completion of 180-deg. rotation, the cam is stopped by the limit switch. Plunger 1 keeps the gear precisely in the new position.

Positioning of the gear is assured accurately by spring-loaded plungers and, because of the small wedge angle of 24 deg., the two plungers are self-locking.

If a limit switch is not employed, the drive will produce a continuous rotary start-and-stop action of the type normally employed in automated operations.

1. **Starting position.**
Plunger 2 determines and locks Gear 3.
Plunger 1 is disengaged.
Cam position: 0°

2. **Cam rotates 60° CW.**
Plunger 1 proceeds halfway to the left.
Plunger 2 does not move.
Cam position: 60°

3. **Cam rotates another 60°**
Plunger 1 does not move.
Plunger 2 is withdrawn to left.
Plunger 1 prevents gear from freewheeling.
Cam position: 120°

4. **Cam rotates 60° more CW.**
Plunger 1 completes stroke moving gear a/2 CCW.
Plunger 1 now holds gear.
Cam position: 180°.

Each clockwise, 180-deg. rotation of the groove cam forces the special-toothed outer gear to step counterclockwise half a tooth space. The device then is brought to a full stop by the limit switch and switching cam shown in top drawing.

Cycloid gear mechanism controls stroke of pump

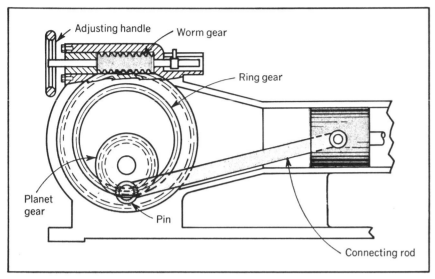

Adjustable ring gear meshes with planet gear of half its diameter to provide an infinitely variable stroke in a pump. Adjustment in ring gear is made by engaging other teeth. In U.S. design below, yoke replaces connecting rod.

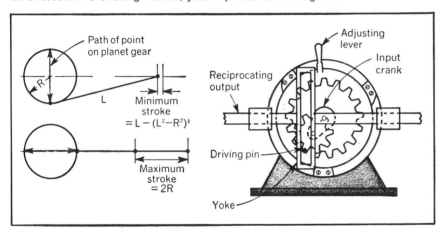

Converting rotary to linear motion

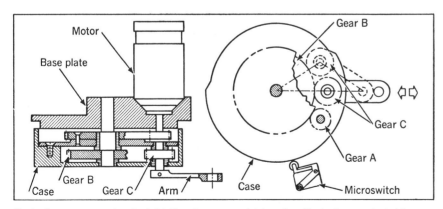

End of arm moves in straight line because of triangle effect shown at right.

A metering pump for liquid or volatile media designed and produced in Czechoslovakia employs an adjustable ring gear, meshing with a special-size planet gear to provide an infinitely variable stroke in the pump. The stroke can be set either manually or automatically through a servo motor. Flow control from 180 to 1200 liter/hr. (48 to 317 gal./hr.) is possible while the pump is at a standstill or running.

Two engineers, B. Dolezal and L. Kurzveil, both of Prague, obtained a Czech patent #112,865 for this design; a similar principle has been used previously in this country.

Straight-line motion is key. The mechanism makes use of a planet gear whose diameter is half that of the ring gear. As the planet is rotated to roll on the inside of the ring, a point on the pitch diameter of the planet will describe a straight line (instead of the usual hypocycloid curve). This line is a diameter of the ring gear. The left end of the connecting rod is pinned to the planet at this point. The ring gear is made shiftable by machining in its outer surface a second set of gear teeth, and then meshing this set with a worm gear for control. Shifting the ring gear alters the slope of the straight-line path. The two extreme positions are shown in the diagram. In the mechanism's present position, the pin will reciprocate vertically to produce the minimum stroke for the piston. Rotating the ring gear 90 deg will cause the pin to reciprocate horizontally to produce the maximum piston stroke.

The diagram at the bottom illustrates the American version which uses a yoke instead of a connecting rod. This permits the length of the stroke to be reduced to zero. Also, the length of the pump can be substantially reduced. □

A compact gear system to provide linear motion from a rotating shaft has been designed, under a NASA contract, by Allen G. Ford of Jet Propulsion Laboratory in California. It neatly uses a planetary gear system so that the end of an arm attached to the planet gear always moves in a linear path (drawing).

The gear system is set in motion by a motor attached to the base plate. Gear A, attached to the motor shaft, turns the case assembly, causing Gear C to rotate along Gear B, which is fixed. The arm is the same length as the center distance between Gears B and C. Lines between the centers of Gear C, the end of the arm, and the case axle form an isosceles triangle, the base of which is always along the plane through the center of rotation. So the output motion of the arm attached to Gear C will be in a straight line.

When the end of travel is reached, a microswitch causes the motor to reverse, returning the arm to its original position. □

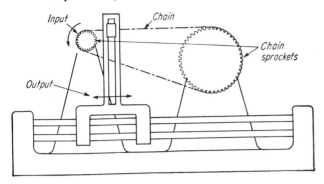

A drive pin, fastened to a chain, rides in a vertical slot to reciprocate a carriage in a horizontal direction. Velocity in both directions is constant and depends on the angle of slope that the chain makes with the vertical slot. Thus, velocity to the left is higher than to the right because the top of the chain is horizontal.

New star wheels challenge geneva drives for indexing

With circular-arc slots, they can be analyzed mathematically and manufactured more easily. These new star wheels are speeding up the operation of automatic machines and assembly lines

A new family of star wheels with circular instead of the usual epicyclic slots (drawings below) are producing fast start-and-stop indexing with relatively low acceleration forces.

Such rapid, jar-free cycling is vital in a wide variety of production machines and automatic assembly lines, in which parts must move from one station to another for drilling, cutting, milling, and other process operations.

The new devices are the brain child of Martin Zugel, a Cleveland engineer who has worked with Dr. R. L. Fox, a Case Institute professor, to put the design of star wheels on a firm mathematical basis.

New design works. After several successful applications of the devices, Zugel formed Cyclo-Index Corp. about 18 months ago to design and manufacture circular-slot star wheels. Applications include the use of one-stop star wheels in Bell & Howell's high-speed film cutter, indexing at 220 rpm.

Older star wheels with epicyclic slots have remained pretty much a

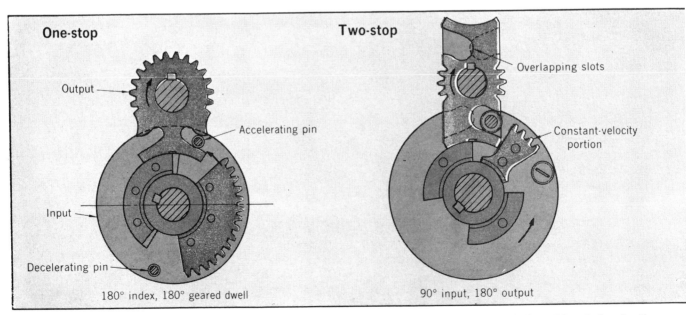

One-stop

Output

Accelerating pin

Input

Decelerating pin

180° index, 180° geared dwell

Two-stop

Overlapping slots

Constant-velocity portion

90° input, 180° output

Star wheels vary in shape, depending on degree of indexing that must be done during one revolution of input. Accelerating

curiosity, difficult to analyze and predict and equally hard to make. Some mention of them is made in the literature from time to time, but the commercially favored indexing drive is the geneva wheel (graph, page 80). Its straight slots make it easier than star wheels to analyze and to make.

The new star wheels with their circular-arc slots are no problem to fabricate. And, because the slots are true circular arcs, they can be visualized for mathematical analysis as four-bar lingages during the entire period of pin-slot engagement (drawings, page 80).

Strong points. With this approach, changes in the radius of the slot can be analyzed and the acceleration curve varied to provide inertia loads below those of the genevas for any design requirement.

Another advantage of the star wheels is that they can index a full 360 deg. in a relatively short period (180 deg.). Such one-stop operation is not possible with genevas. In fact, genevas cannot handle two-stop operations and are hard-pressed to produce three stops per index. The few two-stop indexing devices on the market are cam-operated, which means they require greater input angles for indexing.

Operating sequence. In operation, the input wheel rotates continuously. A sequence starts (drawing, page 80) when the accelerating pin en-

Geared star sector indexes smoothly a full 360 deg. during a 180-deg. rotation of wheel, then pauses during the other 180 deg. to allow wheel to catch up.

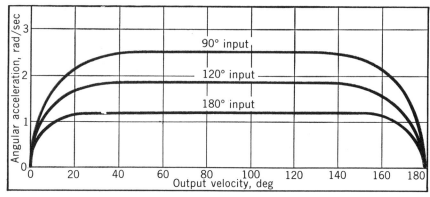

One-stop index motion of unit shown in photo above can easily be designed to take longer to complete its indexing, thus reducing its index velocity.

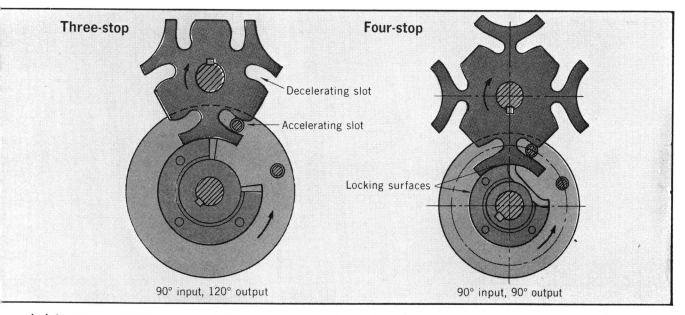

Three-stop

Decelerating slot

Accelerating slot

90° input, 120° output

Four-stop

Locking surfaces

90° input, 90° output

pin brings output wheel up to speed. Gear sectors come into mesh to keep the output rotating beyond 180 deg.

gages the curved slot to start indexing the output wheel clockwise. Simultaneously, the locking surface clears the right side of the output wheel to permit the indexing.

Pin C in the drawings continues to accelerate the output wheel past even the midpoint, where a geneva wheel would start deceleration. Not until the pins are symmetrical (drawing, below right) does acceleration end and deceleration begin. Pin D then takes the brunt of the deceleration force.

Adaptable. The angular velocity of the output wheel, at this point of exit of the acceleration roller from Slot 1, can be varied to suit design requirements. At this point, for example, it is possible either to engage the deceleration roller as described or to start engagement of a constant-velocity portion of the cycle. Many more degrees of output index can be obtained by interposing gear-element segments between the acceleration and deceleration rollers, as shown on page 78.

The device below at left will stop and start four times in making one revolution, while the input turns four times in the same period. In the starting position, the output link has zero angular velocity, which is a prerequisite condition for any device intended for applications at speeds above a near standstill.

In the disengaged position, the angular velocity ratio between the output and input shafts (the "gear" ratio), is entirely dependent upon the design angles α and β and independent of the slot radius, r.

Design comparisons. The slot radius, however, does play an important role in the mode of the acceleration forces. A four-stop geneva provides a good basis for comparison with a four-stage "Cyclo-Index" system.

Assume, for example, that $\alpha = \beta = 22.5$ deg. A little trigonometry leads to:

$$R = A\left[\frac{\sin \beta}{\sin (\alpha+\beta)}\right]$$

which yields $R=0.541A$. The only restriction on r is that it be large enough to allow the device to pass through its mid-position. This is satisfied if:

$$r > \frac{RA(1-\cos \alpha)}{A-2R-A \cos \alpha} \approx 0.1 A$$

There is no upper limit on r, so the slot can be straight.

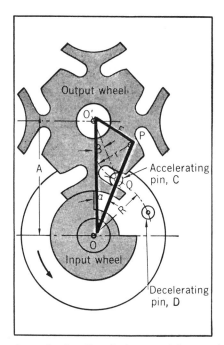
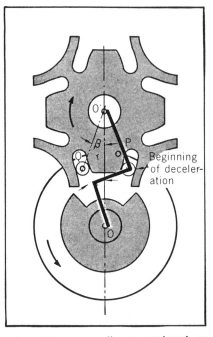

Star-wheel action is improved by curving the slots over radius **r**, centered on initial-contact line OP. Units then act as four-bar linkages OO'PQ.

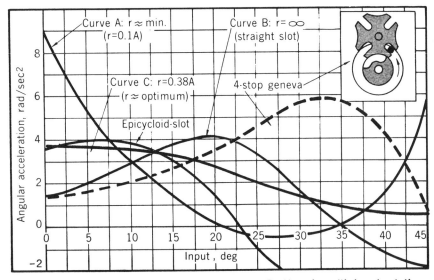

Accelerating force of star wheels (curves A, B, C) varies with input rotation. With optimum slot (curve C), it is lower than for four-stop geneva.

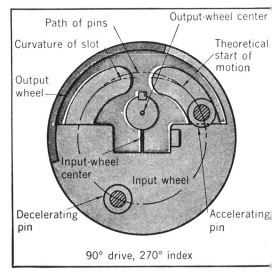

Internal star wheel, still experimental, uses radius difference to cushion indexing shock.

GENEVA MECHANISMS

Locking-arm geneva

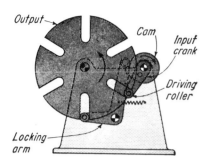

As with most genevas, the driving follower on the rotating input crank enters a slot and rapidly indexes the output. In this version, the roller of the locking-arm (shown leaving the slot) enters slot to prevent the geneva from shifting when not indexing.

Planetary gear geneva

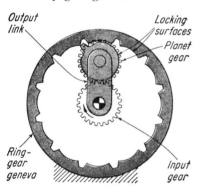

The output link remains stationary while the input gear drives the planet gear with single tooth on the locking disk, which is part of the planet gear, and which meshes with the ring-gear geneva to index the output link one position.

Symbol indicates a pivot point that is fixed to frame

Four-bar geneva

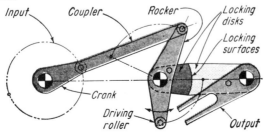

Twin geneva drive

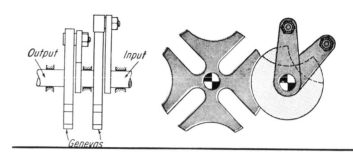

The driven member of the first geneva acts as the driver for the second geneva. This produces a wide variety of output motions including very long dwells between rapid indexes.

Groove cam geneva

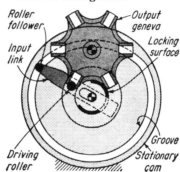

When a geneva is driven by a roller rotating at a constant speed it tends to have very high acceleration and deceleration characteristics. In this modification the input link, which contains the driving roller, can move radially while rotating by means of the groove cam. Thus, as the driving roller enters the geneva slot it moves radially inward, which reduces the geneva acceleration force.

Locking-slide geneva

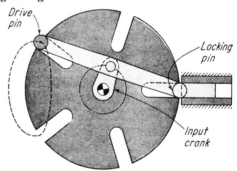

One pin locks and unlocks the geneva; the second rotates the geneva during the unlocked portion. In the position shown the drive pin is about to enter the slot to index the geneva, whereas the locking pin is just clearing the slot.

For obtaining a long-dwell motion form an oscillating output. Rotation of the input causes a driving roller to reciprocate in and out of the slot of the output link. The two disk surfaces keep the output in the position shown during the dwell period.

Rapid-transfer geneva

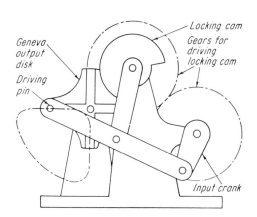

Long-dwell geneva

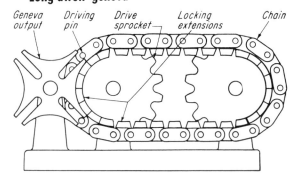

This arrangement employs a chain with an extended pin in combination with a standard geneva. This permits a long dwell between each 90-deg shift in the position of the geneva. The spacing between the sprockets determines the length of dwell. Note that some of the links have special extensions to lock the geneva in place between stations.

The coupler point at the extension of the connecting link of the 4-bar mechanism describes a curve with two approximately straight lines, 90 deg apart. This provides a good entry situation in that there is no motion in the geneva while the driving pin moves deeply into the slot — then there is an extremely rapid index. A locking cam which prevents the geneva from shifting when it is not indexing, is connected to the input shaft through gears.

Modified motion geneva

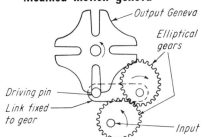

With a normal geneva drive the input link rotates at constant velocity, which restricts flexibility in design. That is, for given dimensions and number of stations the dwell period is determined by the speed of the input shaft. Use of elliptical gears produces a varying crank rotation which permits either extending or reducing the dwell period.

Dual-track geneva

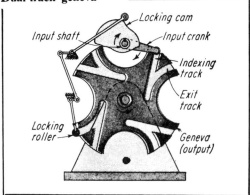

The key factor when designing genevas is to have the input roller enter and leave the geneva slots tangentially (as the crank rapidly indexes the output). This is accomplished in the novel device above by employing two tracks. The roller enters one track, indexes the geneva 90 deg (in a four-stage geneva), and then automatically follows the exit slot to leave the geneva. (Make a model to see how nicely it indexes.)

Purpose of the associate linkage mechanism is to lock the geneva when it is not indexing. In the position shown, the locking roller is just about to exit from the geneva.

The output in this simple device is prevented from turning in either direction—unless actuated by the input motion. In operation, the drive lever indexes the output disk by bearing on the pin. The escapement is cammed out of the way during indexing because the slot in the input disk is in the position to permit the escapement tip to enter into it. But as the lever leaves the pin, the input disk forces the escapement tip out of its slot and into the notch to lock output in both directions.

Internal-groove geneva

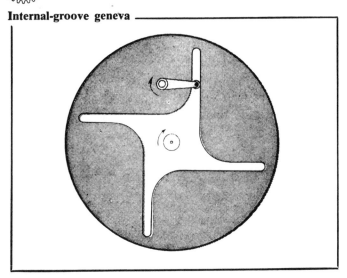

This arrangement again permits the roller to exit and enter the driving slots tangentially. In the position shown, the driving roller has just completed indexing the geneva and is about to coast for 90 deg as it goes around the curve. (During this time a separate locking device may be necessary to prevent an external torque from reversing the geneva.)

Controlled-output escapement

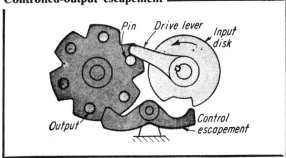

Progressive oscillating drive

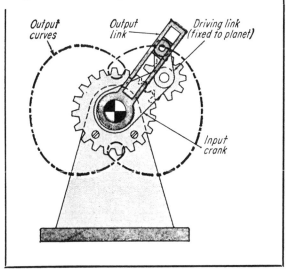

By fixing a crank to the planet gear, a point *P* can be made to describe the double loop curve illustrated. The slotted output crank oscillates briefly at the vertical portions.

Parallel-guidance mechanisms

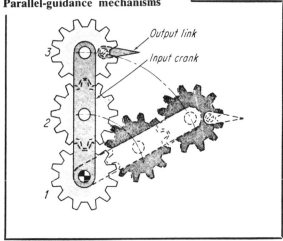

The input crank contains two planet gears. The center sun-gear is fixed as in the previous epicycloid mechanisms. By making the three gears equal in diameter and having gear 2 serve as an idler, any member fixed to gear 3 will remain parallel to its previous positions throughout the rotation of the input ring crank.

Sinusoidal reciprocator

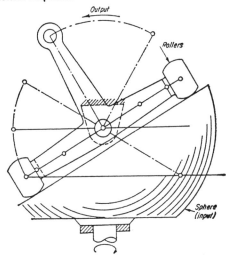

This transforms rotary motion into a reciprocating motion in which the oscillating output member is in the same plane as the input shaft. The output member has two arms with rollers which contact the surface of the truncated sphere. Rotation of the sphere causes the output to oscillate.

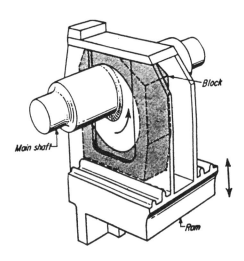

The high-volume 2500-ton press is designed to shape such parts as connecting rods, tractor track links, and wheel hubs. Simple automatic-feed mechanism raises production to a possible 2400 forgings per hour. Produced by Erie Foundry Co, Erie, Penna.

Cross-bar reciprocator

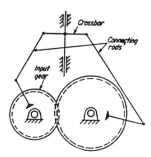

Although complex-looking, this device has been successful in high speed machines for transforming rotary motion into a high-impact linear motion. Both gears contain cranks connected to the cross bar by means of connecting rods.

New drive mechanism

... for hot-gas (Stirling cycle) engines has been developed by Philips of Eindhoven (Netherlands). Essentially, it's a multiple four-bar linkage system that makes it possible to separate power and gas-transfer mechanisms and eliminate the need for pressurizing the crankcase. With previous systems, in which power and gas-transfer functions were combined, the crankcase had to be pressurized to limit leakage of gas past the piston and to limit downward forces on the drive mechanism.

In the new system, called a rhombic drive (see diagram) there are twin cranks and control-rod mechanisms, identical in design and offset from the central axis of the engine. The cranks rotate in opposite directions and are coupled by twin gear wheels. Says R J Meijer of Philips: "The symmetry of the system and the coaxial arrangement of power-piston and displacer-piston rods make it an easy matter to avoid putting the crankcase under high pressure . . . The power-piston rod stuffing-box is subject to no lateral thrust, and the friction forces are low." Only the buffer space need be filled with gas under pressure; and this, he says, is much more easily done than pressurizing the whole crankcase.

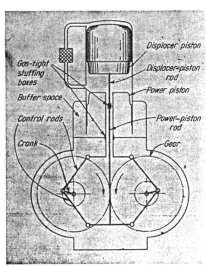

Rhombic drive takes its name from crankcase and control-rod arrangement. Note that power-piston rod is hollow and displacer-piston rod runs through it.

Modified Geneva Drives

These sketches were selected as practical examples of uncommon but often useful mechanisms. Most of them serve to add a varying velocity component to the conventional Geneva motion. The data

Fig. 1—(Below) In the conventional external Geneva drive, a constant-velocity input produces an output consisting of a varying velocity period plus a dwell. In this modified Geneva, the motion period has a constant-velocity interval which can be varied within limits. When spring-loaded driving roller *a* enters the fixed cam *b*, the output-shaft velocity is zero. As the roller travels along the cam path, the output velocity rises to some constant value, which is less than the maximum output of an unmodified Geneva with the same number of slots; the duration of constant-velocity output is arbitrary within limits. When the roller leaves the cam, the output velocity is zero; then the output shaft dwells until the roller re-enters the cam. The spring produces a variable radial distance of the driving roller from the input shaft which accounts for the described motions. The locus of the roller's path during the constant-velocity output is based on the velocity-ratio desired.

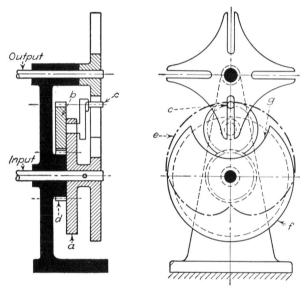

Fig. 2—(Above) This design incorporates a planet gear in the drive mechanism. The motion period of the output shaft is decreased and the maximum angular velocity is increased over that of an unmodified Geneva with the same number of slots. Crank wheel *a* drives the unit composed of plant gear *b* and driving roller *c*. The axis of the driving roller coincides with a point on the pitch circle of the planet gear; since the planet gear rolls around the fixed sun gear *d*, the axis of roller *c* describes a cardioid *e*. To prevent the roller from interfering with the locking disk *f*, the clearance arc *g* must be larger than required for unmodified Genevas.

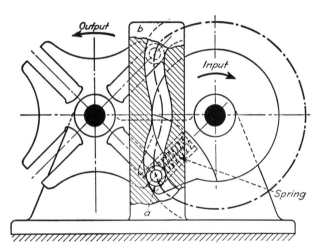

Saxonian Carton Machine Co., Dresden, Germany

Fig. 3—A motion curve similar to that of Fig. 2 can be derived by driving a Geneva wheel by means of a two-crank linkage. Input crank *a* drives crank *b* through link *c*. The variable angular velocity of driving roller *d*, mounted on *b*, depends on the center distance *L*, and on the radii *M* and *N* of the crank arms. This velocity is about equivalent to what would be produced if the input shaft were driven by elliptical gears.

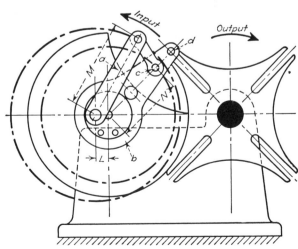

were based in part on material and figures in AWF and VDMA Getriebeblaetter, published by Ausschuss fuer Getriebe beim Ausschuss fuer wirtschaftiche Fertigung, Leipzig, Germany.

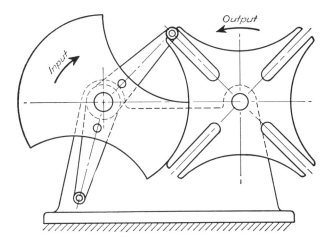

Fig. 4—(Left) The duration of the dwell periods is changed by arranging the driving rollers unsymmetrically around the input shaft. This does not affect the duration of the motion periods. If unequal motion periods are desired as well as unequal dwell periods, then the roller crank-arms must be unequal in length and the star must be suitably modified; such a mechanism is called an "irregular Geneva drive."

Fig. 5—(Below) In this intermittent drive, the two rollers drive the output shaft as well as lock it during dwell periods. For each revolution of the input shaft the output shaft has two motion periods. The output displacement ϕ is determined by the number of teeth; the driving angle, ψ, may be chosen within limits. Gear a is driven intermittently by two driving rollers mounted on input wheel b, which is bearing-mounted on frame c. During the dwell period the rollers circle around the top of a tooth. During the motion period, a roller's path d relative to the driven gear is a straight line inclined towards the output shaft. The tooth profile is a curve parallel to path d. The top land of a tooth becomes the arc of a circle of radius R, the arc approximating part of the path of a roller.

Fig. 6—Intermittent drive with cylindrical lock. Shortly before and after engagement of two teeth with driving pin d at the end of the dwell period, the inner cylinder f is not adequate to effect positive locking of the driven gear. A concentric auxiliary cylinder e is therefore provided with which only two segments are necesary to get positive locking. Their length is determined by the circular pitch of the driven gear.

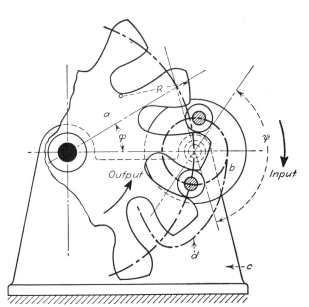

Mechanism for transmitting intermittent motion between two skewed shafts. The shafts need not be at right angles to one another. Angular displacement of the output shaft per revolution of input shaft equals the circular pitch of the output gear wheel divided by its pitch radius. The duration of the motion period depends on the length of the angular join a of the locking disks b. This drive was used extensively on motion picture projectors.

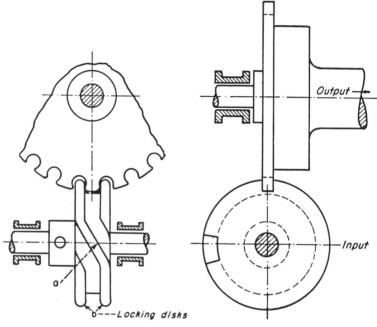

"Multilated tooth" intermittent drive. Driver b is a circular disk of width w with a cut-out d on its circumference and carries a pin c close to the cutout. The driven gear, a of width $2w$ has standard spur gear teeth, always an even number, which are alternately of full and of half width (mutilated). During the dwell period two full width teeth are in contact with the circumference of the driving disk, thus locking it; the multilated tooth between them is behind the driver. At the end of the dwell period pin c comes in contact with the mutilated tooth and turns the driven gear for one circular pitch. Then, the full width tooth engages the cutout d and the driven gear moves one more pitch, whereupon the dwell period starts again and the cycle is repeated. Used only for light loads primarily because of high accelerations encountered.

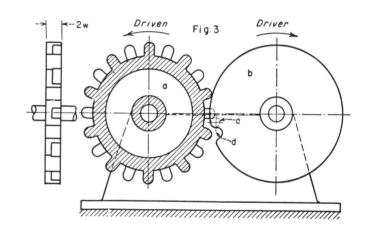

An operating cycle of 180 deg motion and 180 deg dwell is produced by this mechanism. The input shaft drives the rack which is engaged with the output shaft gear during half the cycle. When the rack engages, the lock teeth at the lower end of the coulisse are disengaged and, conversely, when the rack is disengaged, the coulisse teeth are engaged, thereby locking the output shaft positively. The change-over points occur at the dead-center positions so that the motion of the gear is continuously and positively governed. By varying R and the diameter of the gear, the number of revolutions made by the output shaft during the operating half of the cycle can be varied to suit requirements.

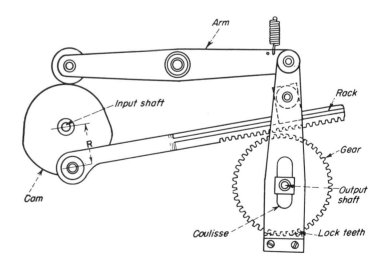

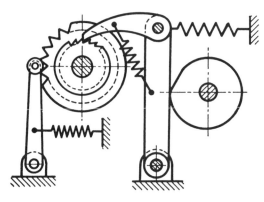

FIG. 1- CAM DRIVEN RATCHET

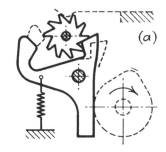

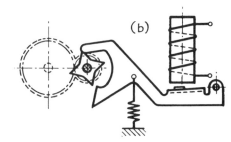

FIG. 3- (a) CAM OPERATED ESCAPEMENT ON A TAXIMETER
(b) SOLENOID OPERATED ESCAPEMENT

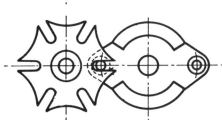

FIG. 2- SIX-SIDED MALTESE CROSS AND
DOUBLE DRIVER GIVES 3:1 RATIO

FIG. 4- ESCAPEMENT
USED ON AN
ELECTRIC METER

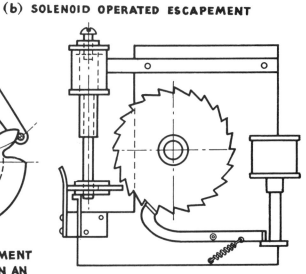

FIG. 5- SOLENOID-OPERATED RATCHET
WITH SOLENOID RESETING MECHAN-
ISM. A SLIDING WASHER ENGAGES
TEETH

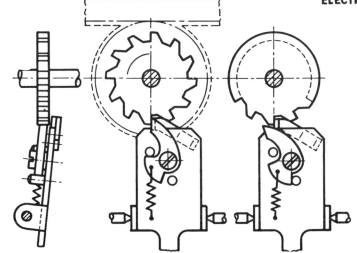

FIG. 6- PLATE OSCILLATING ACROSS PLANE OF RATCHET
GEAR ESCAPEMENT CARRIES STATIONARY AND
SPRING HELD PAWLS

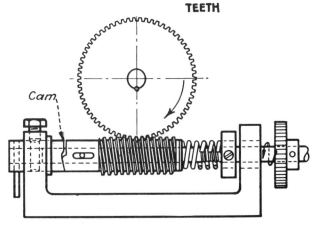

Cam

FIG. 7- WORM DRIVE COMPENSATED BY CAM ON
WORK SHAFT, PRODUCES INTERMITTENT
MOTION OF GEAR

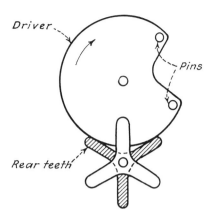

INTERMITTENT COUNTER MECHANISM. One revolution of driver advances driven wheel 120 degrees. Driven wheel rear teeth locked on cam surface during dwell.

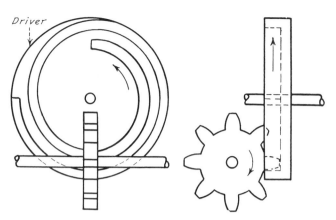

SPIRAL AND WHEEL. One revolution of spiral advances driven wheel 1 tooth. Driven wheel tooth locked in driver groove during dwell.

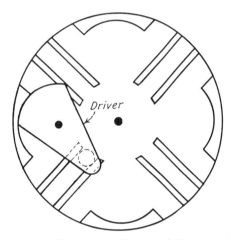

INTERNAL GENEVA MECHANISM. Driver and driven wheel rotate in same direction. Duration of dwell is more than 180 deg of driver rotation.

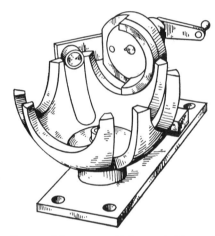

SPHERICAL GENEVA MECHANISM. Driver and driven wheel are on perpendicular shafts. Duration of dwell is exactly 180 deg of driver rotation.

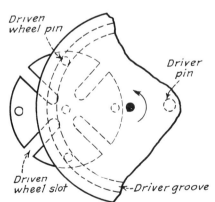

EXTERNAL GENEVA MECHANISM. Driver grooves lock driven wheel pins during dwell. During movement, driver pin mates with driven wheel slot.

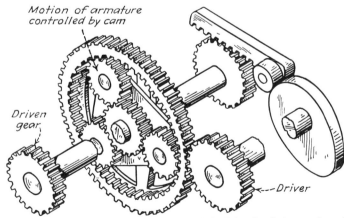

SPECIAL PLANETARY GEAR MECHANISM. Principle of relative motion of mating gears illustrated in method can be applied to spur gears in planetary system. Motion of normally fixed planet centers produces intermittent motion of sum gear.

HYPOCYCLOID MECHANISMS

PREBEN W. JENSEN

THE appeal of cycloidal mechanisms is that they can easily be tailored to provide one of these three common motion requirements:

• **Intermittent motion**—with either short or long dwells

• **Rotary motion with progressive oscillation**—where the output undergoes a cycloidal motion during which the forward motion is greater than the return motion

• **Rotary-to-linear motion with a dwell period**

All the cycloidal mechanisms covered in this article are geared; this results in compact positive devices capable of operating at relatively high speeds with little backlash or "slop." The mechanisms can also be classified into three groups:

Hypocycloid—where the points tracing the cycloidal curves are located on an external gear rolling inside an internal ring gear. This ring gear is usually stationary and fixed to the frame.

Epicycloid—where the tracing points are on an external gear which rolls in another external (stationary) gear

Pericycloid—where the tracing points are located on an internal gear which rolls on a stationary external gear.

Basic hypocycloid curves

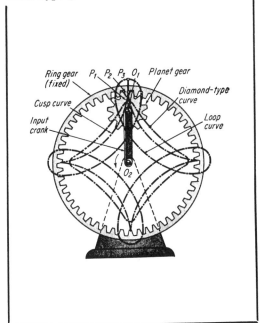

Double-dwell mechanism

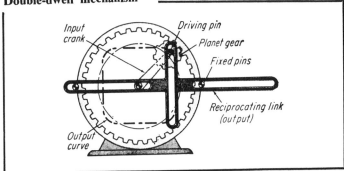

Coupling the output pin to a slotted member produces a prolonged dwell in each of the extreme positions. This is another application of the diamond-type hypocycloidal curve.

Input drives a planet in mesh with a stationary ring gear. Point P_1 on the planet gear describes a diamond-shape curve, point P_2 on the pitch line of the planet describes the familiar cusp curve, and point P_3, which is on an extension rod fixed to the planet gear, describes a loop-type curve. In one application, an end miller located at P_1 was employed in production for machining a diamond-shape profile.

Long-dwell geneva drive

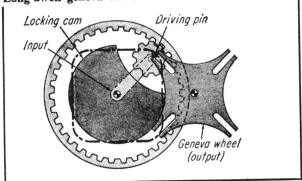

As with standard four-station genevas, each rotation of the input indexes the slotted geneva 90 deg. By employing a pin fastened to the planet gear to obtain a rectangular-shape cycloidal curve, a smoother indexing motion is obtained because the driving pin moves on a noncircular path.

Internal-geneva drive

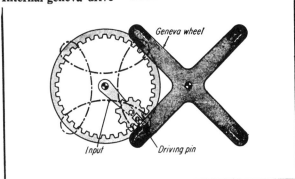

Loop-type curve permits driving pin to enter slot in a direction that is radially outward from the center, and then loop over to rapidly index the cross member. As with the previous geneva, the output rotates 90 deg, then goes into a long dwell period during each 270-deg rotation of the input.

Cycloidal motion is becoming popular for mechanisms in feeders and automatic machines.

Cycloidal parallelogram

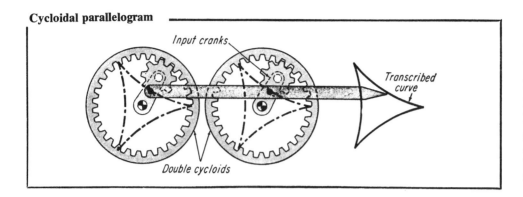

Two identical hypocycloid mechanisms guide the point of the bar along the triangularly shaped path. They are useful also in cases where there is limited space in the area where the curve must be described. Such double-cycloid mechanisms can be designed to produce other types of curves.

Cycloidal rocker

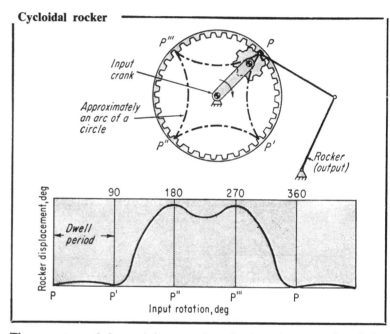

Short-dwell rotary

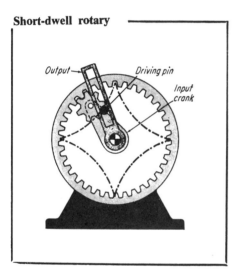

Here the pitch circle of the planet gear is exactly one-quarter that of the ring gear. A pin on the planet will cause the slotted output member to have four instantaneous dwells for each revolution of the input shaft.

The curvature of the cusp is approximately that of an arc of a circle. Hence the rocker comes to a long dwell at the right extreme position while point P moves to P'. There is then a quick return from P' to P'', with a momentary dwell at the end of this phase. The rocker then undergoes a slight oscillation from point P'' to P''', as shown in the displacement diagram.

Rectangular-motion drive

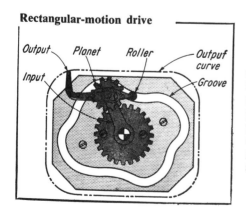

For producing closed curves consisting of several sections of straight lines. Rectangular-shaped curve is shown, but the device is capable of producing many sided curves. The output member is eccentrically mounted on a planet gear and simultaneously guided by the roller which runs in a stationary cam groove.

Cycloidal reciprocator

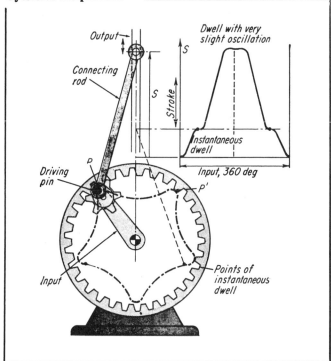

Portion of curve, *P-P'*, produces the long dwell (as in previous mechanism), but the five-lobe cycloidal curve avoids a marked oscillation at the end of the stroke. There are also two points of instantaneous dwell where the curve is perpendicular to the connecting rod.

By making the pitch diameter of the planet equal to half that of the ring gear, every point on the planet gear (such as points P_2 and P_3) will describe elliptical curves which get flatter as the points are selected closer to the pitch circle. Point P_1, at the center of the planet, describes a circle; point P_4 at the pitch circle describes a straight line. When a cutting tool is placed at P_3, it will cut almost-flat sections from round stock, as when machining a bolt. The other two sides of the bolt can be cut by rotating the bolt, or the cutting device, 90 deg. (Reference: H. Zeile, *Unrund- und Mehrkantdrehen*, VDI-Berichte, Nr. 77, 1965.)

Adjustable harmonic drive

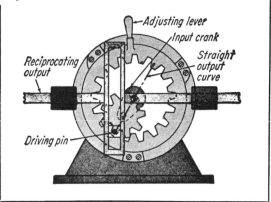

By making the planet-gear half that of the internal gear, a straight-line output curve is produced by the driving pin which is fastened to the planet gear. The pin engages the slotted member to cause the output to reciprocate back and forth with harmonic (sinusoidal) motion. The position of the fixed ring gear can be changed by adjusting the lever, which in turn rotates the straight-line output-curve. When the curve is horizontal, the stroke is at a maximum; when the curve is vertical, the stroke is zero.

Elliptical-motion drive

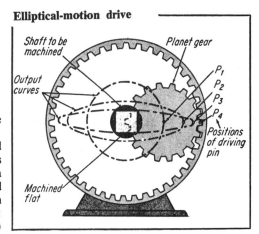

EPICYCLOID MECHANISMS

Epicycloid reciprocator

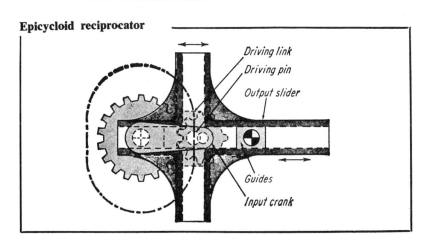

Here the sun gear is fixed and the planet gear driven around it by means of the input link. There is no internal ring gear as with the hypocycloid mechanisms. Driving pin P on the planet describes the curve shown which contains two almost-flat portions. By having the pin ride in the slotted yoke, a short dwell is produced at both the extreme positions of the output member. The horizontal slots in the yoke ride the end-guides, as shown.

Oscillating-output geneva

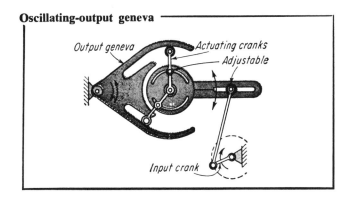

Three adjustable output-links provide a wide variety of oscillating motions. The input crank oscillates the central member which has an adjustable slot to vary the stroke. The oscillation is transferred to the two actuating rollers, which alternately enter the geneva slots to index it, first in one direction and then another. Additional variation in output motion can be obtained by adjusting the angular positions of the output cranks.

Plunger-actuated indexer

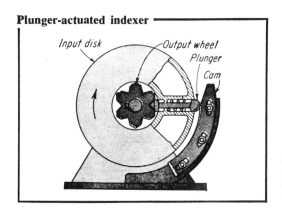

Single tooth indexer

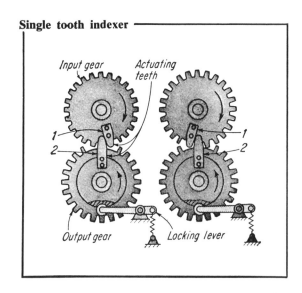

Key factor in this device is the use of an input gear which is smaller than the output gear—hence it can complete its circuit faster than the output when both are in mesh. In the left diagram, the actuating tooth of the input, tooth 1, strikes that of the output, tooth 2, to roll both gears into mesh. After one circuit of the input (right diagram), tooth 1 is now ahead of tooth 2, the gears go out of mesh, and the output comes to a stop (kept in position by the bottom locking detent) for almost 360 deg of the input.

Here the output rotates only when the plunger, which is normally kept in the outer position by means of its spring, is cammed into the toothed wheel attached to the output. Hence, for every revolution of the input disk, the output is driven approximately 60 deg, and then comes to a stop for the remaining 300 deg.

ROTARY TO RECIPROCATING DEVICES

Track-switching scotch yoke

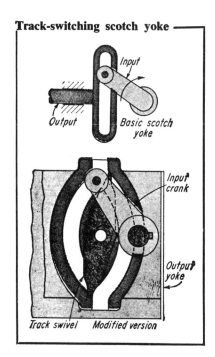

In a typical scotch yoke, the motion of the rotating input crank is translated into the reciprocating motion of the yoke. But this provides only an instantaneous dwell at each end. To obtain sought for long dwells, the left slot (in the modified version) is made curved with a radius equal to that of the input crank radius. This causes a 90-deg dwell at the left end of the stroke. For the right end, the crank pushes aside the springloaded track swivel as it comes around the bend and is shunted into the second track to provide also a 90-deg dwell at the right end.

In-line reciprocator

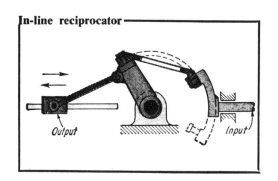

This is a simple way of converting rotary motion to reciprocating motion in which both input and output shafts are in line with each other. The right half of the device is the well known three-dimensional reciprocator. Rotating the input crank causes its link to oscillate. A second, connecting link then converts it into the desired in-line motion.

Elliptical-gear planetary

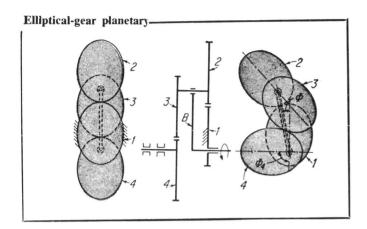

By employing elliptical gears instead of the usual circular gears, a planetary drive is obtained which can provide extra-large variations in the angular speed output.

This is a normal parallel-gear speed reducer, but with cam actuation to provide a desired variation in the output speed. If the center of the idler shaft were stationary, the output motion would be uniform, but the cam attached to the idler shaft gives the shaft an oscillating motion which varies the final output motion.

Cammed-gear speed variator

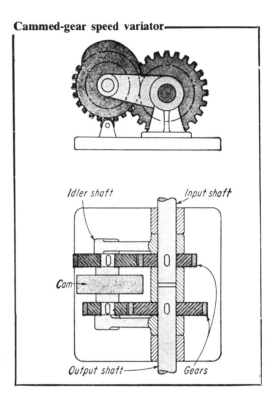

Idler shaft

Input shaft

Cam

Output shaft

Gears

Output of this novel device is varied infinitely by changing the distance that the balls will operate from the main shaft line. The unit employs multiple disks, free to rotate on a common shaft, except for the extreme left and right disks which are keyed to the input and output shafts respectively. Every second disk carries three uniformly spaced balls which can be shifted closer to or away from the center by means of the adjustment lever. When disk 1 rotates the first group of balls, disk 3 will rotate slower because of the different radii, r_{x1} and r_{x2}. Disk 3 will then drive disk 5, and disk 5 will drive disk 7, all with the same speed ratios, thus compounding the ratios to get the final speed reduction.

The effective radii can be calculated from

$$r_{1x} = R_x - \tfrac{1}{2} D \cos \psi$$
$$r_{2x} = R_x + \tfrac{1}{2} D \cos \psi$$

where R_x is the distance from the shaft center to the ball center, D is the diameter of the ball, and ψ is one half the cone angle.

Multicone drive

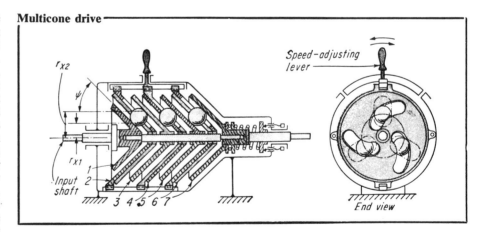

Speed-adjusting lever

End view

Shifting the controlled rod linearly twists the propellers around on the common axis by means of the rack and gear arrangement. Note the use of a double rack, one above and on either side of the other to obtain an *opposing* twisting motion required for propellers.

Adjustable-pitch device

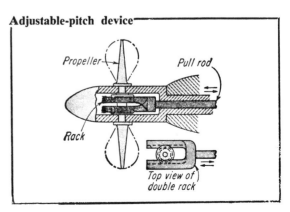

Propeller

Pull rod

Rack

Top view of double rack

ROTARY TO RECIPROCATING

Four-bar slider mechanism

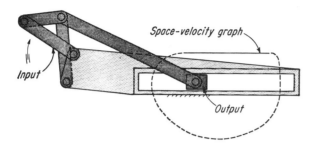

With proper proportions, the rotation of the input link can impart an almost-constant velocity motion to the slider

Three-gear stroke multiplier

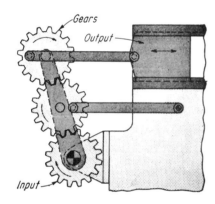

Rotation of the input gear causes the connecting link, attached to the machine frame, to oscillate. This produces a large-stroke reciprocating motion in the output slider.

Rotary motion of the input is translated into linear motion of the linkage end. The linkage is fixed to the smaller sprocket, and the larger sprocket fixed to the frame. For instrumentation.

Oscillating-chain mechanism

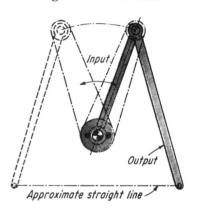

Rack and gear sector

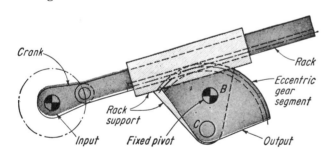

Rotary motion of the input is translated into oscillating motion of the output. The rack support and gear sector are pinned at C, but the gear itself oscillates around B.

Right angle oscillator

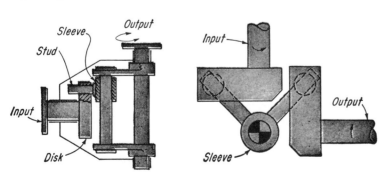

The input shaft and disk (extreme left) drives the stud and sleeve in a circular path. This causes the sleeve to move up and down and imparts an oscillating motion to the output shaft. A second variation is also shown.

Linear reciprocator

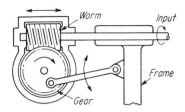

The objective here is to convert a rotary motion into a reciprocating motion that is *in line* with the input shaft. Rotation of the shaft drives the worm gear which is attached to the machine frame by means of a rod. Thus input rotation causes the worm gear to draw itself (and the worm) to the right—thus providing a back and forth motion. Employed in connection with a color-transfer cylinder in printing machines.

Disk and roller drive

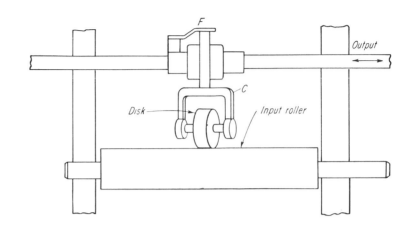

Here a hardened disk, riding at an angle to an input roller, transforms the rotary motion into linear motion *parallel to the axis of the input*. The roller is pressed against the input shaft with the help of a flat spring, *F*. Feed rate is easily varied by changing the angle of the disk. Arrangement can produce an extremely slow feed with a built-in safety factor in case of possible jamming.

Bearings and roller drive

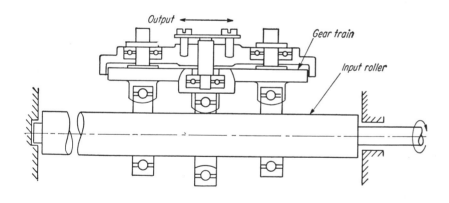

Similar to the previous one, this arrangement avoids large Hertzian stresses between disk and roller by employing three ball bearings in place of the single disk. The inner races of the bearings make contact on one side or the other. Hence a gearing arrangement is required to alternate the angle of the bearings. This arrangement also reduces the bending moment on the shaft.

Reciprocating space crank

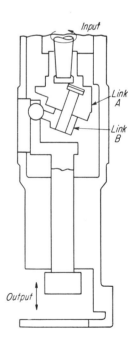

Rotary input causes the bottom surface of link *A* to wobble in reference to the center line. Link *B* is free of link *A* but restrained from rotating by the slot. This causes the output member to reciprocate linearly. Employed for filing machine.

LONG DWELL MECHANISMS

Oscillating crank and planetary drive

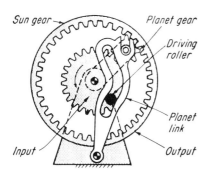

Chain-slider drive

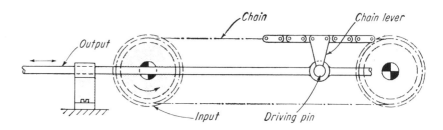

The output reciprocates back and forth with a constant velocity but comes to a long dwell at either end as the chain lever, whose length is equal to the radius of the sprockets, goes around either sprocket.

Here the planet is driven with a stop-and-go motion. The driving roller is shown entering the circular-arc slot on the planet link, hence the link and the planet remain stationary while the roller travels this portion of the slot. Result: a rotating output motion with a progressive oscillation.

Chain and slider drive

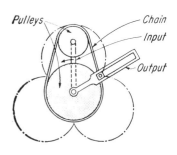

Chain-oscillating drive

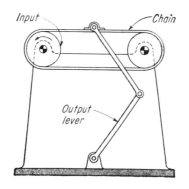

Same principle as before except that the chain link drives a lever which oscillates. The slowdown-dwell occurs as the chain pin goes around the left sprocket.

The input crank causes the small pulley to orbit around the stationary larger pulley. A pivot point in the chain slides inside the groove of the output link. In the position shown the output is about to enter a long dwell period of about 120 deg.

Epicyclic dwell mechanism

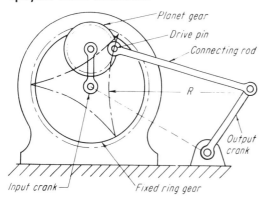

Cam-worm dwell mechanism

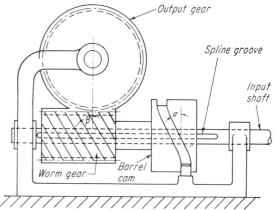

Here the output crank pulsates back and forth with a long dwell at the extreme right position. Input is to the planet gear by means of the rotating crank. The pin on the gear traces the epicyclic three-lobe curve shown. The right portion of the curve is almost a circular arc of radius R. If the connecting rod is made equal to R, the output crank comes virtually to a standstill during a third of the total rotation of the input. The crank then reverses, comes to a stop at left position, reverses, and repeats dwell.

Without the barrel cam, the input shaft would drive the output gear by means of the worm gear at constant speed. The worm and the barrel cam, however, are permitted to slide linearly on the input shaft. Rotation of the input shaft now causes the worm gear to be cammed back and forth, thus adding or subtracting motion to the output. If barrel cam angle α is equal to the worm angle β, the output comes to a stop during the limits of rotation illustrated, then speeds up to make up for lost time.

Cam-helical dwell mechanism

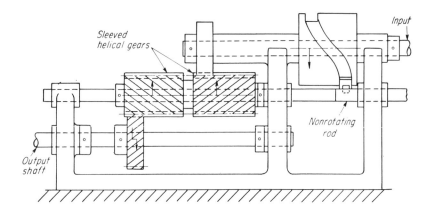

6-Bar dwell mechanism

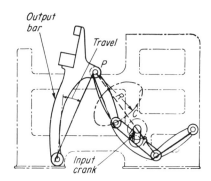

When one helical gear is shifted linearly (but prevented from rotating) it will impart rotary motion to the mating gear because of the helix angle. This principle is used in the mechanism illustrated. Rotation of the input shaft causes the intermediate shaft to shift to the left, which in turn adds or subtracts from the rotation of the output shaft.

Rotation of the input crank causes the output bar to oscillate with a long dwell at the extreme right position. This occurs because point C describes a curve that is approximately a circular arc (from C to C') with center at P. The output is practically stationary during that portion of curve.

Cam-roller dwell mechanism

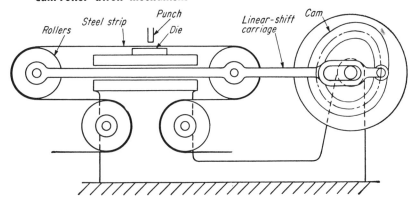

Three-gear drive

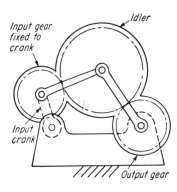

A steel strip is fed at constant linear velocity. But at the die station (illustrated), it is desired to stop the strip so that the punching operation can be performed. The strip passes over movable rollers which, when shifted to the right, cause the strip to move to the right. Since the strip is normally fed to the left, proper design of the cam can nullify the linear feed rates so that the strip stops, and then speeds to catch up to the normal rate.

This is actually a four-bar linkage combined with three gears. As the input crank rotates it takes with it the input gear which drives the output gear by means of the idler. Various output motions are possible. Depending on proportions of the gears, the output gear can pulsate, or come to a short dwell—or even reverse briefly.

Double-crank dwell mechanism

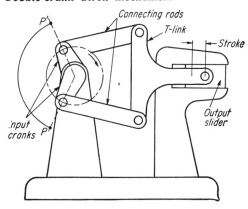

Both cranks are connected to a common shaft which also acts as the input shaft. Thus they always remain a constant distance apart from each other. There are only two frame points—the center of the input shaft and the guide for the output slider. As the output slider reaches the end of its stroke (to the right), it remains practically at standstill while one crank rotates through angle PP'. Used in textile machinery for cutting.

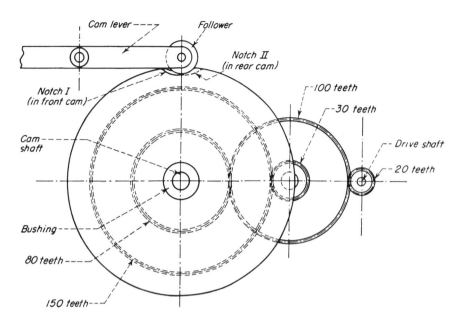

Fast Cam-Follower Motion

Fast cam action every *n* cycles when *n* is a relatively large number, can be obtained with this manifold cam and gear mechanism. A single notched cam geared 1/*n* to a shaft turning once a cycle moves relatively slowly under the follower. The double notched-cam arrangement shown is designed to operate the lever once in 100 cycles, imparting to it a rapid movement. One of the two identical cams and the 150-tooth gear are keyed to the bushing which turns freely around the cam shaft. The latter carries the second cam and the 80-tooth gear. The 30- and 100-tooth gears are integral, while the 20-tooth gear is attached to the one-cycle drive shaft. One of the cams turns in the ratio of 20/80 or 1/4; the other in the ratio 20/100 times 30/150 or 1/25. The notches therefore coincide once every 100 cycles (4 x 25). Lever movement is the equivalent of a cam turning in a ratio of 1/4 in relation to the drive shaft. To obtain fast cam action, *n* must be broken down into prime factors. For example, if 100 were factored into 5 and 20, the notches would coincide after every 20 cycles.

Intermittent Motion

This mechanism, developed by the author and to his knowledge novel, can be adapted to produce a stop, a variable speed without stop or a variable speed with momentary reverse motion. Uniformly rotating input shaft drives the chain around the sprocket and idler, the arm serving as a link between the chain and the end of the output shaft crank. The sprocket drive must be in the ratio N/n with the cycle of the machine, where n is the number of teeth on the sprocket and N the number of links in the chain. When point P travels around the sprocket from point A to position B, the crank rotates uniformly. Between B and C, P decelerates; between C and A it accelerates;; and at C there is a momentary dwell. By changing the size and position of the idler, or the lengths of the arm and crank, a variety of motions can be obtained. If in the sketch, the length of the crank is shortened, a brief reverse period will occur in the vicinity of C; if the crank is lengthened, the output velocity will vary between a maximum and minimum without reaching zero.

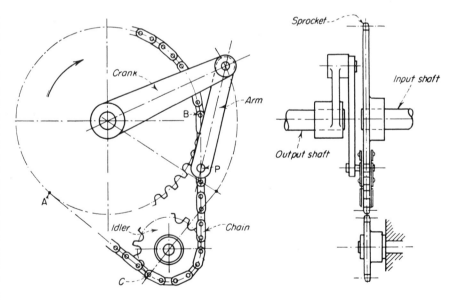

SHORT DWELL MECHANISMS

Gear-slider crank

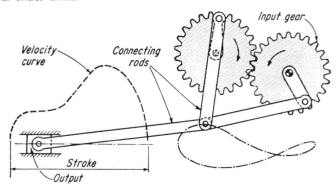

Employed in metal-drawing presses where the piston must move with a low constant velocity. The input drives both gears which in turn drive the connecting rods to produce the velocity curve shown.

Curve slider drive

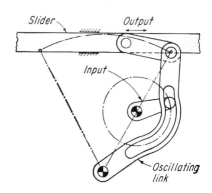

Here, too, the circular arc on the oscillating link permits the link to come to a dwell during the right position of the output slider.

The input drives three gears with connecting rods. A wide variety of reciprocating motions of the output can be obtained by selective proportioning of the linkages, including one to several dwells per cycle.

Gear oscillating crank

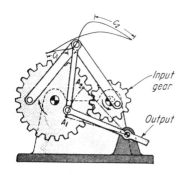

Similar arrangement to the one previously shown, but the curve described by the pin connection has two portions, C_1 and C_2, which are very close to circular arc with centers at A_1 and A_2. Hence the driven link will have a dwell at both extreme positions.

Triple-harmonic drive

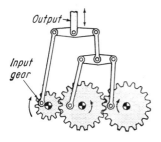

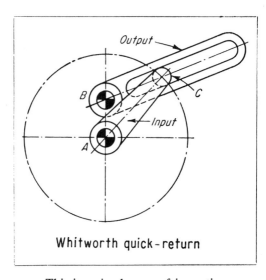

Whitworth quick-return

This is a simple way of imparting a varying motion to the output shaft B. However, the axes, A and B, are not colinear.

Wheel and slider drive

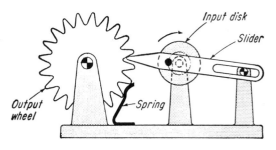

For each revolution of the input disk the slider moves in to engage the wheel and index it one tooth. A flat spring keeps the wheel locked while stationary.

FRICTION DEVICES
For Intermittent Rotary Motion

W. M. HALLIDAY
Southport, England

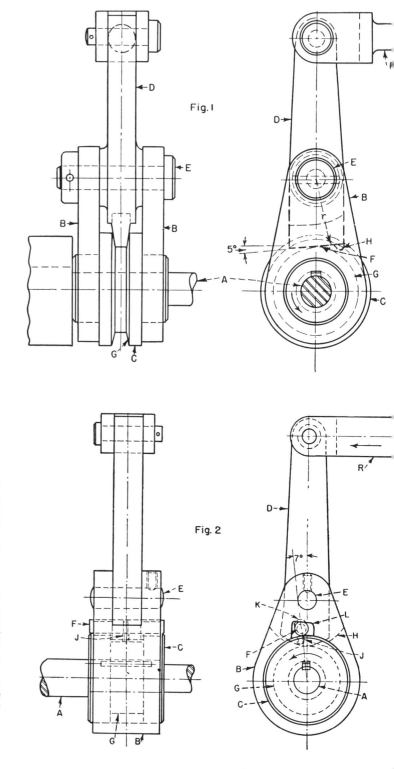

Fig. 1

Fig. 2

FIG. 1—WEDGE AND DISK. Consists of shaft *A* supported in bearing block *J*; ring *C* keyed to *A* and containing an annular groove *G*; body *B* which can pivot around the shoulders of *C*; lever *D* which can pivot about *E*; and connecting rod *R* driven by an eccentric (not shown). Lever *D* is rotated counterclockwise about *E* by the connecting rod moving to the left until surface *F* wedges into groove *G*. Continued rotation of *D* causes *A*, *B* and *D* to rotate counterclockwise as a unit about *A*. Reversal of input motion instantly swivels *F* out of *G*, thus unlocking the shaft which remains stationary during return stroke because of friction induced by its load. As *D* continues to rotate clockwise about *E*, node *H*, which is hardened and polished to reduce friction, bears against bottom of *G* to restrain further swiveling. Lever *D* now rotates with *B* around *A* until end of stroke.

FIG. 2—PIN AND DISK. Lever *D*, which pivots around *E*, contains pin *F* in an elongated hole *K* which permits slight vertical movement of pin but prevents horizontal movement by means of set screw *J*. Body *B* can rotate freely about shaft *A* and has cut-outs *L* and *H* to allow clearances for pin *F* and lever *D*, respectively. Ring *C*, which is keyed to shaft *A*, has an annular groove *G* for clearance of the tip of lever *D*. Counterclockwise motion of lever *D* actuated by the connecting rod jams pin between *C* and the top of cut-out *L*. This occurs about seven degrees from the vertical axis. *A*, *B* and *D* are now locked together and rotate about *A*. Return stroke of *R* pivots *D* around *E* clockwise and unwedges pin until it strikes side of *L*. Continued motion of *R* to the right rotates *B* and *D* clockwise around *A* while the uncoupled shaft remains stationary because of its load.

Friction devices are free from the common disadvantages inherent in conventional pawl and ratchet drives such as: (1) noisy operation; (2) backlash needed for engagement of pawl; (3) load concentrated on one tooth of the ratchet; and (4) pawl engagement dependent on an external spring.

The five mechanisms presented here convert the reciprocating motion of a connecting rod into intermittent rotary motion. The connecting rod stroke to the left drives a shaft counterclockwise; shaft is uncoupled and remains stationary during the return stroke of connecting rod to the right.

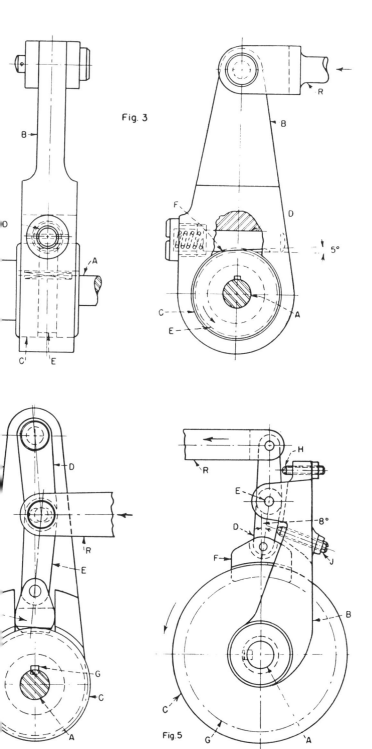

Fig. 3

Fig. 5

FIG. 3—SLIDING PIN AND DISK. Counterclockwise movement of body *B* about shaft *A* draws pin *D* to the right with respect to body *B*, aided by spring pressure, until the flat bottom *F* of pin is wedged against annular groove *E* of ring *C*. Bottom of pin is inclined about five degrees for optimum wedging action. Ring *C* is keyed to *A* and parts *A, C, D* and *B* now rotate counterclockwise as a unit until end of connecting rod's stroke. Reversal of *B* draws pin out of engagement so that *A* remains stationary while body completes its clockwise rotation.

FIG. 4—TOGGLE LINK AND DISK. Input stroke of connecting rod *R* to the left wedges block *F* in groove *G* by straightening toggle links *D* and *E*. Body *B*, toggle links and ring *C* which is keyed to shaft *A*, rotate counterclockwise together about *A* until end of stroke. Reversal of connecting rod motion lifts block, thus uncoupling shaft, while body *B* continues clockwise rotation until end of stroke.

FIG. 5—ROCKER ARM AND DISK. Lever *D*, activated by the reciprocating bar *R* moving to the left, rotates counterclockwise on pivot *E* thus wedging block *F* into groove *G* of disk *C*. Shaft *A* is keyed to *C* and rotates counterclockwise as a unit with body *B* and lever *D*. Return stroke of *R* to the right pivots *D* clockwise about *E* and withdraws block from groove so that shaft is uncoupled while *D*, striking adjusting screw *H*, travels with *B* about *A* until completion of stroke. Adjusting screw *J* prevents *F* from jamming in groove.

NO TEETH ON THESE RATCHETS

L KASPER
design consultant
Philadelphia

With springs, rollers and other devices they keep motion going one way.

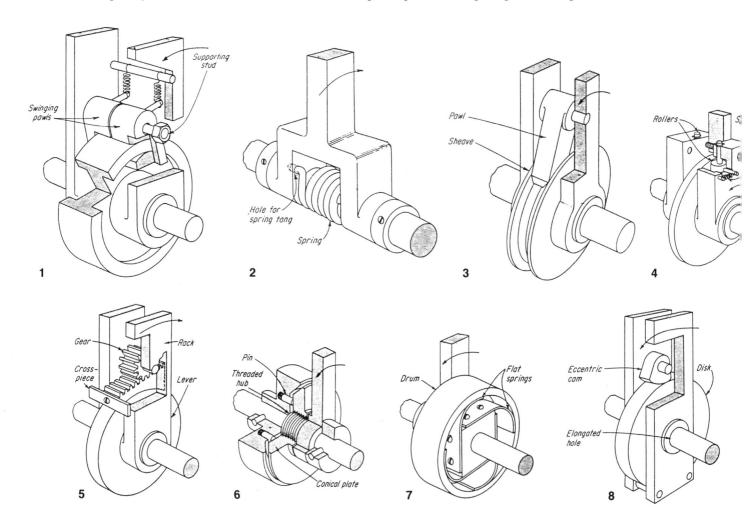

1 **SWINGING PAWLS** lock on rim when lever swings forward, and release on return stroke. Oversize holes for supporting stud make sure both top and bottom surfaces of pawls make contact.

2 **HELICAL SPRING** grips shaft because its inner diameter is smaller than the outer diameter of shaft. During forward stroke, spring winds tighter; during return stroke, it expands.

3 **V-BELT SHEAVE** is pushed around when pawl wedges in groove. For a snug fit, bottom of pawl is tapered like a V-belt.

4 **ECCENTRIC ROLLERS** squeeze disk on forward stroke. On return stroke, rollers rotate backwards and release their grip. Springs keep rollers in contact with disk.

5 **RACK** is wedge-shape so that it jams between the rolling gear and the disk, pushing the shaft forward. When the driving lever makes its return stroke, it carries along the unattached rack by the crosspiece.

6 **CONICAL PLATE** moves as a nut back and forth along the threaded center hub of the lever. Light friction of spring-loaded pins keeps the plate from rotating with the hub.

7 **FLAT SPRINGS** expand against inside of drum when lever moves one way, but drag loosely when lever turns drum in opposite direction.

8 **ECCENTRIC CAM** jams against disk during motion half of cycle. Elongated holes in the levers allow cam to wedge itself more tightly in place.

Escapement Mechanisms

FEDERICO STRASSER

Escapements, a familiar part of watches and clocks, can also be put to work in various control devices, and for the same purpose—to interrupt the motion of a gear train at regular intervals.

Each escapement shown here contains an energy absorber. Whether balance wheel or pendulum, it allows the gear train to advance one tooth at a time, locking the wheel momentarily after each half-cycle. Energy from the escape wheel is transferred to the escapement by the advancing teeth, which push against the locking surfaces (pallets).

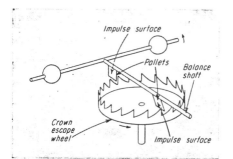

VERGE. Pallets are alternately lifted and cleared by diametrically opposed teeth on escape wheel, thus rocking balance shaft. One pallet or the other is always in contact with a tooth of the escape wheel.

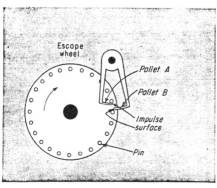

PIN WHEEL. Pallet *A* is about to unlock escape wheel. Pin will give a push to pallet *A* while sliding down impulse surface. As pendulum continues to swing, pallet *B* will stop pin, thus locking escape wheel. After pendulum reverses direction and swings toward dead center, pallet *B* releases wheel and receives an impulse from pin.

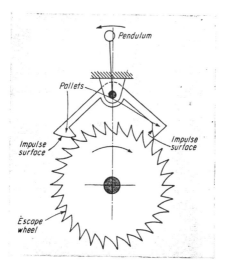

ANCHOR—RECOIL. As pendulum swings to one side, pallet engages tooth of escape wheel. When wheel is gripped, it recoils slightly. Pendulum starts in opposite direction by push of wheel transmitted through pallet. As pendulum swings to other side, the other pallet engages escape-wheel tooth, recoils, and repeats cycle.

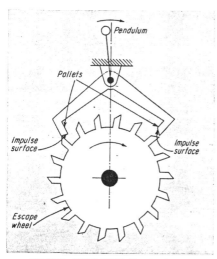

ANCHOR—DEAD BEAT. For greater accuracy, recoil escapement is modified by changing angle of impulse surfaces of pallets and teeth so that recoil is eliminated. The teeth of escape wheel fall "dead" upon pallets; as pendulum completes its swing, escape wheel does not turn backward.

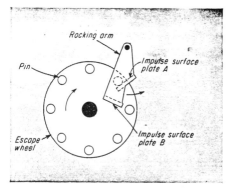

LANTERN WHEEL. Escape wheel is stopped when pin hits plate *A* of pallet, attached to rocking arm. As arm swings to right, pin pushes against impulse surface of plate until pin is clear. Escape wheel then turns freely until pin is stopped by plate *B* of pallet. Arm starts back toward left, receiving impulse from pin until latter clears plate.

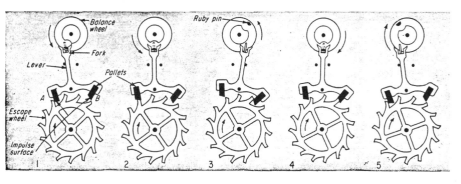

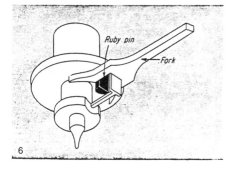

MUDGE'S LEVER. As balance wheel swings counterclockwise (1), the ruby pin enters fork of lever and pushes it to the right. The lever releases tooth of escape wheel, which starts turning and gives a tiny push (2) to impulse surface of pallet *A*. This is transmitted to balance wheel through fork and ruby pin. Balance wheel continues rotating counterclockwise (3) while lever swings to right and pallet *B* stops the escape wheel by engaging tooth. When balance wheel reaches its hairspring's limit, it reverses direction and re-enters fork (4). Ruby pin pushes lever to left, and receives impulse when escape wheel starts to move. Balance wheel continues clockwise until it reaches opposite limit of its hairspring. Lever continues moving to left until it again stops escape wheel (5) and awaits return of ruby pin. Another view of Mudge's lever is given in (6). This escapement is very precise because of the small portion of each half-cycle that escape wheel and balance wheel are in contact. Friction drag is at a minimum.

continued, next page

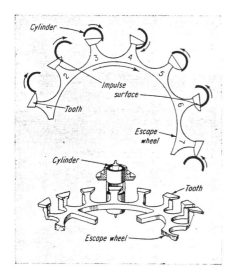

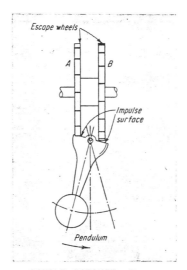

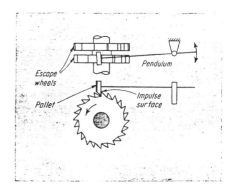

CYLINDER. Tooth of escape wheel and cylinder (balance-wheel shaft) are shown in seven positions. Tooth has been locked (1) and is about to enter cycle. Tooth is providing impulse (2) by sliding action along cylinder lip. Escape wheel is locked while cylinder rotates clockwise (3) to its limit (4), then starts back (5). As tooth starts forward, impulse is again imparted (6) to cylinder lip. Second tooth is locked (7) during period of backswing.

DOUBLE RATCHET. Two escape wheels are mounted on a common axis and pinned together so teeth are aligned with a half-pitch distance between them. Drawing shows pendulum in its extreme left position, with its impulse surface locking a tooth of the escape wheel *A*. As pendulum starts to right, escape wheels start to turn, imparting a push to the pendulum. As it approaches its extreme right position, pendulum again stops the escape wheels, by engaging a tooth on wheel *B*, then receives a second impulse as it starts toward the left.

HIGH-SPEED DOUBLE RATCHET. Same principle as in 2, but used with a small-mass pendulum. Speeds to 50 beats per second can be obtained.

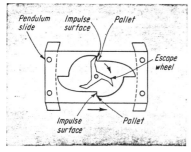

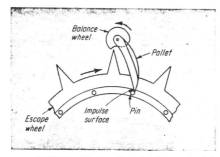

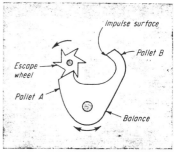

THREE-LEGGED. The three teeth of escape wheel work alternately upon top and bottom pallets. Top pallet is shown being driven to right. When tooth clears pallet, escape wheel will turn until tooth engages bottom pallet. When pendulum slide starts back toward left, bottom pallet will receive push from escape wheel, until tooth clears pallet.

DUPLEX. Single pallet of balance wheel is shown receiving impulse from pin of escape wheel. As balance wheel rotates counterclockwise, tooth of escape wheel clears notch, permitting escape wheel to turn clockwise. Next tooth of escape wheel is stopped by balance wheel until balance wheel reaches limit, reverses direction, and pallet swings back to position shown. Escapement receives but one impulse per cycle.

STAR WHEEL. Dead-beat escapement in which pallets alternately act upon diametrically opposed teeth. Illustration shows balance in extreme clockwise position with escape wheel locked. Balance starts turning counterclockwise, releasing escape wheel. At other extreme, pallet *B* locks escape wheel for a moment before direction of balance is again reversed and push is imparted.

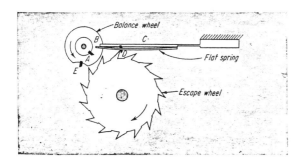

CHRONOMETER. As balance wheel turns counterclockwise, jewel *A* pushes flat spring *B* and raises bar *C*. Tooth clears jewel *D* and escape wheel turns. A tooth imparts push to jewel *E*. As jewel *A* clears flat spring, jewel *D* returns to position and catches next tooth. On return of balance wheel (clockwise), jewel *A* passes flat spring with no action. Thus, escapement receives a single impulse per cycle.

CAM-CONTROLLED PLANETARY GEAR SYSTEM

JOSEPH KAPLAN

By incorporating a grooved cam you get a novel mechanism that's able to produce a wide variety of output motions.

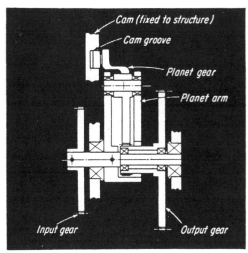 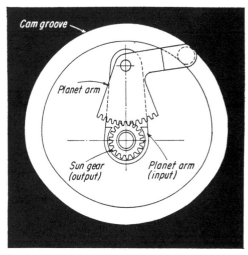

Construction details of cam-planetary mechanism employed in film drive.

Do you want more variety in the kinds of output motion given by a planetary gear system? You can have it by controlling the planet with a grooved cam. The method gives the mechanism these additional features:

• Intermittent motion, with long dwells and minimum acceleration and deceleration.

• Cyclic variations in velocity.

• Two levels, or more, of constant speed during each cycle of the input.

The design is not simple because of need to synchronize the output of the planetary system with the cam contour. However, such mechanisms are now at work in film drives and should prove useful in many automatic machines. Here are equations, tables and a step-by-step sequence that will make the procedure easier.

How the Mechanism Works

The planet gear need not be cut in full—a gear sector will do because the planet is never permitted to make a full revolution. The sun gear is integral with the output gear. The planet arm is fixed to the input shaft, which is coaxial with the output shaft. Attached to the planet is a follower roller which rides in a cam groove. The cam is fixed to the frame.

The planet arm (input) rotates at constant velocity and makes one revolution with each cycle. Sun gear (output) also makes one revolution during each cycle. Its motion is modified, however, by the oscillatory motion of the planet gear relative to the planet arm. It is this motion that is controlled by the cam (a constant-radius cam would not affect the output, and the drive would give only a constant one-to-one ratio).

Comparison with Other Devices

A main feature of this cam-planetary mechanism is its ability to produce a wide range of non-homogeneous functions. These functions can be defined by no less than two mathematical expressions, each valid for a discrete portion of the range. This feature is not shared by the more widely known intermittent mechanisms: the external and internal Genevas, the three-gear drive, and the cardioid drive.

Either three-gear or cardioid can provide a dwell period —but only for a comparatively short period of the cycle. With the cam-planetary, one can obtain over 180° of dwell during a 360° cycle by employing a 4-to-1 gear ratio between planet and sun.

And what about a cam doing the job by itself? This has the disadvantage of producing reciprocating motion. In other words, the output will always reverse during the cycle —a condition unacceptable in many applications.

Design Procedure

Basic equation for an epicyclic gear train is:

$$d\theta_S = d\theta_A - n\, d\theta_{P\text{-}A}$$

where: $d\theta_S$ = rotation of sun gear (output), deg
$\quad\quad d\theta_A$ = rotation of planet arm (input), deg
$\quad\quad d\theta_{P\text{-}A}$ = rotation of planet gear with respect to arm, deg
$\quad\quad n$ = ratio of planet to sun gear.

The required output of the system is usually specified in the form of kinematic curves. Design procedure then is to:

• Select the proper planet-sun gear ratio

• Develop the equations of the planet motion (which also functions as a cam follower)

• Compute the proper cam contour

Inertial drive with swinging mass propels vehicle

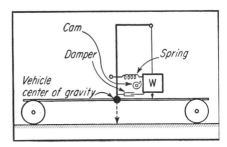

A new idea for vehicle propulsion depends on inertial forces rather than ground traction, thus may find use as an auxiliary drive in situations where wheels only spin helplessly in mud or ice. Prof Arthur W. Farrall of Michigan State Univ demonstrated his design with the scale model in the picture. The model, about 1 ft long and weighing 6½ lb, dragged a ½-lb weight up a moderate incline.

Motive force is provided by three movable weights that are, in sequence, thrust forward by cam action and then snapped back by springs. The backward-swinging weights supply the inertial force to propel the vehicle forward—which Dr Farrall sees as useful in military, construction, and agricultural equipment—or the process can be reversed to provide a negative push, for example, in braking on ice.

A bow to Newton. The system operates under Newton's three basic laws: A body in motion tends to stay in motion; momentum is changed at a rate proportional to the force that's

acting on it, and to every action there is an equal and opposite reaction.

Translated to the model in the picture, Newton's laws result in a two-cycle operation. During the cocking stroke, a small electric motor turns a cam that pushes the weight slowly forward while a spring attached to the weight is tightened (energized). In the power stroke, the cam releases the weight, which is snapped back by the spring. This pull of the swinging weight is strong enough to draw the vehicle toward it. This means the

vehicle must move forward until the force of the weight is taken up by its shock absorber. Calculations indicate that a 60-lb weight accelerated to 500 ft/sec^2 would provide 900 lb of push.

To avoid backsliding, the scale model relies on two sources of friction. There is some friction between the wheels and the ground, though the model is mounted on ball bearings, and the shock absorber slows the impact of the weight, converting the energy of its swing into heat.

Cam-and-wedge device averts truck jackknifing

A British designer has come up with a new idea for preventing the jackknifing of articulated trucks. The method, developed by Reginald Goold, Downend, Bristol, England, with backing of the National Research Development Corp., London SW 1, seems to mark a significant advance on other attempts to solve the problem.

Goold's invention (drawing) is a means of locking the swivel connections between the truck cab and trailer as soon as the trailer swings out of steering control. The device consists of moving wedges that, says NRDC, should be cheap to manufacture and easy to maintain.

When wheels lock. Jackknifing is caused by the locking of the cab's rear wheels. Another form of jackknifing, called trailer swing, is caused by the locking of the trailer wheels. Both forms involve rotation at the cab/trailer coupling, and the angle between the cab and trailer is largely dependent upon the steering angle.

Goold's device, linked to the steering, will lock the turntable coupling

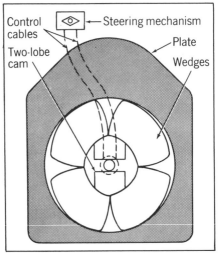

Two-lobe cam is linked to steering; the plate is part of trailer's fifth wheel.

(known as the fifth wheel) whenever the cab/trailer angle ceases to be related to the steering angle.

Cam and wedges. Driven from the steering system is a two-lobe cam that controls the positions of four wedges. These wedges taper radially downward toward the center. The trailer kingpin has a control surface that sits above, and is contoured to mesh with, the tops of the wedges. It

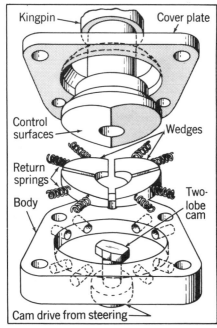

Control surfaces of trailer mesh with wedges unless cab/trailer angle deviates.

is arranged so its rotation relative to the trailer fifth-wheel plate must follow the rotation of the steering-controlled cam, else the bottom of the control surface cams out of the wedges. Under such conditions, it locks the vehicle in its cab/trailer angle.

CHAPTER 3.
SPECIAL-PURPOSE
MECHANISMS

Whiffletree linkage whisks
a load to a new position

Electromechanical system employs simple on-off circuitry to snap a tape head to any of 28 positions from any other position in random-access-memory storage unit

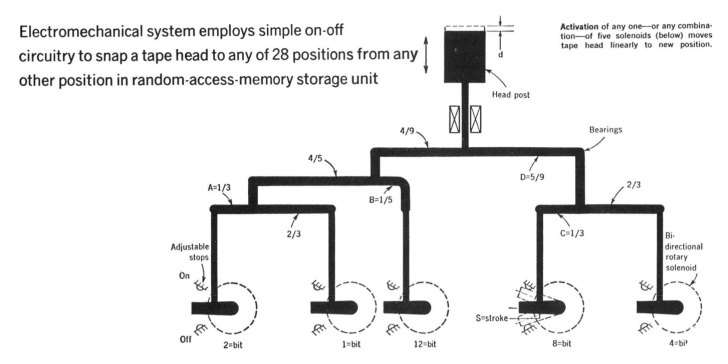

Activation of any one—or any combination—of five solenoids (below) moves tape head linearly to new position.

A simple "whiffletree" linkage arrangement actuated by on-off solenoids is providing fast and accurate linear positioning of a load or a machine member.

This new electromechanical device (drawing above) is designed for Potter Instrument Co.'s random-access memory-storage unit (photo, page 109). It performs a mechanical binary-to-decimal conversion to set the RAM's recording tape head in any of 28 positions.

John A. Altonji, engineering section manager for random-access memories at Potter Instrument, collaborated with Dr. Andrew Gabor in designing the mechanism and working out the mathematics necessary to selecting the proper linkage proportions. Altonji predicts that the whiffletree device can be used in many positioning applications, including those in machine tools and hydraulic control valves.

Why binary? In designing the mechanical positioner, Potter Instrument aimed at the same simplicity and speed that is inherent in electrical binary-number systems. In such systems, simple on-off circuitry can differentiate, for example, among 32 numbers.

Five binary digits would, in this instance, be controlled by five on-off solenoids or relays in which the "off" position stands for zero and the "on" position represents one. Thus:

00000 = 0	01000 = 8
00001 = 1	01001 = 9
00010 = 2	01010 = 10
00011 = 3	01011 = 11
00100 = 4	01100 = 12
00101 = 5	10000 = 16
00110 = 6	10100 = 20
00111 = 7	11111 = 31

Thus, changing a position instantly from, say, from number 2 to number 16 means switching only two solenoids. The same concept is at the heart of the whiffletree design, in which the activation of a given group of solenoids will move the respective linkages, snapping the tape-head post directly from one position to another, with an accuracy within 0.001 in.

Linkage relationships. Five bi-directional rotary solenoids are arranged to provide equal input strokes between fixed stops. The 22-deg. strokes are converted to a linear output through the five intermediate linkages. The solenoids are always energized to either the "on"

or "off" positions; the neutral positions shown in the drawing occur only when the power is turned off. The solenoids are directly coupled to hydraulic dashpots (drawing, next page) to provide terminal damping at the end of stroke and to balance the spring action.

Altonji notes that the number of discrete positions obtainable with n input solenoids is 2^n; in this case, 32 positions because five solenoids are used. Actuation of each solenoid will result in output motion of d, $2d$, $4d$, $8d$, and $16d$, where d is a unity deflection determined by the design (0.017 in. in this example).

In the Potter memory unit, the 16-bit output is modified to provide a deflection of only $12d$ to satisfy a specific design requirement for only 28 discrete positions. For a displacement of, say, $11d$, (or 11×0.017 in. $= 0.187$ in.), three solenoids must be turned "on"—the d, $2d$, and $8d$ solenoids.

In the schematic, A, B, C, and D are the unknown fractional parts of the linkage arms. The complementary sections are therefore $(1-A)$, $(1-B)$, $(1-C)$, etc. letting S equal the unknown length of input stroke for each solenoid, measured at the point where a link is joined to the solenoid link, equations can be written for these binary weights:

1: $S(ABD) = 1d$
2: $S(1-A)BD = 2d$
4: $SC(1-D) = 4d$
8: $S(1-C)(1-D) = 8d$
12: $S(1-B)D = 12d$

In random-access memory unit, tape-head can be positioned so quickly and accurately that data can be stored magnetically on 28 channels across tape.

Adding these equations yields:
$S = 27d$

Solving them simultaneously produces the fractional arm ratios:

$A = \frac{1}{3}, B = \frac{1}{5}, C = \frac{1}{3}, D = \frac{5}{9}$.

Thus, for a desired unit deflection of $d = 0.017$ in., the solenoid stroke is $S = (27)(0.017) = 0.459$ in., for all solenoids.

How accurate a device? Because such positioning mechanisms must operate unfailingly with a high degree of position accuracy, Altonji has also analyzed the output error as a function of the stroke error. (The stroke error can be minimized by adjusting the stops.)

Designating the output errors associated with each bit as e_1, e_2, e_4, e_8, e_{12}, and the input stroke error as ΔS, Altonji modified the first equation to obtain

$(S \pm \Delta S)(ABD) = d + e_1$
$\frac{1}{27}(S \pm \Delta S) = d + e_1$

Combining with $S = 27d$ yields

$e_1 = \pm \frac{1}{27}(\Delta S)$
$e_2 = \pm \frac{2}{27}(\Delta S)$

etc., to $e_{12} = \frac{12}{27}(\Delta S)$

These equations predict that the largest error attributable to the most significant bit is less than half the input error. The total error when all solenoids are turned on will be equal to ΔS.

Speeding the response. When positioning time must be minimized, as in a random-access memory unit, acceleration and settling time become critical factors. The acceleration characteristics are improved through the use of biasing springs attached to each actuator arm (drawing below, left).

At initial command, the acceleration torque in a rotary solenoid is low, because the magnetic gap is so large. This torque increases with travel as the gap closes (drawing). In normal operation the actuator arms are at either one stop or the other. Thus, one spring always exerts a tensile force in the direction of the other stop. The total accelerating force on command is, therefore, the sum of the tensile force plus the solenoid torque.

The mechanical stops are isolated from ground and are wired to provide contact signals that logically fire a 20-millisec single-shot multivibrator (MMV) to allow contact bounce to settle. When the MMV has run out, the head is considered to be in position for reading and writing.

The whiffletree mechanism helps the Potter RAM (photo left) to store a total of 50 million bits of information on 16 2-in.-wide tape loops that fly continuously on foil bearings at 600 in./sec. Eight tape loops are housed in each of two rugged, interchangeable cartridges. Data can be retrieved at random in an average time of 96 millisec.

—*Nicholas P. Chironis*

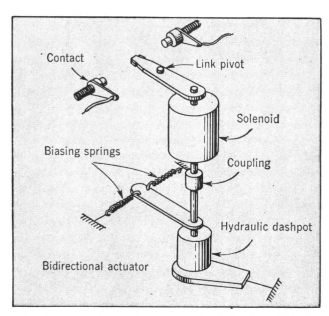

Solenoid assembly employs hydraulic dashpot and biasing springs to minimize the settling time for contact bounce and thus further to speed the response of the solenoids.

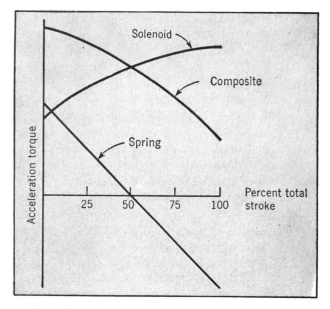

Initial acceleration torque of the solenoid is boosted by spring system to provide a high torque at the beginning of the stroke (composite curve), speeding unit's response.

Unique linkage produces precise straight-line motion

Patented family of straight-line mechanisms promises to serve many demands for movement without guideways and with low friction

A new mechanism for producing, without guideways, a straight-line motion very close to being true has been invented by James A. Daniel, Jr., a mechanism consultant at Newton, N.J. A patent has been granted, and the linkage is being applied on a camera to replace slides and telescoping devices.

Linkages, with their minimal friction of pivots, serve many purposes in today's machinery, in place of sliding and rolling parts that need guideways of one type or another.

James Watt, who developed the first such mechanism in 1784, is said to have been prouder of it than of his steam engine. Other well-known linkages include those of Evans, Tchebicheff, Roberts, and Scott-Russell.

Four-bar arrangement. Like other mechanisms that aim at straight-line motion, the Daniel design is based on the common four-bar linkage. Usually it is the selection of a certain point on the center link—the "coupler," which may extend past its pivot points—and of the location and proportions of the links that is the key to a straight-line device.

According to Daniel, the deviation of his mechanism from a straight line is "so small it cannot easily be measured." Also, the linkage has the ability to support a weight from the moving point of interest with an equal balance as the point moves along. "This gives the mechanism powers of neutral equilibrium," says Daniel.

Patented action. The basic version of Daniel's mechanism (Fig. 1) consists of the four-bar $ABCD$. The coupler link BC is extended to P (the proportions of the links must be selected according to a rule). Rotation of link CD about D (Fig. 2) causes BA to rotate about A and point P to follow approximately a straight line as it moves to P^1. Another point, Q, will move along a straight path to Q^1, also without need for a guide. A weight hung on P would be in equilibrium.

"At first glance," says Daniel, "the Evans linkage [Fig. 4] may look similar to mine, but link CD, being offset from the perpendicular at A, prevents the path of P from being a straight line."

Watt's mechanism EFGD (Fig. 5) is another four-bar mechanism that will produce a path of C that is roughly a straight line as EF or GD is rotated. Tchebicheff combined the Watt and Evans mechanisms to create a linkage in which point C will move almost perpendicularly to the path of P.

Steps in layout. Either end of the coupler can be redundant when only one straight-line movement is required (Fig. 6). Relative lengths of the links and placement of the pivots are critical, although different proportions are easily obtained for design purposes (Fig. 7). One proportion, for example, allows the path of P to pass below the lower support pivot, giving complete clearance to the traveling member. Any Daniel mechanism can be laid out straightforwardly:

- Lay out any desired right triangle PQF (Fig. 3). Best results are with angle A approximately 75 to 80 deg.
- Pick point B on PQ. For greatest straight-line motion, B should be at or near the midpoint of PQ.
- Lay off length PD along FQ from F to find point E.
- Draw BE and its perpendicular bisector to find point A.
- Pick any point C. Lay off length PC on FQ from F to find point G.
- Draw CG and its perpendicular bisector to find D. The basic mechanism is $ABCD$ with PQ as the extension of BC.

Multilinked versions. A "gang" arrangement (Fig. 8) is useful for, say, stamping or punching five evenly spaced holes at one time. Two basic linkages are joined, and the Q points used to obtain short, powerful strokes.

An extended dual arrangement (Fig. 9) can support the traveling point at both ends and can permit a long stroke with no interference. A doubled-up parallel arrangement (Fig. 10 and photo below) provides a rigid support and two pivot points for the straight-line motion of a horizontal bar.

When the traveling point is allowed to clear the pivot support (Fig. 11), the ultimate path will curve upward to provide a handy "kick" action. A short kick is obtained by adding a stop (Fig. 12) to reverse the direction of the frame links while the long coupler continues its stroke. Daniel suggests this curved path is useful in engaging or releasing an object on a straight path.

Nicholas P. Chironis

Inventor James Daniel demonstrates the self-balancing characteristics of a multi-link, doubled-up parallel arrangement of links for a versatile mechanism.

Daniel's basic straight-line linkage...

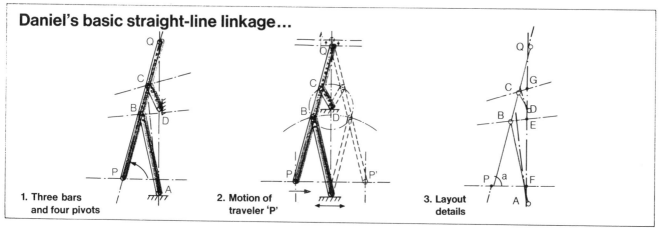

1. Three bars
   and four pivots

2. Motion of
   traveler 'P'

3. Layout
   details

...differs from competing types and leads to many variations

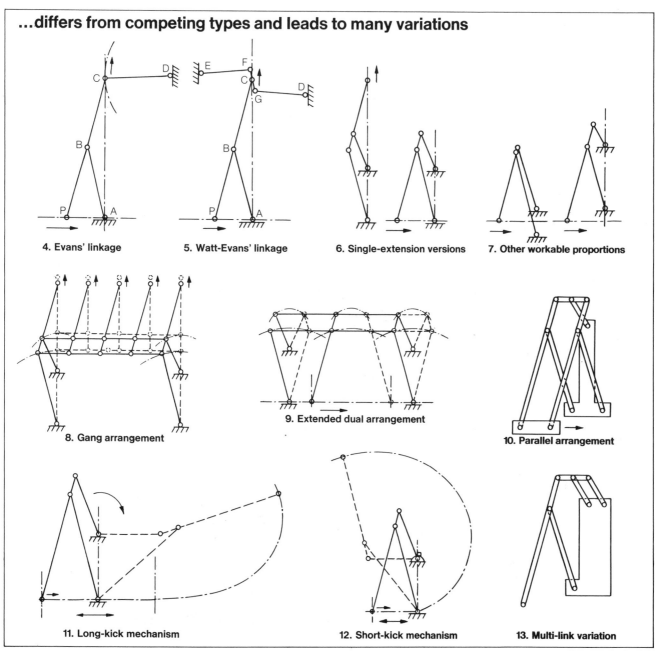

4. Evans' linkage

5. Watt-Evans' linkage

6. Single-extension versions

7. Other workable proportions

8. Gang arrangement

9. Extended dual arrangement

10. Parallel arrangement

11. Long-kick mechanism

12. Short-kick mechanism

13. Multi-link variation

Linkages pivot two ways for straight-line extension

New family of expanding and contracting devices
produces a greater distance of true straight-line motion
than ever before, to support protruding machine parts

Travel of link

The first camera with no sliding parts (photo, p.114; drawing above) is only one application of a new type of straight-line mechanism that can guide a reciprocating part along a truly straight path. All of this camera's moving parts are restrained and guided by linkages.

Extension of the mechanism, invented by J. A. Daniel, Jr., of Newton, N.J., who also designed the camera, produces a greater distance of straight-line movement than other linkage or telescoping systems. Hence the system can replace slides, telescoping devices, tracks, and rollers wherever friction or excess play is a problem. The mechanism requires no housing and can be operated in any position.

In the camera, an array of 12 links keeps lenses in line and parallel with each other as the camera is focused through the viewfinder. It replaces a telescoping tube that tended to bind. Several other variations of the mechanism are possible also (drawings right).

Search for true paths. Linkage designers have long been fascinated by the problem of obtaining a truly straight output motion. Some well-known designers and mathematicians have tackled the problem, with solutions including Roberts' 4-bar coupler linkage, Chebyshev's cross-bar linkage, Scott-Russel's long-link mechanism, Watt's modified linkage, and Peaucellier's cell linkage.

All such mechanisms reveal inherent faults that severely limit their application. Of the five just mentioned, for example, only Peaucellier's linkage will produce an exact straight-line motion and, even then, will guide only a point—the part itself may have to be constrained to move along a guideway. Moreover, all produce a stroke that is short for their size.

Series of linkages. Daniel's patented "straight-line carrier device" is an ingenious arrangement of a series of linkages. joined so the counterclockwise movement of one set of links will produce a clockwise

movement, of identical amount and speed, in a second set of links.

Daniel's theory can be explained by considering two identical pendulums placed on a level plane and swinging the same horizontal dis-

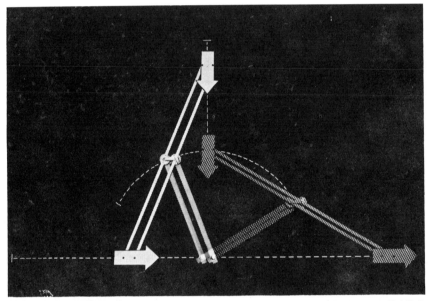

In geared version, extending links in one parallelogram products parallel-link setup in which a set of points moves down while keeping horizontal orientation.

Groups of straight-line linkages can be joined in an expanding unit that, unlike the familiar scissor-type gate, needs no end slots to prevent binding.

Initial design had sliding pin . . .

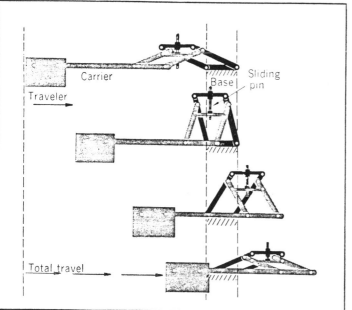

Traveler

Carrier

Base

Sliding pin

Total travel

. . . New design uses accordion arrangement . . .

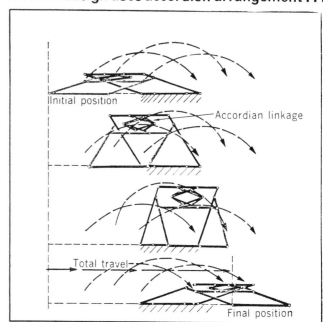

Initial position

Accordian linkage

Total travel

Final position

. . . Future units will have pinned gears . . .

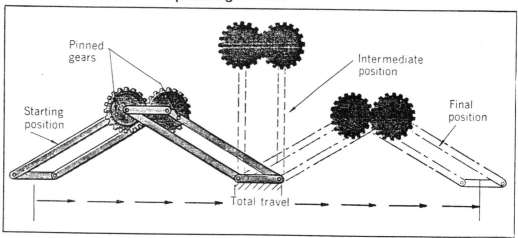

Pinned gears

Starting position

Intermediate position

Final position

Total travel

. . . Or relocated linkages for the accordion arrangement

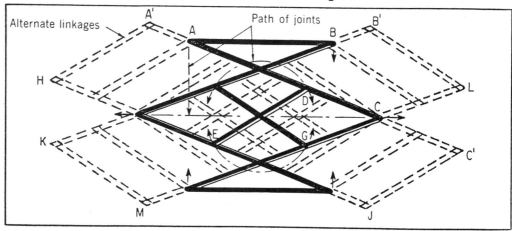

Alternate linkages

Path of joints

A' A B B'

H L

K C'

M J

D C

E G

tance. Any device that will hold a point on one pendulum to exactly the same horizontal level as the equivalent point on the other pendulum will suffice to limit the movement of one pendulum relative to the other when one is swinging clockwise and the other, counterclockwise.

Counter motion. To obtain this counter motion, Daniel first tried a sliding pin (first drawing, page 113) and a set of links of carefully selected lengths. The pin arrangement allowed the vertical links to swing like pendulums while keeping the pivot points in line vertically. Use of a sliding pin, however, prevented Daniel from claiming a slideless camera, so he came up with the

accordion linkage (second drawing, page 113).

This arrangement in turn can be changed into still another variation by relocation of the links (drawing, bottom right on page 113). This variation also illustrates the "magic ratio" that Daniel found essential in selecting the proper linkage proportions. The links, he discovered, must be in the ratio of:

$$AB = 16, \quad AC = 23, \quad DE = 12$$
$$AD = 14, \quad DC = 9$$

Any variation from these proportions—as little as 1/64 in. in a foot-long mechanism—will prevent the linkage from operating properly.

Finally, a set of gears was pinned on one prototype which improved the length of the stroke.

Miniature straight-line device (drawing) in camera allows rear link (in color) to be pushed past front link.

No mast on this lift truck

... Yet the load moves straight up instead of swinging through an arc. It's done with compensating links that shift the pivots as load arm is raised.

Coils are strapped ...

to 52-in.-sq pallets that are loaded in boxcars. Truck had to be designed to enter the 8 x 8-ft boxcar door, pick up pallets as near as 1 ft, 7 in. to the doors and maneuver inside the car's 9-ft, 2-in. width. Coils weigh 15,000 lb each and are stacked three-high when unloaded. Stacking required lifting forks to tilt from 10° back to 5° forward and have an 8-in. side-shifting motion, for stack alignment. Maneuverability was obtained with rear-wheel steering.

Straight-line lifting action is result of adding links A and B to the lifting mechanism. The load arm is raised by link A which arcs rearward as the hydraulic cylinder pushes it upward. Relative lengths of links A and B and the load arm let this action counteract the forward arcing motion of the swinging load arm. Load actually moves to rear as it is lifted.

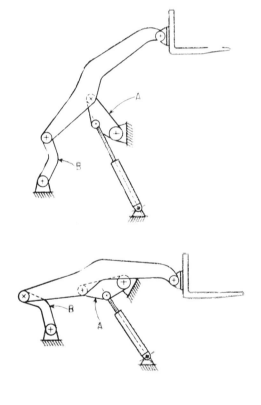

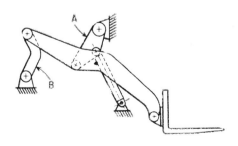

Designed by Automatic Transportation Co.

LINKAGE ARRANGEMENTS FOR ENGINE VALVES

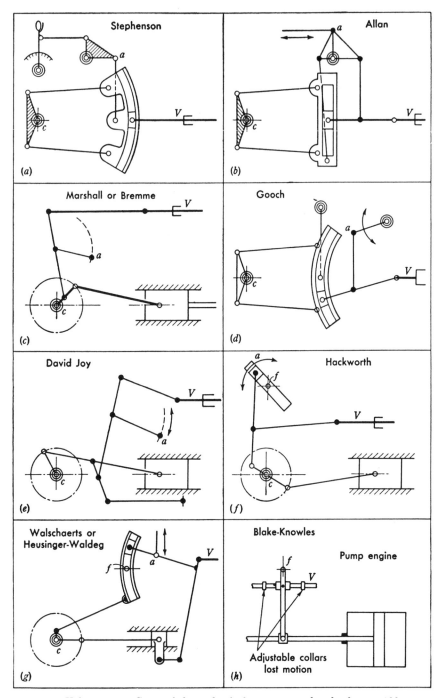

Stephenson

(a)

Allan

(b)

Marshall or Bremme

(c)

Gooch

(d)

David Joy

(e)

Hackworth

(f)

Walschaerts or Heusinger-Waldeg

(g)

Blake-Knowles

Pump engine

Adjustable collars
lost motion

(h)

Valve gears. Some of these classical movements date back over 100 years. They are of general interest because of the possibility of modifying them for other purposes. The following notation is used: f denotes a pivot fixed in the frame. a denotes an adjustable pivot for reversing or for changing cutoff. c denotes crankshaft center. V denotes valve rod.

(From *Mechanisms* by J. S. Beggs)

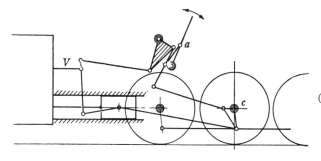

The Baker valve gear applied to a New York Central (fast freight) locomotive. This gear is popular because of the absence of slides exposed to dirt. (*The Pilliod Company*.)

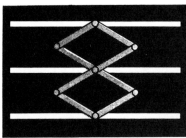

12 EXPANDING and

Parallel bars, telescoping slides and other mechanisms that can spark answers to man design problems.

FEDERICO STRASSER

Expanding grilles . . .
are often put to work as a safety feature. Single parallelogram (1) requires slotted bars; double parallelogram (2) requires none—but middle grille-bar must be held parallel by some other method.

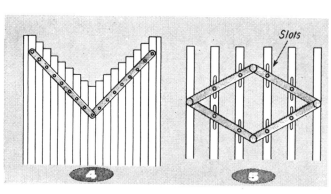

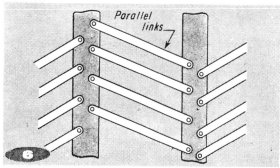

Multi-bar . . .
shutters, gates, etc. (4) can take various forms. Slots (5) allow for vertical adjustment. Space between bars may be made adjustable (6) by connecting the vertical bars with parallel links.

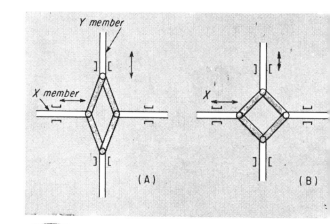

Variable motion . . .
can be produced with this arrangement. In (A) position, Y member is moving faster than X member. In (B), speeds both members are instantaneously equal. If the motion is c tinued in same direction, speed of X will become the greate

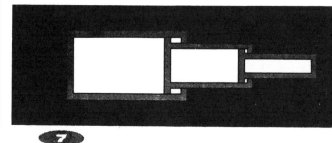

Telescoping devices . . .
are basis for many expanding and contracting mechanisms. arrangement shown, nested tubes can be sealed and filled a highly temperature-responsive medium such as a volatile liq

CONTRACTING DEVICES

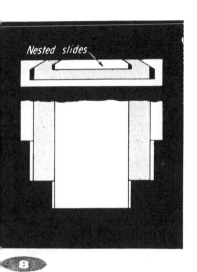

8

Nested slides . . .

can provide extension for machine-tool
table or other designs where accurate
construction is necessary. In this design
adjustments to obtain smooth sliding must
be made first, then the table surface must
be levelled.

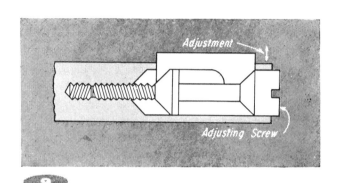

9

Expanding mandrels . . .
of the circular type are well-known. Shown here is a less common mandrel-type adjustment. Parallel member, adjusted by
two tapered surfaces on screw, can exert powerful force if
taper is small.

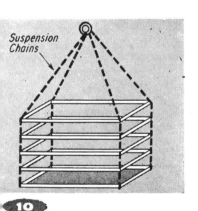

10

Expanding basket . . .
is opened when suspension chains are
lifted. Baskets take up little space when
not in use. Typical use for these baskets
is for conveyor systems. As tote baskets
they also allow easy removal of contents
because they collapse clear of the load.

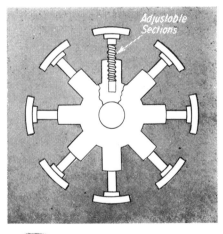

11

Expanding wheel . . .
has various applications besides acting as
pulley or other conventional wheel. Examples: electrical contact on wheel surfaces
allow many repetitive electrical functions to
be performed while wheel turns; dynamic
and static balancing is simplified when expanding wheel is attached to non-expanding
main wheel. As a pulley, expanding wheel
may have a steel band fastened to only one
section and passing twice around the circumference, thereby allowing adjustment.

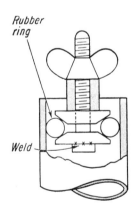

12

Pipe stopper . . .
relies on bulging of rubber ring
for its action—soft rubber will
allow greater adjustment than
hard rubber, and also conform
more easily to rough pipe-surfaces.
Hard rubber, however, withstands
higher pressures. Weld screw
head to washer for leaktight joint.

5 LINKAGES for

These devices convert rotary to straight-line motion without the need for guides.

SIGMUND RAPPAPORT

Evans' linkage . . .

has oscillating drive-arm that should have a maximum operating angle of about 40°. For a relatively short guide-way, the reciprocating output stroke is large. Output motion is on a true straight line in true harmonic motion. If an exact straight-line motion is not required, however, a link can replace the slide. The longer this link, the closer does the output motion approach that of a true straight line—if link-length equals output stroke, deviation from straight-line motion is only 0.03% of output stroke.

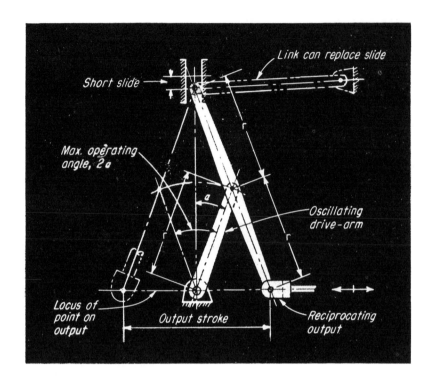

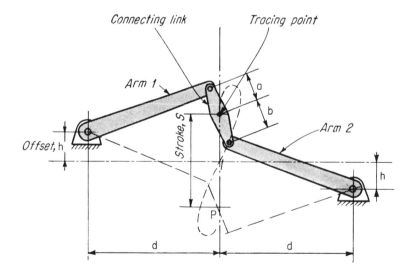

Simplified Watt's linkage . . .

generates an approximate straight-line motion. If the two arms are equally long, the tracing point describes a symmetrical figure 8 with an almost straight line throughout the stroke length. The straightest and longest stroke occurs when the connecting-link length is about ⅔ of the stroke, and arm length is 1.5S. Offset should equal half the connecting-link length. If the arms are unequal, one branch of the figure-8 curve is straighter than the other. It is straightest when a/b equals (arm 2)/(arm 1).

STRAIGHT-LINE MOTION

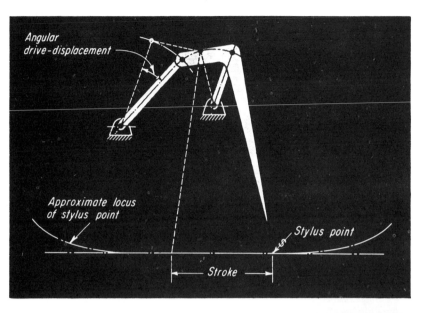

Angular drive-displacement

Approximate locus of stylus point

Stylus point

Stroke

Four-bar linkage . . .
produces approximately straight-line motion. This arrangement provides motion for the stylus on self-registering measuring instruments. A comparatively small drive-displacement results in a long, almost-straight line.

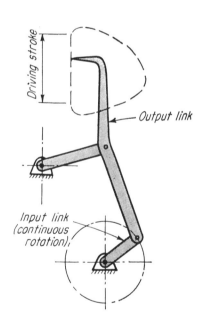

Driving stroke

Output link

Input link (continuous rotation),

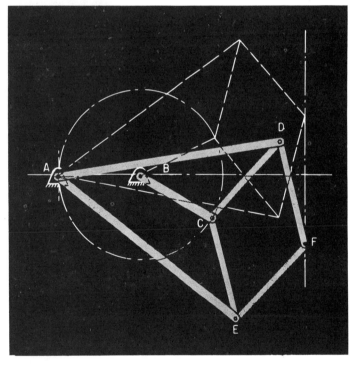

D-drive . . .
results when linkage arms are arranged as shown here. Output-link point describes a path resembling the letter D, thus it contains a straight portion as part of its cycle. Motion is ideal for quick engagement and disengagement before and after a straight driving-stroke. Example, the intermittent film-drive in movie-film projectors.

The "Peaucellier cell" . . .
was first solution to the classical problem of generating a straight line with a linkage. Its basis: within the physical limits of the motion, AC x AF remains constant. Curves described by C and F are, therefore, inverse; if C describes a circle that goes through A, then F will describe a circle of infinite radius—a straight line, perpendicular to AB. The only requirements are: AB=BC; AD=AE; and CD, DF, FE, EC are all equal. The linkage can be used to generate circular arcs of large radius by locating A outside the circular path of C.

Linkage Ratios for Straight-Line Mechanisms

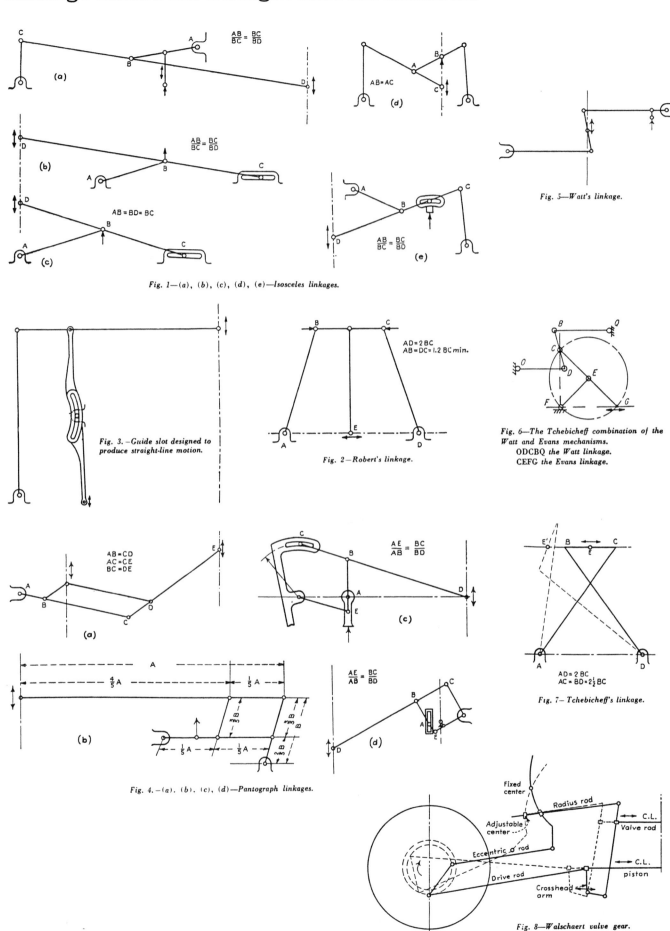

$\frac{AB}{BC} = \frac{BC}{BD}$

(a)

$AB = AC$

(d)

$\frac{AB}{BC} = \frac{BC}{BD}$

(b)

$AB = BD = BC$

(c)

$\frac{AB}{BC} = \frac{BC}{BD}$

(e)

Fig. 5—Watt's linkage.

Fig. 1—(a), (b), (c), (d), (e)—Isosceles linkages.

Fig. 3. —Guide slot designed to produce straight-line motion.

$AD = 2 BC$
$AB = DC = 1.2\ BC\ min.$

Fig. 2—Robert's linkage.

Fig. 6—The Tchebicheff combination of the Watt and Evans mechanisms.
ODCBQ the Watt linkage.
CEFG the Evans linkage.

$AB = CD$
$AC = CE$
$BC = DE$

(a)

$\frac{AE}{AB} = \frac{BC}{BD}$

(c)

$AD = 2\ BC$
$AC = BD = 2\frac{1}{2}\ BC$

Fig. 7— Tchebicheff's linkage.

A

$\frac{4}{5}A$ $\frac{1}{5}A$

(b)

$\frac{1}{5}A$ $\frac{1}{5}A$

$\frac{AE}{AB} = \frac{BC}{BD}$

(d)

Fig. 4. —(a), (b), (c), (d)—Pantograph linkages.

Fixed center
Radius rod
C.L.
Adjustable center
Valve rod
Eccentric rod
C.L.
Drive rod
piston
Crosshead arm

Fig. 8—Walschaert valve gear.

120

FIG. 1—No linkages or guides are used in this modified hypocyclic drive which is relatively small in relation to the length of its stroke. The sun gear of pitch diameter D is stationary. The drive shaft, which turns the T-shaped arm, is concentric with this gear. The idler and planet gears, the latter having a pitch diameter of $D/2$, rotate freely on pivots in the arm extensions. Pitch diameter of the idler is of no geometrical significance, although this gear does have an important mechanical function. It reverses the rotation of the planet gear, thus producing true hypocyclic motion with ordinary spur gears only. Such an arrangement occupies only about half as much space as does an equivalent mechanism containing an internal gear. Center distance R is the sum of $D/2$, $D/4$ and an arbitrary distance d, determined by a particular application. Points A and B on the driven link, which is fixed to the planet, describe straight-line paths through a stroke of $4R$. All points between A and B trace ellipses, while the line AB envelopes an astroid.

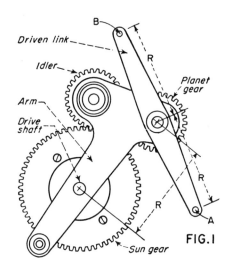

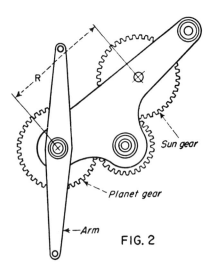

FIG.1

FIG. 2

Parallel Motion

FIG. 2—A slight modification of the mechanism in Fig. 1 will produce another type of useful motion. If the planet gear has the same diameter as that of the sum gear, the arm will remain parallel to itself throughout the complete cycle. All points on the arm will thereby describe circles of radius R. Here again, the position and diameter of the idler gear are of no geometrical importance. This mechanism can be used, for example, to cross-perforate a uniformly moving paper web. The value for R is chosen such that $2\pi R$, or the circumference of the circle described by the needle carrier, equals the desired distance between successive lines of perforations. If the center distance R is made adjustable, the spacing of perforated lines can be varied as desired.

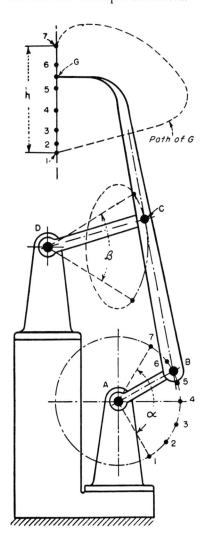

—The task of describing a "D" curve.

Hint for designing: Start with a straight line portion of the path of G, replace the oval arc of C by an arc of the osculating circle, thus determining the length of link DC.

This mechanism is intended to accomplish the following: (1) Film hook, while moving the film strip, must describe very nearly a straight line; (2) Engagement and disengagement of the hook with the perforation of the film must take place in a direction approximately normal to the film; (3) Engagement and disengagement should be shock free. Slight changes in the shape of the guiding slot f enable the designer to vary the shape of the output curve as well as the velocity diagram appreciably.

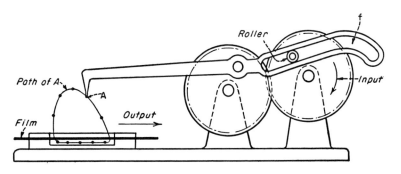

5 CARDAN-GEAR

SIGMUND RAPPAPORT,

These gearing arrangements convert rotation into straight-line motion, without need for slideways.

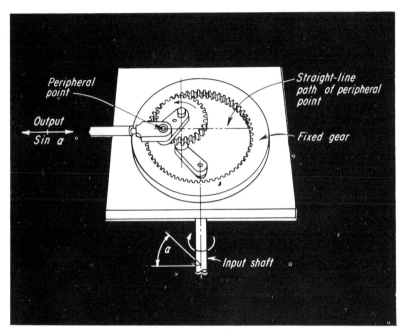

●1 Cardan gearing . . .

works on the principle that any point on the periphery of a circle rolling on the inside of another circle describes, in general, a hypocyloid. This curve degenerates into a true straight line (diameter of the larger circle) if diameters of both circles are in the ratio of 1:2. Rotation of input shaft causes small gear to roll around the inside of the fixed gear. A pin located on pitch circle of the small gear describes a straight line. Its linear displacement is proportional to the theoretically true sine or cosine of the angle through which the input shaft is rotated. Among other applications, Cardan gearing is used in computers, as a component solver (angle resolver).

●2 Cardan gearing and Scotch yoke . . .

in combination provide an adjustable stroke. Angular position of outer gear is adjustable. Adjusted stroke equals the projection of the large dia, along which the drive pin travels, upon the Scotch-yoke centerline. Yoke motion is simple harmonic.

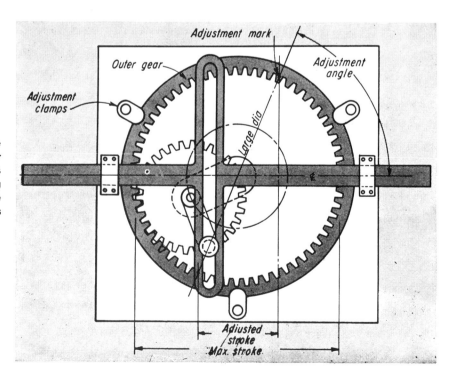

MECHANISMS

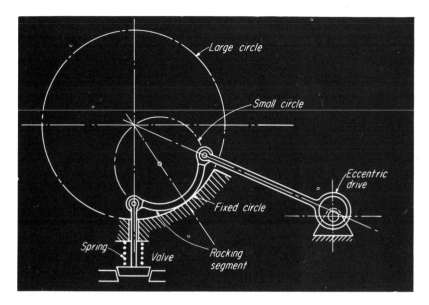

Large circle

Small circle

Eccentric drive

Fixed circle

Spring

Valve

Rocking segment

3

Valve drive . . .
exemplifies how Cardan principle may be applied. A segment of the smaller circle rocks to and fro on a circular segment whose radius is twice as large. Input and output rods are each attached to points on the small circle. Both these points describe straight lines. Guide of the valve rod prevents the rocking member from slipping.

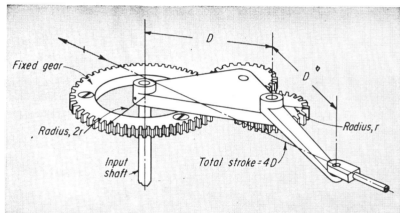

Fixed gear

D

D

Radius, 2r

Radius, r

Input shaft

Total stroke = 4D

4

Simplified Cardan principle . . .
does away with need for the relatively expensive internal gear. Here, only spur gears may be used and the basic requirements should be met, i.e. the 1:2 ratio and the proper direction of rotation. Latter requirement is easily achieved by introducing an idler gear, whose size is immaterial. In addition to cheapness, this drive delivers a far larger stroke for the comparative size of its gears.

5

Rearrangement of gearing . . .
in (4) results in another useful motion. If the fixed sun-gear and planet pinion are in the ratio of 1:1, then an arm fixed to the planet shaft will stay parallel to itself during rotation, while any point on it describes a circle of radius R. An example of application: in conjugate pairs for punching holes on moving webs of paper.

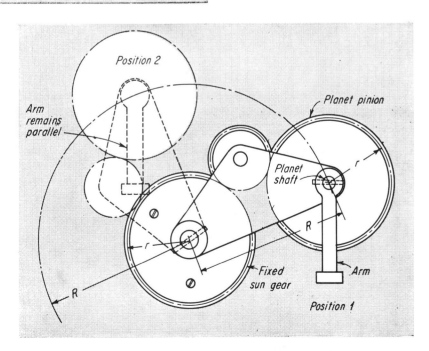

Position 2

Planet pinion

Arm remains parallel

Planet shaft

r

R

Fixed sun gear

Arm

R

Position 1

10 ways to change
STRAIGHT-LINE DIRECTION

Arrangements of linkages, slides, friction drives and gears that can be the basis of many ingenious devices.

FEDERICO STRASSER

Linkages

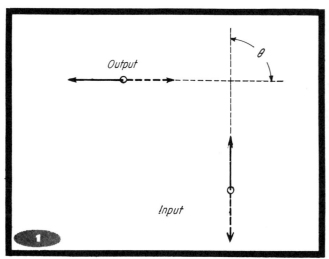

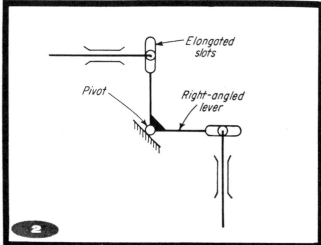

Basic problem (θ is generally close to 90°)

Slotted lever

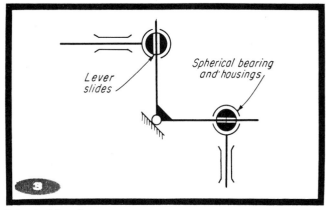

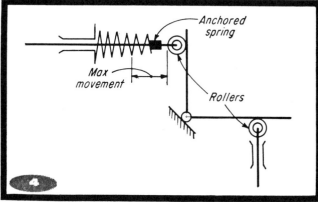

Spherical bearings

Spring-loaded lever

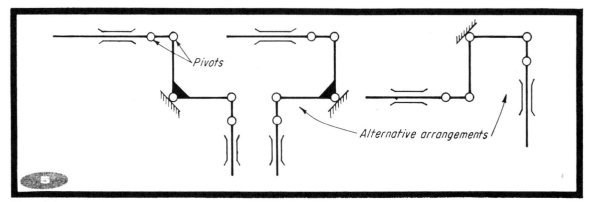

Pivoted levers with alternative arrangements

Guides

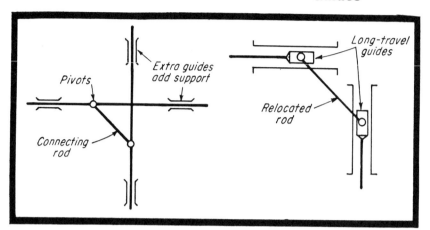

Single connecting rod (left) is relocated (right) to get around need for extra guides

Friction Drives

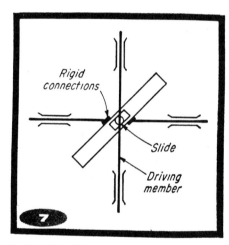

Inclined bearing-guide

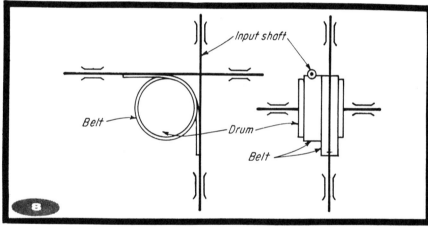

Belt, steel band, or rope around drum, fastened to driving and driven members; sprocket-wheels and chain can replace drum and belt

Gears

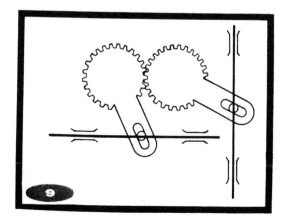

Matching gear-segments

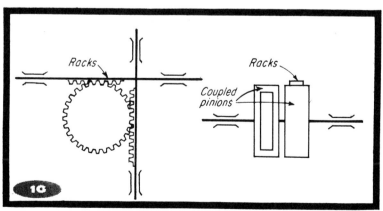

Racks and coupled pinions (can be substituted by friction surfaces for low-cost setup)

9 more ways to

CHANGE STRAIGHT-LINE DIRECTION

These devices, using gears, cams, pistons, and solenoids, supplement 10 similar arrangements employing linkages, slides, friction drives, and gears.

FEDERICO STRASSER

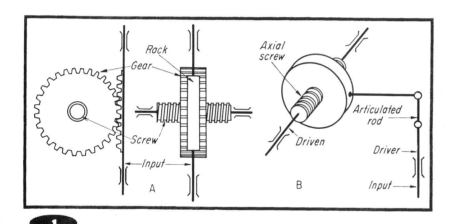

1

Axial screw with rack-actuated gear (A) and articulated driving rod (B) are both irreversible movements, i.e. driver must always drive.

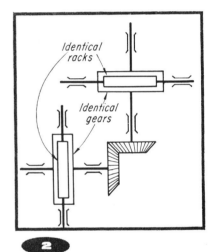

2

Rack-actuated gear with associated bevel gears is reversible.

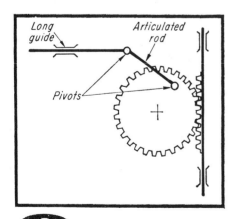

3

Articulated rod on crank-type gear with rack driver. Action is restricted to comparatively short movements.

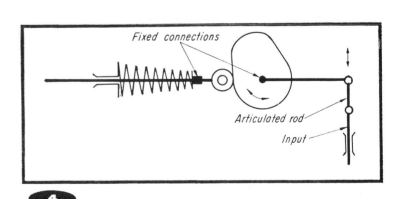

4

Cam and spring-loaded follower allow input/output ratio to be varied according to cam rise. Movement is usually irreversible.

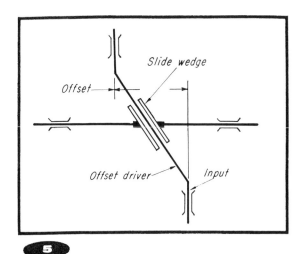

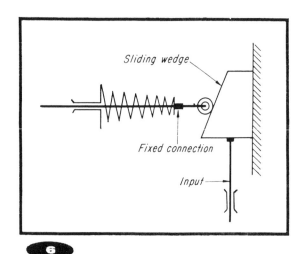

5

Offset driver actuates driven member by wedge action. Lubrication and low coefficient of friction help to allow max offset.

6

Sliding wedge is similar to previous example but requires spring-loaded follower; also, low friction is less essential with roller follower.

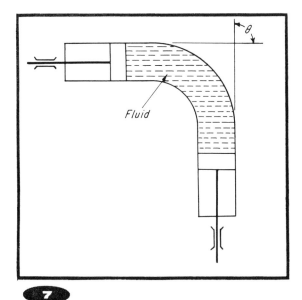

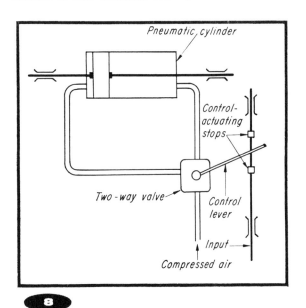

7

Fluid coupling is simple, allows motion to be transmitted through any angle. Leak problems and accurate piston-fitting can make method more expensive than it appears to be. Also, although action is reversible it must always be a compressive one for best results.

8

Pneumatic system with two-way valve is ideal when only two extreme positions are required. Action is irreversible. Speed of driven member can be adjusted by controlling input of air to cylinder.

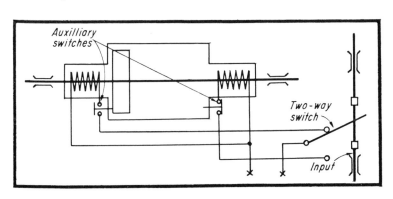

9

Solenoids and two-way switch are here arranged in analogous device to previous example. Contact to energized solenoid is broken at end of stroke. Again, action is irreversible.

Linkages for Accelerating and Decelerating Linear Strokes

When ordinary rotary cams cannot be conveniently applied, the mechanisms here presented, or adaptations of them, offer a variety of interesting possibilities for obtaining either acceleration or deceleration, or both

JOHN E. HYLER

Fig. 1—Slide block moves at constant rate of reciprocating travel and carries both a pin for mounting link *B* and a stud shaft on which the pinion is freely mounted. Pinion carries a crankpin for mounting link *D* and engages stationary rack, the pinion may make one complete revolution at each forward stroke of slide block and another in opposite direction on the return—or any portion of one revolution in any specific instance. Many variations can be obtained by making the connection of link *F* adjustable lengthwise along link that operates it, by making crankpin radially adjustable, or by making both adjustable.

Fig. 2—Drive rod, reciprocating at constant rate, rocks link *BC* about pivot on stationary block and, through effect of toggle, causes decelerative motion of driven link. As drive rod advances toward right, toggle is actuated by encountering abutment and the slotted link *BC* slides on its pivot while turning. This lengthens arm *B* and shortens arm *C* of link *BC*, with decelerative effect on driven link. Toggle is spring-returned on the return stroke and effect on driven link is then accelerative.

Fig. 3—Same direction of travel for both the drive rod and the driven link

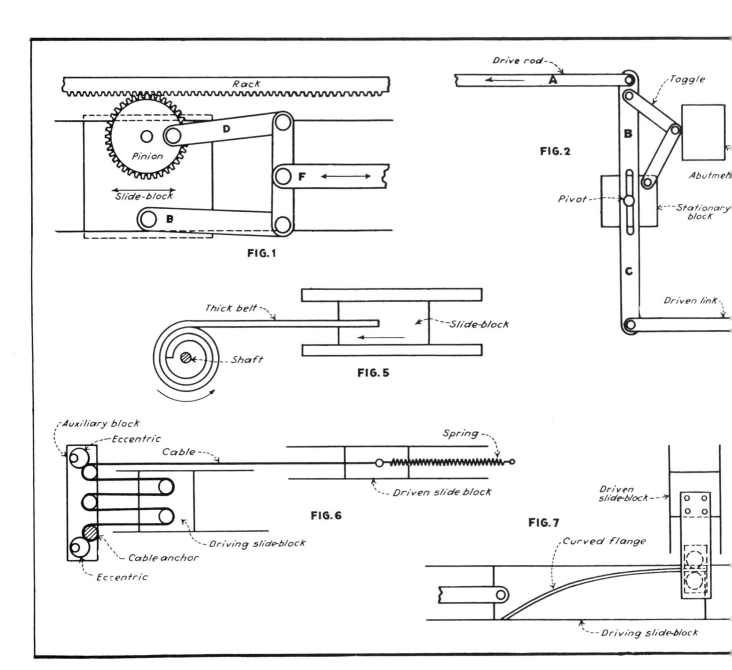

FIG. 1

FIG. 2

FIG. 5

FIG. 6

FIG. 7

s provided by this variation of the preceding mechanism. Here, acceleration is n direction of arrows and deceleration occurs on return stroke. Accelerative effect becomes less as toggle flattens.

ɪɢ. 4—Bellcrank motion is accelerated s rollers are spread apart by curved member on end of drive rod, thereby in urn accelerating motion of slide block. Driven elements must be spring-returned to close system.

ɪɢ. 5—Constant-speed shaft winds up hick belt, or similar flexible member, and ncrease in effective radius causes ac-

celerative motion of slide block. Must be spring or weight-returned on reversal.

Fɪɢ. 6 — Auxiliary block, carrying sheaves for cable which runs between driving and driven slide blocks, is mounted on two synchronized eccentrics. Motion of driven block is equal to length of cable paid out over sheaves resulting from additive motions of the driving and the auxiliary blocks.

Fɪɢ. 7—Curved flange on driving slide block is straddled by rollers pivotally mounted in member connected to driven slide block. Flange can be curved to

give desired acceleration or deceleration, and mechanism is self-returned.

Fɪɢ. 8—Stepped acceleration of the driven slide block is effected as each of the three reciprocating sheaves progressively engages the cable. When the third acceleration step is reached, the driven slide block moves six times faster than the drive rod.

Fɪɢ. 9—Form-turned nut, slotted to travel on rider, is propelled by reversing screw shaft, thus moving concave roller up and down to accelerate or decelerate slide block.

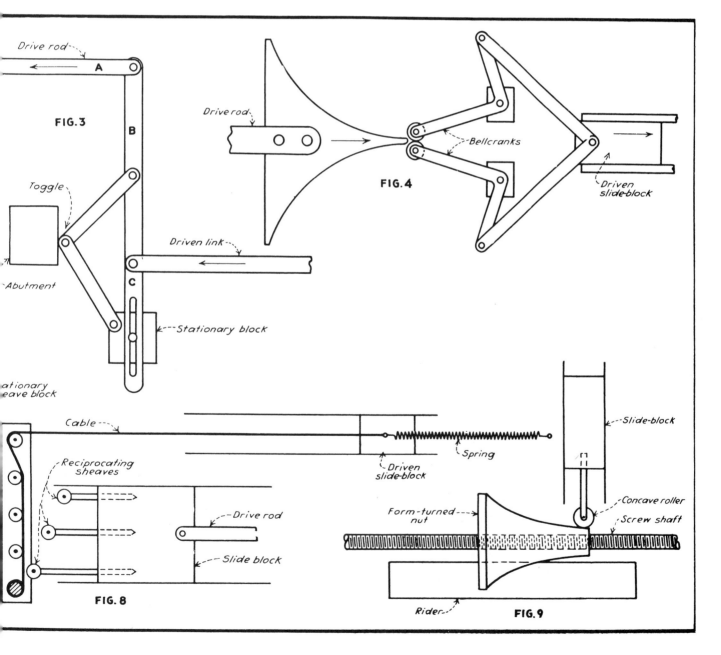

Linkages for Multiplying Short Motions

THE ACCOMPANYING SKETCHES show typical mechanisms for multiplying short linear motions, usually converting the linear motion into rotation. Although the particular mechanisms shown are designed to multiply the movements of diaphragms or bellows, the same or similar constructions have possible applications wherever it is required to obtain greatly multiplied motions. These patented transmissions depend on cams, sector gears and pinions, levers and cranks, cord or chain, spiral or screw feed, magnetic attraction, or combinations of these devices.

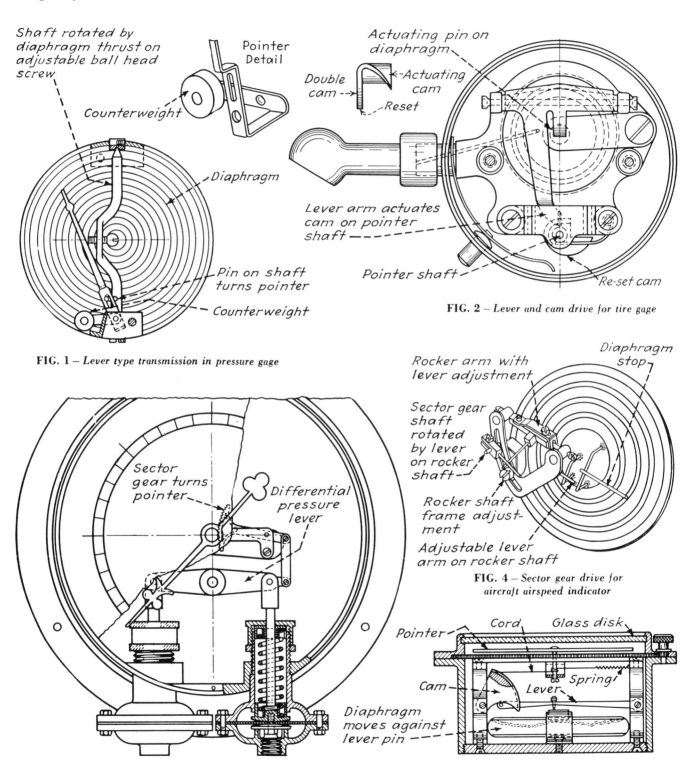

Shaft rotated by diaphragm thrust on adjustable ball head screw

Pointer Detail

Counterweight

Diaphragm

Pin on shaft turns pointer

Counterweight

FIG. 1 – Lever type transmission in pressure gage

Actuating pin on diaphragm

Double cam

Actuating cam

Reset

Lever arm actuates cam on pointer shaft

Pointer shaft

Re-set cam

FIG. 2 – Lever and cam drive for tire gage

Sector gear turns pointer

Differential pressure lever

FIG. 3 – Lever and sector gear in differential pressure gage

Rocker arm with lever adjustment

Diaphragm stop

Sector gear shaft rotated by lever on rocker shaft

Rocker shaft frame adjustment

Adjustable lever arm on rocker shaft

FIG. 4 – Sector gear drive for aircraft airspeed indicator

Pointer

Cord

Glass disk

Cam

Spring

Lever

Diaphragm moves against lever pin

FIG. 5 – Lever, cam and cord transmission in barometer

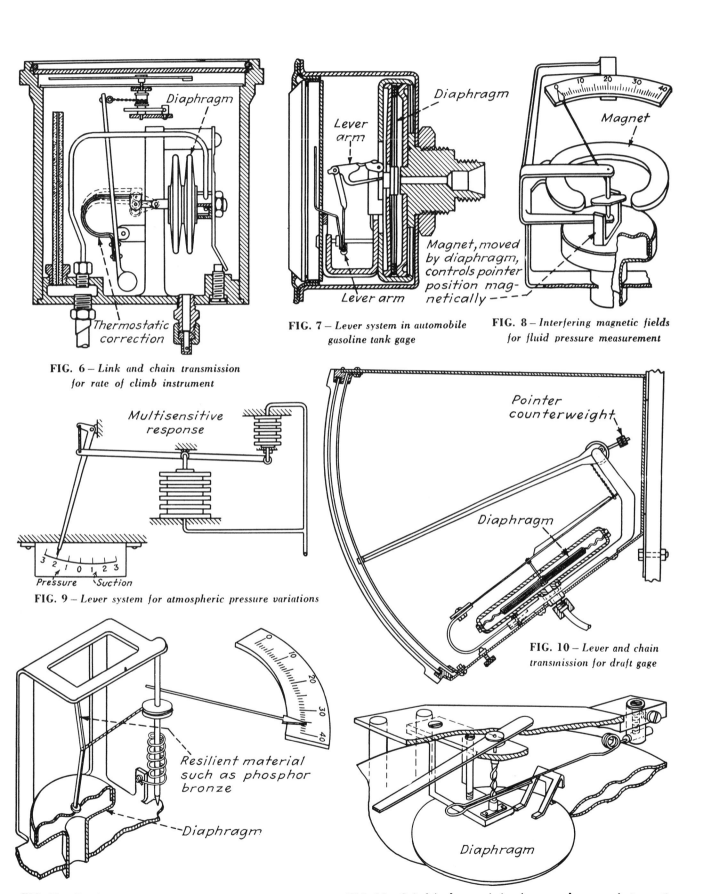

FIG. 6 – *Link and chain transmission for rate of climb instrument*

Diaphragm

Thermostatic correction

FIG. 7 – *Lever system in automobile gasoline tank gage*

Diaphragm

Lever arm

Lever arm

FIG. 8 – *Interfering magnetic fields for fluid pressure measurement*

Magnet

Magnet, moved by diaphragm, controls pointer position magnetically - - - -

Multisensitive response

Pressure   Suction

FIG. 9 – *Lever system for atmospheric pressure variations*

Pointer counterweight

Diaphragm

FIG. 10 – *Lever and chain transmission for draft gage*

Resilient material such as phosphor bronze

Diaphragm

FIG. 11 – *Toggle and cord drive for fluid pressure instrument*

Diaphragm

FIG. 12 – *Spiral feed transmission for general purpose instrument*

131

Converting Impulses to Mechanical Movements

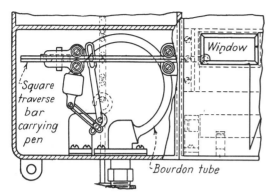

FIG. 1 – *Straight line movement produced by linkage from Bourdon tube*

TRANSMISSION OF MECHANICAL MOVEMENTS or impulses and conversion of electrical circuit variations to mechanical motion may employ any of a wide variety of mechanical linkages as well as hydraulic, pneumatic, or electrical forces. Patent records show all of the typical devices illustrated on these two pages. While these were designed for transmission of small forces, it is obvious that the parts can be made heavier and modified to withstand heavier loads.

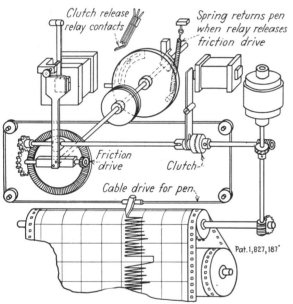

FIG. 2 – *Cable drive on recorder of workman efficiency*

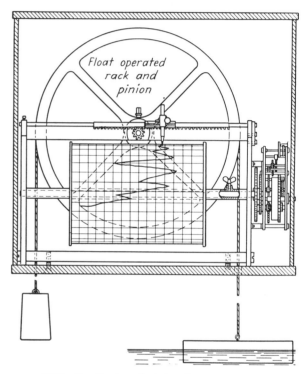

FIG. 3 – *Rack and pinion drive to pen*

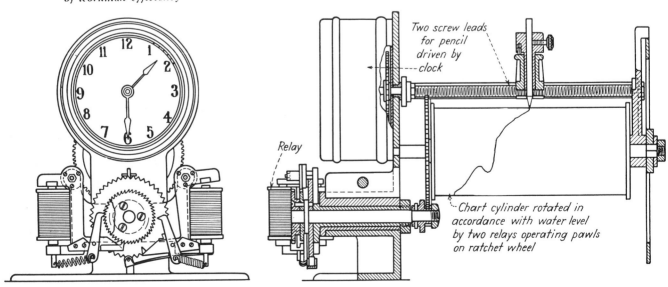

FIG. 4 – *Variable reversing chart movement controlled by relays; clock driven pencil*

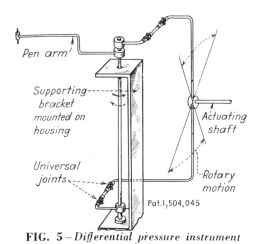

FIG. 5—*Differential pressure instrument with unique pen arm linkage*

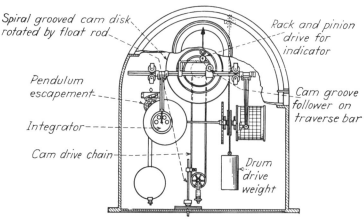

FIG. 6—*Spiral groove cam rotated by float moves pen, integrator arm, and indicator hand*

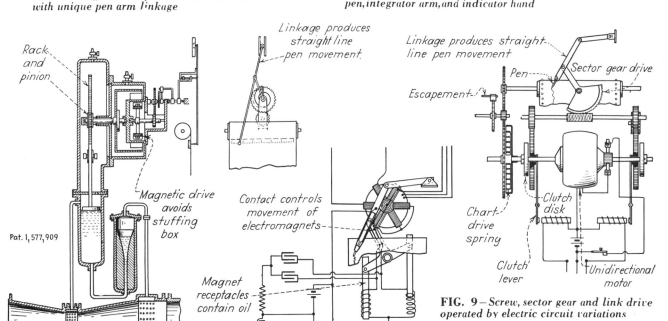

FIG. 7—*Magnetic drive from rack and pinion with straight line movement of pen*

FIG. 8—*Magnetically controlled pen activated by contacts on T-arm and armature*

FIG. 9—*Screw, sector gear and link drive operated by electric circuit variations*

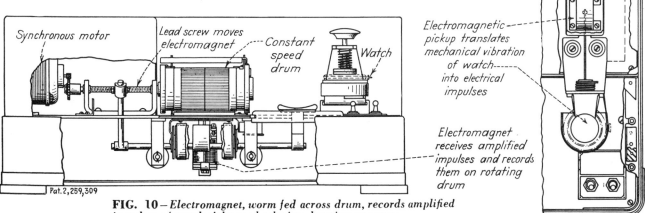

FIG. 10—*Electromagnet, worm fed across drum, records amplified impulses of watch ticks to check time keeping accuracy*

PARALLEL LINK MECHANISMS

PREBEN W. JENSEN

Eight-bar linkage

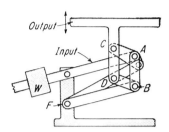

Double-handed screw mechanism

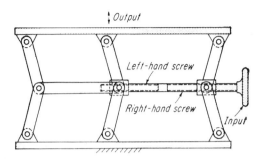

Link AB in this interesting arrangement will always be parallel to EF, and link CD parallel to AB. Hence CD will always be parallel to EF. Also, the linkages are so proportioned that point C moves in approximately a straight line. Final result is that the output plate will be kept horizontal while moving almost straight up or down. The weight permitted this device to function as a disappearing platform in a stage.

A simple parallel-link device that produces needed tensioning in webs, wires, tapes and strip steels. Adjusting the weight varies the drag.

Turning the adjusting screw spreads or contracts the linkage pairs to raise or lower the table. Six parallel links are shown but the mechanism can be built with four, eight, or more links.

Tensioning mechanism

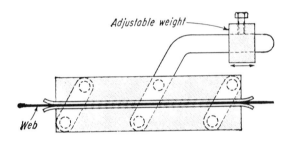

Triple-pivot mechanism

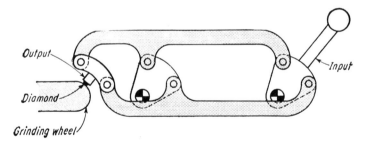

Two triangular plates pivot around fixed points on a machine frame. The output point produces a circular-arc curve. Built for rounding out the ends of the grinding wheels.

STROKE MULTIPLIER

Reciprocating-table drive

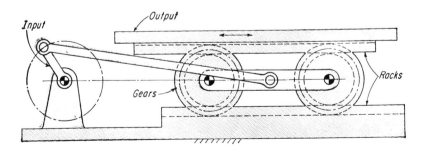

Two gears rolling on stationary bottom rack, drive the movable top rack which is attached to a printing table. When the input crank rotates, the table will move to a distance of four times the crank length.

Parallel link feeder

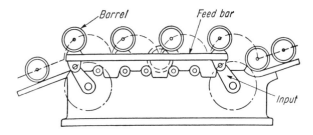

One of the cranks is the input, the other follows to keep the feeding bar horizontal. Employed for moving barrels from station to station.

Parallelogram linkage

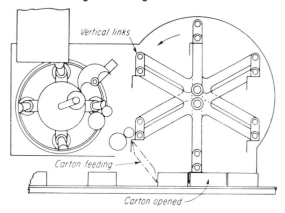

All seven short links are kept in a vertical position while rotating. The center link is the driver. This particular application feeds and opens cartons, but the device is capable of many diverse applications.

Parallel link driller

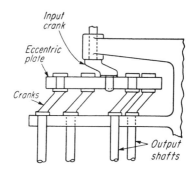

For powering a group of shafts. The input crank drives the eccentric plate. This in turn rotates the output cranks, all of which are of the same length and rotate at the same speed. Gears would require more room between shafts and are usually more expensive.

Parallel plate driver

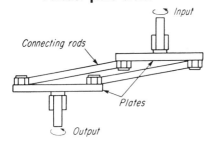

Here again, the input and output rotate with the same angle of relationship. The position of the shafts, however, can vary to suit other requirements without affecting the input-output relationship between shafts.

Curve-scribing mechanism

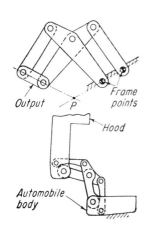

The output link rotates in such a way as to appear revolving around a point moving in space (P). This avoids the need for hinges at distant or inaccessible spots. Known also for hinging the hood of automobiles.

Parallel-link coupling

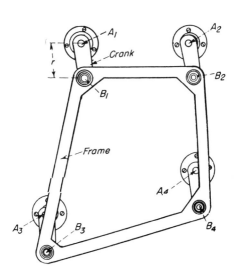

The absence of backlash makes this old but little used mechanism a precision, low-cost replacement for gear or chain drives otherwise used to rotate parallel shafts. Any number of shafts greater than two can be driven from any one of the shafts, provided two conditions are fulfilled: (1) All cranks must have the same length r; and (2) the two polygons formed by the shafts A and frame pivot centers B must be identical. The main disadvantage of this mechanism is its dynamic unbalance, which limits the speed of rotation. To lessen the effect of the vibrations produced, the frame should be made as light as is consistent with strength requirements.

FORCE AND STROKE MULTIPLIERS

Wide angle oscillator

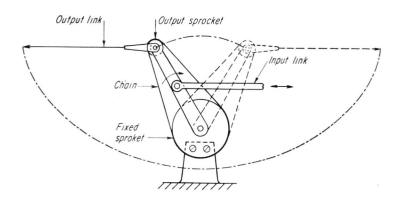

Motion of the input linkage is converted into a wide angle oscillation by means of the two sprockets and chain in the diagram illustrated. An oscillation of 60 deg is converted into 180-deg oscillation.

Angle-doubling drive

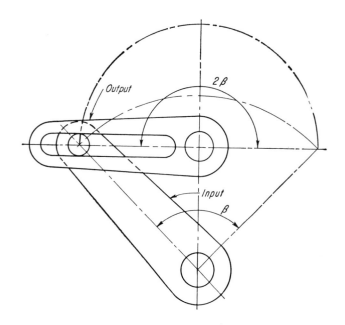

Pulley drive

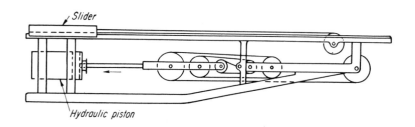

Gear-sector drive

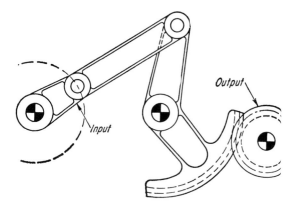

This is actually a four-bar linkage combined with a set of gears. A four-bar linkage usually can get no more than about 120-deg maximum oscillation. The gear segments multiply the oscillation in inverse proportion to the radii of the gears. For the proportions shown, the oscillation is boosted 2½ times.

Frequently it is desired to enlarge the oscillating motion β of one machine member into an output oscillation of, say, 2β. If gears are employed, the direction of rotation cannot be the same unless an idler gear is used, in which case the centers of the intput and output shafts cannot be too close. Rotating the input link clockwise causes the output to also follow in a clockwise direction. For a particular set of link proportions, distance between the shafts determines the gain in angle multiplication.

A simple arrangement which multiplies the stroke of a hydraulic piston, causing the slider to propel rapidly to the right for catapulting.

Typewriter drive

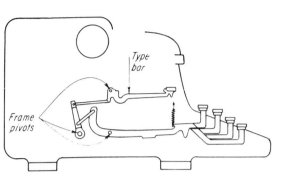

Multiplies the finger force in a typewriter, thus producing a strong hammer action at the roller from a light touch. There are three pivot points attached to the frame. Arrangement of the links is such that the type bar can move in free flight after a key has been struck. The mechanism illustrated is actually two four-bar linkages in series. In certain typewriters, as many as four 4-bar linkages are used in a series.

Double-toggle puncher —

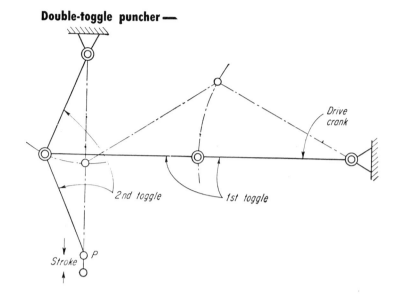

The first toggle keeps point P in the raised position even though its weight may be exerting a strong downward force (as in a heavy punch weight). When the drive crank rotates clockwise (driven, say, by a reciprocating mechanism), the second toggle begins to straighten to create a strong punching force.

Gear-rack drive

This mechanism is frequently employed to convert the motion of an input crank into a much larger rotation of the output (say, 30 to 360 deg). The crank drives the slider and gear rack, which in turn rotates the output gear.

Chain drive

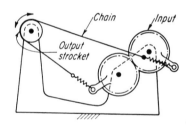

Springs and chains are attached to geared cranks to operate a sprocket output. Depending on the gear ratio, the output will produce a specified oscillation, say two revolutions of output in each direction for each 360 deg of input.

Linkage-train drive

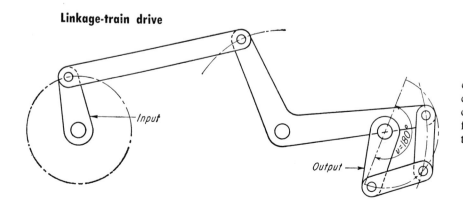

Arranging linkages in series can increase the angle of oscillation. In the case illustrated the oscillating motion of the L-shaped rocker is the input for the second linkage. Final oscillation is 180 deg.

Stroke-amplifying mechanisms

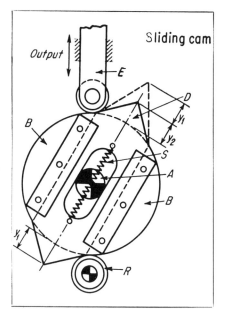

Sliding cam

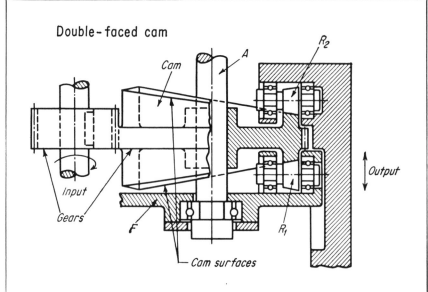

Double-faced cam

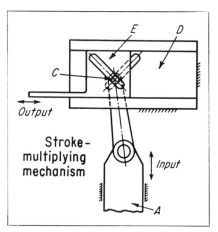

Stroke-multiplying mechanism

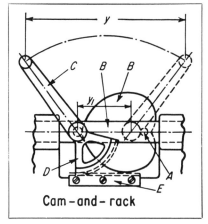

Cam-and-rack

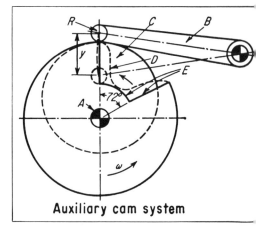

Auxiliary cam system

When the pressure angles are too high to satisfy the design requirements, and it is undesirable to enlarge the cam size, then certain devices can be employed to reduce the pressure angles:

Sliding cam — This device is used on a wire-forming machine. Cam D has a rather pointed shape because of the special motion required for twisting wires. The machine operates at slow speeds, but the principle employed here is also applicable to high-speed cams.

The original stroke desired is $(y_1 + y_2)$ but this results in a large pressure angle. The stroke therefore is reduced to y_2 on one side of the cam, and a rise of y_1 is added to the other side. Flanges B are attached to cam shaft A. Cam D, a rectangle with the two cam ends (shaded), is shifted upward as it cams off stationary roller R. during which the cam follower E

is being cammed upward by the other end of cam D.

Stroke multiplying mechanism — This device is used in power presses. The opposing slots, the first in a fixed member D and the second in the movable slide E, multiply the motion of the input slide A driven by the cam. As A moves upward. E moves rapidly to the right.

Double-faced cam — This device doubles the stroke, hence reduces the pressure angles to one-half of their original values. Roller R_1 is stationary. When the cam rotates, its bottom surface lifts itself on R_1, while its top surface adds an additional motion to the movable roller R_2. The output is driven linearly by roller R_2 and thus is approximately the sum of the rise of both surfaces.

Cam-and-rack — This device increases the throw of a lever. Cam B rotates around A. The roller follower

travels at distances y_1, during which time gear segment D rolls on rack E. Thus the output stroke of lever C is the sum of transmission and rotation giving the magnified stroke y.

Cut-out cam — A rapid rise and fall within 72 deg was desired. This originally called for the cam contour, D, but produced severe pressure angles. The condition was improved by providing an additional cam C which also rotates around the cam center A, but at five times the speed of cam D because of a 5:1 gearing arrangement (not shown). The original cam is now completely cut away for the 72 deg (see surfaces E). The desired motion, expanded over 360 deg (since 72 x 5 = 360), is now designed into cam C. This results in the same pressure angle as would occur if the original cam rise occurred over 360 deg instead of 72 deg.

ADJUSTABLE STROKE MECHANISMS

PREBEN W. JENSEN

Adjustable slider drive

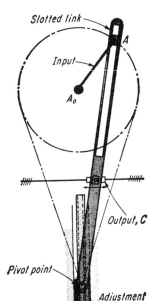

Shifting the pivot point by means of the adjusting screw changes the stroke of the output rod.

Adjustable chain drive

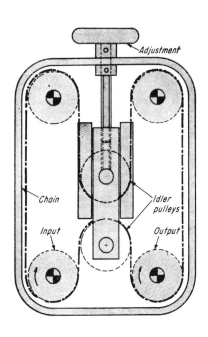

Synchronization between input and output shafts is varied **by** shifting the two idler pulleys by means of the adjusting screw.

Adjustable clutch drive

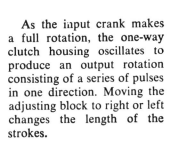

As the input crank makes a full rotation, the one-way clutch housing oscillates to produce an output rotation consisting of a series of pulses in one direction. Moving the adjusting block to right or left changes the length of the strokes.

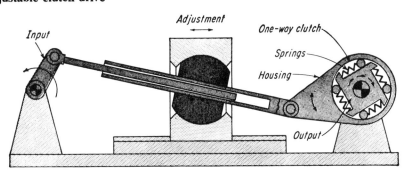

Adjustable-pivot drive

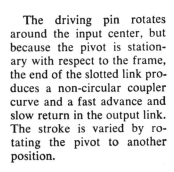

The driving pin rotates around the input center, but because the pivot is stationary with respect to the frame, the end of the slotted link produces a non-circular coupler curve and a fast advance and slow return in the output link. The stroke is varied by rotating the pivot to another position.

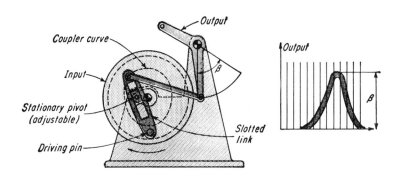

ADJUSTABLE-OUTPUT MECHANISMS

Linkage motion adjuster

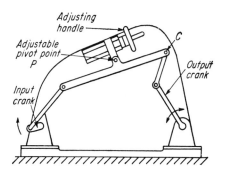

Here the motion and timing of the output link can be varied *during operation* by shifting the pivot point of the intermediate link of the 6-bar linkage illustrated. Rotation of the input crank causes point C to oscillate around the pivot point P. This in turn imparts an oscillating motion to the output crank. Shifting of point P is accomplished by a screw device.

Valve stroke adjuster

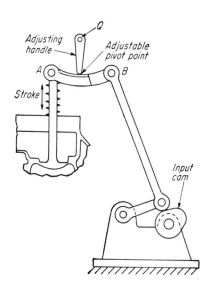

This mechanism adjusts the stroke of valves of combustion-engines. One link has a curved surface and pivots around an adjustable pivot point. Rotating the adjusting link changes the proportion of strokes of points A and B and hence of the valve. The center of curvature of the curve link is at point Q.

Cam motion adjuster

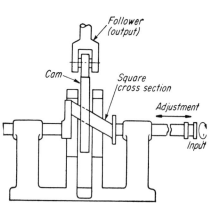

The output motion of the cam follower is varied by linearly shifting the input shaft to the right or left during operation. The cam has a square hole which fits over the square cross section of the crank shaft. Rotation of the input shaft causes eccentric motion in the cam. Shifting the input shaft to the right, for instance, causes the cam to move radially outward, thus increasing the stroke of the follower.

3-D mechanism

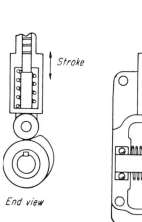

End view

Double-cam mechanism

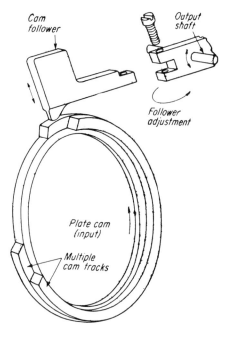

This is a simple but effective technique for changing the timing of a cam. The follower can be adjusted in the horizontal plane but is restricted in the vertical plane. The plate cam contains two or more cam tracks.

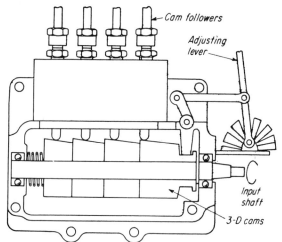

Output motions of four followers are varied during the rotation by shifting the quadruple 3-D cam to the right or left. Linear shift is made by means of adjustment lever which can be released in any of the 6 positions.

Piston-stroke adjuster

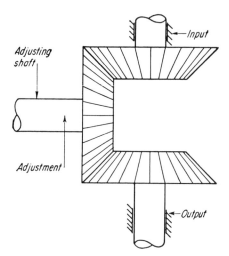

Shaft synchronizer

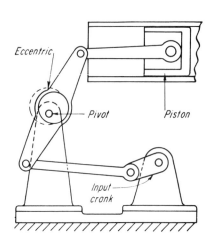

Eccentric pivot point

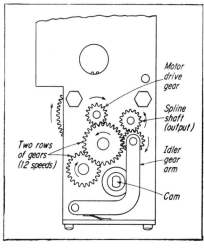

The input crank oscillates the slotted link to drive the piston up and down. The position of the pivot point can be adjusted by means of the screw mechanism even while the piston is under full load.

Actual position of the adjusting shaft is normally kept constant. The input then drives the output by means of the bevel gears. Rotating the adjusting shaft in a plane at right angle to the input-output line changes the relative radial position of the input and output shafts, used for introducing a torque into the system while running, synchronizing the input and output shafts, or changing the timing of a cam on the output shaft.

Rotation of the input crank reciprocates the piston. The stroke depends on position of the pivot point which is easily adjusted, even during rotation, by rotating the eccentric shaft.

Mechanical positioner varies machine element relation

Claimed to be highly accurate, the Phase Variator is a 1:1 ratio mechanical unit for varying the relative position of one machine or machine element to another. Installed between the element to be controlled and its drive, the endless chain device assures positive, simple adjustment of timing, registration, sequencing, indexing, stroking, cam positioning and synchronizing.

It is infinitely adjustable within a 360-degree cycle and is easily installed in any position. Adjustments can be made while unit is running or stopped. Size is 3¼ x 8¼ x 12 in.
Candy Mfg Co.

Gear shift mechanism

Variable-speed strip recorders make the tester's life simpler all round: First off, you can choose a chart speed that will produce easily interpreted data. Then you save on expensive paper since you get a longer recording period. These various advantages are to be had with a new strip chart recorder made by F. L. Mosley Co, Pasadena, Calif — model 7100A — which provides 12 chart speeds at the twist of a knob.

Drive mechanism. With 22 nylon gears in continuous mesh, a series of cam-controlled idler spur gears are used to engage selected speed reduction gear sets. The synchronous motor is connected either to the optional 10 to 1 speed reducer or through transfer gears to the motor drive gear.

Any of the 12 output gears may be coupled to the output spline and chart drive by the manually selected idlers. Mounted on spring-loaded arms, the idlers are controlled by a 12-position detented cam. By rotating the cam the appropriate idler is pivoted into simultaneous mesh with the output of the selected speed reducer and the output drive spline.

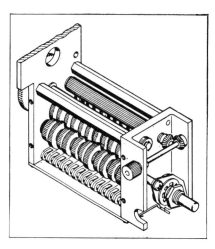

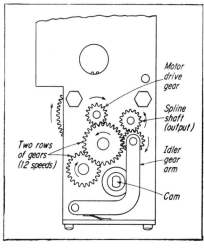

Ways to amplify mechanical movements

FEDERICO STRASSER

How levers, membranes, cams, and
gears are arranged to measure, weigh,
gage, adjust, and govern.

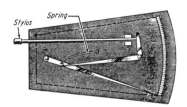

HIGH AMPLIFICATION for
simple measuring instruments
is provided by double lever
action. Accuracy can be as high
as 0.0001 in.

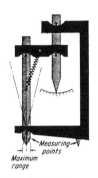

PIVOTED LEVERS allow extremely sensitive action in com-
parator-type measuring device
shown here. The range, however, is small.

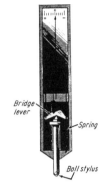

**ULTRA-HIGH AMPLIFICA-
TION**, with only one lever, is
provided in the Hirth-Mini-
meter shown here. Again, the
range is small.

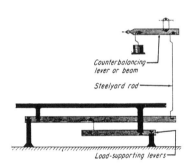

LEVER - ACTUATED weigh-
scale needs no springs to maintain balance. The lever system,
mounted on knife edges, is extremely sensitive.

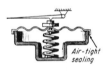

CAPSULE UNIT for gas-
pressure indicators should be
provided with a compression
spring to preload the membrane
for more positive action.

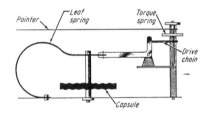

**AMPLIFIED MEMBRANE
MOVEMENT** can be gained by
the arrangement shown here. A
small chain-driven gear links
the lever system.

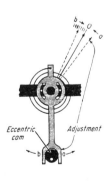

FOR CLOSE ADJUSTMENT,
electrical measuring instruments employ eccentric cams.
Here movement is reduced, not
amplified.

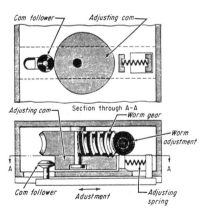

**MICROSCOPIC ADJUST-
MENT** is achieved here by employing a large eccentric-cam
coupled to a worm-gear drive.
Smooth, fine adjustment result.

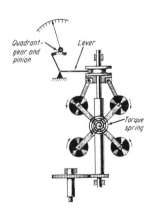

**QUADRANT-GEAR AND
PINION** coupled to an L-lever
provide ample movement of indicator needle for small changes
in governor speed.

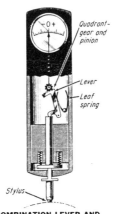

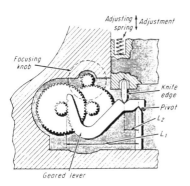

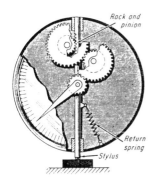

COMBINATION LEVER AND GEARED quadrant are used here to give the comparator maximum sensitivity combined with ruggedness.

LEVER AND GEAR train amplify the microscope control-knob movement. Knife edges provide frictionless pivots for lever.

DIAL INDICATOR starts with rack and pinion amplified by gear train. The return-spring takes out backlash.

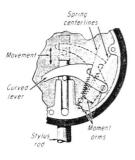

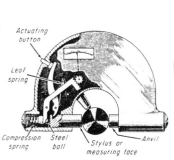

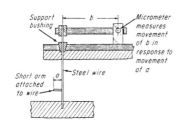

CURVED LEVER is so shaped and pivoted that the force exerted on the stylus rod, and thus stylus pressure, remains constant.

ZEISS COMPARATOR is provided with a special lever to move the stylus clear of the work. A steel ball greatly reduces friction.

"HOT-WIRE"AMMETER relies on the thermal expansion of a current-carrying wire. A relatively large needle movement occurs.

ACCURACY of 90° squares can be checked with a device shown here. The rod makes the error much more apparent.

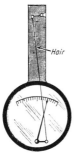

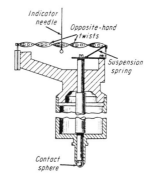

HYGROMETER is actuated by a hair. When humidity causes expansion of the hair, its movement is amplified by a lever.

STEEL RIBBONS transmit movement without the slightest backlash. The movement is amplified by differences in diameter.

METAL BAND is twisted and supported at each end. Small movement of contact sphere produces large needle movement.

TORSIONAL deflection of the short arm is transmitted with low friction to the longer arm for micrometer measurement.

REVERSING MECHANISMS

Double-link reverser

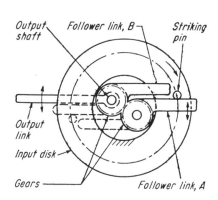

Automatically reverses the output drive every 180-deg rotation of the input. Input disk has press-fit pin which strikes link *A* to drive it clockwise. Link *A* in turn drives link *B* counterclockwise by means of their respective gear segments (or gears pinned to the links). The output shaft and the output link (which may be the working member) are connected to link *B*.

After approximately 180 deg of rotation, the pin slides past link *A* to strike link *B* coming to meet it—and thus reverses the direction of link *B* (and of the output). Then after another 180-deg rotation the pin slips past link *B* to strike link *A* and start the cycle over again.

Toggle-link reverser

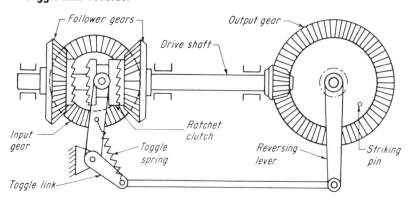

This mechanism also employs a striking pin—but in this case the pin is on the output member. The input bevel gear drives two follower bevels which are free to rotate on their common shaft. The ratchet clutch, however, is spline-connected to the shaft—although free to slide linearly. As shown, it is the right follower gear that is locked to the drive shaft. Hence the output gear rotates clockwise, until the pin strikes the reversing level to shift the toggle to the left. Once past its center, the toggle spring snaps the ratchet to the left to engage the left follower gear. This instantly reverses the output which now rotates counterclockwise until the pin again strikes the reversing level. Thus the mechanism reverses itself every 360-deg rotation of the input.

Modified-Watt's reverser

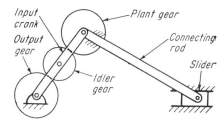

This is actually a modification of the well-known Watt crank mechanism. The input crank causes the planet gear to revolve around the output gear. But because the planet gear is fixed to the connecting rod it causes the output gear to continually reverse itself. If the radii of the two gears are equal then each full rotation of the input link will cause the output gear to oscillate through same angle as rod.

Automatically switching from one pivot point to another in midstroke. by Corpet Louvet & Co,

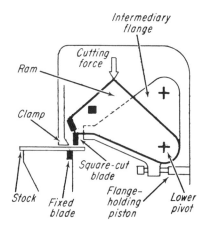

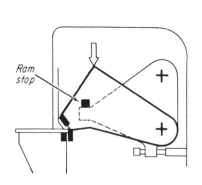

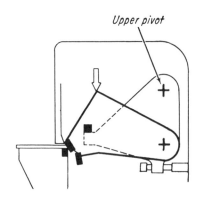

TWO PIVOTS and the intermediary flange govern sequence of cutting action. Flange is connected to the press frame at the upper pivot. The cutting ram is connected to the flange at lower pivot. In first part of cycle, the ram turns around the lower pivot and shears the plate with the square-cut blade; motion of the intermediary flange has been restrained by the flange-holding piston.

After the shearing cut, the ram stop bottoms on the flange. This overcomes the restraining force of the flange-holding piston and the ram turns around upper pivot. This brings the beveling blade into contact with the plate for the bevel cut.

Limited rotation

Here's a ingenious linkage for rotating an object 180 deg around a fixed point without the need for ring bearings. Such rotational requirements are specified for certain large radar antenna today, but are needed for smaller devices as well.

As the schematic shows rotation is controlled by two parallelogram linkages. The geometrical key to this design is quite simple: The outside points of the parallelogram (B_1, B_2) form a triangle with (and rotate around) a third point (B_0). This triangle must be congruent with the second triangle formed by the inside points of the parallelogram (B_3, B_4) and the rotational axis (A_0).

This design was conceived by Preben W. Jensen, an assistant professor of engineering at the Univ of Bridgeport. He believes the linkage is unique and will apply for a patent on it.

The major advantage of such an arrangement, he says, is that large ring bearings for supporting antennas can be eliminated. One such bearing in a radar operating today is about 13 ft in diameter. Jensen's linkage could quite possibly handle even larger antenna, possibly up to 26 ft dia. Another advantage is that the forces are concentrated in the pivots; so the link-

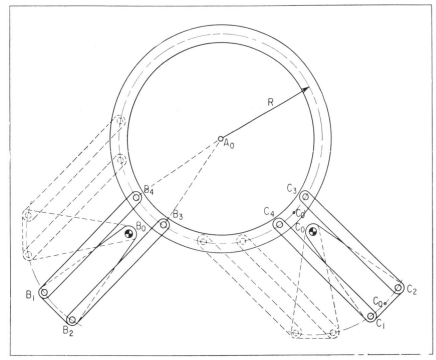

Bearings are out, linkages in

age has relatively small losses from friction when suspended in a vertical position.

The linkage is inspired by the known, but often overlooked, simple parallelogram linkage. In one version of this linkage a pair of parallel arms,

connected by crossmembers, is attached to a ring to be rotated. In all there are six pivot points—four at corners of the parallelogram and two on the ring. A change in the shape of the parallelogram thus rotates the ring.

Mechanical Linkage Varies Compression In Radial Engine

...ough mechanical linkage, the pilot can vary the ...el of the pistons during flight. All connecting rods ...the new design have identical motions and cen-...ugal force can be properly balanced. In the conven-...al radial engine with master connecting rod con-...ction, only one rod has true connecting rod mo-...a and vibrations are necessarily set up.

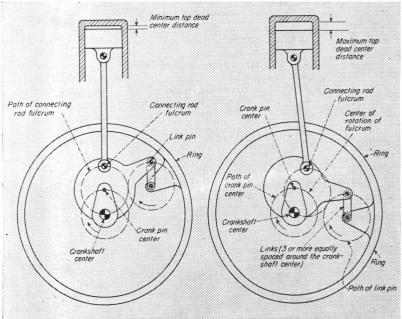

DIAGRAMMATIC VIEW of linkage for one cylinder shows the two extremes to give maximum and minimum compression ratios. Connections for other cylinders are exactly similar. A single link pin connection is shown here; the actual design includes several spaced to maintain balance.

RING IS ROTATED to displace fulcrums angularly around the crankpin center and force the connecting rods to operate in unsymmetrical position to lower compression ratio in cylinders. Linkage shown is for one cylinder only.

Computing mechanisms–I

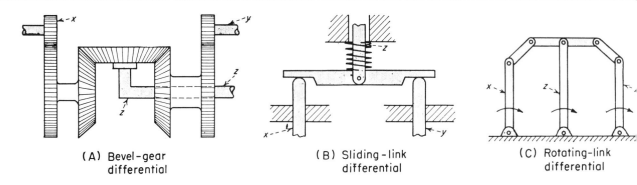

(A) Bevel-gear differential

(B) Sliding-link differential

(C) Rotating-link differential

Fig. 1—ADDITION AND SUBTRACTION. Usually based on the differential principle; variations depend on whether inputs: (A) rotate shafts, (B) translate links, (C) angularly displaced links. Mechanisms solve equation: $z=c\,(x\pm y)$, where c is scale factor, x and y are inputs, and z is the output. Motion of x and y in same direction results in addition; opposite direction—subtraction.

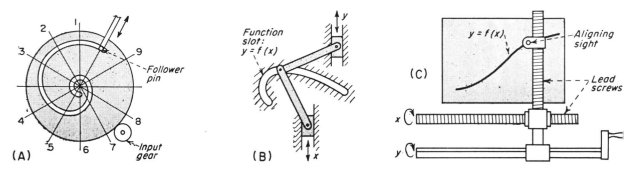

(A)

(B)

(C)

Fig. 2—FUNCTION GENERATORS mechanize specific equations. (A) Reciprocal cam converts a number into its reciprocal. This simplifies division by permitting simple multiplication between a numerator and its denominator. Cam rotated to position corresponding to denominator. Distance between center of cam to center of follower pin corresponds to reciprocal. (B) Function-slot cam. Ideal for complex functions involving one variable. (C) Input table. Function is plotted on large sheet attached to table. Lead screw for x is turned at constant speed by an analyzer. Operator or photoelectric follower turns y output to keep sight on curve.

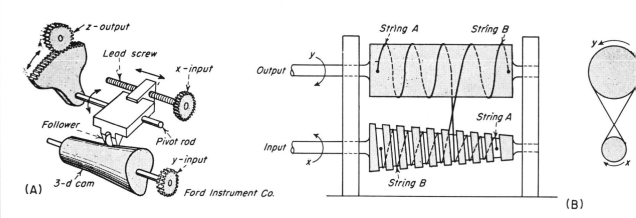

(A)

(B)

Fig. 3—(A) THREE-DIMENSIONAL CAM generates functions with two variables: $z=f\,(x,\,y)$. Cam rotated by y-input; x-inputs shifts follower along pivot rod. Contour of cam causes follower to rotate giving angular displacement to z-output gear. (B) Conical cam for squaring positive or negative inputs: $y=c\,(\pm x)^2$. Radius of cone at any point is proportional to length of string to right of point; therefore, cylinder rotation is proportional to square of cone rotation. Output is fed through a gear differential to convert to positive number.

Analog computing mechanisms are capable of virtually instantaneous response to minute variations in input. Basic units, similar to the types shown, are combined to form the final computer. These mechanisms add, subtract, resolve vectors, or solve special or trigonometric functions.

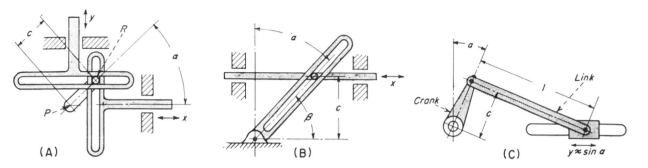

Fig. 4—TRIGONOMETRIC FUNCTIONS. (A) Scotch-yoke mechanism for sine and cosine functions. Crank rotates about fixed point P generating angle α and giving motion to arms: $y = c \sin \alpha$; $x = c \cos \alpha$. (B) Tangent-cotangent mechanism generates: $x = c \tan \alpha$ or $x = c \cot \beta$. (C) Eccentric and follower is easily manufactured but sine and cosine functions are approximate. Maximum error is: $e \max = l - \sqrt{l^2 - c^2}$; error is zero at 90 and 270 deg. l is the length of the link and c is the length of the crank.

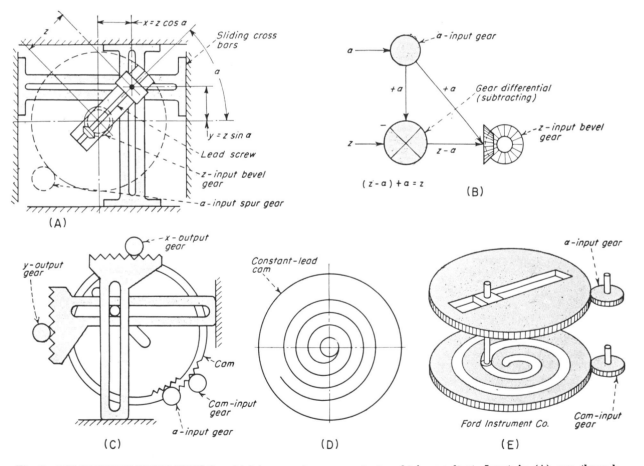

Fig. 5—COMPONENT RESOLVERS for obtaining x and y components of vectors that are continuously changing in both angle and magnitude. Equations are: $x = z \cos \alpha$, $y = z \sin \alpha$ where z is magnitude of vector, and α is vector angle. Mechanisms can also combine components to obtain resultant. Input in (A) are through bevel gears and lead screws for z input, and through spur gears for α-input. Compensating gear differential (B) prevents α-input from affecting z-input. This problem solved in (C) by using constant-lead cam (D) and (E).

Computing mechanisms-II

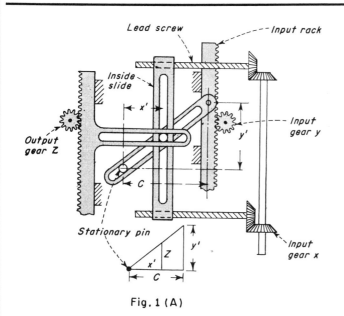

Fig. 1 (A)

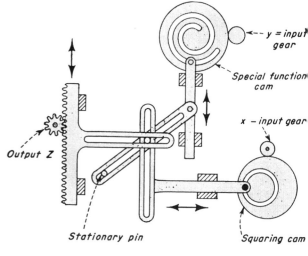

Fig.2 (A)

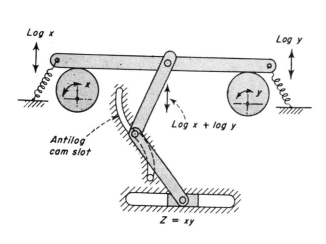

Fig.1 (B)

Fig. 1—MULTIPLICATION OF TWO TABLES x **and** y usually solved by either: (A) Similar triangle method, or (B) logarithmic method. In (A), lengths x' and y' are proportional to rotation of input gears x and y. Distance c is constant. By similar triangles: $z/x = y/c$ or $z = xy/c$, where z is vertical displacement of output rack. Mechanism can be modified to accept negative variables. In (B), input variables are fed through logarithmic cams giving linear displacements of log x and log y. Functions are then added by a differential link giving $z = \log x + \log y = \log xy$ (neglecting scale factors). Result is fed through antilog cam; motion of follower represents $z = xy$.

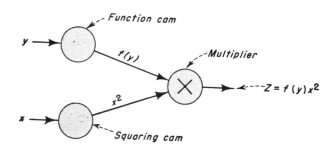

Fig. 2 (B)

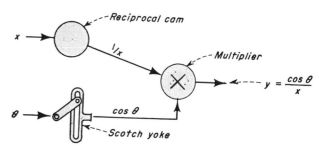

Fig. 2 (C)

Fig. 2—MULTIPLICATION OF COMPLEX FUNCTIONS can be accomplished by substituting cams in place of input slides and racks of mechanism in Fig. 1. Principle of similar triangles still applies. Mechanism in (A) solves the equation: $z = f(y) x^2$. Schematic is shown in (B). Division of two variables can be done by feeding one of the variables through a reciprocal cam and then multiplying it by the other. Schematic in (C) shows solution of $y = \cos \theta / x$.

Several typical computing mechanisms for performnig the mathematical operations of multiplication, division, differentiation, and integration of variable functions.

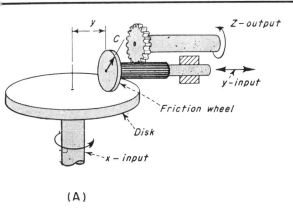

(A)

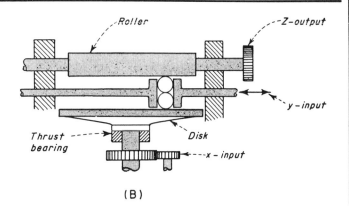

(B)

Fig. 3—INTEGRATORS are basically variable speed drives. Disk in (A) is rotated by x-input which, in turn, rotates the friction wheel. Output is through gear driven by spline on shaft of friction wheel. Input y varies the distance of friction wheel from center of disk. For a wheel with radius c, rotation of disk through infinitesmal turn dx causes corresponding turn dz equal to: $dz=(1/c)\,y\,dx$. For definite x revolutions, total z revolutions will be equal to the integral of $(1/c)\,y\,dx$, where y varies as called for by the problem. Ball integrator in (B), gives pure rolling in all directions.

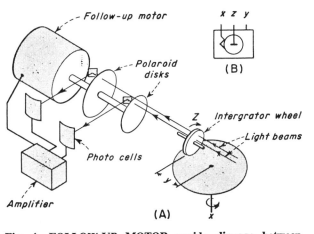

(A)

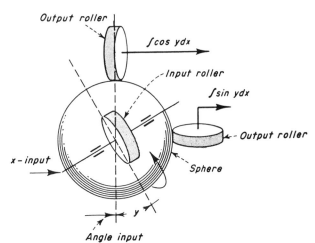

Angle input

Fig. 5—COMPONENT INTEGRATOR uses three disks to obtain x and y components of a differential equation. Input roller x spins sphere; y input changes angle of roller. Output rollers give integrals of components paralleling x and y axes. Ford Instrument Company.

Fig. 4—FOLLOW-UP MOTOR avoids slippage between wheel and disk of integrator in Fig. 3 (A). No torque is taken from wheel except to overcome friction in bearings. Web of integrator wheel is made of polaroid. Light beams generate current to amplifier which controls follow-up motor. Symbol at upper right corner is schematic representation of integrator. For more information see, "Mechanism," by Joseph S. Beggs, McGraw-Hill Book Co., N. Y., 1955.

Fig. 6—DIFFERENTIATOR uses principle that viscous drag force in thin layer of fluid is proportional to velocity of rotating x-input shaft. Drag force counteracted by spring; spring length regulated by servo motor controlled by contacts. Change in shaft velocity causes change in viscous torque. Shift in housing closes contacts causing motor to adjust spring length and balance system. Total rotation of servo gear is proportional to dx/dt.

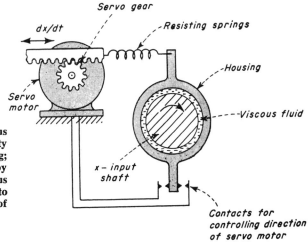

18 Variations of the Differential Mechanism

ALFRED KUHLENKAMP

Fig. 1 (below) shows modifications of the differential linkage shown in Fig. 2(A), based on variations in the triple-jointed intermediate link 6. The links are designated as follows: Frame links: links 2, 3 and 4; two-jointed intermediate links; links 5 and 7; three-jointed intermediate links: link 6.

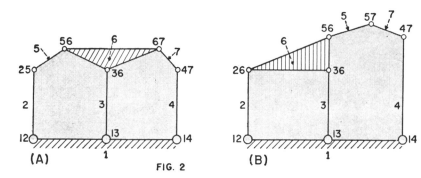

FIG. 2

Input motions to be added are *a* and *b*; their sum *s* is equal to $c_1 a + c_2 b$ where and c_2 are scale factors. The links are numbered in the same manner as in Fig. 2(A

Fig. 1

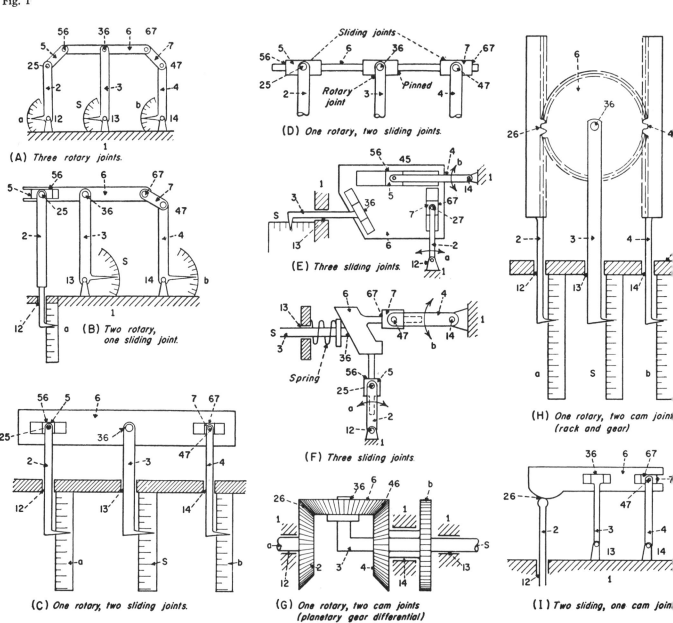

(A) *Three rotary joints.*

(B) *Two rotary, one sliding joint.*

(C) *One rotary, two sliding joints.*

(D) *One rotary, two sliding joints.*

(E) *Three sliding joints.*

(F) *Three sliding joints.*

(G) *One rotary, two cam joints (planetary gear differential)*

(H) *One rotary, two cam joint (rack and gear)*

(I) *Two sliding, one cam join*

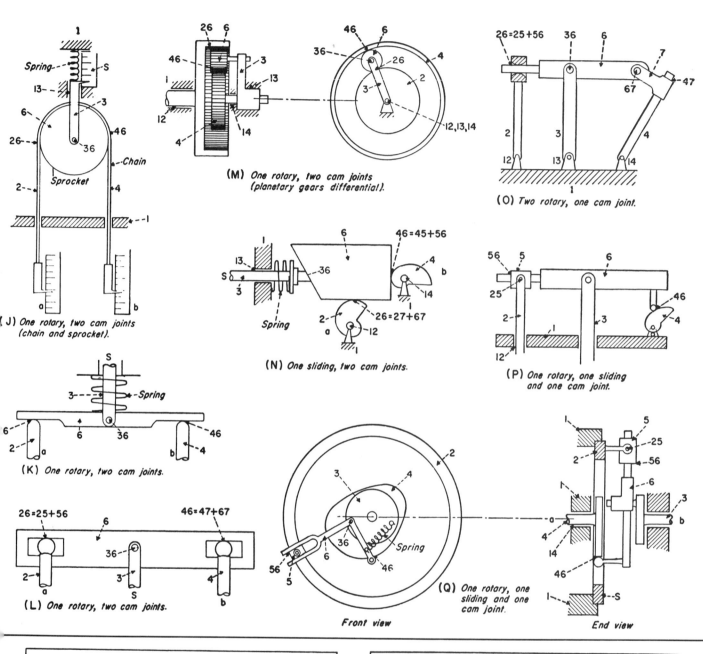

(J) One rotary, two cam joints (chain and sprocket).

(K) One rotary, two cam joints.

(L) One rotary, two cam joints.

(M) One rotary, two cam joints (planetary gears differential).

(N) One sliding, two cam joints.

(O) Two rotary, one cam joint.

(P) One rotary, one sliding and one cam joint.

(Q) One rotary, sliding and one cam joint.

Front view

End view

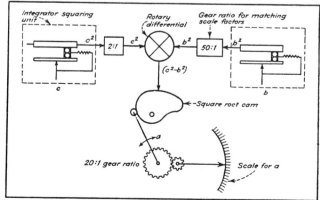

The integrator method of mechanizing the equation $a = \sqrt{c^2 - b^2}$, shown in schematic form. It requires an excessive number of parts.

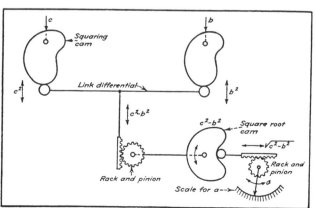

Cam method of mechanizing $a = \sqrt{c^2 - b^2}$ uses function generators for squaring and a link differential for subtraction. Note the reduction in parts.

Computing mechanism for sine function

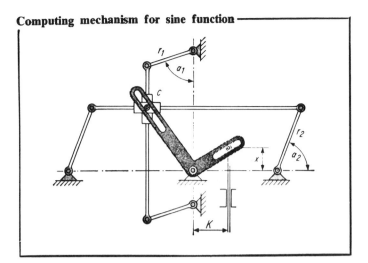

This device makes use of two four-bar linkages which restrain a block but permit it to slide along the linkage rods. The block causes the L shaped link to pivot which, in turn, raises or lowers the vertical rod. For any desired input rotation of a_1, and a_2, the device will solve the equation

$$x = K \frac{\sin a_1}{\sin a_2}$$

where K is a constant put into the device by locating the output rod an appropriate distance.

Grinding wheel dresser

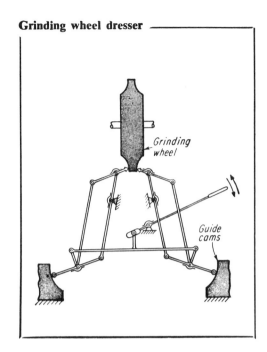

Two pantographs follow fixed guide cams to produce the desired curves at the dressing wheel. The cams are easily changed to produce different contours.

FUNCTION MECHANISMS

Precision-function mechanism

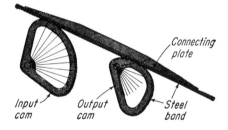

A steel band wrapped around two cams and fastened to a connecting plate can transfer the precise angular rotation of one shaft on to another over long distances. With proper cam design, the mechanism also produces a mathematical-function rotation for a given constant input rotation; for example, it can square, or give a multiple of the input angle.

Sinusoidal-adding mechanism

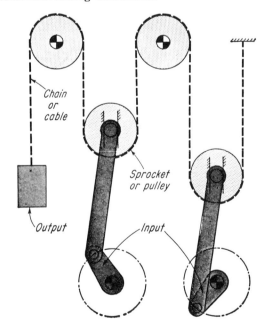

A simple arrangement which can quickly be set up in a lab to superimpose two input motions on to an output. This permits adding sinusoidal motions which have different amplitudes, phase angles, and frequencies.

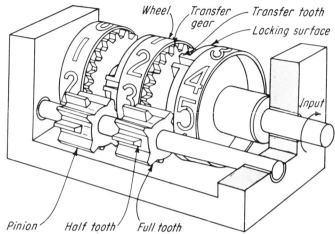

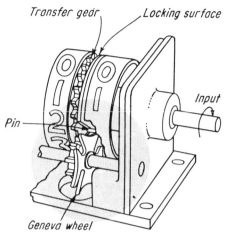

EXTERNAL PINION COUNTER carries the numbered wheels on one shaft, the pinions on a countershaft. The first wheel is direct drive; the rest driven by the pinions. Once per revolution, the transfer tooth (kicker) on the first wheel picks up a full tooth on the first pinion, and turns the second wheel one number. When the next full tooth hits the locking surface, the pinion stops. Until the next transfer the pinion cradles the first wheel between two successive full teeth, locking itself and the second wheel. Because of this locking method, accumulation of tolerances misaligns each number slightly more than the preceding one. If tolerances are too large, the numbers spiral.

The pinion rotates the second wheel one number while the first wheel turns 36°. Thus, if a 2-wheel counter reads 09, it changes to 10 between 9.5 and 10.5 Start and finish of a transfer can be shifted about, though still taking place over 36° rotation. For instance, one manufacturer designs his counter to transfer between 9.0 and 10.0; another has the transfer occur between 9.1 and 10.1.

MODIFIED GENEVA TRANSFER turns second wheel one number while the first turns 72° whereas standard 4-slot Geneva wheel turns 90° while the driver turns 90° (with both methods, pin on first wheel slides in and out of slot as it turns the Geneva wheel). Unlike usual Geneva transfer, modified version does not lock with its cutout against the first wheel. Instead, the pinion locks as in the external-pinion type. This way, diameter of the locking surface is larger and positioning more accurate.

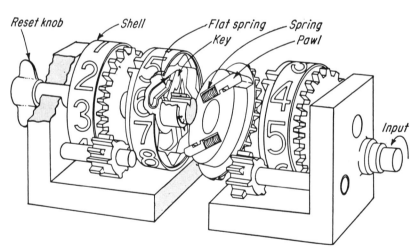

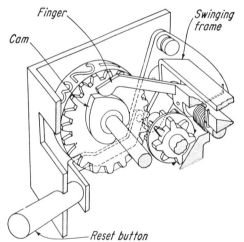

RESET, MOVING SHELL is separate part of wheel. Other half of wheel which carries the transfer gear drives the shell through spring-loaded pawls. Except during reset, wheel-shaft does not turn; therefore, input to first wheel must be through gearing.

At the start of count, with a zero reading, all keys sit in the keyway. Movement of wheels drags them out. During reset, the shells move and transfer gears stand still. A single turn of the wheel-shaft drops all keys back into the keyway; then a slight additional turn brings up all zeros.

RESET, HEART-SHAPE CAM has steadily decreasing radius from point to cleft. Thus the pressing fingers always turn wheels cleft upward so that zeros face out the window.

The swinging frame that carries pinions is linked to fingers by springs. When a push of the reset button forces down the fingers, springs swing the frame, disengaging the pinions. Center disks keep pinions aligned during the reset cycle. Counting method is same as for an ordinary external pinion counter.

A look at what's new in
Space mechanisms

There are potentially hundreds of them, but
only a few have been discovered so far. Here are the
best of one class—the four-bar space mechanisms

NICHOLAS P. CHIRONIS

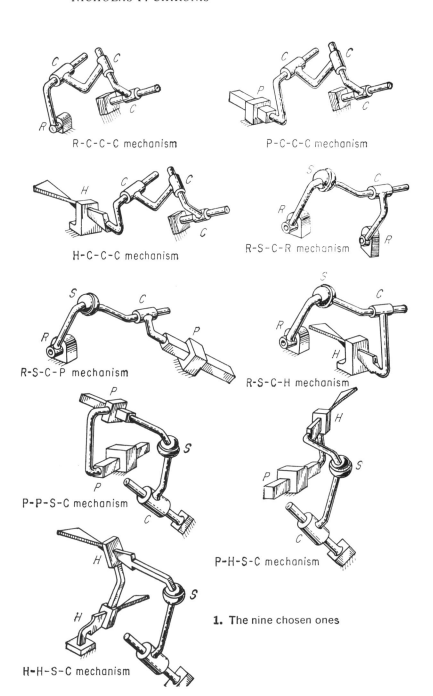

R-C-C-C mechanism

P-C-C-C mechanism

H-C-C-C mechanism

R-S-C-R mechanism

R-S-C-P mechanism

R-S-C-H mechanism

P-P-S-C mechanism

P-H-S-C mechanism

H-H-S-C mechanism

1. The nine chosen ones

A virtually unexplored area of mechanism research is the vast domain of the three-dimensional linkage, frequently called the space mechanism. Only a comparatively few kinds have been investigated or described, and little has been done to classify those that are known. Result: Many engineers do not know much about them and applications of space mechanisms have not been as widespread as might be.

One researcher, however, is making headway in this field. As part of a long-range study of space mechanisms at Oklahoma State University, Prof. Lee Harrisberger set out to discover and identify the kinds of space mechanisms possible within the existing mobility criteria and to study those which have practical characteristics.

Because a space mechanism can exist with a wide variety of connecting joints or "pair" combinations, Harrisberger identifies the mechanisms by the type and sequence of their joints. He established a listing of all the physically realizable kinematic pairs (opposite page) based on the number of degrees of freedom a joint may have. *These pairs are all the known ways of connecting two bodies together for every possible freedom of relative motion between them.*

The practical nine

The next step was to find the combination of pairs and links that would produce practical mechanisms. Using the "Kutzbach criterion" (the only known mobility criterion—it determines the degree of freedom of a mechanism due to the constraints imposed by the pairs), Harrisberger came up with 417 different kinds of space

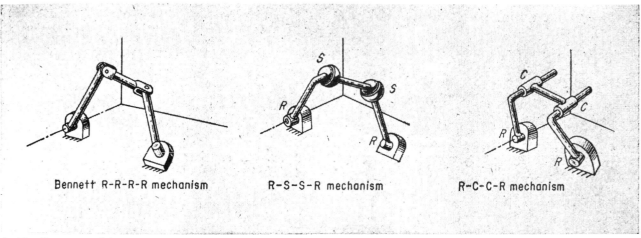

Bennett R-R-R-R mechanism   R-S-S-R mechanism   R-C-C-R mechanism

2. The three mavericks

mechanisms. Detailed examination showed many of these to be mechanically complex and of limited adaptability. But the four-link mechanisms had particular appeal because of their mechanical simplicity. At the recent ASME Mechanism Conference held at Purdue University, Harrisberger reported that he found 138 kinds of four-bar space mechanisms, nine of which have particular merit (Fig 1).

These nine, four-link mechanisms have superior physical realizability because they contain only those joints which have area contact and are self-connecting. In the table, these joints are the five closed, lower pair types:

R = Revolute joint, which permits rotation only

P = Prism joint, which permits sliding motion only

H = Helix or screw type of joint

C = Cylinder joint, which permits both rotation and sliding (hence has two degrees of freedom)

S = Sphere joint, which is the common ball joint permitting rotation in any direction (three degrees of freedom)

All these mechanisms can produce rotary or sliding output motion from a rotary input—the most common mechanical requirement for which linkage mechanisms are designed.

The type letters of the kinematic pairs in the table are used to identify the mechanism by ordering the letter symbols consecutively around the closed kinematic chain. The first letter identifies the pair connecting the input link and the fixed link; the last letter identifies the output link, or last link, with the fixed link. Thus a

mechanism labeled *R-S-C-R* is a double-crank mechanism with a spherical pair between the input crank and the coupler, and a cylinder pair between the coupler and output crank.

The mavericks

The Kutzbach criterion, Harrisberger notes, is inadequate for the job since it cannot predict the existence of such mechanisms as the Bennett *R-R-R-R* mechanism, the double-ball joint *R-S-S-R* mechanism, and the *R-C-C-R* mechanism (Fig 2). These are "special" mechanisms which require special geometric conditions to have a single degree of freedom. The *R-R-R-R* mechanism requires a particular orientation of the revolute axes and a particular ratio of link lengths to function as a single degree of freedom space mechanism. The *R-S-S-R* configuration, when functioning as a single degree of freedom mechanism, will have a passive degree of freedom of the coupler link. When properly constructed, the configuration *R-C-C-R* also will have a passive degree of freedom of the coupler and will function as a single degree mechanism.

Of these three special four-link mechanisms, Harrisberger feels that the *R-S-S-R* mechanism is an outstanding choice as the most versatile and practical configuration for meeting double-crank motion requirements. As we go to press, however, our editors have dug up another maverick—the Space Crank (Fig 3), described exclusively in PRODUCT ENGINEERING (see also page 159). It would seem this also is an *R-R-R-R* mechanism. One wonders how many other mavericks are running around.

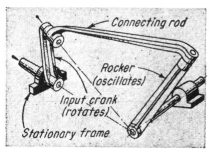

3. Space crank—another maverick?

Classification of kinematic pairs

Degree of freedom	Type number*	Type of joint	
		Symbol	Name
1	100	R	Revolute
	010	P	Prism
	001	H	Helix
2	200	T	Torus
	110	C	Cylinder
	101	T_H	Torus-helix
	020	..	....
	011	..	....
3	300	S	Sphere
	210	S_s	Sphere-slotted cylinder
	201	S_{SH}	Sphere-slotted helix
	120	P_L	Plane
	021	..	....
	111	..	....
4	310	S_G	Sphere-groove
	301	S_{GH}	Sphere-grooved helix .
	220	C_p	Cylinder-plane
	121	..	....
	211	..	....
5	320	S_p	Sphere-plane
	221	..	....
	311		....

*Number of freedoms, given in the order of N_R, N_T, N_H.

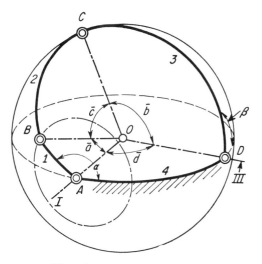

The Spherical Crank

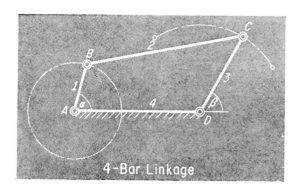

Seven
popular types
of

THREE-
DIMENSIONAL
DRIVES

DR W MEYER ZUR CAPELLEN

Main advantage of
three-dimensional drives is their
ability to transit motion
between nonparallel shafts.
They can also generate other
types of helpful motion.
With this roundup
are descriptions of industrial
applications.

1 spherical crank drive

This type of drive is the basis for most 3-D linkages, much as the common 4-bar linkage is the basis for the two-dimensional field. Both mechanisms operate on similar principles. (In the accompanying sketches, a is the input angle, and β the output angle. This notation has been used throughout the article.)

In the 4-bar linkage, the rotary motion of driving crank *1* is transformed into an oscillating motion of output link *3*. If the fixed link is made the shortest of all, then you have a double-crank mechanism, in which both the driving and driven members make full rotations.

In the spherical crank drive, link *1* is the input, link *3* the output. The axes of rotation intersect at point O; the lines connecting AB, BC, CD and DA can be thought of as part of great circles of a sphere. The length of the link is best represented by angles a, b, c and d.

156

2 spherical-slide oscillator

The two-dimensional slider crank is obtained from a 4-bar linkage by making the oscillating arm infinitely long. By making an analogous change in the spherical crank, you can obtain the spherical slider crank shown at right.

The uniform rotation of input shaft I is transferred into a nonuniform oscillating or rotating motion of output shaft III. These shafts intersect at an angle δ corresponding to the frame link 4 of the spherical crank. Angle γ corresponds to length of link 1. Axis II is at right angle to axis III.

The output oscillates when γ is smaller than δ; the output rotates when γ is larger than δ.

Relation between input angle α and output angle β is (as designated in skewed Hook's joint, below)

$$\tan \beta = \frac{(\tan \gamma)(\sin \alpha)}{\sin \delta + (\tan \gamma)(\cos \delta)(\cos \alpha)}$$

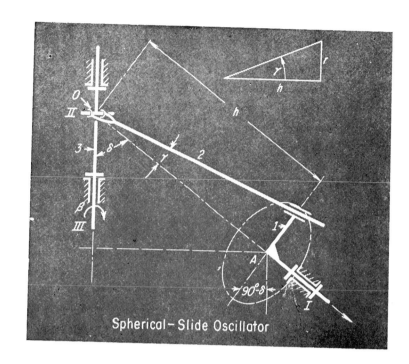

Spherical-Slide Oscillator

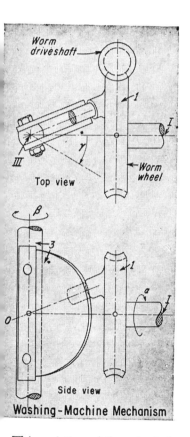

Worm driveshaft

Top view

Side view

Washing-Machine Mechanism

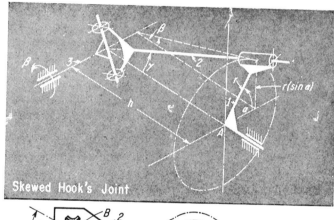

Skewed Hook's Joint

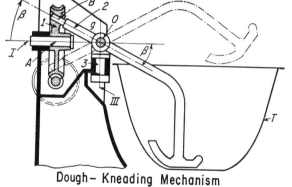

Dough- Kneading Mechanism

3 skewed hook's joint

This variation of the spherical crank is often used where an almost linear relation is desired between input and output angles for a large part of the motion cycle.

The equation defining the output in terms of the input can be obtained from the above equation by making = 90°. Thus sin δ = 1, cos δ = 0, and

$$\tan \beta = \tan \gamma \sin \alpha$$

The principle of the skewed Hook's joint has been recently applied to the drive of a washing machine (see sketch at left).

Here, the driveshaft drives the worm wheel 1 which has a crank fashioned at an angle γ. The crank rides between two plates and causes the output shaft III to oscillate in accordance with the equation above.

The dough-kneading mechanism at right is also based on the Hook's joint, but utilizes the path of link 2 to give a wobbling motion that kneads dough in the tank.

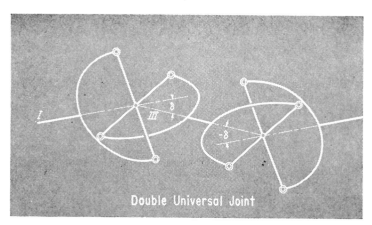

Double Universal Joint

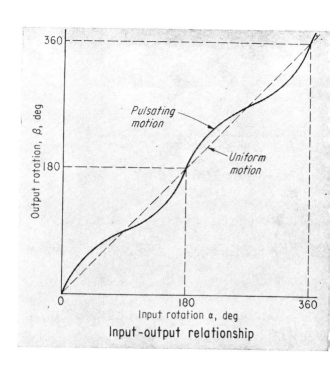

Input-output relationship

4 the universal joint

The universal joint is a variation of the spherical-slide oscillator, but with angle $\gamma = 90°$. This drive provides a totally rotating output and can be operated as a pair, as shown above.

Equation relating input with output for a single universal joint, where δ is angle between connecting link and shaft I:

$$\tan \beta = \tan \alpha \cos \delta$$

Output motion is pulsating (see curve) unless the joints are operated as pairs to provide a uniform motion.

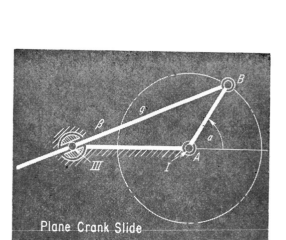

Plane Crank Slide

5 the 3-D crank slide

The three-dimensional crank slide is a variation of a plane crank slide (see sketch), with a ball point through which link g always slides, while a point B on link g describes a circle. A 3-D crank is obtained from this mechanism by making output shaft III not normal to the plane of the circle; another way is to make shafts I and III nonparallel.

A practical variation of the 3-D crank slide is the agitator mechanism (right). As input gear I rotates, link g swivels around (and also lifts) shaft III. Hence, vertical link has both an oscillating rotary motion and a sinusoidal

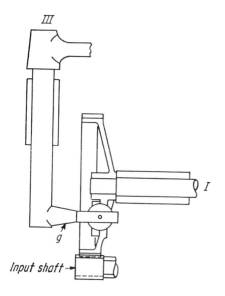

Agitator Mechanism

harmonic translation in the direction of its axis of rotation. The link performs what is essentially a screw motion in each cycle.

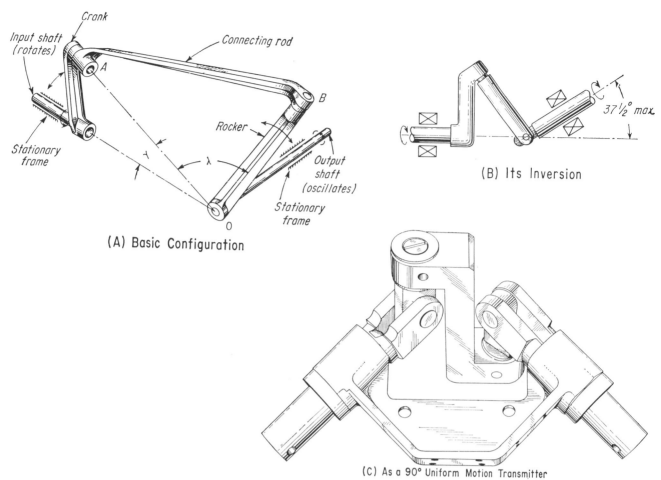

(A) Basic Configuration

(B) Its Inversion

(C) As a 90° Uniform Motion Transmitter

6 the space crank

One of the most recent developments in 3-D linkages is the space crank shown in (A) [see also page 155]. It resembles the spherical crank discussed on page 156, but has different output characteristics. Relationship between input and output displacements is:

$$\cos \beta = (\tan \gamma)(\cos \alpha)(\sin \beta) - \frac{\cos \lambda}{\cos \gamma}$$

Velocity ratio is:

$$\frac{\omega_o}{w_i} = \frac{\tan \gamma \sin \alpha}{1 + \tan \gamma \cos \alpha \cot \beta}$$

where ω_o is the output velocity and ω_i is the constant input velocity.

An inversion of the space crank is shown in (B). It can couple intersecting shafts, and permits either shaft to be driven with full rotations. Motion is transmitted up to 37½° misalignment.

By combining two inversions, (C), a method for transmitting an exact motion pattern around a 90° bend is obtained. This unit can also act as a coupler or, if the center link is replaced by a gear, it can drive two output shafts; in addition, it can be used to transmit uniform motion around two bends.

(Continued on next page)

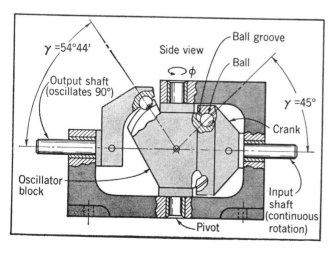

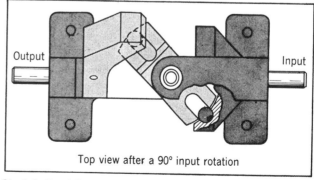

Top view after a 90° input rotation

Steel balls riding within spherical grooves convert a continuous rotary input motion into an output motion that oscillates the shaft back and forth.

VARIATIONS OF THE SPACE CRANK

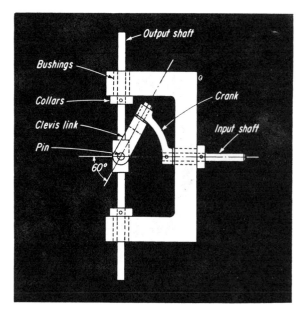

Oscillating motion . . .

powered at right angles. Input shaft making full rotations causes output shaft to oscillate 120°.

Constant-speed-ratio . . .

universal is obtained by using two "inversions" back-to-back. Motion transmitted up to 75° misalignment.

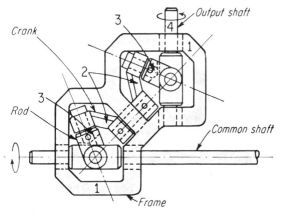

Right-angle . . .

limited-stroke drive transmits exact motion pattern. A multiplicity of fittings can be operated from common shaft.

7 the elliptical slide

The output motion, β, of a spherical slide oscillator, can be duplicated by means of a two-dimensional "elliptical slide." The mechanism has a link g which slides through a pivot point D and is fastened to a point P moving along an elliptical path. The ellipse can be generated by a Cardan drive, which is a planetary gear system with the planet gear half the diameter of the internal gear. The center of the planet, point M, describes a circle; any point on its periphery describes a straight line, and any point in between, such as point P, describes an ellipse.

There are certain relationships between the dimensions of the 3-D spherical slide and the 2-D elliptical slide: $\tan \gamma / \sin \delta = a/d$ and $\tan \gamma / \cot \delta = b/d$, where a is the major half-axis, b the minor half-axis of the ellipse, and d is the length of the fixed link DN. The minor axis lies along this link.

If point D is moved within the ellipse, a completely rotating output is obtained, corresponding to the rotating spherical crank slide.

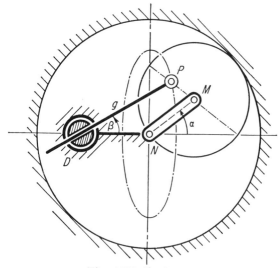

The Elliptical Slide

Walking-link drive

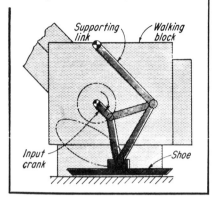

This is actually a four-bar linkage with a triangular extension which supports a shoe which can pivot. The crank and supporting link are connected to a block which is moved forward by the linkage. The drive motor can be inside the block and powered by flexible cables. Used in earth-moving equipment.

If you want to design a fork truck or tractor that won't bog down in soft earth, or new vehicle to explore the moon, take a look at an old idea: walking plates.

In England, Lansing Bagnall, Ltd has just designed and patented what it claims is the first application of this type of mechanism to a moving vehicle. In this case, the vehicle is a fork lift truck. The mechanism (see diagrams) consists of a pair of plates mounted between the front and rear wheels of a truck. The plates, attached to a crankpin, are mounted so that normally they do not interfere with operation of the wheels. But, when they are brought into play, they can be pushed down so the wheels are lifted from the ground and a walking motion is created. They can also be used to assist the wheels when needed, since power can be applied to either system, or to both.

In the US, one of the latest systems for lunar locomotion, proposed by J. D. McKenney of Space-General Corp at the last ARS meeting, is a remarkably similar arrangment, the plates in this case being attached to jointed legs (see sketch).

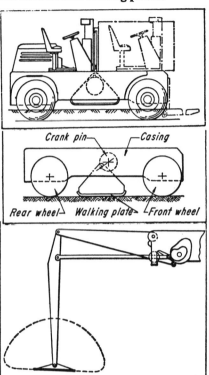

Helical-groove mechanism controls vibration intensity

A novel design of a vibrator, developed by a Czech engineer, employs a mechanical linkage capable of continuously controlling the centrifugal forces within the vibrator, hence the amplitude of vibration, while the vibrator is running.

Vibrators are employed for countless jobs in manufacturing and testing. The mechanical design that produces a wide range of vibration intensities, without need for fluids, shows a number of advantages: faster response, self-contained design, and more reliable operation.

The device employs two sets of eccentric masses, one fixed and the other capable of swiveling around the axis of rotation of the unit (drawing below). The centrifugal forces exerted by the two sets of masses remain constant and equal; what alters is the mutual angular position of the masses and thus the degree of imbalance throughout the entire system.

The unit has a split shaft, whose solid ends carry fixed eccentric weights that are interconnected by a hollow section containing the pivotable eccentric weight. The pivoted weight, in turn, can be rotated by the control rod.

How it's done. The rod has a pair of rollers that travel in two mutually opposed helical grooves within a sleeve. The sleeve, free to slide axially, can be shifted by means of the lever or by a servo system. Shifting the lever swivels the pivoted weight.

In one application, two counter-running units of this type, which are fitted to the frame of a vibratory table, produce controlled centrifugal forces, infinitely variable during full-load running, from 0 to 2500 kg at 1425 to 2650 rpm. The simplicity is paying off in production capabilities.

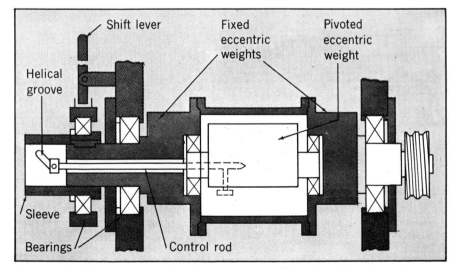

A new principle in creating vibrations and controlling their intensity makes use of fixed and pivotable eccentric weights that vary the resultant centrifugal force.

Five typewriter mechanisms

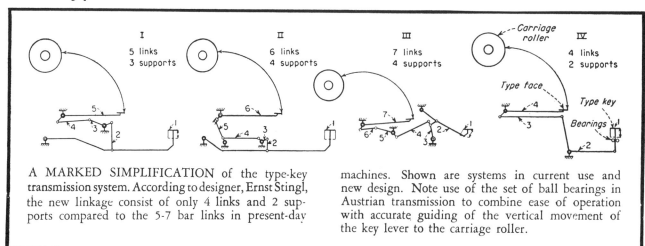

A MARKED SIMPLIFICATION of the type-key transmission system. According to designer, Ernst Stingl, the new linkage consist of only 4 links and 2 supports compared to the 5-7 bar links in present-day machines. Shown are systems in current use and new design. Note use of the set of ball bearings in Austrian transmission to combine ease of operation with accurate guiding of the vertical movement of the key lever to the carriage roller.

Stroke of oscillating, rotating shaft is varied while turning

In designing a golf-ball winding machine, Meyer Bloom of Babox Co. (Brooklyn, N.Y.) devised a mechanism for adjusting the stroke length of an oscillating, rotating shaft while the shaft is in motion.

He did it by pivoting a pair of parallel-plate cams that control the shaft motion through action of a pin follower fixed in a disk mounted on the shaft. Unbalancing forces on the cams and disk are equalized by a second pin that is free to slide in a hole in the opposite side of the disk.

Spherical bearing controls rotations

For engineers working with bearings and linkages, there is now a design that provides weight and cost savings in the actuating joints of aircraft flight control. It is an efficient, compact, spherical bearing.

Bearing does it. The new design, proposed by Forest Johnson of the Lanseair Development Corp. (Hurst, Tex.), limits the self-aligning action of a spherical bearing to two axes by a feature within the bearing itself.

This TAF bearing, as it is called, has a continuous semi-circular groove around the periphery of the spherical surface of the inner race.

Two small balls inserted in holes through the outer race engage the

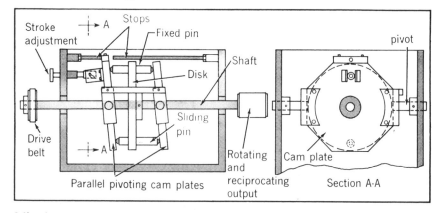

Adjusting the cam plates on the pivots varies stroke on shaft. Pin fixed in disk follows surfaces of cam plates as shaft rotates; a sliding pin balances the forces.

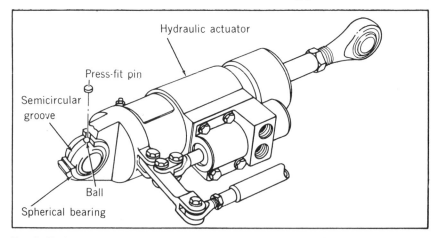

Two balls, located 180 deg. apart in outer race, restrict rotation of aircraft actuator about its longitudinal axis. The groove allows cross-axis rotation.

groove at points 180 deg. apart and are retained by press-fitted pins. The balls act as pivots that permit bearing rotation about their axis.

The bearing also allows movement about the axis of the groove, but it restricts self-alignment about an axis normal to both axes.

CHAPTER 4

SPRING, BELLOW, FLEXURE, SCREW AND BALL DEVICES

Fiendishly simple roller-band device challenges established mechanisms

NICHOLAS P. CHIRONIS

Donald Wilkes (above), inventor of the revolutionary roller-band concept, explains the importance of band tension in a basic configuration. One application of the new principle is to trigger a circuit in an acceleration switch, a model of which is shown at left, with a single-roller version at right. Any acceleration over a preset value will cause the roller to move to right and snap-close an electric circuit.

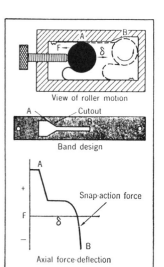

View of roller motion

Band design

Axial force-deflection

How rolamite works

Take a metal band

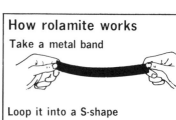

Loop it into a S-shape

Put between parallel guides and r

Draw taut and fasten ends

To make it act as a spring, taper the band

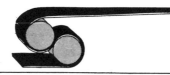

How it works. The rolamite concept is absurdly simple. In its basic form, the device consists of a metal band looped around two rollers (drawing). The resilient band seeks to stay flat, so when it is looped, it stores elastic energy. When the band ends are pulled taut and are fastened, the rolamite assumes its basic S-configuration.

In this arrangement, each loop seeks to straighten itself out. If the band width is constant, the forces are equalized and the roller cluster stays where it is, unless moved. When it is pushed, the cluster rolls with extremely little effort—less, its developers claim, than if bearings were used.

There is no sliding of surfaces,

hence no sliding friction. Tests at Sandia have shown a coefficient of friction of about 0.0005—about 10% that of a ball or roller bearing. Thus, a rolamite needs no lubrication and is almost wear-free.

For the rolamite to function as a spring, the band is tapered, perforated, or otherwise modified from the basic parallel-edge shape.

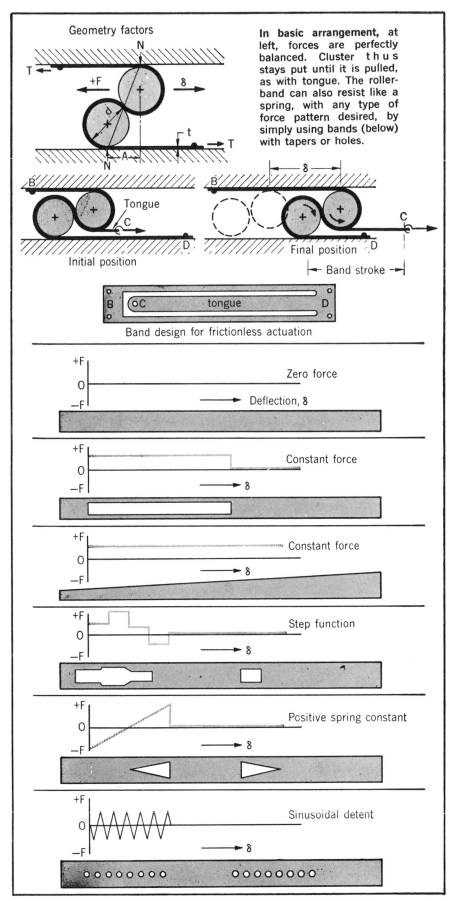

Geometry factors

In basic arrangement, at left, forces are perfectly balanced. Cluster t h u s stays put until it is pulled, as with tongue. The roller-band can also resist like a spring, with any type of force pattern desired, by simply using bands (below) with tapers or holes.

Initial position

Final position

Band stroke

Band design for frictionless actuation

Zero force

Deflection, δ

Constant force

Constant force

Step function

Positive spring constant

Sinusoidal detent

Loop a long, flexible steel band around a couple of rollers, slip the assembly between parallel surfaces, pull the band ends taut, and anchor them. You now have a unique mechanical element (photo, top left) for an almost endless variety of functions and applications.

The Atomic Energy Commission's Sandia Laboratory, Albuquerque, N. M., where the mechanism was invented by Donald F. Wilkes, calls it "rolamite." The device is so simple that you can hardly believe it's brand-new, but a thorough search of technical literature by Sandia has failed to turn up any similar mechanism.

AEC has filed a patent application in Wilkes' name. However, devices covered by AEC patents may be manufactured under royalty-free, nonexclusive licenses. So the roller-band concept is wide open for the taking.

Simplicity itself. The almost frictionless roller-band device can be rolled from one end of its assembly to the other with very little force. Tilt it slightly, and it will roll with only 1% to 10% of the friction in ball and roller bearings.

Even in its rudimentary form, the roller-band can perform useful tasks —as a linear bearing, for example. But its fascination lies in its ability to assume all sorts of shapes and to be worked into countless useful jobs.

It can be designed to produce any type of force profile simply by variation of the band shape (drawings left). It can replace any type of spring or any grouping of springs, precisely reproducing their characteristics. Another variation of the band produces the G-switch (far left), which will roll and snap-close a switch as soon as a specified level of acceleration force is exceeded.

A roller-band unit can be miniaturized to a remarkable degree, needs no lubrication in service, is inexpensive to produce, demands no precise tolerances, and is relatively insensitive to dirt.

In its various forms, it can function as a thermostat, machine way, pivot, bearing, speed reducer, snap-

Rolamite, continued

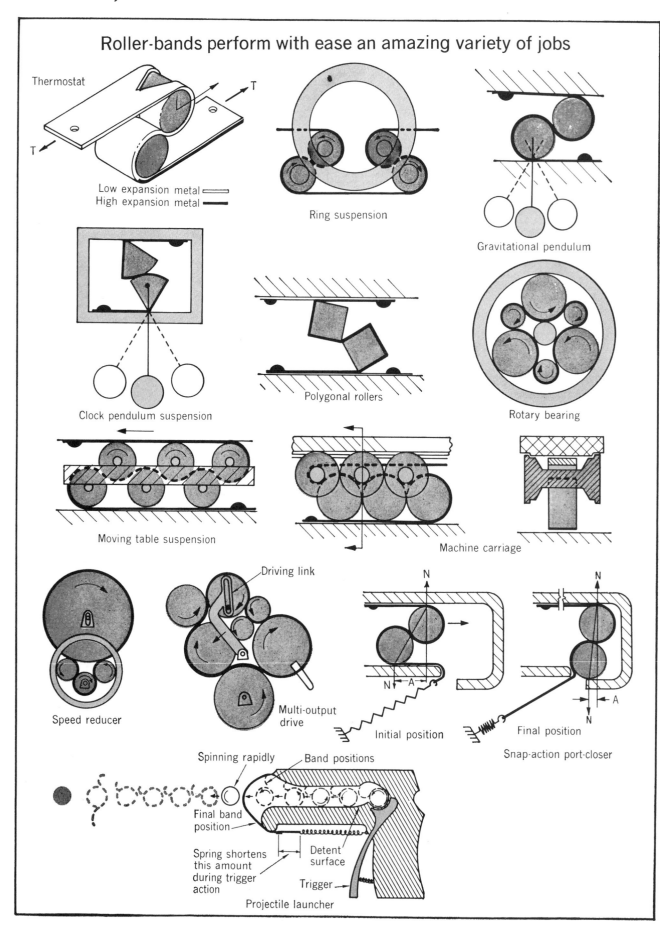

Roller-bands perform with ease an amazing variety of jobs

Thermostat

Low expansion metal ▭
High expansion metal ▬

Ring suspension

Gravitational pendulum

Clock pendulum suspension

Polygonal rollers

Rotary bearing

Moving table suspension

Machine carriage

Speed reducer

Driving link

Multi-output drive

N

N

A

Initial position

N

N

A

Final position

Snap-action port-closer

Spinning rapidly

Band positions

Final band position

Spring shortens this amount during trigger action

Detent surface

Trigger

Projectile launcher

action switch, valve, or projectile launcher (drawings left). Potentially, it could also act as a shock absorber, damper, hinge, lock, relay, piston, or light switch.

Generating force. The key to a roller-band's operation is that each loop seeks to straighten itself out at all times. As long as the band width is constant, these forces are equalized, and the cluster of rollers stays where it is unless moved, as in the zero-force condition.

If, however, the band is tapered, with the top portion thinner than the bottom portion, the stronger loop below dominates the weaker loop above. Then the band unwinds, creating a constant driving force, F exerted axially according to the following formula:

$$F = \frac{Et^3\,(b-a)}{6\,(d+t)^2}$$

where E is modulus of elasticity, d is roller diameter, t is band thickness, and $b-a$ is the difference in band widths. If the difference in band widths remains constant, as from a taper or a stamped triangular slot, the roller-band will function as a constant-force spring. To obtain a linearly increasing spring characteristic, as in a helical compression spring, two opposing tapers or triangular cutouts must be employed so $b-a$ factor varies with travel (next-to-bottom graph, page 165).

There are other ways, too, to get the roller-band to produce useful forces. The bands can be made of prestressed steel, as in the Neg'ator spring, or of bimetallic materials, as in the thermostat application in drawing at top left.

In the thermostat, the cutout on the top keeps the band normally tensioned to the left, against a stop. As temperature rises, the cluster moves to the right, and the low coefficient of rolling friction, about 0.0005, permits a substantial output.

Force amplification. Roller-band geometry can also be designed to produce a rapid release of mechanical force, as in the snap-action port-closer shown at left.

The normal reaction force, N, is a function of the band tension T, the spacing between the guide surfaces s, and the axial component of the distance between centers of rolling elements A, in equation:

$$N = \frac{T\,(s-t)}{A}$$

Note that as A approaches zero, N approaches infinity. Thus, when the bottom roller in the port-closer slips off to a lower surface and A rapidly diminishes, the normal force N, acting to close the port against pressure, is suddenly and sharply amplified. A similar principle provides the propulsion force in the launcher shown at bottom left.

Moreover, because the invention is open to all manufacturers on a royalty-free license from AEC, Hamilton Watch Co. (Lancaster, Pa.), which makes thin metal bands for rolamite applications, believes some companies are keeping their rolamite-based products under wraps until ready for full-scale production.

On the band wagon. In addition to Hamilton Watch, companies that have shown a strong interest in rolamite products include Rolamite Technology, Inc. (RTI, Albuquerque, N.M.), Technar, Inc. (Pasadena, Calif.), Statham Instruments, Inc. (Los Angeles), Hughes Engineering Co. (Wayland, Mich.), and Sandia Laboratories (Albuquerque, N.M.), which has established a new Rolamite Laboratory.

Thermostats. Hamilton Watch is experimenting with the use of laminated rolamite bands for thermostats (drawing 1, left). The bands consist of two materials with different coefficients of thermal expansion.

As the ambient temperature is decreased, the different expansion rates in the portion of the band encircling the top roller create a strong tendency for the cluster to move upward. This tendency is opposed by a triangular band cutout operating on the other roller. (See also Fig. 1, p. 168).

An electrical circuit may be completed, depending on the pre-selected

position of the temperature selector. Advantages over existing thermostats, says Hamilton Watch, include fewer parts, ruggedness, and enhanced accuracy without adjustment. Also, the band contour can be designed to provide high sensitivity at normal room temperatures and less sensitivity at temperatures not normally desired.

Meter dials. The rolamite principle can be applied to instruments to produce a compact meter-dial arrangement with good resolution, according to Fred Duimstra of Sandia. The needle is attached to one of the two rollers of the rolamite (Fig. 2, p.168). A portion of the trochoidal curve approximates the arc of a circle with a large radius, permitting the center of action to be closer to the point of the needle. Hughes Engineering has applied this principle to amplify the motion of the dial of a thread gage (drawing 3).

In another application, RTI has used rolamite successfully as a door hinge (drawing 4). The door opens a maximum of 120 deg., and the rollers also rotate 120 deg.

Mechanisms. Statham Instruments is applying the concept to convert rotary motion from a torque motor to linear motion for a pump (drawing 5). One of the company's products (drawing 6) pumps fluid at a rapid rate of 160 strokes per minute against a pressure head of 200 mm of mercury.

The amplification capabilities of rolamite arrangements also can amplify linear-to-rotary or linear-to-linear motion (drawing 7).

Donald Wilkes, inventor of rolamite and now with RTI, has built a bandless rolamite (drawing 8) in which the magnetic field causes the two rollers to automatically assume a rolamite geometry and provides friction-free travel of the rollers.

Technar, Inc., has applied the magnetic concept to a fluid valve (drawing 9). Energizing the left magnet causes the rollers to close the outlet port. Fluid pressure self-seals the port. The use of shaped rollers (drawing 10) minimizes the air gap during roller travel.

continued, next page

Rolamite principle is applied in a broad spectrum of devices entering hardware stage

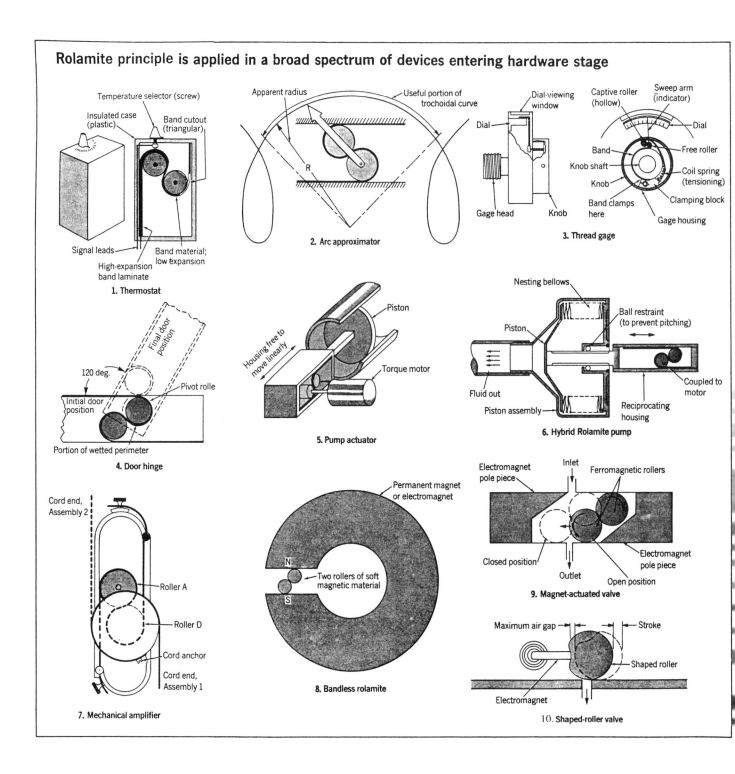

1. Thermostat
- Temperature selector (screw)
- Insulated case (plastic)
- Band cutout (triangular)
- Signal leads
- High-expansion band laminate
- Band material; low expansion

2. Arc approximator
- Apparent radius
- Useful portion of trochoidal curve
- R

3. Thread gage
- Dial-viewing window
- Captive roller (hollow)
- Sweep arm (indicator)
- Dial
- Band
- Free roller
- Knob shaft
- Coil spring (tensioning)
- Knob
- Clamping block
- Band clamps here
- Gage housing
- Gage head
- Knob

4. Door hinge
- Final door position
- 120 deg.
- Initial door position
- Pivot rolle
- Portion of wetted perimeter

5. Pump actuator
- Piston
- Housing free to move linearly
- Torque motor

6. Hybrid Rolamite pump
- Nesting bellows
- Ball restraint (to prevent pitching)
- Piston
- Fluid out
- Coupled to motor
- Piston assembly
- Reciprocating housing

7. Mechanical amplifier
- Cord end, Assembly 2
- Roller A
- Roller D
- Cord anchor
- Cord end, Assembly 1

8. Bandless rolamite
- Permanent magnet or electromagnet
- N
- S
- Two rollers of soft magnetic material

9. Magnet-actuated valve
- Electromagnet pole piece
- Inlet
- Ferromagnetic rollers
- Closed position
- Electromagnet pole piece
- Outlet
- Open position

10. Shaped-roller valve
- Maximum air gap
- Stroke
- Shaped roller
- Electromagnet

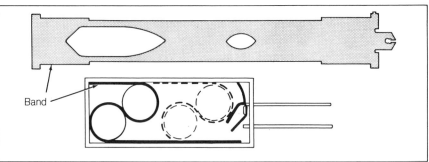

Coiled band in tiny rolamite acts as a spring to keep the rollers tightly bunched at one end. The triangular cutout in the band permits the rollers to roll to the right when an exact acceleration force is exceeded.

- Band

7 APPLICATIONS FOR THE CONSTANT-FORCE SPRING

This spring, known as the Neg'ator, finds some up-to-date applications in a magazine feed, one-way brake, motion transfer device, mechanical servo and lifting jack, etc.

HARRY E. NANKONEN

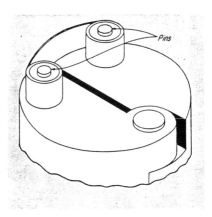

Strip of spring material . . .

prestressed, formed into a tight coil, and mounted on a freely rotating drum (A) resists withdrawal from the coil with a force that remains constant throughout any extension. In this form the Neg'ator is widely used as a long-deflection spring to perform such functions as counterbalancing, constant-force tensioning and retracting. Spring motor (B) is a second basic form of spring. Here the band is extended and reverse wound about a larger dia output drum. Motor produces almost constant torque throughout entire rundown—50-60 turns or more.

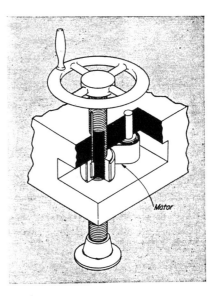

Governor spring . . .

provides constant resistance to centrifugal force when hinged governor opens. Pins holding equal coils of spring may either be capped or fitted with retaining rings to prevent spring from working off during rotation.

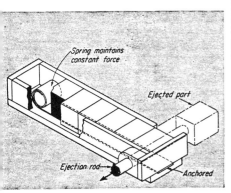

Magazine feed . . .

is powered by the self-adjusting coil of extended spring pushing against material being fed. Long feed has no force variation. Device advances work in machine tools and feeds products in vending machines. Assembly stations, where parts must be fed one at a time, can often be made more efficient with these feeds.

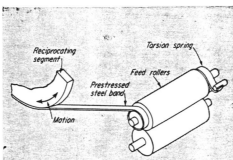

Motion transfer . . .

is quiet and accurate; band replaces gears, transmits oscillating motion of cam-operated segment, produces accurately registered rotation of feed rollers. Steel band is prestressed to insure dimensional stability. Torsion spring returns rollers to starting position.

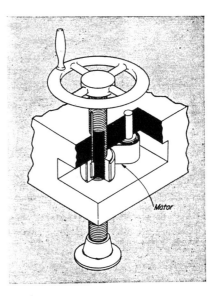

Spring-assisted jack . . .

reduces manual effort required to raise heavy equipment. Without added torque supplied by spring motor a larger handwheel is needed for easy operation.

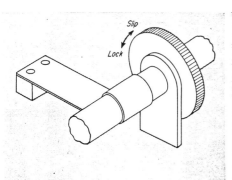

One-way brake . . .

prevents counter clockwise rotation of this mechanism, as spring tends to grip more tightly. Clockwise rotation expands the spring coil, lets shaft slip. There is no backlash.

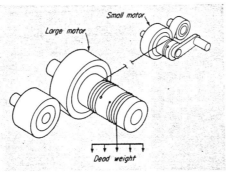

Mechanical servo . . .

is formed by two opposed, cable-connected motors. Larger motor provides an output torque sufficient to support both dead weight and force of smaller motor. Friction keeps system static until the remotely located smaller motor is manipulated. Then, the larger motor and load follow.

Flat springs in mechanisms

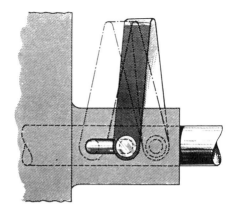

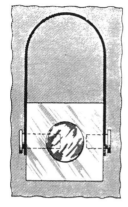

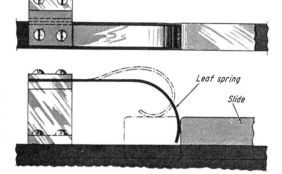

CONSTANT FORCE is approached because of the length of this U-spring. Don't align studs or spring will fall.

FLAT-WIRE SPRAG is straight until the knob is assembled; thus tension helps the sprag to grip for one-way clutching.

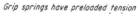

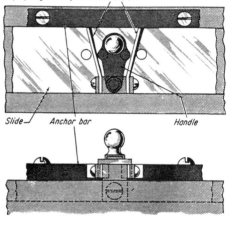

SPRING-LOADED SLIDE will always return to its original position unless it is pushed until the spring kicks out.

INCREASING SUPPORT AREA as the load increases on both upper and lower platens is provided by a circular spring.

EASY POSITIONING of the slide is possible when the handle pins move a grip spring out of contact with the anchor bar.

CONSTANT TENSION in the spring, and thus force required to activate slide, is (almost) provided by this single coil.

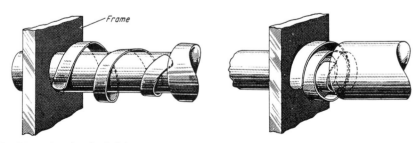

VOLUTE SPRING here lets the shaft be moved closer to the frame, thus allowing maximum axial movement.

These devices all rely on a flat spring for
their efficient actions, which would otherwise need more complex configurations.

L. KASPER

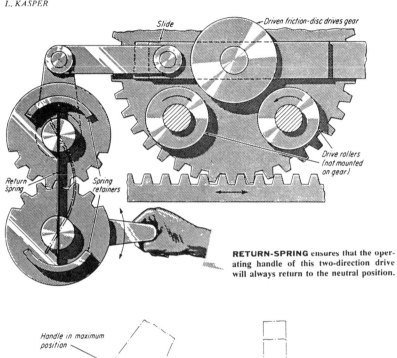

RETURN-SPRING ensures that the operating handle of this two-direction drive
will always return to the neutral position.

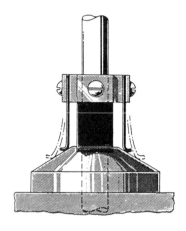

CUSHIONING device features rapid increase of spring tension because of the small
pyramid angle. Rebound is minimum, too.

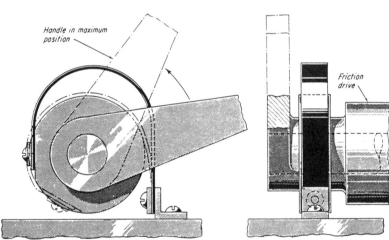

SPRING-MOUNTED DISK changes center position as handle is rotated to move
friction drive, also acts as built-in limit stop.

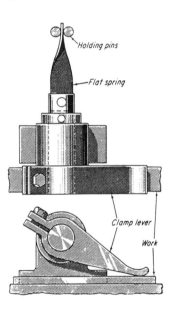

HOLD-DOWN CLAMP has flat spring assembled with initial twist to provide clamping force for thin material.

INDEXING is accomplished simply, efficiently, and at low cost by the flat-spring
arrangement shown here.

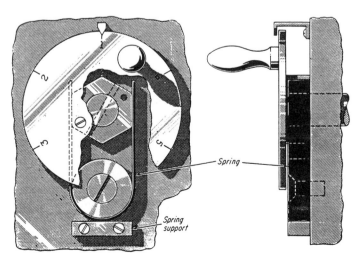

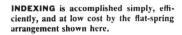

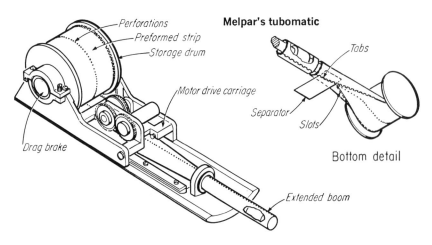

Perforations
Preformed strip
Storage drum
Motor drive carriage
Drag brake
Extended boom

Melpar's tubomatic

Tabs
Separator
Slots

Bottom detail

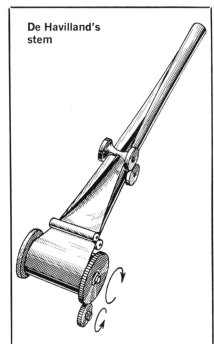

De Havilland's stem

Pop-up springs get new backbone

An addition to the new family of retractable coil springs, initially popular for use as antennas, holds promise of solving one problem in such applications: lack of torsional and flexural rigidity when extended. Melpar Inc, under contract to the Goddard Space Flight Center at Greenbelt, Md, has developed a pop-up boom that locks itself into a stiffer tube than any made so far.

In two previous versions—De Havilland Aircraft's Stem and Hunter Spring's Helix (*PE*—Feb 14 '66, p 53)—rigidity was obtained by permitting the material to overlap. In Melpar's design, the strip that unrolls from the drum to form the cylindrical mast has tabs and slots that interlock to produce a strong tube.

Melpar has also added a row of perforations along the center of the strip to aid in accurate control of the spring's length during extension or contraction. This adds to the spring's attractiveness as a positioning device, besides its established uses as antennas for spacecraft and portable equipment and as gravity gradient booms and sensing probes.

Curled by heat. Retractable, prestressed coil springs have been in the technical news for the past year, yet most manufacturers have been rather closemouthed about exactly how they convert a strip of beryllium copper or stainless steel into such a spring.

In its Helix, Hunter induces the prestressing at an angle to the axis of the strip, so the spring uncoils helically; De Havilland and Melpar prestress the material along the axis of the strip.

Melpar uses a prestressing technique worked out by John J. Park of the Goddard Center. Park found early in his assignment that technical papers are lacking on just how a metal strip can be given a new "memory" that makes it curl longitudinally unless restrained.

Starting from scratch, Park ran a series of experiments using a glass tube, 0.65 in. ID, and strips of beryllium copper alloy, 2 in. wide and 0.002 in. thick. He found it effective to roll the alloy strip lengthwise into the glass tube and then to heat it in a furnace. Test strips were then allowed to cool at room temperature.

Experience showed that the longer and hotter the furnace time, the more tightly the strip would curl along its length, producing a smaller tube. For example, a test strip heated at 920 F for 5 min would produce a tube that remained at the 0.65-in. inside diameter of the glass holder; at 770 F, heating for even 15 min produced a tube that would expand to an 0.68-in. dia.

By proper correlation of time and temperature in the furnace, Park suggests a continuous tube-forming process could be set up, cutting off segments of the completed tube at the lengths desired. □

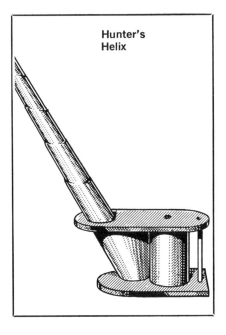

Hunter's Helix

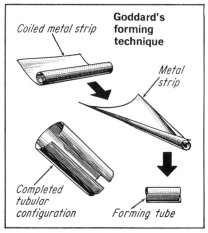

Goddard's forming technique

Coiled metal strip

Metal strip

Completed tubular configuration

Forming tube

12 ways to put springs to work

Variable-rate arrangements, roller
positioning, space saving, and other
ingenious ways to get the most from springs.

L. KASPER

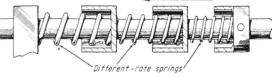

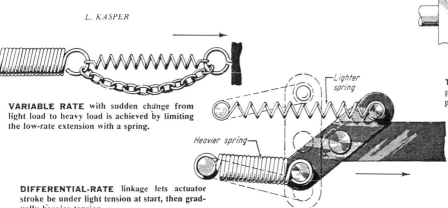

VARIABLE RATE with sudden change from light load to heavy load is achieved by limiting the low-rate extension with a spring.

THREE-STEP RATE change at predetermined positions. The lighter springs will always compress first regardless of their position.

DIFFERENTIAL-RATE linkage lets actuator stroke be under light tension at start, then gradually heavier tension.

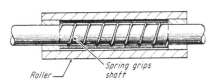

ROLLER POSITIONING by tight-wound spring on shaft obviates necessity for collars. Roller will slide under excess end thrust.

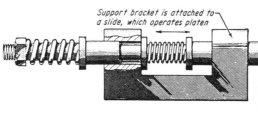

COMPRESSING MECHANISM has dual rate for double-action compacting. In one direction pressure is high, in reverse pressure is low.

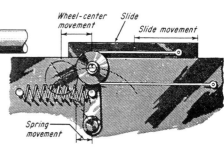

SHORT EXTENSION of spring for long movement of slide keeps tension change between maximum and minimum low.

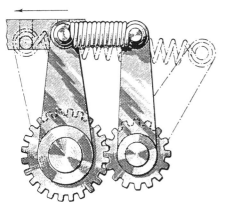

INCREASED TENSION for same movement is gained by providing a movable spring mount and gearing it to the other movable lever.

CLOSE-WOUND SPRING is attached to a hopper and will not buckle when used as a movable feed-duct for nongranular material.

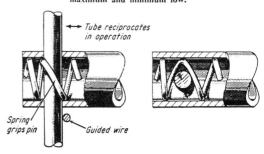

PIN GRIP is spring that holds pin by friction against end movement or rotation, but lets pin be repositioned without tools.

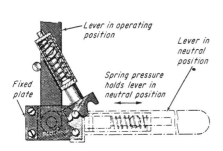

TOGGLE ACTION here is used to make sure the gear-shift lever will not inadvertently be thrown past neutral.

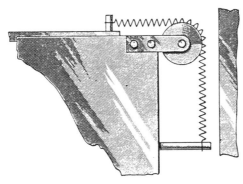

SPRING WHEEL helps distribute deflection over more coils than if spring rested on corner. Less fatigue and longer life result.

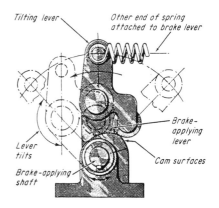

TENSION VARIES at different rate when brake-applying lever reaches the position shown. Rate is reduced when tilting lever tilts.

173

Overriding Spring Mechanism

Extensive use is made of overriding spring mechanisms in the design of instruments and controls. Anyone of the arrangements illustrated allows an incoming motion to override the outgoing moti[on] whose limit has been reached. In an instrument, [for] example, the spring device can be placed betwe[en]

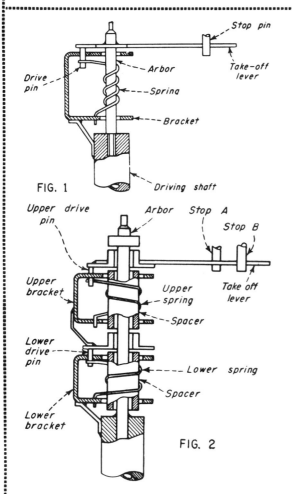

FIG. 1

Fig. 1—Unidirectional Override. The take-off lever of this mechanism [can] rotate nearly 360 deg. It's movement is limited by only one stop pin. In [one] direction, motion of the driving shaft also is impeded by the stop pin. But [in] the reverse direction the driving shaft is capable of rotating approximately [360] deg past the stop pin. In operation, as the driving shaft is turned clockw[ise] motion is transmitted through the bracket to the take-off lever. The spr[ing] serves to hold the bracket against the drive pin. When the take-off lever [has] traveled the desired limit, it strikes the adjustable stop pin. However, [the] drive pin can continue its rotation by moving the bracket away from the dr[ive] pin and winding up the spring. An overriding mechanism is essential [in] instruments employing powerful driving elements, such as bimetallic eleme[nts] to prevent damage in the overrange regions.

FIG. 2

Fig. 2—Two-directional Override. This mechanism is similar to that [de]scribed under Fig. 1, except that two stop pins limit the travel of the take[-off] lever. Also, the incoming motion can override the outgoing motion in ei[ther] direction. With this device, only a small part of the total rotation of [the] driving shaft need be transmitted to the take-off lever and this small part [may] be anywhere in the range. The motion of the driving shaft is transmi[tted] through the lower bracket to the lower driv[e] pin, which is held against [the] bracket by means of the spring. In turn, the lower drive pin transfers the [mo]tion through the upper bracket to the upper drive pin. A second spring h[olds] this pin against the upper drive bracket. Since the upper drive pin is attac[hed] to the take-off lever, any rotation of the drive shaft is transmitted to the le[ver] provided it is not against either stop A or B. When the driving shaft turn[s in] a counterclockwise direction, the take-off lever finally strikes against the [ad]justable stop A. The upper bracket then moves away from the upper d[rive] pin and the upper spring starts to wind up. When the driving shaft is rot[ated] in a clockwise direction, the take-off lever hits adjustable stop B and the lo[wer] bracket moves away from the lower drive pin, winding up the other spr[ing.] Although the principal uses for overriding spring arrangements are in [the] field of instrumentation, it is feasible to apply these devices in the drive[s of] major machines by beefing up the springs and other members.

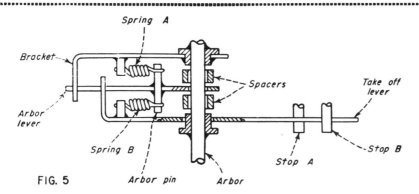

FIG. 5

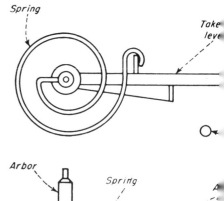

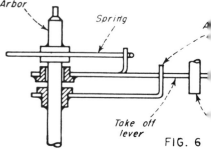

FIG. 6

Fig. 5—Two-directional, 90 Degree Override. This double overriding mechanism allows a maximum overtravel of 90 deg in either direction. As the arbor turns, the motion is carried from the bracket to the arbor lever, then to the take-off lever. Both the bracket and the take-off lever are held against the arbor lever by means of springs A and B. When the arbor is rotated counterclockwise, the take-off lever hits stop A. The arbor lever is held stationary in contact with the take-off lever. The bracket, which is soldered to the arbor, rotates away from the arbor lever, putting spring A in tension. When the arbor is rotated in a clockwise direction, the take-off lever comes against stop B and the bracket picks up the arbor lever, putting spring B in tension.

for Low-Torque Drives HENRY L. MILO, JR.

the sensing and indicating elements to provide over-range protection. The dial pointer is driven positively up to its limit, then stops; while the input shaft is free to continue its travel. Six of the mechanisms described here are for rotary motion of varying amounts. The last is for small linear movements.

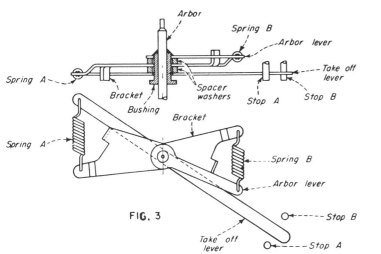

FIG. 3

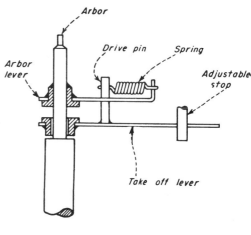

FIG. 4

Fig. 3—Two-directional, Limited-Travel Override. This mechanism performs the same function as that shown in Fig. 2, except that the maximum override in either direction is limited to about 40 deg, whereas the unit shown in Fig. 2 is capable of 270 deg movement. This device is suited for uses where most of the incoming motion is to be utilized and only a small amount of travel past the stops in either direction is required. As the arbor is rotated, the motion is transmitted through the arbor lever to the bracket. The arbor lever and the bracket are held in contact by means of spring B. The motion of the bracket is then transmitted to the take-off lever in a similar manner, with spring A holding the take-off lever and the bracket together. Thus the rotation of the arbor is imparted to the take-off lever until the lever engages either stops A or B. When the arbor is rotated in a counterclockwise direction, the take-off lever eventually comes up against the stop B. If the arbor lever continues to drive the bracket, spring A will be put in tension.

Fig. 4—Unidirectional, 90 Degree Override. This is a single overriding unit, that allows a maximum travel of 90 deg past its stop. The unit as shown is arranged for over-travel in a clockwise direction, but it can also be made for a counterclockwise override. The arbor lever, which is secured to the arbor, transmits the rotation of the arbor to the take-off lever. The spring holds the drive pin against the arbor lever until the take-off lever hits the adjustable stop. Then, if the arbor lever continues to rotate, the spring will be placed in tension. In the counterclockwise direction, the drive pin is in direct contact with the arbor lever so that no overriding is possible.

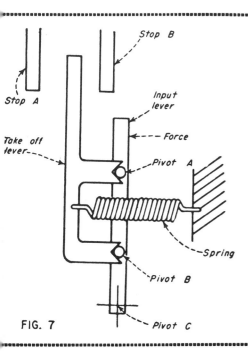

FIG. 7

Fig. 6—Unidirectional, 90 Degree Override. This mechanism operates exactly the same as that shown in Fig. 4. However, it is equipped with a flat spiral spring in place of the helical coil spring used in the previous version. The advantage of the flat spiral spring is that it allows for a greater override and minimizes the space required. The spring holds the take-off lever in contact with the arbor lever. When the take-off lever comes in contact with the stop, the arbor lever can continue to rotate and the arbor winds up the spring.

Fig. 7—Two-directional Override, Linear Motion. The previous mechanisms were overrides for rotary motion. The device in Fig. 7 is primarily a double override for small linear travel although it could be used on rotary motion. When a force is applied to the input lever, which pivots about point C, the motion is transmitted directly to the take-off lever through the two pivot posts A and B. The take-off lever is held against these posts by means of the spring. When the travel is such the take-off lever hits the adjustable stop A, the take-off lever revolves about pivot post A, pulling away from pivot post B and putting additional tension in the spring. When the force is diminished, the input lever moves in the opposite direction, until the take-off lever contacts the stop B. This causes the take-off lever to rotate about pivot post B, and pivot post A is moved away from the take-off lever.

SPRING MOTORS AND TYPICAL

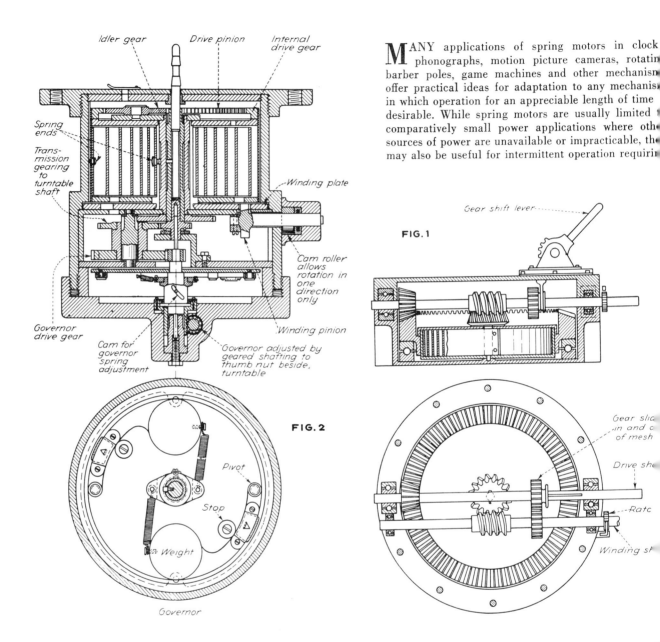

Idler gear • Drive pinion • Internal drive gear

Spring ends

Transmission gearing to turntable shaft

Winding plate

Cam roller allows rotation in one direction only

Governor drive gear

Winding pinion

Cam for governor spring adjustment

Governor adjusted by geared shafting to thumb nut beside turntable

MANY applications of spring motors in clock phonographs, motion picture cameras, rotatin barber poles, game machines and other mechanism offer practical ideas for adaptation to any mechanis in which operation for an appreciable length of time desirable. While spring motors are usually limited comparatively small power applications where oth sources of power are unavailable or impracticable, the may also be useful for intermittent operation requiri

FIG.1

Gear shift lever

FIG.2

Pivot

Stop

Weight

Governor

Gear slid un and o of mesh

Drive sh

Ratc

Winding sh

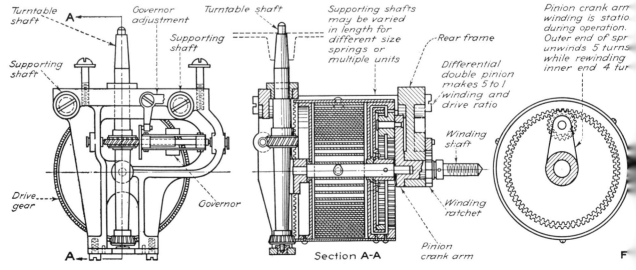

Turntable shaft

Governor adjustment

Turntable shaft

Supporting shafts may be varied in length for different size springs or multiple units

Supporting shaft

Rear frame

Supporting shaft

Differential double pinion makes 5 to 1 winding and drive ratio

Pinion crank arn winding is statio during operation. Outer end of spr unwinds 5 turns while rewinding inner end 4 tur

Drive gear

Governor

Winding shaft

Winding ratchet

Section A-A

Pinion crank arm

F

ASSOCIATED MECHANISMS

paratively high torque or high speed, using a low
ver electric motor or other means for building up
rgy.

he accompanying patented spring motor designs
w various methods of transmission and control of
ing motor power. Flat-coil springs, confined in
ms, are most widely used because they are com-
t, produce torque directly, and permit long angular
placement. Gear trains and feed-back mechanisms
d down excess power so that it can be applied for a
ger time, and governors are commonly used to regu-
speed.

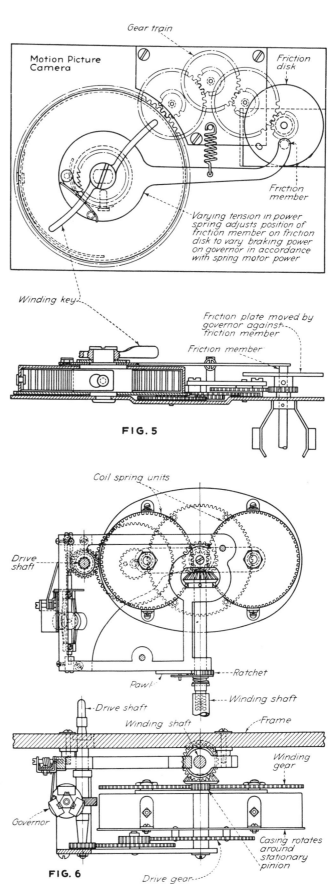

Motion Picture Camera

Gear train

Friction disk

Friction member

Varying tension in power
spring adjusts position of
friction member on friction
disk to vary braking power
on governor in accordance
with spring motor power

Winding key

Friction plate moved by
governor against
friction member

Friction member

FIG. 5

Coil spring units

Drive shaft

Ratchet

Pawl

Winding shaft

Drive shaft

Winding shaft

Frame

Winding gear

Governor

Casing rotates around stationary pinion

FIG. 6

Drive gear

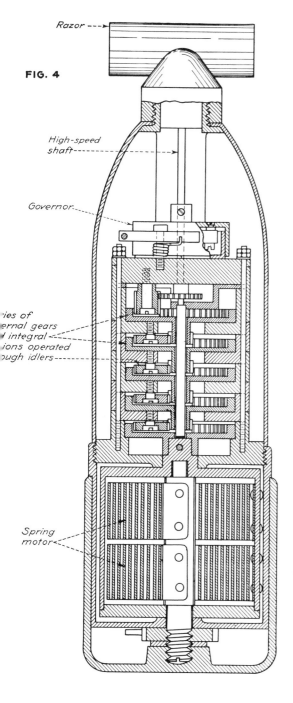

Razor

FIG. 4

High-speed shaft

Governor

ries of
ernal gears
integral
ions operated
ough idlers

Spring motor

Flexures accurately support pivoting mechanisms and instruments

Often bypassed by the various rolling bearings, flexures are now making steady progress—often getting the nod for applications in space and industry where their numerous assets outweigh the fact that they cannot give the full rotation that bearings offer.

Flexures, or flexible suspensions as they are officially called, lie between the worlds of rolling bearings—such as the ball and roller bearings—and of sliding bearings—which include sleeve and hydrostatic bearings. Neither rolling nor sliding, flexures simply cross-suspend a part and then flex to allow the necessary movement.

There are many applications today in which parts or components must reciprocate or oscillate, so flexures are being made increasingly available as off-the-shelf parts with precise characteristics.

Flexures to the moon. Flexures are being selected over bearings in space applications, because they do not wear out, have better lubrication requirements, and are less subject to backlash.

One aerospace flexure—scarcely more than 2 in. high—will be used for a key task on the Apollo Applications Program (AAP), in which Apollo spacecraft and hardware are to be employed for scientific research. The flexures' job will be to keep a 5000-lb. telescope pointed at the sun with unprecedented accuracy, so that solar phenomena can be viewed.

Demands met. Lockheed's answer, recommended by Dr. George G. Herzl, a consultant to Lockheed, was a flexure pivot.

The pivot contains thin connecting beams that have flexing action, so they perform like a combination spring and bearing.

Unlike a true bearing, however, it has no rubbing surfaces. Unloaded, or with a small load, a flexure pivot acts as a positive—or center-seeking—spring; loaded above a certain amount, it acts as a negative spring.

A consequence of this duality is that in space, the AAP telescope always returns to a central position, while during ground testing it drifts away from center. The Lockheed design takes advantage of this phenomenon of flexure pivots: By attaching a balancing weight to the telescope during ground tests, Lockheed can closely simulate the dynamic conditions of space.

Potential of flexures. Lockheed has adapted flexure pivots to other situations as well. In one case, a flexure is used for a gimbal mount in a submarine. Another operates a

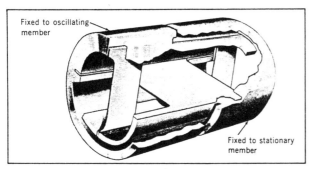

Frictionless flexure pivot, which resembles a bearing, is made of flat, angular crossed springs that support rotating sleeves in a variety of structural designs.

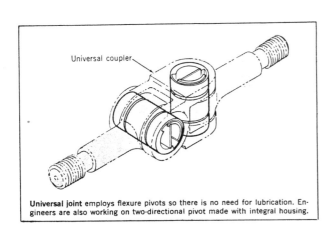

Universal joint employs flexure pivots so there is no need for lubrication. Engineers are also working on two-directional pivot made with integral housing.

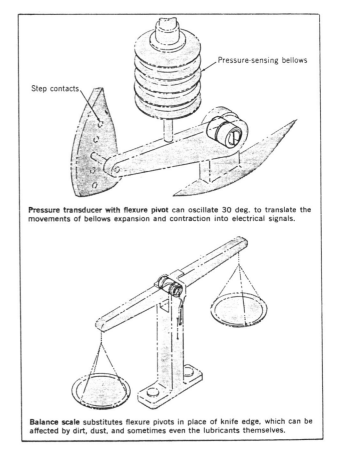

Pressure transducer with flexure pivot can oscillate 30 deg. to translate the movements of bellows expansion and contraction into electrical signals.

Balance scale substitutes flexure pivots in place of knife edge, which can be affected by dirt, dust, and sometimes even the lubricants themselves.

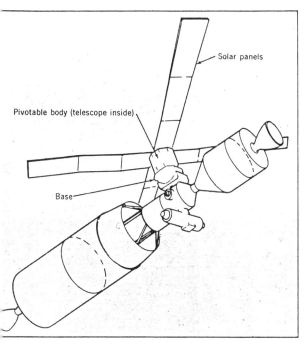

Solar panels

Pivotable body (telescope inside)

Base

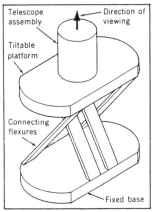

Telescope assembly

Direction of viewing

Tiltable platform

Connecting flexures

Fixed base

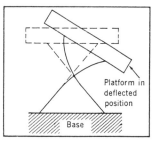

Platform in deflected position

Base

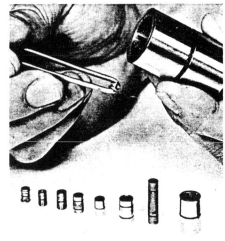

Apollo telescope-mount cluster (far left) employed flexures for tilting X-ray telescope. Platform (left) is tilted without break-away torque. Photo above shows typical range of flexure sizes.

safety shutter to protect delicate sensors in a satellite.

Realizing the potential of flexure pivots, Bendix Corp. (Utica, N.Y.) has developed an improved type of bearing flexure, commonly known as "flexure pivot." It is designed to be compliant around one axis and rigid around the cross axes. The flexure pivots have the same type of flat, crossed springs as the rectangular type, but they are designed as a simple package that can be easily installed and integrated into a design (photo, above). The compactness of the flexure pivot makes it a natural to replace ordinary bearings in many oscillating applications (drawings left).

The Bendix units are built around three elements: flexures, a core or inner housing, and an outer housing or mounting means. They permit angular deflections of 7½ deg., 15 deg., or 30 deg.

The cantilever type, (drawing at left) can support an overhung load. There is also a double-ended type that supports central loads. The width of each cross member of the outer flexure is equal to one-half that of the inner flexure, so that when assembled at 90 deg. from each other, the total flexure width in each plane is the same.

Key point. The heart of any flexure pivot is the flexure itself. F. A. Seelig, an engineer at Bendix, points out that a key factor in using a flexure is the torsional-spring constant of the assembly—in other words, the resisting restoring torque per angle of twist, which can be predicted from an equation:

$$K = C\,\frac{NEbt^3}{12L}$$

where K = spring constant, in.-lb./deg.

N = number of flexures of width b

E = modulus of elasticity, lb./in.^{2}

b = flexure width, in.

t = flexure thickness, in.

L = flexure length, in.

C = summation of constants resulting from variations in tolerances and flexure shape.

Flat springs serve as a frictionless pivot

A flexible mount, suspended by a series of flat vertical springs that converge spoke-like from a hub, is capable of pivoting through small angles without any friction. The device, developed by C. O. Highman of Ball Bros. Research Corp. under contract to Marshall Space Flight Center, Huntsville, Ala., is free also of any hysteresis when rotated (it will return exactly to its position before being pi-

voted). Moreover, its rotation is smooth and linearly proportional to torque.

The pivot mount, which in a true sense acts as a pivot bearing without need for any lubrication, was developed with the aim of improving the pointing accuracies of telescopes, radar antennas, and laser ranging systems. It has other interesting potential applications, however. When the pivot mount is supported by springs

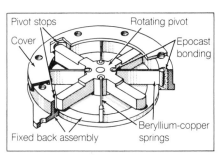

Pivot stops

Rotating pivot

Cover

Epocast bonding

Fixed back assembly

Beryllium-copper springs

Assembly of flat springs gives accurate, smooth pivoting with no starting friction

that have different thermal expansion coefficients, for example, heat applied to one spring segment produces an angular rotation independent of external drive.

Flexing springs. The steel pivot mount is supported by beryllium-copper springs attached to the outer frame. Stops limit the thrust load. The flexure spring constant is about 4 ft-lb/radian.

The flexible pivot mount can be made in tiny sizes, too, and can be driven by a dc torque motor or a mechanical linkage. In general, the mount can be used in any application requiring small rotary motion with zero chatter. ∎

Flexure devices—for economic action

Advantages: Often simpler, friction and wear are virtually nil, no lubrication.

JAMES F. MACHEN

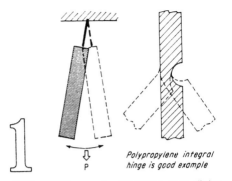

Polypropylene integral hinge is good example

1 BASIC FLEXURE connection (single-strip pivot) eliminates need for bearing in oscillatory linkages such as relay armatures

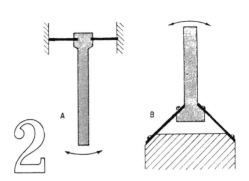

2 TWO EXAMPLES of two-strip pivots

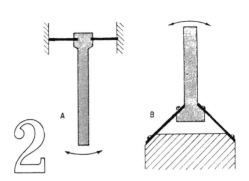

3 CROSS-STRIP PIVOT combines flexibility with some load carrying capacity

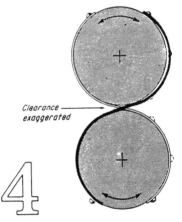

Clearance exaggerated

4 CROSS-STRIP ROLLING PIVOT maintains "geared" rolling contact between two cylinders — different diameters give "gear" ratio

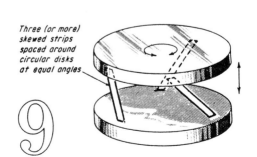

5 "RACK AND PINION" equivalent of rolling pivots

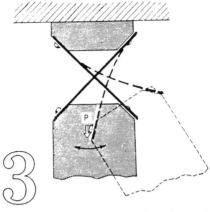

Travel limited by strip clearance through side walls

6 120° Y CROSS-STRIP pivot holds center location to pro frictionless bearing with angular spring rate

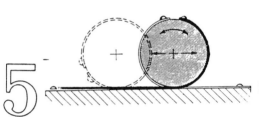

8 IN THIS PARALLEL-MOTION linkage, platform A remains level and its height does not change with sideways oscillation

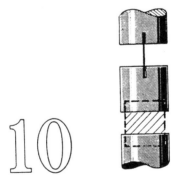

10 LIGHT-DUTY UNIVERSAL JOINT is ideal for many sealed ins ment actions

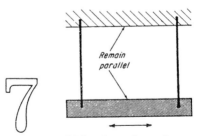

Remain parallel

7 PARALLEL-MOTION linkage has varying spacing

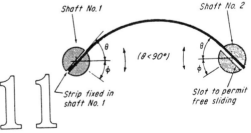

Three (or more) skewed strips spaced around circular disks at equal angles

9 SKEWED STRIP converts angular motion into linear motion or vice versa

Shaft No. 1   *Shaft No. 2*

$(\theta < 90°)$

Strip fixed in shaft No. 1   *Slot to permit free sliding*

11 FLEXURE TRANSMITS equal but opposite, low-torque, angular motion between parallel shafts

12 SINE SPRING, straight line mechanism lets point A approximately a straight line for short distances.

Flexible bands substitute for worm gear in a precisely repeatable rotary mechanism used as a star tracker. The tracker instrumentation, mounted on the platform, rotates by means of an input rotation to the lead screw.

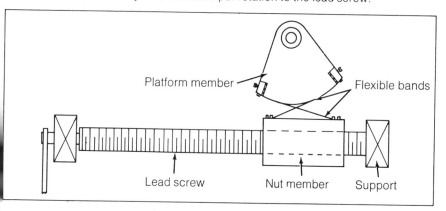

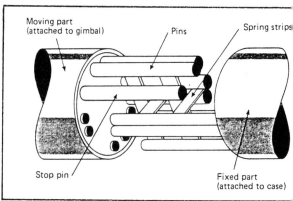

Flexure pivot boasts high mechanical stability for use in precision instruments

Taut bands and lead screw provide accurate rotary motion

A pair of opposed, taut, flexible bands in combination with a lead screw is providing an extremely accurate technique for converting rotary motion in one plane to rotary in another plane. Normally, a worm-gear set would be employed for such motion. The new technique, however, developed by Kenneth G. Johnson of Jet Propulsion Laboratory, Pasadena, Calif. under a NASA sponsored project, is providing repeatable, precise positioning within two seconds of an arc for a star tracker mechanism (drawing, photo).

Crossed bands. In the new mechanism, a precision-finished lead screw and a fitted mating nut member are employed to produce linear translatory motion. This motion is then transformed to a rotary movement of a pivotally platform member. The transformation is achieved by coupling the nut member and the platform member through a pair of crossed flexible phosphor bronze bands.

The precision lead screw is journaled at its ends in the two supports.

With the bands drawn taut, it is apparent that, as the lead screw is rotated to translate the nut member, the platform member will be drawn about its pivot without any lost motion or play. Since the nut member is accurately fitted to the lead screw, and since precision-ground lead screws have a minimum of lead error, the uniform linear translation pro-

duced by rotation of the lead screw will result in a uniform angular rotation of the platform member.

Johnson believes there is a good chance the device will be granted a patent, since a literature and patent search failed to uncover a similar design. Royalty-free, nonexclusive licenses for its commercial use will be granted by NASA.

As to whether or not the bands would transmit a true fidelity of rotary motion (as worm gears would), Johnson points out that points on the radial periphery of the sector are governed by the relationship $S = R\Theta$, which states that rotation is directly proportional to distance as measured at the circumference. The nut that translates on the lead screw is directly related to the rotary input by reason of an accurately ground and lapped lead screw. Also, 360 deg. of rotation of lead screw translates the saddle nut a distance of one thread pitch. This translation results in rotation of the sector through an angle equal to S/R.

The relationship is true at any point within the operating range of the instrument, provided that R remains constant. Two other necessary conditions for maintaining the relationship are that the saddle nut be constrained against rotation, and that there be a zero gap between sector and saddle nut.

Pivots with a twist

A multi-pin flexure type pivot by the British firm of Smiths Industries has combined high radial and axial stiffness with the inherent advantages of a cross-spring pivot—which it is.

The pivot (see above) provides non-sliding, non-rolling radial and axial support without the need for lubrication. The design combines high radial and axial stiffness with a relatively low and controlled angular stiffness. Considerable attention has been applied to solving the practical problems of mounting the pivot in a precise and controlled way.

The finished pivot is substantially free from residual mechanical stress to achieve stability in service. Maraging steel is used throughout the assembly to avoid any differential expansion due to material mismatch. The blades of the flexure pivot are free from residual braze so as to avoid any bimetallic movements when the temperature of the pivot changes.

The comparatively open construction of the pivot makes it less susceptible to jamming by loose particles. Further, the simple geometric arrangement of the support pins and flexure blade allows blade anchor points to be defined with greater accuracy. The precision ground integral mounting flanges simplify installation.

Advantages, according to the designer, include frictionless, stictionless and negligible hysteresis characteristics. The bearing is radiation-resistant and can be used in high vacuum conditions or in poor environments where there is dirt and contamination. □

Air Spring Mechanisms

Eight ways to actuate mechanisms with

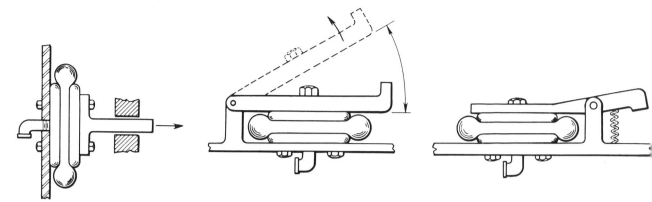

Linear force link—One- or two-convolution air spring drives guide rod. Rod returned by gravity, opposing force, metal spring or, at times, internal stiffness of air spring.

Rotary force link—Pivoted plate can be driven by one-convolution or two-convolution spring to 30 deg of rotation. Limitation on angle is based on permissible spring misalignment.

Clamp—Jaw is normally held open b means of metal spring. Actuation of a spring then closes clamp. Amount opening in jaws of clamp can be up t 30 deg of arc.

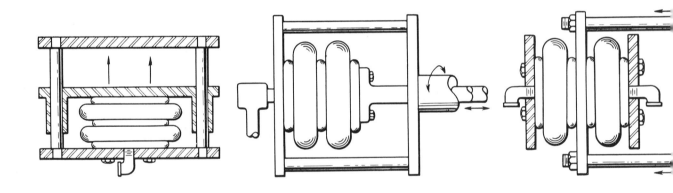

Direct-acting press—One-, two-, or three-convolution air springs used singly or in gangs. Naturally stable when used in groups. Gravity returns platform to starting position.

Rotary shaft actuator — Shifts shaft longitudinally while the shaft is rotating. Air springs with one, two, or three convolutions can be used. Standard rotating air fitting is required.

Reciprocating linear force lin —With one-, two-, or three-convol tion air springs in back-to-back arrang ment. Two- and three-convoluti springs may need guides for force roc

Popular types of air springs

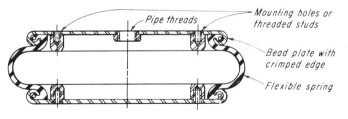

ONE-CONVOLUTION BELLOW

A IR is an ideal load-carrying medium. It is highly elastic, its spring rate can be easily varied, it is not subject to permanent set.

Air springs are elastic devices that employ compressed air as the spring. They maintain a soft ride and a constant vehicle height under varying load. In industrial applications they control vibration (isolate or amplify it) and actuate linkages to provide either rotary or linear movement. Three kinds of air springs (bellows, rolling sleeve, rolling diaphragm) are illustrated.

Bellows type

A single-convolution spring looks like a tire lying on its side. It has limited stroke and relatively high spring rate. Natural frequency is about 150 cpm without auxiliary volume for

most sizes, and as high as 240 cpm smallest size. Lateral stiffness is hi (about half the vertical rate); therefc the spring is quite stable latera when used for industrial vibrati isolation. It can be filled manually

air springs

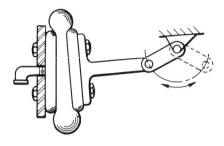

Pivot mechanism—Rotates rod through 145 deg of rotation. Can take 30-deg misalignment owing to circular path of connecting-link pin. Metal spring or opposing force retracts link.

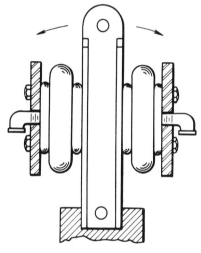

Reciprocating rotary motion—With one-convolution and two-convolution springs. Arc up to 30 deg is possible. Can pair large air spring with smaller one or lengthen lever.

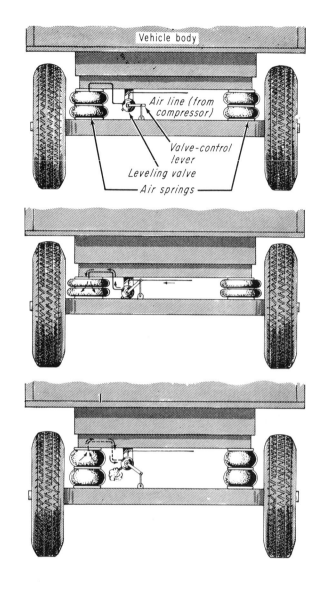

AIR SUSPENSION ON VEHICLE: (a) Normal static conditions—air springs at desired height, height-control valve closed. (b) Load added to vehicle—valve opens to admit air to springs and restore height, but at higher pressure. (c) Load removed from vehicle—valve permits bleeding off excess air pressure to atmosphere and restores design height.

kept inflated to constant height if connected to factory air supply through a pressure regulator. This spring will also actuate linkages where short axial length is desirable. It is seldom used in vehicle suspensions.

Rolling-sleeve type

This is sometimes called the reversible-sleeve or rolling-lobe type. It has telescoping action—the lobe at the bottom of the air spring rolls up and down along the piston. The spring is used primarily in vehicle suspensions because lateral stiffness is almost zero.

Rolling-diaphragm type They are laterally stable and can be used as vibration isolators, actuators, or con-stant-force springs. But because of the negative effective-area curve, they are not generally supplied by pressure regulators

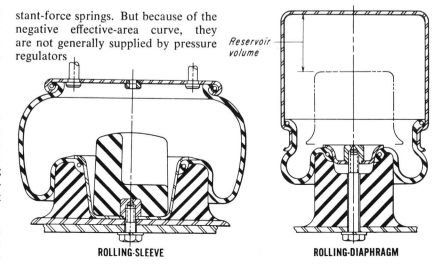

ROLLING-SLEEVE **ROLLING-DIAPHRAGM**

Obtaining variable rates from springs

How stops, cams, linkages and other arrangements can vary the load/deflection ratio during extension or compression.

JAMES F MACHEN

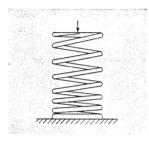

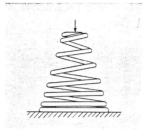

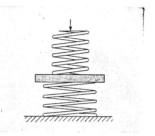

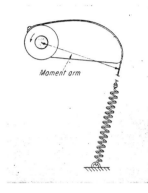

WITH TAPERED-PITCH SPRINGS the number of effective coils changes with deflection—the coils "bottom" progressively. Tapered

O.D. and pitch combine to produce similar effect except spring with tapered O.D. will have shorter solid height.

IN DUAL SPRINGS one spring closes solid before the other.

CAM-AND-SPRING DEVICE causes torque relationship to vary during rotation as moment arm changes.

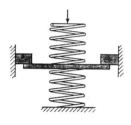

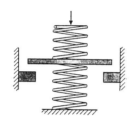

STOPS can be used with either compression or extension springs.

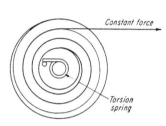

TORSION SPRING combined with variable-radius pulley gives constant force.

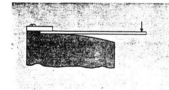

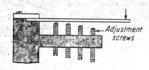

LEAF SPRINGS can be arranged so that their effective lengths change with deflection.

4-BAR MECHANISM in conjunction with a spring has a great variety of load/deflection characteristics.

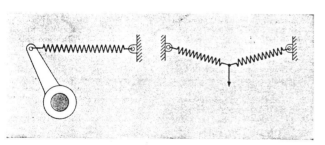

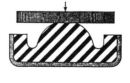

LINKAGE-TYPE ARRANGEMENTS are often used in instruments where torque control or anti-vibration suspension is required.

MOLDED-RUBBER SPRING has deflection characteristics that vary with its shape.

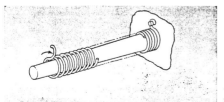

TAPERED MANDREL AND TORSION SPRING. Effective number of coils decreases with torsional deflection.

ARCHED LEAF-SPRING gives almost constant force when shaped like the one illustrated.

Belleville springs

According to H. K. Metalcraft, a variety of belleville traits are possible:

For h/t of about 0.4—A linear spring rate and high load resistance with small deflections.

For h/t ratios between 0.8 and 1.0—An almost linear spring rate for fasteners and bearings and in stacks.

For h/t around 1.6—A constant (flat) spring rate starting about 60% of the deflection (relative to the fully compressed flat position) and proceeding to the flat position and, if desired, on to the flipped side to a deflection of about 140%. In most applications, the flat position is the limit of travel, and for deflections beyond the flat, the contact elements must be allowed unrestricted travel.

One use of bellevilles with constant spring rate is on so-called live spindles on the tailpiece of a lathe. The work can be loaded on the lathe, and as the piece heats up and begins to expand, the belleville will absorb this change in length without adding any appreciable load.

For high h/t ratios exceeding about 2.5—The spring is stiff and as the stability point (high point on the curve) is passed the spring rate becomes negative causing resistance to drop rapidly. If allowed, the belleville will snap through the flat position. In other words, it will turn itself inside out. One application of snap-through bellevilles is for land mines where a considerable resisting force is first built up, after which the washer snaps through to actuate a firing mechanism. By careful choice of dimensions, a land mine will resist being triggered by pressure from a person's weight but will explode under a vehicle.

Working in groups. Belleville washers stacked in the parallel arrangement have been used successfully in a variety of applications.

One is a small-arms buffer mechanism (Fig 4) designed to absorb repeated, high-energy shock loads. A preload nut pre-deflects the washers to stiffen their resistance. The stacked washers are guided by a central shaft, an outside guide cylinder, guide rings, or a combination.

A wind-up starter mechanism for diesel engines (shown in Fig 5 and in a booklet by Bauer Springs) replaces a heavy-duty electric starter or auxiliary gas engine. To turn over the engine, energy is manually stored in a stack of bellevilles compressed by a hand crank. When released, the expanding spring pack rotates a pinion meshed with the flywheel ring gear to start the engine.

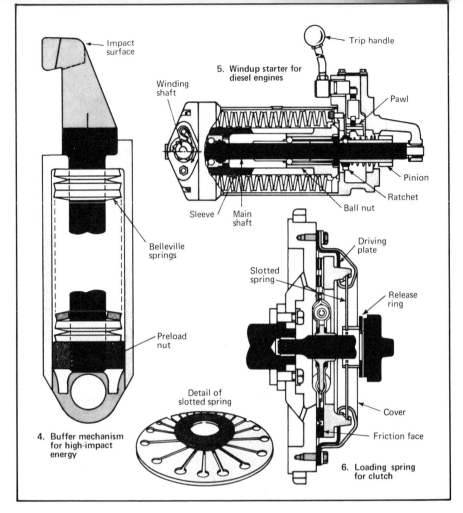

4. Buffer mechanism for high-impact energy

Detail of slotted spring

5. Windup starter for diesel engines

6. Loading spring for clutch

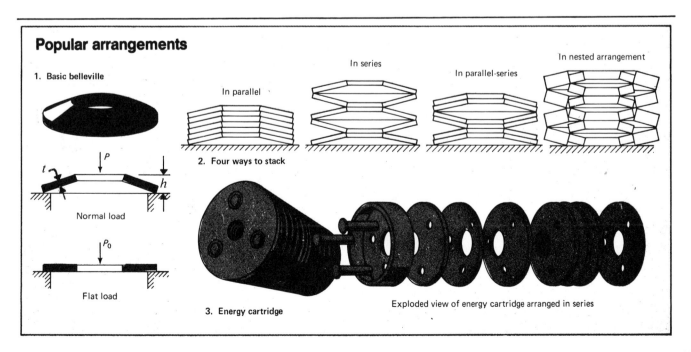

Popular arrangements

1. Basic belleville

2. Four ways to stack

In parallel

In series

In parallel-series

In nested arrangement

Normal load

Flat load

3. Energy cartridge

Exploded view of energy cartridge arranged in series

185

Spring and Linkage Arrangements for Vibration Control

Need a buffer between vibrating machinery and the surrounding structure? These isolators, like a capable fighter, absorb the light jabs and stand firm against the forces that are haymakers.

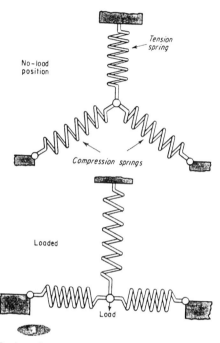

Basic spring arrangement . . .

has zero stiffness, is "soft as a cloud" when compression springs are in line, as illustrated in the loaded position. But change the weight or compression-spring alignment, and stiffness increases greatly. Such a support is adequate for vibration isolation because zero stiffness gives a greater range of movement than the vibration amplitude—generally in the hundredths-of-an-inch range. Arrangements shown here are highly absorbent when required, yet provide a firm support when large force changes occur. By contrast, isolators depending upon very "soft" springs, such as the sine spring, are unsatisfactory in many applications—they allow large movement of the supported load with any slight weight change or large-amplitude displacing force.

FROM ENGLAND

This new form of spring arrangement is described in "Supports for Vibration Isolation," a British Aeronautical Research Council paper by W. G. Molyneux. He explains and illustrates the basic arrangement, as well as alternatives. From his paper come the applications shown here, together with a brief description of the principles involved.

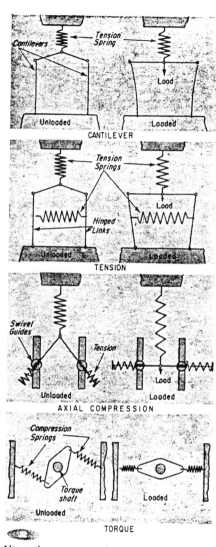

CANTILEVER

TENSION

AXIAL COMPRESSION

TORQUE

Alternative arrangements . . .

illustrate adaptability of basic design. Here, instead of the inclined, helical compression springs, either tension or cantilever springs can serve. Similarly, different types of springs can replace the axial, tension spring. Zero torsional stiffness can also be provided.

Various applications . . .

of the principle to vibration isolation show how versatile the design is. Coil spring (4), as well as cantilever and torsion-bar suspension of automobiles can be all reduced in stiffness by adding an inclined spring; stiffness of tractor seat (5) and, consequently, transmitted shocks can be similarly reduced. Mechanical tension meter (6) provides sensitive indication of small variations in tension: as a weighing device, for example, it could detect small variations in nominally identical objects. Nonlinear torque meter (7) provides sensitive indication of torque variations about a predetermined level.

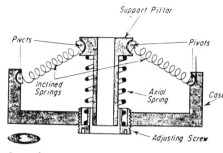

General-purpose support . . .

is based on basic spring arrangement except that axial compression spring is substituted for tension spring. Inclined compression springs, spaced around a central pillar, carry the component to be isolated. When load is applied, adjustment may be necessary to bring inclined springs to zero inclination. Load range that can be supported with zero stiffness on a specific support is determined by the adjustment range and physical limitations of the axial spring.

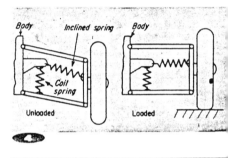

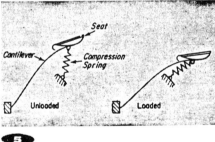

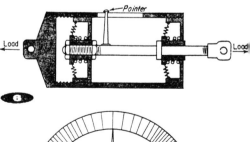

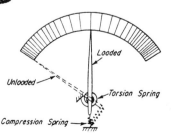

Twenty screw devices

You need a threaded shaft, a nut . . . plus some way for one of these members to rotate without translating and the other to translate without rotating. That's all. Yet these simple components can do practically all of the adjusting, setting, or locking used in design.

Most such applications have low-precision requirements. That's why the thread may be a coiled wire or a twisted strip; the nut may be a notched ear on a shaft or a slotted disk. Standard screws and nuts right off your supply shelves can often serve at very low cost.

Here are the basic motion transformations possible with screw threads (Fig 1):
- transform rotation into linear motion or reverse (A),
- transform helical motion into linear motion or reverse (B),
- transform rotation into helical motion or reverse (C).

Of course the screw thread may be combined with other components: in a 4-bar linkage (Fig 2), or with multiple screw elements for force or motion amplification.

continued, next page

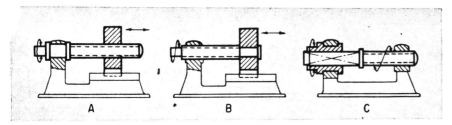
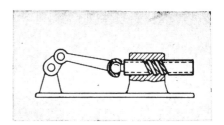

1 MOTION TRANSFORMATIONS of a screw thread include: rotation to translation (A), helical to translation (B), rotation to helical (C). Any of these is reversible if the thread is not self-locking (see screw-thread mathematics on following page—thread is reversible when efficiency is over 50%).

2 STANDARD 4-BAR LINKAGE has screw thread substituted for slider. Output is helical rather than linear.

Rotation to Translation

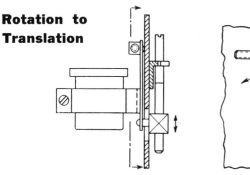
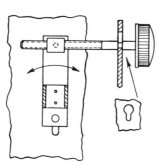
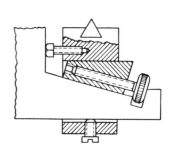

3 TWO-DIRECTIONAL LAMP ADJUSTMENT with screwdriver to move lamp up and down. Knob adjust (right) rotates lamp about pivot.

4 KNIFE-EDGE BEARING is raised or lowered by screw-driven wedge. Two additional screws locate the knife edge laterally and lock it.

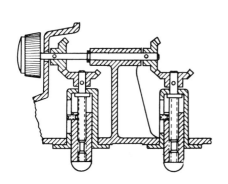
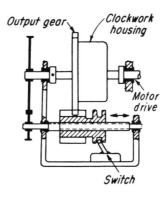
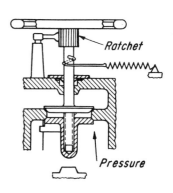

5 SIDE-BY-SIDE ARRANGEMENT of tandem screw threads gives parallel rise in this height adjustment for projector.

6 AUTOMATIC CLOCKWORK is kept wound tight by electric motor turned on and off by screw thread and nut. Note motor drive must be self-locking or it will permit clock to unwind as soon as switch turns off.

7 VALVE STEM has two oppositely moving valve cones. When opening, the upper cone moves up first, until it contacts its stop. Further turning of the valve wheel forces the lower cone out of its seat. The spring is wound up at the same time. When the ratchet is released, spring pulls both cones into their seats.

Translation to Rotation

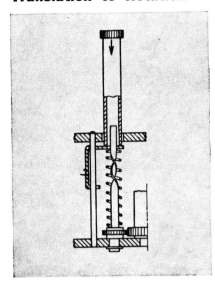

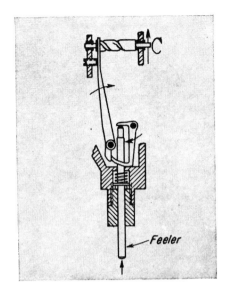

8 A METAL STRIP or square rod may be twisted to make a long-lead thread, ideal for transforming linear into rotary motion. Here a push-button mechanism winds a camera. Note that the number of turns or dwell of output gear is easily altered by changing (or even reversing) twist of the strip.

9 FEELER GAGE has its motion amplified through a double linkage and then transformed to rotation for dial indication.

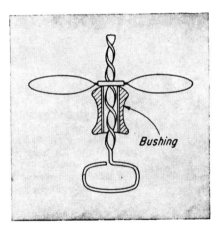

10 THE FAMILIAR flying propeller-toy is operated by pushing the bushing straight up and off the thread.

Self-Locking

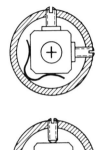

11 HAIRLINE ADJUSTMENT for a telescope, with two alternative methods of drive and spring return.

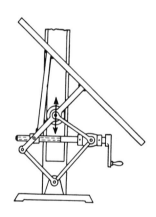

12 SCREW AND NUT provide self-locking drive for a complex linkage.

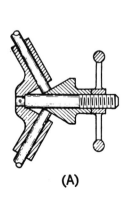

(A)

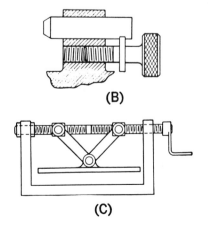

(B)

(C)

FORCE TRANSLATION. Threaded handle in (A) drives coned bushing which thrusts rods outwardly for balanced pressure. Screw in (B) retains and drives dowel pin for locking applications. Right- and left-handed shaft (C) actuates press.

Double Threading

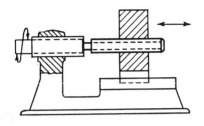

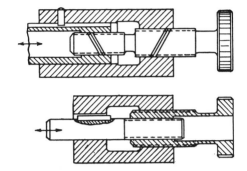

14 D I F F E R E N T I A L **SCREWS** can be made in dozens of forms. Here are two methods: above, two opposite-hand threads on a single shaft; below, same hand threads on independent shafts.

13 DOUBLE-THREADED SCREWS, when used as differentials, provide very fine adjustment for precision equipment at relatively low cost.

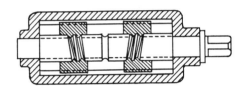

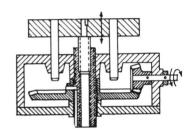

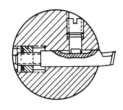

15 OPPOSITE-HAND THREADS make a high-speed centering clamp out of two moving nuts.

16 MEASURING TABLE rises very slowly for many turns of the input bevel gear. If the two threads are 1½—12 and ¾—16, in the fine-thread series, table will rise approximately 0.004 in. per input-gear revolution.

17 LATHE TURNING TOOL in drill rod is adjusted by differential screw. A special double-pin wrench turns the intermediate nut, advancing the nut and retracting the threaded tool simultaneously. Tool is then clamped by setscrew.

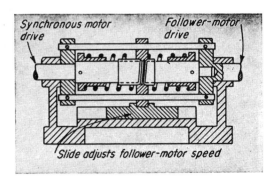

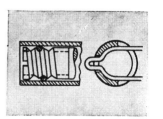

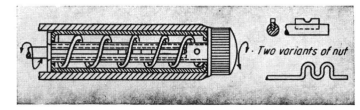

19 (left) A WIRE FORK is the nut in this simple tube-and-screw design.

20 (below) A MECHANICAL PENCIL includes a spring as the screw thread and a notched ear or a bent wire as the nut.

18 ANY VARIABLE-SPEED MOTOR can be made to follow a small synchronous motor by connecting them to the two shafts of this differential screw. Differences in number of revolutions between the two motors appear as motion of the traveling nut and slide so an electrical speed compensation is made.

189

10 WAYS TO EMPLOY SCREW MECHANISMS

FEDERICO STRASSER

Three basic components of screw mechanisms are: actuating member (knob, wheel, handle), threaded device (screw-nut set) and sliding device (plunger-guide set).

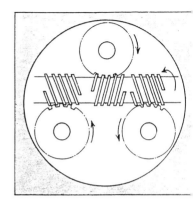

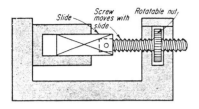

Nut can rotate . . .

but will not move longitudinally. Typical applications: screw jacks, heavy vertically-moved doors; floodgates, opera-glass focusing, vernier gages, Stillson wrenches.

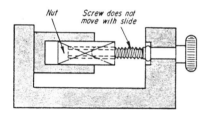

Screw can rotate . . .

but only the nut moves longitudinally. Typical applications: lathe tailstock feed, vises, lathe apron.

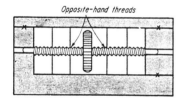

Opposing movement . . .

of lateral slides; adjusting members or other screw-actuated parts can be achieved with opposite-hand threads.

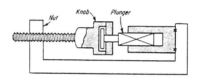

Differential movement . . .

is given by threads of different pitch. When screw is rotated the nuts move in same direction but at different speeds.

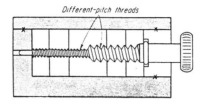

Screw and plunger . . .

are attached to knob. Nut and guide are stationary. Used on: screw presses, lathe steady-rest jaws for adjustment, shaper slide regulation.

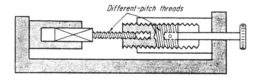

Concentric threading . . .

also gives differential movement. Such movements are useful wherever rotary mechanical action is required. Typical example is a gas-bottle valve, where slow opening is combined with easy control.

One screw actuates three gears . . .

simultaneously. Axes of gears are at right angles to that of screw. This type of mechanism can sometimes replace more expensive gear setups where speed reduction and multi-output from single input is required.

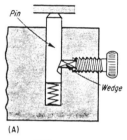

(A)

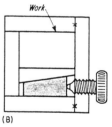

(B)

Screw-actuated wedges . . .

lock locating pin (A) and hold work in fixture (B). These are just two of many tool and diemaking applications for these screw actions.

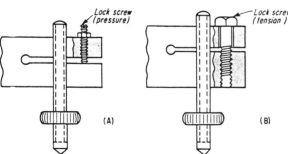

(A)   (B)

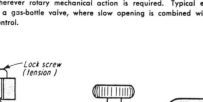

(C)

Adjustment screws . . .

are effectively locked by either a pressure screw (A) or tension screw (B). If the adjusting screw is threaded into a formed sheet-metal component (C) a setscrew can be used to lock the adjustment.

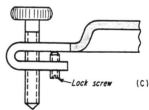

Locking nuts . . .

are often placed on opposite sides of a panel, to prevent axial screw movement and simultaneously lock against vibrations. Drill-press depth stops, and adjustable stops for shearing and cutoff dies are some examples.

7 Special screw arrangements

How differential, duplex, and other types of
screws can provide slow and fast feeds,
minute adjustments, and strong clamping action.

LOUIS DODGE

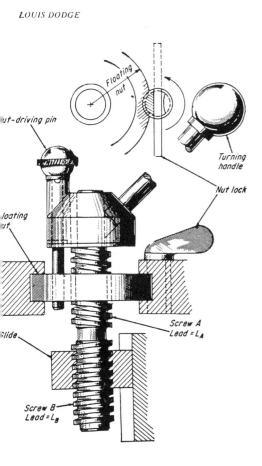

RAPID AND SLOW FEED. With left- and right-
hand threads, slide motion with nut locked equals
L_A plus L_B per turn; with nut floating, slide motion
per turn equals L_B. Get extremely fine feed with
rapid return motion when threads are differential.

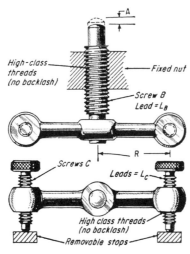

EXTREMELY SMALL MOVEMENTS.
Microscopic measurements, for example, are characteristic of this arrange-
ment. Movement A is equal to $N(L_B \times L_1)/2\pi R$, where N equals number of
turns of screw C.

SHOCK ABSORBENT SCREW. When
springs coiled as shown are used as
worm drives for light loads, they have
the advantage of being able to absorb
heavy shocks.

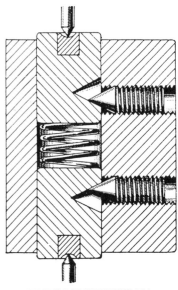

BEARING ADJUSTMENT. This screw
arrangement is a handy way of providing for bearing adjustment and overload protection.

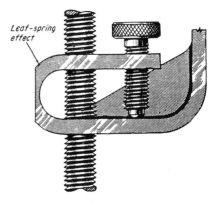

BACKLASH ELIMINATION. The large
screw is locked and all backlash is eliminated when the knurled screw is tightened — finger torque is sufficient.

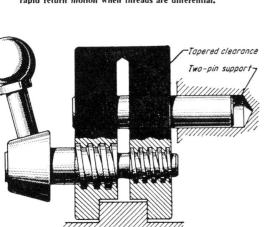

DIFFERENTIAL CLAMP. This method of using a
differential screw to tighten clamp jaws combines
rugged threads with high clamping power. Clamp-
ing pressure, $P = Te\ [R\ (\tan \varphi + \tan \alpha]$, where
T = torque at handle, R = mean radius of screw
threads, φ = angle of friction (approx. 0.1), α =
mean pitch angle of screw, and e = efficiency of
screw (generally about 0.8).

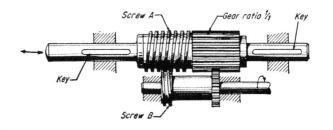

HIGH REDUCTION of rotary motion to fine linear
motion is possible here. Arrangement is for low
forces. Screws are left and right hand. $L_1 = L_B$
plus or minus a small increment. When $L_B = 1/10$
and $L_1 = 1/10.05$ the linear motion of screw A
will be 0.05 in. per turn. When screws are the same
hand, linear motion equals $L_1 + L_B$.

14 ADJUSTING DEVICES

FEDERICO STRASSER

Here is a selection of some basic arrangements that provide and hold mechanical adjustment.

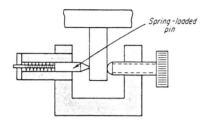

Spring-loaded pin

1

Spring-loaded pin . . .
supplies counterforce against which adjustment force must always act. A levelling foot would work against gravity—but for most other setups a spring is needed to give counterforce.

Dual adjusting screws

2

Dual screws . . .
provide inelastic counterforce. Backing-off one screw and tightening the other allows extremely small adjustments to be made. Also, once adjusted, the position remains solid against any forces tending to move device out of adjustment.

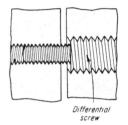

Differential screw

3

Differential screw . . .
has same-hand thread but with different pitches. Relative distance between two components can be adjusted with high precision by differential screws.

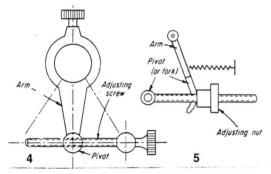

Arm
Pivot (or fork)
Arm
Adjusting screw
Arm
Adjusting nut
Pivot

4   **5**

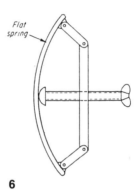

Flat spring

6

Arc-drafting guide . . .
is example of adjusting device where one of its own components, the flat spring, both supplies the counterforce and performs the mechanism's main function—guiding the pencil.

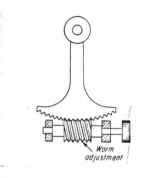

Worm adjustment

7

Worm adjustment . . .
is shown here in device for varying position of an arm. Measuring instruments, and other designs requiring fine adjustments, usually need this type of adjusting device.

Swivel motion . . .
is necessary in (4) between adjusting screw and arm because of circular locus of female thread in the actuated member. Similar action (5) requires either the screw to be pivoted or the arm to be forked.

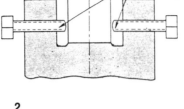

8

9

Tierods . . .
with opposite-hand threads at ends require (8) only a similarly threaded nut to provide simple, axial adjustment. Flats on rod ends (9) make it unnecessary to restrain both rods against rotation when adjusting screw is turned—restraining one rod is enough.

13

Toothed stop
Rack
Worm
Indexing holes
Pin

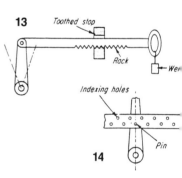

14

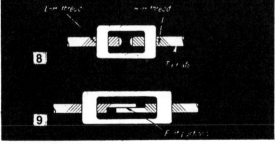

Split leg

10

Split-leg caliper . . .
is example of simple but highly efficient adjusting-device design. Tapered screw forces split leg apart, thus enlarging opening between the two legs.

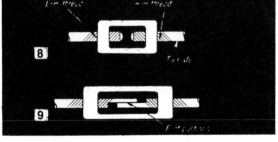

Torsion spring
Shaft
Torsion spring
Slots
Adjusting dog

11

12

Shaft torque . . .
is adjusted (11) by rotating the spring-holding collar relative to shaft, and locking collar at position of desired torque. Adjusting slots (12) accommodate torsion-spring arm after spring is wound to desired torque.

Rack and toothed stop . . .
(13) are frequently used to adjust heavy louvers, boiler doors and similar equipment. Adjustment is not continuous, depends on the rack pitch. Large counter-adjustment forces may necessitate weighted rack to prevent tooth disengagement. Indexing holes (14) provide similar adjustment to rack. Pin makes position less liable to be accidentally moved.

Ball-bearing screws—*converters of rotary to linear motion*

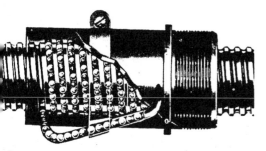

1. CONTINUOUS RECIRCULATING BALLS, rolling in helical grooves, provide ability to carry heavy loads with low friction and high efficiency.

In many applications a single circuit design—in which one tubular ball guide is used—is adequate. In critical applications where constant or very frequent operation is required the multiple circuit —utilizing two or more tubular ball guides—is used.

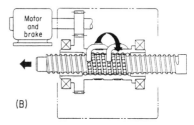

(A)

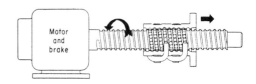

(B)

BASIC OPERATIONS: Rotary motion to (A) screw, or (B) nut, causes other member to travel linearly or provide linear force.

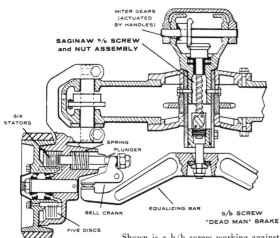

Shown is a b/b screw working against spring tension in a dead man brake. The screw is backdriven by the spring to set the brakes when the handle is released.

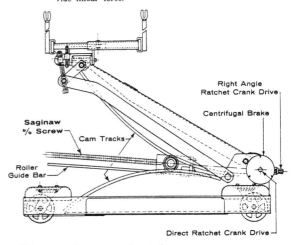

This automotive transmission jack uses a single b/b screw to move rollers between scissor-like cams. Drive speed of the screw is relatively slow and the load is in tension. The brake prevents backdriving under load.

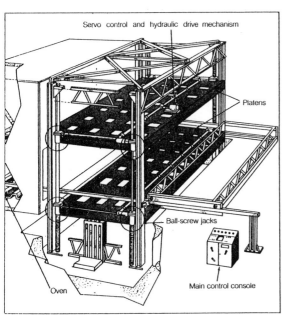

Three ball-screw jacks, operated in unison by 20-hp motor, control pilothouse as boat (right) passes under low bridge. World's largest thermoforming machine (above) forms plastic auto bodies, boats, other parts between two movable platens.

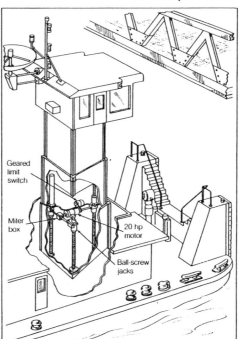

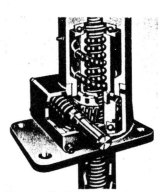

Worm gear drives ball screw to produce high lifting force with little effort.

More Ball-Bearing Screw Applications

Cartridge-operated rotary actuator quickly retracts webbing to forcibly separate a pilot from his seat as the seat is ejected in emergencies. Tendency of pilot and seat to tumble together after ejection prevented opening of chute. Gas pressure from ejection device fires the cartridge in the actuator to force ball-bearing screw to move axially. Linear motion of screw is translated into rotary motion of ball nut. This rapidly rolls up the webbing (stretching it as shown) which snaps the pilot out of his seat. *Talley Industries.*

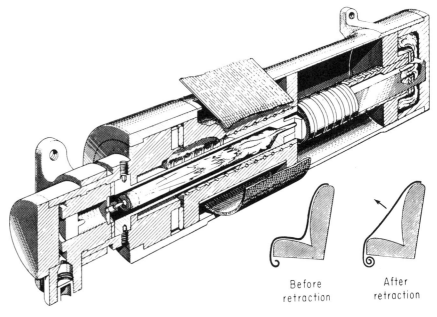

Before retraction

After retraction

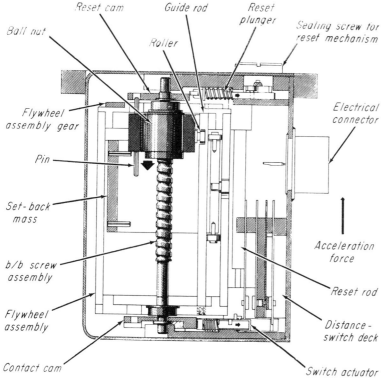

Time-delay switching device integrates time function with missile's linear travel. Purpose is to safely arm the warhead. A strict "minimum G-time" 'system may arm a slow missile too soon for adequate protection of own forces; a fast missile may arrive before warhead is fused. Weight of nut, plus inertia under acceleration will rotate the ball-bearing screw which has a fly wheel on the end. Screw pitch is such that a given number of revolutions of flywheel represents distance traveled. *Globe Industries.*

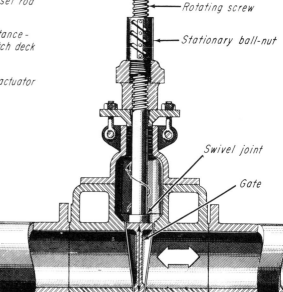

Speedy, easily operated, but more accurate control of flow through valve obtained by rotary motion of screw in stationary ball nut. Screw produces linear movement of gate. The swivel joint eliminates rotary motion between screw and gate.

Rotary to linear motion

RUSSELL C BALL Jr.

Three typical applications

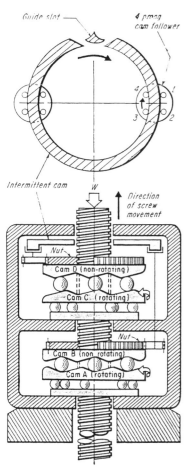

Guide slot — 4 prong cam follower

Intermittent cam

W

Direction of screw movement

Nut

Cam D (non-rotating)

Cam C (rotating)

Nut

Cam B (non rotating)

Cam A (rotating)

CONVERSION OF ROTARY TO LINEAR MOTION. Two cam assemblies operate in phased relationship. When one pair is driving, the second pair is retracting. The nuts are alternately rewound by the intermittent cam, shown also in the top view, which is driven continuously with cams A and C. As the deflector strikes prong 1 of the 4-prong cam follower at right (which is meshed with one of the nuts), the follower is rotated one quarter turn clockwise. Prong 4 then moves into the position of prong 1. This motion rewinds the nut to the top of cam B. A half revolution later, the operation is repeated for the follower on the left side.

SLUICE-GATE OPERATION requires lifting and lowering forces of about 75,000 lb. Actuator must have self-locking characteristics to prevent weight of gate from reversing the drive, and accurate positioning control. With new actuator, long stem need not be rigidly restrained to prevent twisting.

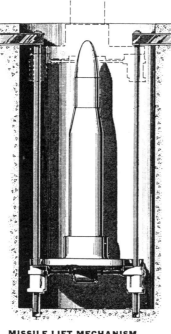

MISSILE LIFT MECHANISM has four actuators powered from a single source to move the platform uniformly at all points. Screw is locked in place while the actuator travels with its prime mover along the screw. The platform can be positioned with precision and very high lifts can be accomplished.

RUDDER-CONTROL UNIT employs actuator to provide continuous back and forth movement, usually of small displacement. Friction forces must be kept to a minimum. Power source can be a reversible hydraulic motor. Because of self-locking characteristics, no backloading from the rudder can move the threaded stem.

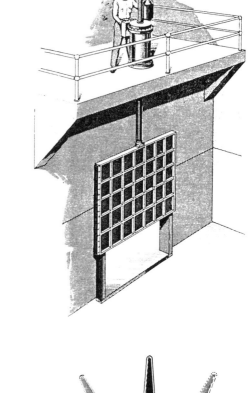

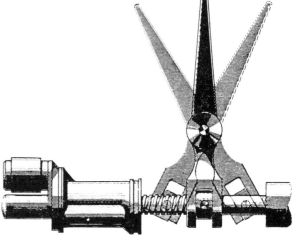

The "Roll-Ramp" actuator is the result of 15 years of investigation by its inventor, Arthur Maroth. A commercial version has been under development and test by Philadelphia Gear Corp for several years and the actuator is now in the production stage.

Principle of operation

The mechanism is enclosed in a stationary housing; an Acme-threaded shaft moves through the housing as the thrusting member. The thread, however, provides only a gripping surface for the internal mechanisms of the actuator.

The basic driving element is an inclined plane—actually two inclined planes mounted in opposition which are wrapped around a disk. This forms a face cam with a succession of long inclines and short declines. Rollers are placed on each incline between two such opposed cams and the assembly is set on a thrust bearing. Rotating the lower cam, while keeping the upper cam from rotating, will cause the upper cam to generate a straight-line, up-and-down motion. This system, however, will yield a very limited stroke (twice the use of one wedge) before the cams must be reset to lift again.

To generate a continuous lifting motion, two such cam assemblies work in phased relationship. A nut with gear teeth cut in its periphery is set on top of each cam assembly. A screw of a length suitable for the desired travel is threaded through the two nuts. The nuts are free to rotate until they snub down against the upper cam.

In the position shown, rollers in the lower cam assembly are at the bottoms of the inclines. As cam A rotates, cam B lifts the nut and screw. The rollers in the upper cam assembly are almost at the crests of the inclines. As cam C rotates further (driven by the same input as A) its rollers go over the crests and down the cam declines so cam D falls away from the top nut. The top nut is unloaded but it is carried upward by the advance of the screw (still being driven by the bottom rollers). Now, the nut is counter-rotated on the screw by the intermittent motion mechanism.

The counter-rotation returns the nut until it again meets cam D. This allows cam D to re-engage the nut when the rollers of the upper assembly are at the bottoms of the inclines and are about to climb. At this time, the rollers of the lower assembly pass the crests of the inclines and begin to return. The bottom nut is then counter-rotated by the intermittent mechanism to its new starting position in the same manner as the top nut.

Ball and groove convert motion to suit design

A unique mechanical-control concept employs steel balls for converting linear or rotary input motion to the other form of output motion, in any combination desired, and for remote actuation around corners. Developed by Alexander B. Hulse, who is known to PRODUCT ENGINEERING readers for his invention of the space crank (described on page 159), the concept is being produced in several configurations (drawings, right) by Martin Schall, Inc., Staten Island, N.Y., a tool and machinery manufacturer.

"While translation of motion in grooves or channels, using balls or rollers, is by no means a new art, a recent thorough patent search has indicated that these configurations are new and patentable," states Hulse.

Hydraulic sender. Hulse's mechanisms consist of balls routed in special-path grooves and pushed or pulled by various means to effect the change in the type of motion desired.

In one application, the actuator is made in the form of a hydraulic sender in which fluid entering a cylinder drives a plunger to rotate a sector. Such a sender can, among

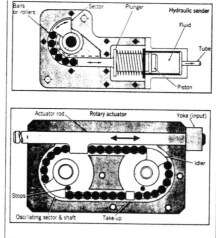

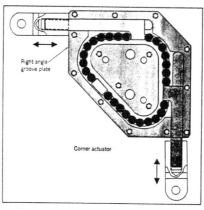

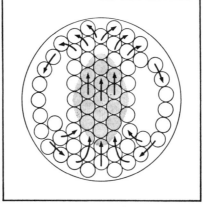

other things, be used as a master brake cylinder. Moreover, if the compression spring on the plunger is replaced by a torsion spring on the shaft, the mechanism becomes a hydraulic or pneumatic rotary actuator.

Rotary actuator. Here an idler section is employed, and the balls are confined between both sides of the sectors and both ends of the reciprocating bar. When the shaft is rotated in either direction, the bar will move in the same direction.

"While the same kinematics can be effected with a gear rack and pinion, the balls and rollers can take loading uncommon to gear teeth, backlash can readily be controlled by an adjustable extension at one end of the reciprocating bar, lubrication can readily be supplied to the moving members, and a neat compact package results," points out Hulse.

Corner actuator. This device is a push-pull mechanism with the two rods disposed at an angle to each other (90 deg. in the drawing). The device has a closed groove circuit; backlash is controlled with a cylindrical filler piece.

Compact ball transfer units roll loads every which way

An improved design of an oft-neglected device for moving loads—ball transfers—is opening up new applications in air cargo planes (photo below) and other materials handling jobs. It can serve in production lines to transfer sheets, tubes, bars, and parts.

Uses of established ball transfer units have been limited largely to furniture (in place of casters) and other prosaic duties. With new design that takes fuller advantage of their multiple-axis translation and instantaneous change of direction, ball transfer units can be realistically considered as another basic type of anti-friction bearing.

How they work. Essentially, ball transfers (photo below) are devices that translate omnidirectional linear motion into rolling motion to provide an unlimited number of axes of movement in any given plane. In such a unit, a large main ball rotates on its own center within a housing. This ball is supported by a circular group of smaller balls (drawing below) that roll under load and, in so doing, recirculate within the housing in endless chains.

Main ball shown, 1 in. dia., is supported by 70 smaller balls, hidden.

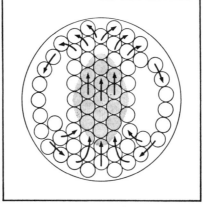

Shaded area seen from above shows load; arrows show ball circulation.

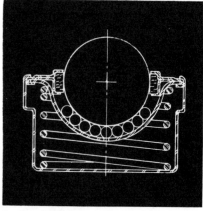

Spring loading assures even distribution of the load on the small balls.

CHAPTER 5.

CAM, TOGGLE, CHAIN AND BELT MECHANISMS

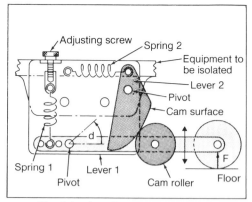

Cam and memory rollers can take up rotational shock

A new energy-absorbing device takes up the rotational shock load when a powered shaft is forced to an abrupt stop by causing resilient rollers to be compressed or deformed.

Developed by Dura Corp., Toledo, Ohio, a subsidiary of Walter Kidde & Co., the overload device is already being used on electric-powered windows for automobiles. In this application it absorbs the energy of a rotating armature when the window comes to stall at either the up or down position.

Says the inventor. "The armature carries about three times the energy at the stall position which means that without such a device, it may break up the gear teeth and cause failure in other components of the mechanism," says J.C. Littmann, chief engineer, Electro-Mechanical Devices, Dura Corp., and inventor of the overload stopping device.

The new device is interspersed between the motor shaft and the output which in this case is a sector gear that actuates the window. The motor shaft drives the worm gear and torque is transmitted through the rollers to the three-lobe cam and hence to the pinion and sector gear.

As the window comes to the top of its position and abruptly stops, stopping also the pinion and cam, the worm continues to rotate until the rollers, made of polyurethane, are compressed into an oblong shape, corresponding to the triangular shape of the opening and the configuration of the cam.

Because they are made of polyurethane, the rollers have a tremendous memory capability because they can regain their initial shape even after considerable deformation. The rollers thus retain the torque in their deflected position, holding the window snugly against the seals in the up position even after the window button is released.

When the motor is reversed, the rollers return to their neutral state to drive the pinion in the reverse direction. The rollers are always in slight compression.

Wide applications. The device is being used in all the Ford and Chrysler models now in production for power side-door and rear-gate windows. Other

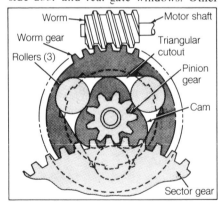

Key elements (drawing) are a set of deformable rollers. The input motor drives the outer worm gear. Mirror (photo) shows the output pinion at rear of device

applications include overload mechanisms for washing-machine drives.

"When the spin-dry cycle of a washing machine is clutched in," says Littmann, "the initial shock load boosts the stress on the gear teeth. Our device helps distribute the stress to a much lower level so that the load can be picked up much more gradually."

Other applications are in drapery drives where the motor drives a mechanism to the end of travel and stalls as long as the operator holds the button.

"Automatic machinery is another potential area," says Littmann. Many of the gear drives are heavy and have high inertia. The overload device would eliminate shock damage in abrupt stops."

Each model of the device has been specifically developed for an application and cannot be classified as a shelf item. ∎

Cam-and-spring mount tailors shock absorption

A simple, yet highly effective technique for varying the energy-absorption characteristics of shock mounts brings together in one compact arrangement several springs and a cam. Developed by Dr. J.S. Turton of the Univ. of Sussex, Sussex, England, the shock mounting can be designed to produce a wide variety of force-deflection relationships.

The device, which Dr. Turton calls a Camspring, employs two springs, both fixed at one end and attached to pivotable levers at the other end. One of the levers contains a cam surface that bears against a roller, which is mounted on the other lever.

How it works. In operation, four Camsprings are mounted under the equipment to be shock-isolated, one under each corner. A shock force F pivots lever l and extends spring l. At the same time, the cam roller bears on the cam surface to extend spring 2 in a non-linear fashion.

If the angular deflection of lever l is designated as d measured from the horizontal, the cam profile and the spring rates can be selected to produce a specific force-deflection relationship (between F and d). For example, the Camspring can be made to resist with a constant force during most of the deflection of lever l, in which case the Camspring is functioning as a constant-force spring of zero rate. Such spring characteristics are frequently sought in suspension systems for cars in railroad trains.

Applications. Moreover, the cam profile and springs can be selected to produce an Fsd spring rate that will give a specific natural frequency for the entire system being supported or a spring rate that increases with an increase in the deflection imparted.

Dr. Turton has developed a computer program to obtain any Fsd relationship, and he believes the technique should find wide use in vehicle suspensions, anti-shock mountings, and in supporting hydrofoil ships, floating rafts, and breakwater jetties. ∎

Your guide to mechanisms for
Generating cam curves

It usually doesn't pay to design a complex cam
curve if it can't be easily machined—so check these
mechanisms before starting your cam design

PREBEN W. JENSEN

IF you have to machine a cam curve into the metal
blank without using a master cam, how accurate
can you expect it to be? That depends primarily on
how precisely the mechanism you use can feed the
cutter into the cam blank. The mechanisms described
here have been carefully selected for their practicabil-
ity. They can be employed directly to machine the
cams, or to make master cams for producing others.

The cam curves are those frequently employed in
automatic-feed mechanisms and screw machines. They
are the circular, constant-velocity, simple-harmonic,
cycloidal, modified cycloidal, and circular-arc cam
curve, presented in that order.

Circular cams

This is popular among machinists because of the
ease in cutting the groove. The cam (Fig 1A) has a
circular groove whose center, A, is displaced a distance a from the cam-plate center, A_0, or it may simply be a plate cam with a spring-loaded follower (Fig
1B).

Interestingly, with this cam you can easily duplicate
the motion of a four-bar linkage (Fig 1C). Rocker
BB_0 in Fig 1C, therefore, is equivalent to the motion of
the swinging follower in Fig 1A.

The cam is machined by mounting the plate eccentrically onto a lathe. The circular groove thus can
be cut to close tolerances with an excellent surface
finish.

If the cam is to operate at low speeds you can replace the roller with an arc-formed slide. This permits the transmission of high forces. The optimum
design of such "power cams" usually requires time-
consuming computations.

continued, next page

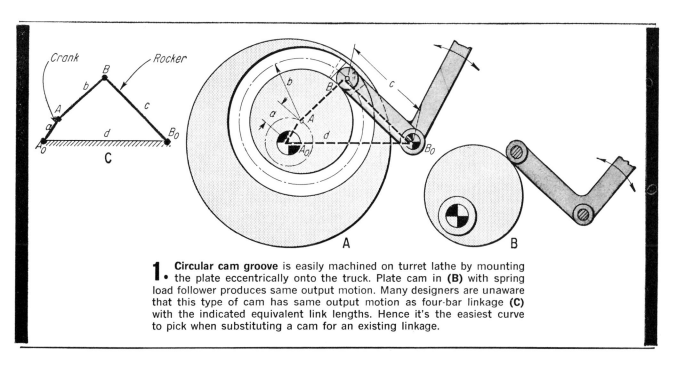

1. **Circular cam groove** is easily machined on turret lathe by mounting
the plate eccentrically onto the truck. Plate cam in **(B)** with spring
load follower produces same output motion. Many designers are unaware
that this type of cam has same output motion as four-bar linkage **(C)**
with the indicated equivalent link lengths. Hence it's the easiest curve
to pick when substituting a cam for an existing linkage.

The disadvantage (or sometimes, the advantage) of the circular-arc cam is that, when traveling from one given point, its follower reaches higher speed accelerations than with other equivalent cam curves.

Constant-velocity cams

A constant-velocity cam profile can be generated by rotating the cam plate and feeding the cutter linearly, both with uniform velocity, along the path the translating roller follower will travel later (Fig 2A). In the case of a swinging follower, the tracer (cutter) point is placed on an arm equal to the length of the actual swinging roller follower, and the arm is rotated with uniform velocity (Fig 2B).

Simple-harmonic cams

The cam is generated by rotating it with uniform velocity and moving the cutter with a scotch yoke geared to the rotary motion of the cam. Fig 3A shows the principle for a radial translating follower; the same principle is, of course, applicable for offset translating and swinging roller follower. The gear ratios and length of the crank working in the scotch yoke control the pressure angles (the angles for the rise or return strokes).

For barrel cams with harmonic motion the jig in Fig 3B can easily be set up to do the machining. Here, the barrel cam is shifted axially by means of the rotating, weight-loaded (or spring-loaded) truncated cylinder.

The scotch-yoke inversion linkage (Fig 3C) replaces the gearing called for in Fig 3A. It will cut an approximate simple-harmonic motion curve when the cam has a swinging roller follower, and an exact curve when the cam has a radial or offset translating roller follower. The slotted member is fixed to the machine frame 1. Crank 2 is driven around the center 0. This causes link 4 to oscillate back and forward in simple harmonic motion. The sliding piece 5 carries the cam to be cut, and the cam is rotated around the center of 5 with uniform velocity. The length of arm 6 is made equal to the length of the swinging roller follower of the actual cam mechanism and the device adjusted so that the extreme positions of the center of 5 lie on the center line of 4.

The cutter is placed in a stationary spot somewhere along the centerline of member 4. In case a radial or offset translating roller follower is used, the sliding piece 5 is fastened to 4.

The deviation from simple harmonic motion when the cam has a swinging follower causes an increase in acceleration ranging from 0 to 18% (Fig 3D), which depends on the total angle of oscillation of the follower. Note that for a typical total oscillating angle of 45 deg, the increase in acceleration is about 5%.

Cycloidal motion

This curve is perhaps the most desirable from a designer's viewpoint because of its excellent acceleration characteristic. Luckily, this curve is comparatively easy to generate. Before selecting the mechanism it is worthwhile looking at the underlying theory of the cycloids because it is possible to generate not only cycloidal motion but a whole family of similar curves.

The cycloids are based on an offset sinusoidal wave (Fig 4). Because the radii of curvatures in points C, V, and D are infinite (the curve is "flat" at these points), if this curve was a cam groove and moved in the direction of line CVD, a translating roller follower, actu-

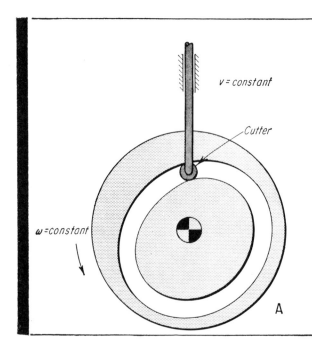

ated by this cam, would have zero acceleration at points C, V, and D no matter in what direction the follower is pointed.

Now, if the cam is moved in the direction of CE and the direction of motion of the translating follower is lined perpendicular to CE, the acceleration of the follower in points C, V, and D would still be zero.

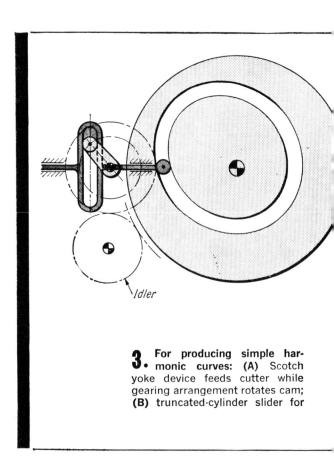

3. For producing simple harmonic curves: (A) Scotch yoke device feeds cutter while gearing arrangement rotates cam; (B) truncated-cylinder slider for

2. **Constant-velocity** cam is machined by feeding the cutter and rotating the cam at constant velocity. Cutter is fed linearly **(A)** or circularly **(B)**, depending on type of follower.

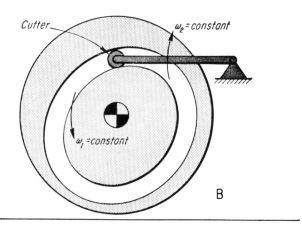

B

This has now become the basic cycloidal curve, and it can be considered as a sinusoidal curve of a certain amplitude (with the amplitude measured perpendicular to the straight line) superimposed on a straight (constant-velocity) line.

The cycloidal is considered the best *standard* cam contour because of its low dynamic loads and low

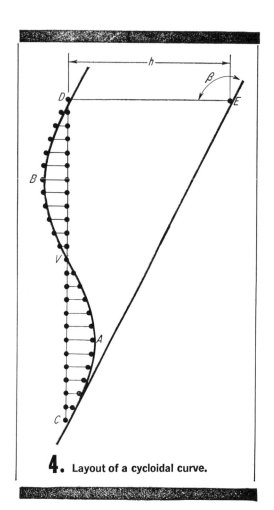

4. **Layout of a cycloidal curve.**

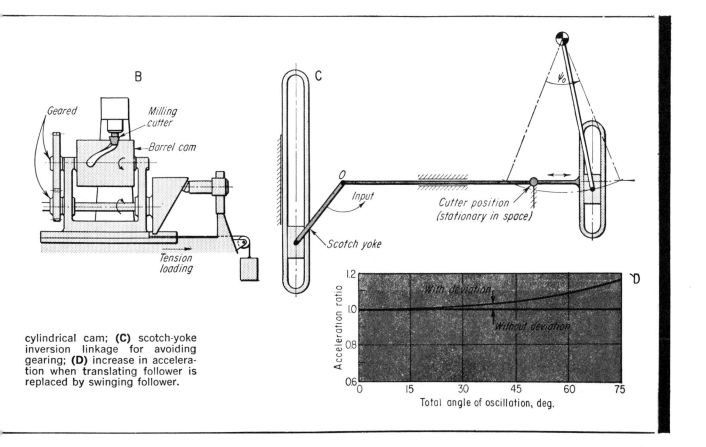

B

cylindrical cam; **(C)** scotch-yoke inversion linkage for avoiding gearing; **(D)** increase in acceleration when translating follower is replaced by swinging follower.

shock and vibration characteristics. One reason for these outstanding attributes is that it avoids any sudden change in acceleration during the cam cycle. But improved performances are obtainable with certain modified cycloidals.

Modified cycloids

To get a modified cycloid, you need only change the direction and magnitude of the amplitude, while keeping the radius of curvature infinite at points *C, V,* and *D*.

Comparisons are made in Fig 5 of some of the modified curves used in industry. The true cycloidal is shown in the cam diagram of *A*. Note that the sine amplitudes to be added to the constant-velocity line are perpendicular to the base. In the Alt modification shown in *B* (after Hermann Alt, German kinematician, who first analyzed it), the sine amplitudes are perpendicular to the constant-velocity line. This results in improved (lower) velocity characteristics (see *D*), but higher acceleration magnitudes (see *E*).

The Wildt modified cycloidal (after Paul Wildt) is constructed by selecting a point *w* which is 0.57 the distance *T/2*, and then drawing line *wp* through *yp* which is midway along *OP*. The base of the sine curve is then constructed perpendicular to *yw*. This modification results in a maximum acceleration of 5.88 h/T^2, whereas the standard cycloidal curve has a maximum acceleration of 6.28 h/T^2. This is a 6.8% reduction in acceleration.

(It's quite a trick to construct a cycloidal curve to go through a particular point *P*—where *P* may be anywhere within the limits of the box in *C*—and with a specific slope at *P*. There is a growing demand for this type of modification, and a new, simple, graphic technique developed for meeting such requirements will be shown in the next issue.)

Generating the modified cycloidals

One of the few devices capable of generating the family of modified cycloidals consists of a double carriage and rack arrangement (Fig 6A).

The cam blank can pivot around the spindle, which in turn is on the movable carriage *I*. The cutter center is stationary. If the carriage is now driven at constant speed by the lead screw, in the direction of the arrow, the steel bands 1 and 2 will also cause the cam blank to rotate. This rotation-and-translation motion to the cam will cause a spiral type of groove.

For the modified cycloidals, a second motion must be imposed on the cam to compensate for the deviations from the true cycloidal. This is done by a second steel band arrangement. As carriage *I* moves, the bands 3 and 4 cause the eccentric to rotate. Because of the stationary frame, the slide surrounding the eccentric is actuated horizontally. This slide is part of carriage *II*, with the result that a sinusoidal motion is imposed on to the cam.

Carriage *I* can be set at various angles β to match angle β in Fig 5B and C. The mechanism can also be modified to cut cams with swinging followers.

Circular-arc cams

Although in recent years it has become the custom to turn to the cycloidal and other similar curves even when speeds are low, there are many purposes for which

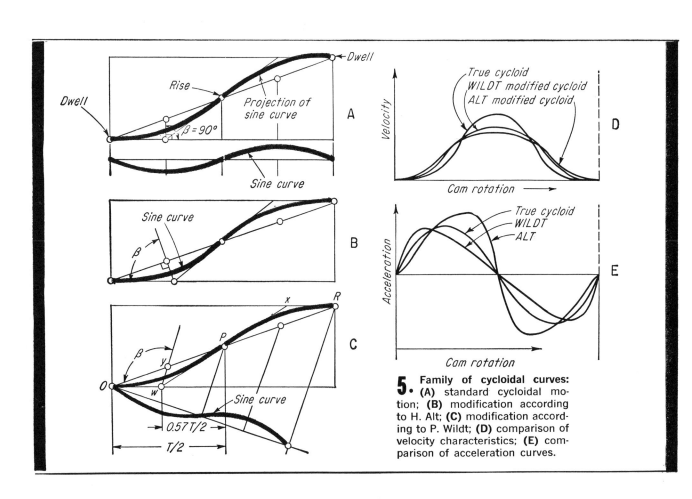

5. **Family of cycloidal curves:** **(A)** standard cycloidal motion; **(B)** modification according to H. Alt; **(C)** modification according to P. Wildt; **(D)** comparison of velocity characteristics; **(E)** comparison of acceleration curves.

circular-arc cams suffice. Such cams are composed of circular arcs, or circular arcs and straight lines. For comparatively small cams the cutting technique illustrated in Fig 7 produces good accuracy.

Assume that the contour is composed of circular arc *1-2* with center at O_2, arc *3-4* with center at O_3, arc *4-5* with center at O_1, arc *5-6* with center at O_4, arc *7-1* with center at O_1, and the straight lines *2-3* and *6-7*. The method involves a combination of drilling, lathe turning, and template filing.

First, small holes about 0.1 in diameter are drilled at O_1, O_3, and O_4, then a hole is drilled with the center at O_2 and radius of r_2. Next the cam is fixed in a turret lathe with the center of rotation at O_1, and the steel plate is cut until it has a diameter of $2r_5$. This takes care of the larger convex radius. The straight lines *6-7* and *2-3* are now milled on a milling machine.

Finally, for the smaller convex arcs, hardened pieces are turned with radii r_1, r_3, and r_4. One such piece is shown in Fig 7B. The templates have hubs which fit into the drilled holes at O_1, O_3, and O_4. Now the arc *7-1*, *3-4*, and *5-6* are filed, using the hardened templates as a guide. Final operation is to drill the enlarged hole at O_1 to a size that a hub can be fastened to the cam.

This method is frequently better than copying from a drawing or filing the scallops away from a cam where a great number of points have been calculated to determine the cam profile.

Compensating for dwells

One disadvantage with the previous generating devices is that, with the exception of the circular cam, they cannot include a dwell period within the rise-and-fall cam

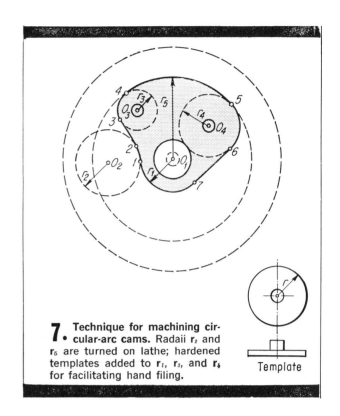

7. **Technique for machining circular-arc cams.** Radaii r_2 and r_5 are turned on lathe; hardened templates added to r_1, r_3, and r_4 for facilitating hand filing.

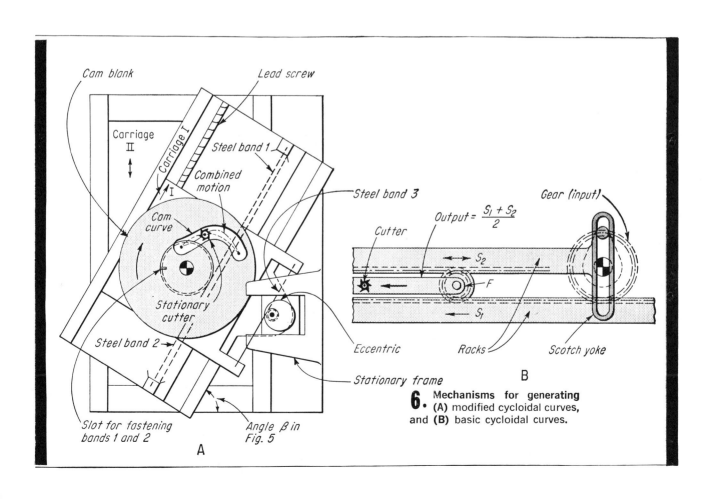

6. **Mechanisms for generating (A)** modified cycloidal curves, and **(B)** basic cycloidal curves.

cycle. The mechanisms must be disengaged at the end of rise and the cam rotated in the exact number of degrees to where the fall cycle begins. This increases the inaccuracies and slows down production.

There are two devices, however, that permit automatic machining through a specific dwell period: the double-geneva drive and the double eccentric mechanism.

Double-genevas with differential

Assume that the desired output contains dwells (of specific duration) at both the rise and fall portions, as shown in Fig 8A. The output of a geneva that is being rotated clockwise will produce an intermittent motion similar to the one shown in Fig 8B—a rise-dwell-rise-dwell . . . etc, motion. These rise portions are distorted simple-harmonic curves, but are sufficiently close to the pure harmonic to warrant use in many applications.

If the motion of another geneva, rotating counter-clockwise as shown in (C), is added to that of the clockwise geneva by means of a differential (D), then the sum will be the desired output shown in (A).

The dwell period of this mechanism is varied by shifting the relative position between the two input cranks of the genevas.

The mechanical arrangement of the mechanism is shown in Fig 8D. The two driving shafts are driven by gearing (not shown). Input from the four-star geneva to the differential is through shaft 3; input from the eight-station geneva is through the spider. The output from the differential, which adds the two inputs, is through shaft 4.

The actual device is shown in Fig 8E. The cutter is fixed in space. Output is from the gear segment which rides on a fixed rack. The cam is driven by the motor which also drives the enclosed genevas. Thus, the entire device reciprocates back and forth on the slide to feed the cam properly into the cutter.

Genevas driven by couplers

When a geneva is driven by a constant-speed crank, as shown in Fig 8D, it has a sudden change in acceleration at the beginning and end of the indexing cycle (as the crank enters or leaves a slot). These abrupt changes can be avoided by employing a four-bar linkage with coupler in place of the crank. The motion of the coupler point C (Fig 9) permits smooth entry into the geneva slot.

Double eccentric drive

This is another device for automatically cutting cams with dwells. Rotation of crank A (Fig 10) imparts an oscillating motion to the rocker C with a prolonged dwell at both extreme positions. The cam, mounted on the rocker, is rotated by means of the chain drive and thus is fed into the cutter with the proper motion. During the dwells of the rocker, for example, a dwell is cut into the cam.

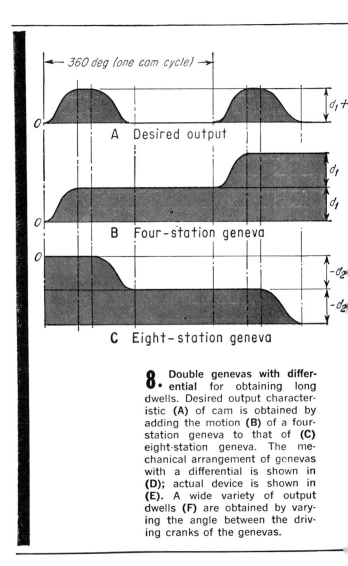

8. Double genevas with differential for obtaining long dwells. Desired output characteristic (A) of cam is obtained by adding the motion (B) of a four-station geneva to that of (C) eight-station geneva. The mechanical arrangement of genevas with a differential is shown in (D); actual device is shown in (E). A wide variety of output dwells (F) are obtained by varying the angle between the driving cranks of the genevas.

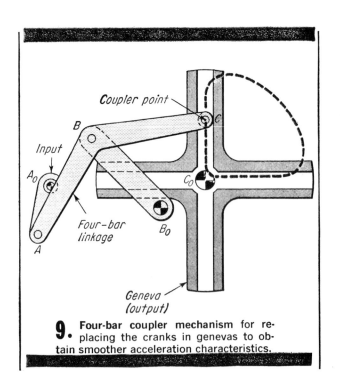

9. Four-bar coupler mechanism for replacing the cranks in genevas to obtain smoother acceleration characteristics.

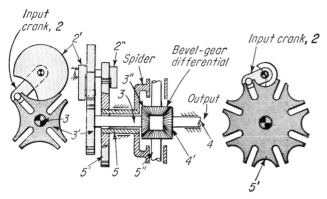

Input crank, 2

2'

2"

Spider

3"

3

Bevel-gear differential

Input crank, 2

Output

3'

3

4

4'

5' 5 5"

5'

D   Double geneva with differential

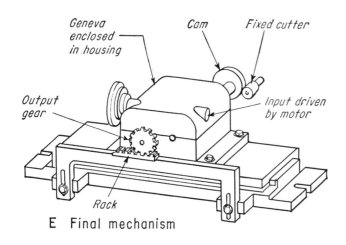

Geneva enclosed in housing

Cam

Fixed cutter

Output gear

Input driven by motor

Rack

E   Final mechanism

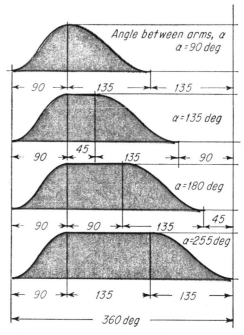

Angle between arms, α

α = 90 deg

90   135   135

α = 135 deg

90   45   135   90

α = 180 deg

90   90   135   45

α = 255 deg

90   135   135

360 deg

F   Various dwell resultants

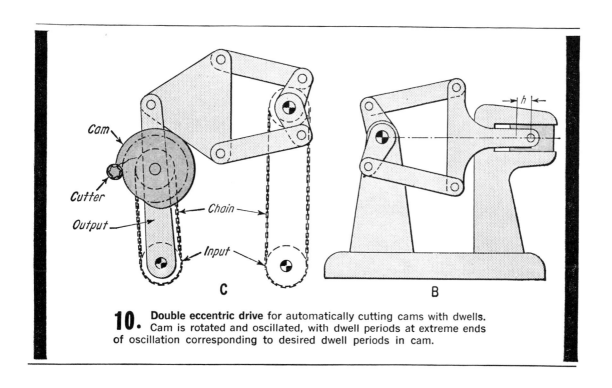

Cam

Cutter

Output

Chain

Input

C

h

B

10. **Double eccentric drive** for automatically cutting cams with dwells. Cam is rotated and oscillated, with dwell periods at extreme ends of oscillation corresponding to desired dwell periods in cam.

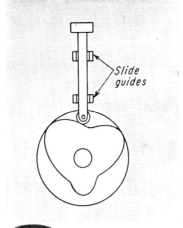

Slide guides

This assortment of devices reflects the variety of ways in which cams can be put to work.

F STRASSER

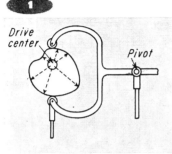

Drive center

Pivot

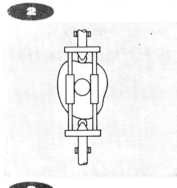

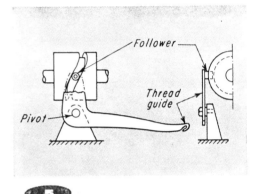

Follower

Pivot

Thread guide

Reciprocating cam

Circular cam

Automatic feed . . .
for automatic machines. There are two cams; one with circular motion, the other with reciprocating motion. This combination eliminates the trouble caused by irregularity of feeding and lack of positive control over stock feed.

Barrel cam . . .
has milled grooves; is used in, for example, sewing machines to guide thread. This type of cam is also used extensively in textile manufacturing machines such as looms and other intricate fabric-making devices.

Constant-speed rotary . . .
motion is converted (1) into a variable, reciprocating motion; into rocking or vibratory motion of simple forked follower (2); or more robust follower (3), which can provide valve-moving mechanism for steam engines. Vibratory-motion cams must be designed so that their opposite edges are everywhere equidistant when measured through drive-shaft center.

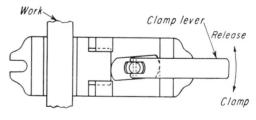

Work

Clamp lever

Release

Clamp

Double eccentric . . .
actuated by a suitable handle, provides powerful clamping action for machine-tool holding fixture.

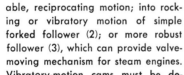

Index arm

Fixed weel

Follower

Cam

Planet

Carrier

Indexing mechanism . . .
combines epicyclic gear and cam. Planetary wheel and cam are fixed relative to one another; the carrier is rotated at uniform speed round the fixed wheel. Index arm has non-uniform motion, with dwell periods.

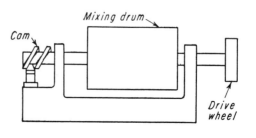

Mixing drum

Cam

Drive wheel

Mixing roller . . .
for paint or candy, etc. Mixing drum has small oscillating motion as well as rotation.

Mechanisms

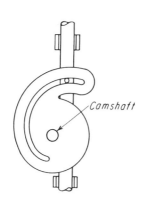

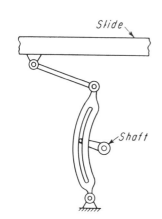

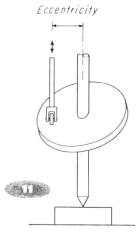

Slot cam . . .
onverts oscillating motion of camshaft to
ariable but straight-line motion of rod.
according to slot shape, rod motion can
e made to suit specific design require-
ents such as straight-line and logarithmic
otion.

Continous rotary motion . . .
of shaft is converted into reciprocating motion of a slide. Device
is used on certain types of sewing
machines and printing presses.

Swash-plate cams . . .
are feasible for light loads only,
such as in a pump. Eccentricity
produces resultant forces that
cause excessive loads. Multiple
followers may ride on plate,
thereby providing smooth pumping action for multipiston pump.

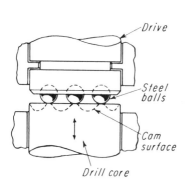

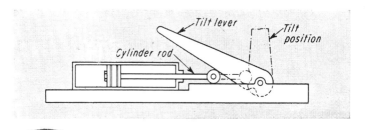

Tilting device . . .
can be designed so that lever remains in tilted position when cylinder
rod is withdrawn, or can be spring-loaded to return with cylinder rod.

teel-ball cam . . .
onverts high-speed rotary motion of, for
xample, an electric drill into high-fre-
uency vibrations that power the drill core
or use as a rotary hammer for masonry,
oncrete, etc. Attachment can be designed
o fit hand drills.

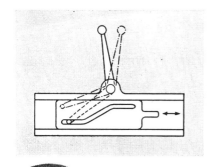

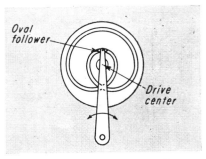

Sliding cam . . .
finds employment in remote control to
shift gears in position that is otherwise
inaccessible on machines.

Groove and oval . . .
follower provide device that requires two
revolutions of cam for one complete follower cycle.

207

Special-Function Cams

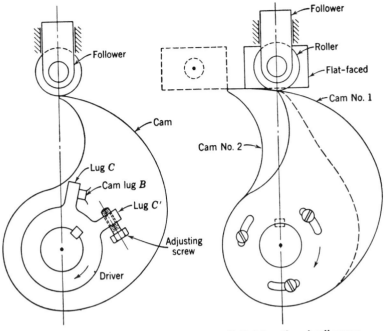

1. Quick-acting floating cam

2. Quick-acting dwell cams

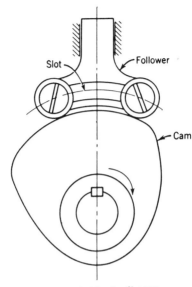

3. Adjustable-dwell cam

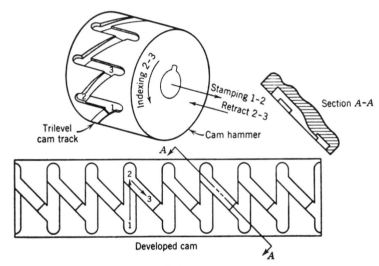

4. Indexing cam

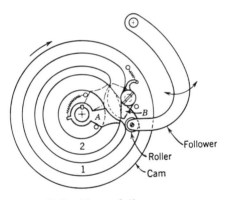

5. Double-revolution cam

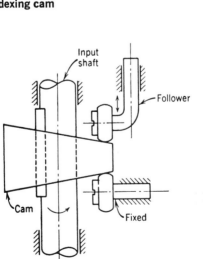

6. Increased-stroke barrel cam

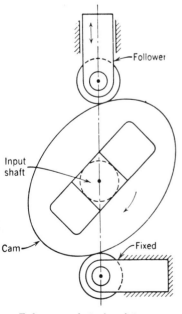

7. Increased-stroke plate cam

Adjustable-Dwell Cams

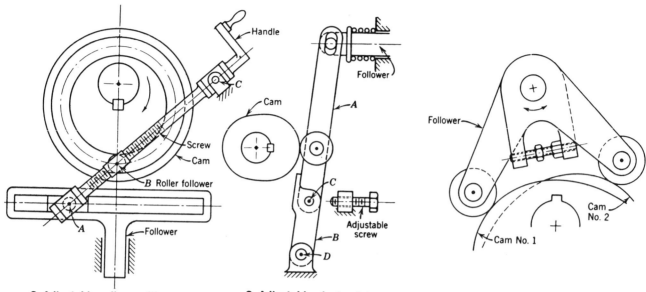

8. Adjustable roller-position cam

9. Adjustable pivot-point cam

Fig 1 — A quick drop of the follower is obtained by permitting the cam to be pushed out of the way by the follower itself as it reaches the edge of the cam. Lugs C and C' are fixed to the cam-shaft. The cam is free to turn (float) on the cam-shaft, limited by lug C and the adjusting screw. With the cam rotating clockwise, lug C drives the cam through lug B. At the position shown, the roller will drop off the edge of the cam which is then accelerated clockwise until its cam lug B hits the adjusting screw of lug C'.

Fig 2 — Instantaneous drop is obtained by the use of two integral cams and followers. The roller follower rides on cam 1. Continued rotation will transfer contact to the flat-faced follower which drops suddenly off the edge of cam 2. After the desired dwell, the follower is restored to its initial position by cam 1.

Fig 3 — Dwell period of cam can be varied by changing the distance between two rollers in the slot.

Fig 4 — Reciprocating pin (not shown) causes the barrel cam to rotate intermittently. Cam is stationary while pin moves from 1 to 2. Groove 2-3 is at a lower level; thus as the pin retracts it cams the barrel cam, then it climbs the incline from 2 to the new position of 1.

Fig 5 — Double groove cam makes two revolutions for one complete movement of the follower. Cam has movable switches, A and B, which direct the follower alternately in each groove. At the instant shown, B is ready to guide the roller follower from slot 1 to slot 2.

Fig 6 and 7 — Increased stroke is obtained by permitting the cam to shift on the input shaft. Total displacement of follower is therefore the sum of the cam

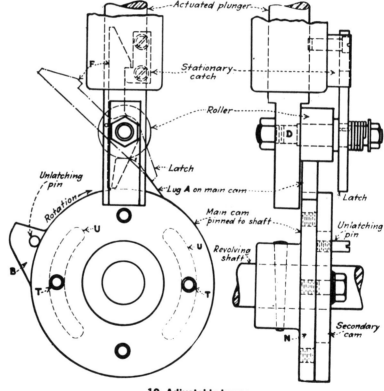

10. Adjustable lug cam

displacement on the fixed roller plus the follower displacement relative to the cam.

Fig 8 — Stroke of follower is adjusted by turning the screw handle which changes distance AB.

Fig 9 — Pivot point of connecting link to follower is changed from point D to point C by adjusting the screw.

Fig 10 — Adjustable dwell is obtained by having the main cam, with lug A, pinned to the revolving shaft. Lug A forces

the plunger up into the position shown and allows the latch to hook over the catch, thus holding the plunger in the up position. The plunger is unlatched by lug B. The circular slots in the cam plate permit shifting of lug B, thereby varying the time that the plunger is held in the latched position.

REFERENCE: Rothbart, H. A. *Cams—Design, Dynamics, and Accuracy*, John Wiley and Sons, Inc., New York.

Roll-Cam Devices

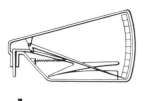 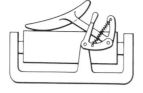  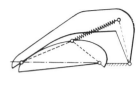

1

Sensitive contact gage uses a rocking pair to decrease the effect of friction and increase the accuracy.

Variable electrical resistor has a rocking surface instead of a sliding brush to reduce wear with smooth operation.

Typewriter linkage has a rocking pair that actuates upper or lower case letters with smoother, quieter action.

Rocking mechanism derived from 4-bar linkage has constant spring length which transmits no force to bearing.

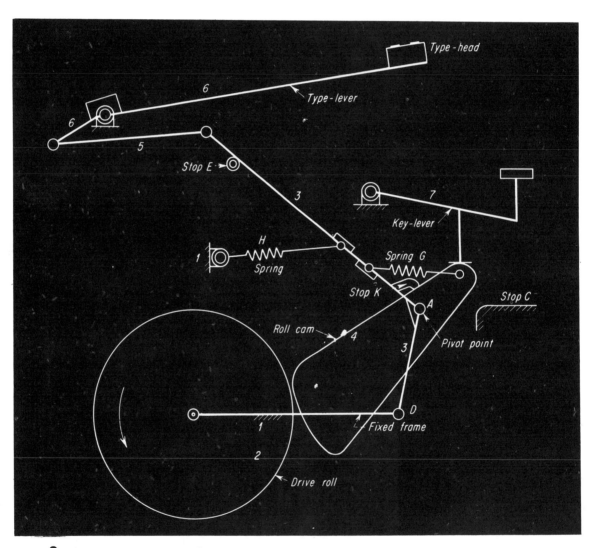

2 Electric-typewriter mechanism . . .
uses roll cam for motion amplification. Here, path of pivot point on roll cam is curvilinear.

Roll cams are also employed in IBM electric typewriters, Fig. 2. Here the cam is triggered by a touch of the typist's finger to power the type heads.

The roll is driven by an electric motor at constant speed. Cycle begins when the typist depresses the key lever which rotates the cam into contact with the drive roll. The cam is connected to link 3 at pin point A. Rotation of drive roll makes link 3 rotate clockwise, causing link 6 to rotate (via link 5) until the type head contacts the platen (not shown).

At end of the cycle, the cam strikes stop C and loses contact with drive roll. Spring G then returns the cam to its position against the stop K.

During this time, while cam and drive are disengaged, the type head continues to approach the platen because of kinetic energy stored in type lever (link 6). After type head strikes platen, spring H returns the linkage to the home position where link 3 contacts stop E.

How a Roll Cam Works   A C DUNK

The roll-cam mechanism shown in Fig. 1 is for a rectilinear type—its job in an RCA 45-rpm record changer is to control tone-arm motion and record changing. Basic components are:

- **Drive roll**—rotates at constant speed about fixed pivot and provides input power.
- **Roll cam**—its cutout allows it to dwell until activated.
- **Slide rod**—is reciprocated by rotation of the roll cam.
- **Frame**—provides a fixed pivot and slot for the rod.

The sketch shows the mechanism in the dwell position --cam and drive not in contact. To start the cycle, finger pressure is employed momentarily (through a lever not shown) to slightly rotate the cam until it comes into contact with the drive roll (which an electric motor is rotating at constant speed). Friction between the surface then rotates the cam, in turn causing the rod to move upward. The cam goes through a complete cycle and returns down to its original position. Here the stop on the frame will arrest the cam, which will again be out of engagement until activated again.

Spring pressure prevents slippage during rotation—especially during the return cycle. However, to make the transfer of motion more positive, one member can be made of rubber and the other of metal with small serrations on its periphery. To prevent any slip whatsoever, conjugate tooth forms could be used.

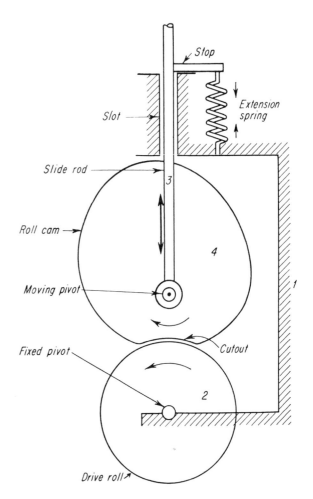

3 Record changer . . .
has this roll-cam mechanism. Path of movable pivot is rectilinear. Cam is triggered by rotating cam slightly until it contacts drive.

Cam Drives for Machine Tools

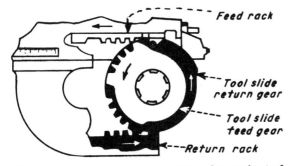

Two-directional rack-and-gear drive for main tool slide combines accuracy, uniform movement and minimum idle time. Mechanism makes a full double stroke per cycle. It approaches fast, shifts smoothly into feed and returns fast. Point of shift is controlled by an adjustable dog on a calibrated gear. Automatic braking action assures smooth shift from approach to feed. Used on Greenlee 4-Spindle Bar Automatics.

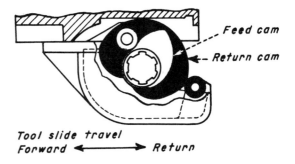

Cam drive for tool slide mechanism is used instead of rack feed when short stroke is required to get fast machining cycle on Greenlee Automatic. Sketch shows cams and rollers with slide in retracted position.

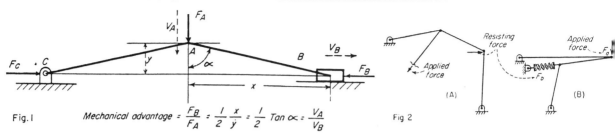

Fig.1   Mechanical advantage $= \dfrac{F_B}{F_A} = \dfrac{1}{2}\dfrac{x}{y} = \dfrac{1}{2}\, Tan\, \alpha = \dfrac{V_A}{V_B}$ Fig 2

MANY MECHANICAL LINKAGES are based on the simple toggle which consists of two links that tend to line-up in a straight line at one point in their motion. The mechanical advantage is the velocity ratio of the input point A to the outpoint point B: or V_A/V_B. As the angle α approaches 90 deg, the links come into toggle and the mechanical advantage and velocity ratio both approach infinity. However, frictional effects reduce the forces to much less than infinity although still quite high.

FORCES CAN BE APPLIED through other link and need not be perpendicular to each other. (A One toggle link can be attached to another li rather than to a fixed point or slider. (B) Tw toggle links can come into toggle by lining up top of each other rather than as an extension each other. Resisting force can be a spring for

HIGH MECHANICAL ADVANTAGE

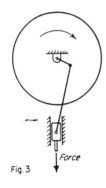

IN PUNCH PRESSES, large forces are needed at the lower end of the work-stroke, however little force is required during the remainder. Crank and connecting rod come into toggle at the lower end of the punch stroke, giving a high mechcanical advantage at exactly the time it is most needed.

Fig.3

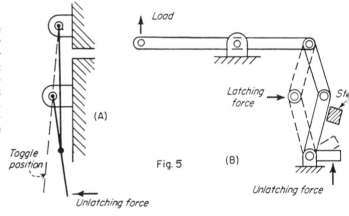

Fig. 5

LOCKING LATCHES produce a high mechanical advantage when in the toggle portion of the stroke. (A) Simple latch exerts a large force in the locked position. (B) For positive locking, closed position of latch is slightly beyond toggle position. Small unlatching force opens linkage.

COLD-HEADING RIVET MACHINE is designed to give each rivet two successive blows. Following the first blow (point 2) the hammer moves upward a short distance (to point 3). Following the second blow (at point 4), the hammer then moves upward a longer distance (to point 1) to provide clearance for moving the workpiece. Both strokes are produced by one revolution of the crank and at the lowest point of each stroke (points 2 and 4) the links are in toggle.

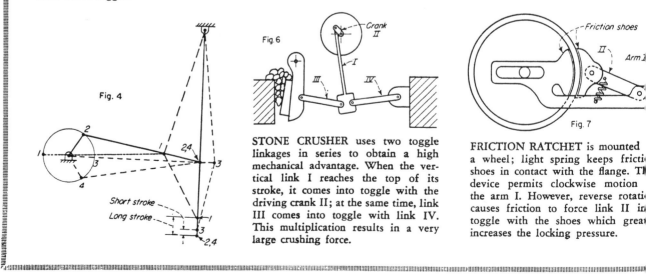

Fig. 4

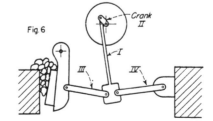

STONE CRUSHER uses two toggle linkages in series to obtain a high mechanical advantage. When the vertical link I reaches the top of its stroke, it comes into toggle with the driving crank II; at the same time, link III comes into toggle with link IV. This multiplication results in a very large crushing force.

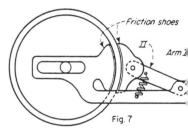

Fig. 7

FRICTION RATCHET is mounted a wheel; light spring keeps fricti shoes in contact with the flange. Th device permits clockwise motion the arm I. However, reverse rotati causes friction to force link II in toggle with the shoes which grea increases the locking pressure.

Different Mechanisms
THOMAS P. GOODMAN

HIGH VELOCITY RATIO

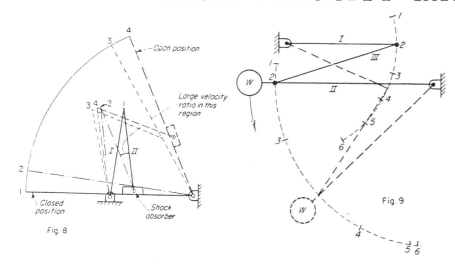

Fig. 8

Fig. 9

DOOR CHECK LINKAGE gives a high velocity ratio at one point in the stroke. As the door swings closed, connecting link I comes into toggle with the shock absorber arm II, giving it a large angular velocity. Thus, the shock absorber is more effective retarding motion near the closed position.

IMPACT REDUCER used on some large circuit breakers. Crank I rotates at constant velocity while lower crank moves slowly at the beginning and end of the stroke. It moves rapidly at the mid stroke when arm II and link III are in toggle. Accelerated weight absorbs energy and returns it to the system when it slows down.

VARIABLE MECHANICAL ADVANTAGE

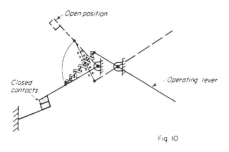

Fig 10

TOASTER SWITCH uses an increasing mechanical advantage to aid in compressing a spring. In the closed position, spring holds contacts closed and the operating lever in the down position. As the lever is moved upward, the spring is compressed and comes into toggle with both the contact arm and the lever. Little effort is required to move the links through the toggle position; beyond this point, the spring snaps the contacts open. A similar action occurs on closing.

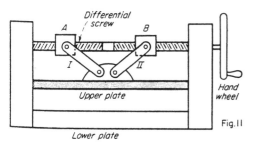

Fig. 11

TOGGLE PRESS has an increasing mechanical advantage to counteract the resistance of the material being compressed. Rotating handwheel with differential screw moves nuts A and B together and links I and II are brought into toggle.

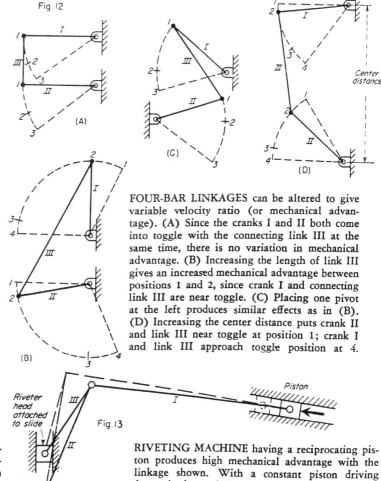

Fig. 12

Fig. 13

FOUR-BAR LINKAGES can be altered to give variable velocity ratio (or mechanical advantage). (A) Since the cranks I and II both come into toggle with the connecting link III at the same time, there is no variation in mechanical advantage. (B) Increasing the length of link III gives an increased mechanical advantage between positions 1 and 2, since crank I and connecting link III are near toggle. (C) Placing one pivot at the left produces similar effects as in (B). (D) Increasing the center distance puts crank II and link III near toggle at position 1; crank I and link III approach toggle position at 4.

RIVETING MACHINE having a reciprocating piston produces high mechanical advantage with the linkage shown. With a constant piston driving force, the force of the head increases to a maximum value when links II and III come into toggle.

16 Latch, Toggle and

Diagrams of basic latching and quick-release mechanisms.

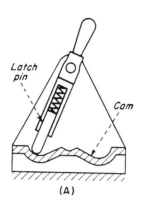

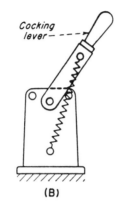

Fig. 1—Cam-guided latch (A) has one cocked, two relaxed positions. (B) Simple over-center toggle action. (C) Over-center toggle with slotted link. (D) Double toggle action is often used in electrical switches.

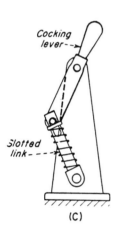

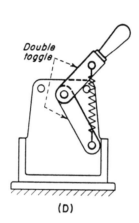

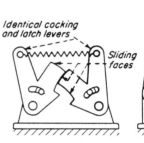

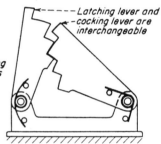

Fig. 2—Identically shaped cocking lever and latch (A) allow functions to be interchangeable. Radii of sliding faces must be dimensioned for mating fit. Stepped latch (B) has several locking positions.

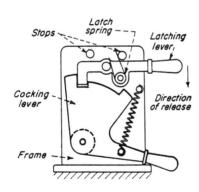

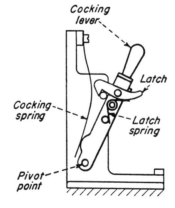

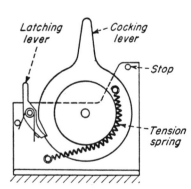

Fig. 3—Latch and cocking lever spring-loaded so latch movement releases cocking lever. Cocked position can be held indefinitely. Studs in frame provide stops, pivots or mounts for springs.

Fig. 4—Latch mounted on cocking lever allows both levers to be reached at same time with one hand. After release, cocking spring initiates clockwise lever movement, then gravity takes over.

Fig. 5—Disk-shaped cocking lever has tension spring resting against cylindrical hub. Spring force thus alway acts at constant radius from lever pivot point.

Trigger Devices

SIGMUND RAPPAPORT

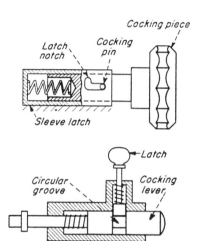

Fig. 7—Geared cocking device has rachet fixed to pinion. Torsion spring exerts clockwise force on spur gear; tension spring holds gear in mesh. Device wound by turning ratchet handle counter-clockwise which in turn winds torsion spring. Moving release-lever permits spur gear to unwind to original position without affecting ratchet handle.

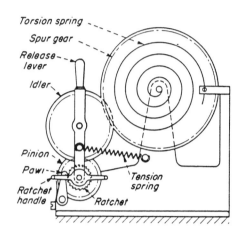

Fig. 6—Sleeve latch (A) has L-shaped notch. Pin in shaft rides in notch. Cocking requires simple push and twist action. (B) Latch and plunger use axial movement for setting and release. Circular groove needed if plunger can rotate.

Fig. 8—In over-center lock (A) clockwise movement of latching lever cocks and locks slide. Release requires counter-clockwise movement. (B) Latching-cam cocks and releases cocking lever with same counter-clockwise movement.

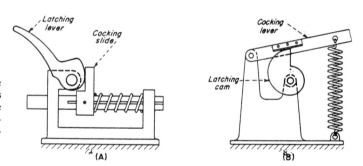

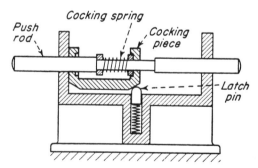

Fig. 9—Spring-loaded cocking piece has chamfered corners. Axial movement of push-rod forces cocking piece against spring-loaded ball or pin set in frame. When cocking builds up enough force to overcome latch-spring, cocking piece snaps over to right. Action can be repeated in either direction.

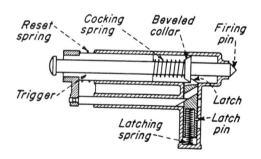

Fig. 10—Firing-pin type mechanism has bevelled collar on pin. Pressure on trigger forces latch down until it releases collar, when pin snaps out under force of cocking spring. Reset spring pulls trigger and pin back. Latch is forced down by bevelled collar on pin until it snaps back after overcoming force of latch spring. (Latch pin retains latch if trigger and firing pin are removed.)

Data based on material and sketches in AWF und VDMA Getrieblaetter, published by Ausschuss fuer Getriebe beim Ausssbuss fuer Wirtschaftiche Fertigung, Leipzig, Germany.

Diagrams show the basic ways to produce mechanical snap action.

PETER C NOY

Snap action results when a force is applied to a device over a period of time; buildu... of this force to a critical level causes a sudden motion to occur. The ideal snap devic... would have no motion until the force reached critical level. This, however, is not poss... ble, and the way in which the mechanism approaches this ideal is a measure of its eff... ciency as a snap device. Some of the designs shown here approach closely to the idea... others less so, but may nevertheless have other good features.

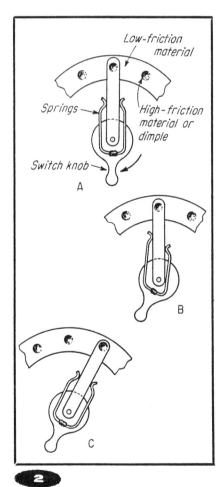

2

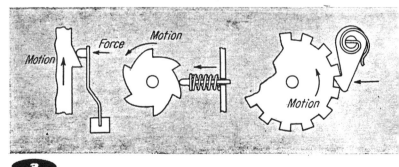

1

Dished disk . . .
is a simple, common method of producing snap action. Snap leaf made fro... spring material may have various-shaped impressions stamped at the point whe... overcentering action occurs. Frog clacker is, of course, a typical application. ... bimetal made this way will reverse itself at a predetermined temperature.

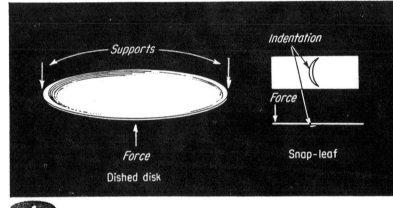

3

Friction override . . .
may be used to hold against an increasing load until friction is suddenly overcome. This is a useful action for small sensitive devices where large forces and movements are undesirable. It is interesting to note that this is the way we snap our fingers, and thus is probably the original snap mechanism. Moisture affects this action.

Ratchet-and-pawl . . .
principle is perhaps the most widely used type of snap mechanism. Its ma... variations are an essential feature in practically every complicated mechanic... device. By definition, however, this movement is not true snap-action.

MECHANISMS

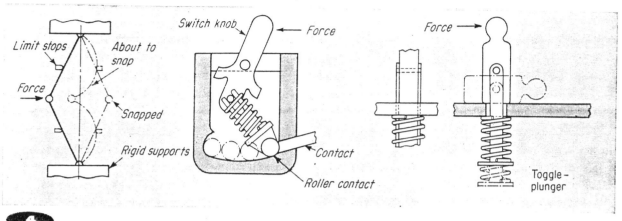

4

Over-centering . . .

devices find their greatest application in electrical switches. Considerable design ingenuity has been used to fit this principle into many different mechanisms. In the final analysis, over-centering is actually the basis of most snap-action devices.

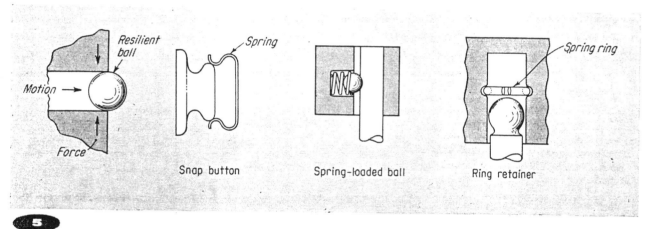

5

Sphere ejection . . .

principle employs snap buttons, spring-loaded balls and catches, and retaining-rings for fastening that will have to be used repeatedly. Action can be adjusted in design to provide either easy or difficult removal. Wear may change force required.

6

Pneumatic dump valve . . .

produces snap action by preventing piston movement until air pressure has built up in front end of cylinder to a relatively high pressure. Dump-valve area in low-pressure end is six times larger than its area on high-pressure side, thus the pressure required on the high-pressure side to dislodge the dump valve from its seat is six times that required on the low-pressure side to keep the valve seated.

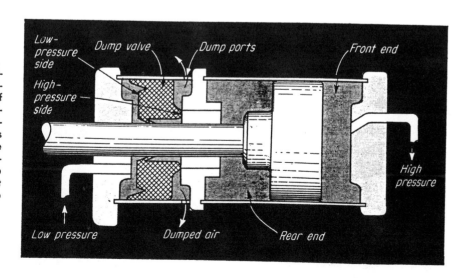

A further selection of basic arrangements for

obtaining sudden motion after gradual buildup of for

PETER C NOY

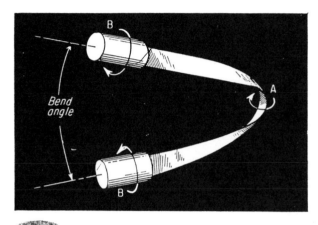

1

Torsion ribbon . . .
bent as shown will turn "inside out" at A with a snap action when twisted at B. Design factors are ribbon width and thickness, and bend angle.

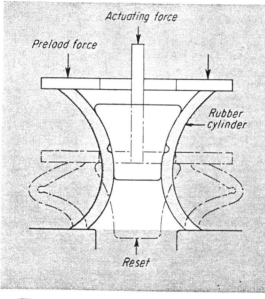

2

Collapsing cylinder . . .
has elastic walls that may be deformed gradually u
their stress changes from compressive to bending with
resultant collapse of the cylinder

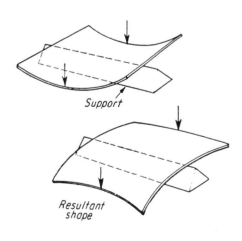

3

Bowed spring . . .
will collapse into new shape when loaded as shown. "Push-pull" type of steel measuring tape illustrates this action; the curved material stiffens the tape so that it can be held out as a cantilever until excessive weight causes it to collapse suddenly.

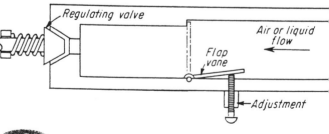

4

Flap vane . . .
is for air or liquid flow cutoff at a limiting velocity. With a regulat
valve, vane will snap shut (because of increased velocity) when press
is reduced below a certain value.

DEVICES

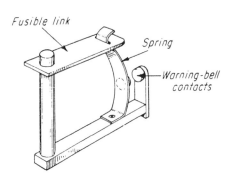

Fusible link

Spring

Warning-bell contacts

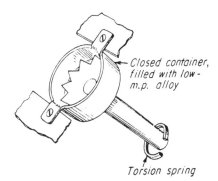

Closed container, filled with low-m.p. alloy

Torsion spring

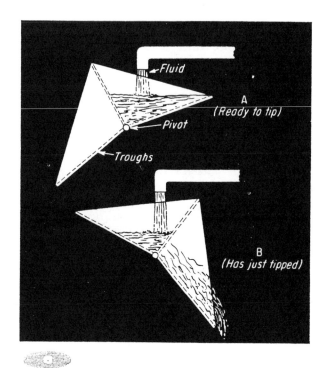

Fluid

A (Ready to tip)

Pivot

Troughs

B (Has just tipped)

acrificing link . . .
s used generally where high heat or chem-
cally corrosive conditions would be hazardous
—if temperature becomes too high, or atmos-
here too corrosive, link will yield at whatever
conditions it is designed for. Usually the device
s required to act only once, although a device
ike the lower one is quickly reset but restricted
o temperature control.

Gravity-tips . . .
although slower acting than most snap mechanisms, can
be called snap mechanisms because they require an accumulation of energy to trigger an automatic release. Tipping-
troughs used to spread sewage exemplify arrangement shown
in A, once overbalanced, action is fast.

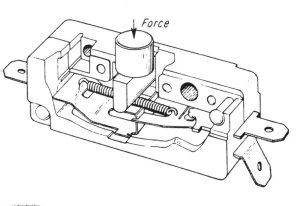

Force

A

B

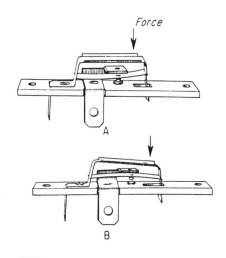

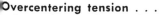

Overcentering tension . . .
pring combined with pivoted contact-strip is o:e arrangement
mong many similar ones used in switches. Arrangement shown
ere is somewhat unusual, since the actuating force bears on
he spring itself.

Overcentering leaf-spring . . .
action is also the basis for many ingenious
snap-action switches used for electrical control.
Sometimes spring action is combined with the
thermostatic action of a bimetal strip to make
the switch respond to heat or cold either for control purposes or as a safety feature.

Applications of Differential

Known for its mechanical advantage, the differential winch is a control mechanism
that can supplement the gear and rack and four-bar linkage systems in changing

ALEXANDER B. HULSE, JR. and ROBERT AYMAR

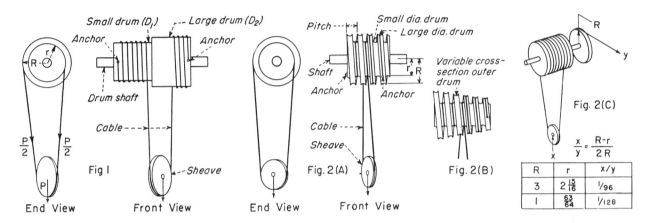

FIG. 1—Standard Differential Winch; consists of two drums D_1 and D_2 and a cable or chain which is anchored on both ends and wound clockwise around one drum and counterclockwise around the other. The cable supports a load carrying sheave and if the shaft is rotated clockwise, the cable, which unwinds from D_1 on to D_2, will raise the sheave a distance

$$\text{Sheave rise/rev} = \frac{2\pi R - 2\pi r}{2} = \pi(R - r)$$

The winch, which is not in equilibrium, exerts a counterclockwise torque.

$$\text{Unbalanced torque} = \frac{P}{2}(R - r)$$

FIG. 2(A)—Hulse Differential Winch*. Two drums, which are in the form of worm threads contoured to guide the cables, concentrically occupy the same longitudinal space. This keeps the cables approximately at right angles to the shaft and eliminates cable shifting and rubbing especially when used with variable cross sections as in Fig. 2(B) where any equation of motion can be satisfied by choosing suitable cross sections for the drums.

Ways for resisting or supporting the axial thrust may have to be considered in some installations. Fig. 2(C) shows typical reductions in displacement.

*Pat. No. 2,590,623

R	r	x/y
3	$2\frac{15}{16}$	$1/96$
1	$\frac{63}{64}$	$1/128$

$$\frac{x}{y} = \frac{R-r}{2R}$$

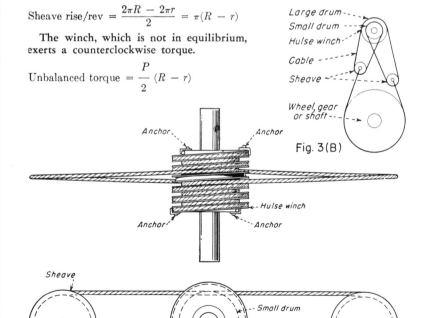

Fig. 3(B)

Fig. 3(A)

FIG. 3(A)—Hulse Winch with Opposing Sheaves. This arrangement which uses two separate cables and four anchor points can be considered as two winches back-to-back using one common set of drums. Variations in motion can be obtained by: (1) restraining the sheaves so that when the system is rotated the drums will travel toward one of the sheaves; (2) restraining the drums and allowing the sheaves to travel. The distance between the sheaves will remain constant and are usually connected by a bar; (3) permitting the drums to move axially while restraining them transversely. When the system is rotated, drums will travel axially one pitch per revolution, and sheaves remain in same plane perpendicular to drum axis. This

Winch to Control Systems

rotary motion into linear. It can magnify displacement to meet the needs of delicate instruments or be varied almost at will to fulfill uncommon equations of motion.

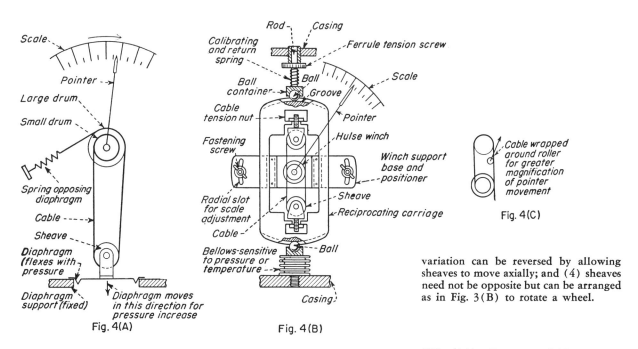

Fig. 4(A)

Fig. 4(B)

Fig. 4(C)

variation can be reversed by allowing sheaves to move axially; and (4) sheaves need not be opposite but can be arranged as in Fig. 3(B) to rotate a wheel.

FIG. 4(A)—Pressure and Temperature Indicators. A pressure change causes the diaphragm and sheave to move vertically and the pointer radially. Equilibrium occurs when spring force balances actuating torque. Replacing diaphragm with a thermal element changes instrument into a temperature indicator. Two sheaves and a reciprocating carriage, Fig. 4(B), are based on the principle shown in Fig. 3(A). Carriage is activated by pressure or temperature and is balanced by a spring force in opposite end. Further magnification can be obtained, Fig. 4(C), by wrapping cable around a roller to which pointer is attached.

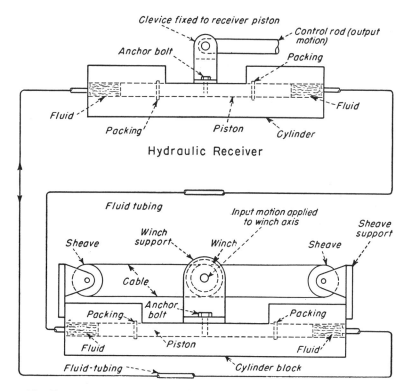

Hydraulic Receiver

Fig. 5   Hydraulic Sender

FIG. 5—Hydraulic Control System Actuated by a Differential Winch. Used for remote precision positioning of a control rod with a minimum of applied torque. The sending piston, retained in a cylinder block, reciprocates back and forth by a torque applied to the winch shaft. Fluid is forced out from one end of the cylinder through the pipe lines to displace the receiving piston which in turn activates a control rod. The receiver simultaneously displaces a similar amount of fluid from the opposite end back to the sender. By suitable valving the sender may be used as a double-acting pump.

6 APPLICATIONS FOR THE CAPSTAN-TYPE POWER AMPLIFIER

Precise positioning and movement of heavy loads are two basic jobs for this all-mechanical torque booster.

L. A. ZAHORSKY

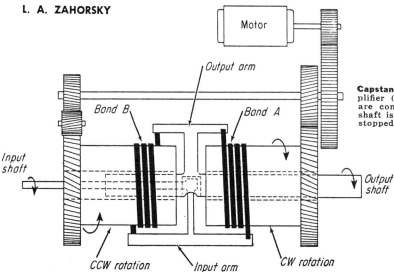

Capstan principle (above) is basis for mechanical power amplifier (below) that combines two counterrotating drums. Drums are continuously rotating but only transmit torque when input shaft is rotated to tighten band on drum A. Overrun of output is stopped by drum B, when overrun tightens band on this drum.

CAPSTAN is simple mechanical amplifier—rope wound on motor-driven drum slips until slack is taken up on the free end. Force needed on free end to lift the load depends on the coefficient of friction and number of turns of rope.

By connecting bands A and B to an input shaft and arm, the power amplifier provides an output in both directions, plus accurate angular positioning. When the input shaft is turned clockwise, the input arm takes up the slack on band A, locking it to its drum. Inasmuch as the load end of locked band A is connected to the output arm, it transmits the CW motion of the driven drum on which it is wound, to the output shaft. Band B therefore slacks off and slips on its drum. When the CW motion of the input shaft stops, tension on band A is released and it slips on its own drum. If the output shaft tries to overrun, the output arm will apply tension to band B, causing it to tighten on the CCW rotating drum and stop the shaft.

From: Control Engineering

This mechanical power amplifier has a fast response. Power from its continuously rotating drums is instantaneously available. When used for position-control applications, such methods as pneumatic, hydraulic, and electrical systems—even with continuously running power sources—require transducers of some kind to change signals from one energy form to another. The mechanical power amplifier, on the other hand, permits direct sensing of the controlled motion.

Four major advantages of this all-mechanical device are:

1. Kinetic energy of the power source is continuously available for rapid response.

2. Motion can be duplicated and power amplified without converting energy forms.

3. Position and rate feedback are inherent design characteristics.

4. Zero slip between input and output eliminates the possibility of cumulative error.

One other important advantage is the ease with which this device can be adapted to perform special functions—jobs for which other types of systems would require the addition of more costly and perhaps less reliable components. The six applications which follow illustrate how these advantages have been put to work in solving widely divergent problems.

1. Nonlinear broaching

Problem: In broaching large bore rifles, the twist given to the lands and grooves represent a nonlinear function of barrel length. Development work on such rifles usually requires some experimentation with this function. At present, rotation of the broaching head is performed by a purely mechanical arrangement consisting of a long heavy wedge-type cam and appropriate gearing. For steep twist angles, however, the forces acting on this mechanism becomes extremely high.

Solution: A suitable mechanical power amplifier, with its inherent position feedback, was added to the existing mechanical arrangement, as shown in sketch. The cam and follower, instead of having to drive the broaching head, simply furnish enough torque to position the input shaft of the amplifier.

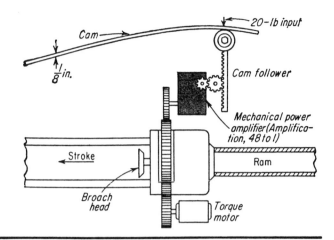

2. Hydraulic winch control

Problem: Hydraulic pump-motor systems represent an excellent method of controlling position and motion at high power levels. In the 10- to 150-hp range, for example, the usual approach is to vary the output of a positive displacement pump in a closed-loop hydraulic circuit. In many of the systems that might be used to control this displacement, however, a force feedback proportional to system pressure can lead to serious errors or even oscillations.

Solution: Sketch shows an external view of the complete package. The output shaft of the mechanical power amplifier controls pump displacement, while its input is controlled by hand. In a more recent development, requiring remote manual control, a servomotor replaces this local handwheel. Approximately 10 lb-in. torque drives a 600 lb-in. load. If this system had to transmit 600 lb-in. the equipment would be more expensive and more dangerous.

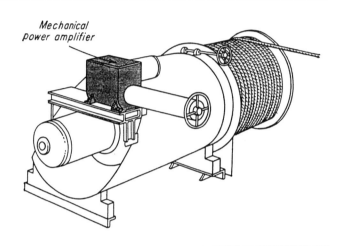

3. Load positioning

Problem: It was necessary for a 750-lb load to be accelerated from standstill in 0.5 sec and brought into speed and position synchronization with a reference linear motion. It was also necessary that the source of control motion be permitted to accelerate more rapidly than the load itself. Torque applied to the load could not be limited by means of a slipping device.

Solution: A system using a single mechanical power amplifier provided the solution, sketch. Here a mechanical memory device preloaded for either rotation is used to drive the input shaft of the amplifier. This permits the input source to accelerate as rapidly as desired. Total control input travel minus the input travel of the amplifier shaft is temporarily stored. After 0.5 sec the load reaches proper speed, and the memory device transmits position information in exact synchronization with the input.

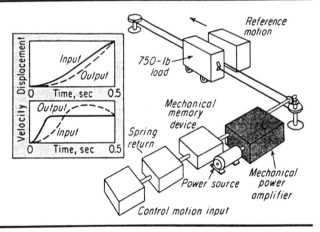

4. Tensile testing machine

Problem: On a hydraulic tensile testing machine, stroke of the power cylinder had to be controlled as a function of two variables: tension in, and extension of, the test specimen. A programming device designed to provide a control signal proportional to these variables, had an output power level of about 0.001 hp—too low to drive the pressure regulator controlling the flow to the cylinder.

Solution: An analysis of the problem revealed three requirements: output of the programmer had to be amplified about 60 times, position accuracy had to be within 2 deg, and acceleration had to be held at a very low value. A mechanical power amplifier satisfied all three requirements. Sketch illustrates the completed system. Its design is based principally on steady-state characteristics.

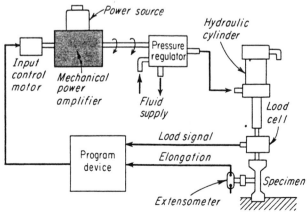

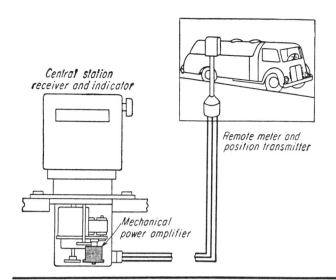

Central station
receiver and indicator

Remote meter and
position transmitter

Mechanical
power amplifier

5. Remote metering & counting

Problem: For a remote liquid metering job, synchro systems had been used to transmit remote meter readings to a central station and repeat this information on local indicating counters. The operation involved a large number of meters and indicators. As new devices were added (ticket printers, for instance), the torque requirement also grew.

Solution: Use of mechanical power amplifiers in the central station indicators not only supplied extra output torque but also made it possible to use synchros even smaller than those originally selected to drive the indicators alone.

The synchro transmitters currently used operate at a maximum speed of 600 rpm and produce only about 3 oz-in. of torque. The mechanical power amplifiers furnish up to 100 lb-in. and are designed to fit in the bottom of the registers as shown in sketch. Total accuracy is within 0.25 gallon, error is noncumulative.

6. Irregular routing

Problem: To remotely control the table position of a routing machine from information stored on a film strip. The servoloop developed to interpret this information produced only about 1 oz-in. of torque. About 20 lb-ft was required at the table feedscrew.

Solution: Sketch shows how a mechanical power amplifier supplied the necessary torque at the remote table location. A position transmitter converts the rotary motion output of the servoloop to a proportional electrical signal and sends it to a differential amplifier at the machine location. A position receiver, geared to the output shaft, provides a signal proportional to table position. The differential amplifier compares these, amplifies the difference, and sends a signal to either counterrotating electromagnetic clutch, which drives the input shaft of the mechanical power amplifier.

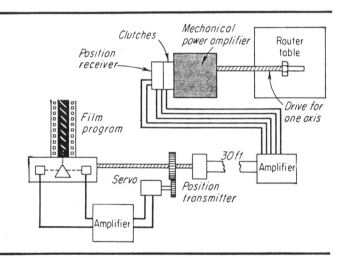

Clutches
Mechanical power amplifier
Router table
Position receiver
Film program
Drive for one axis
30 ft
Amplifier
Servo
Position transmitter
Amplifier

Mechanical power amplifier . . .

that drives crossfeed slide uses the principle of the windlass. By varying the control force, all or any part of power to the drum can be utilized.

Two drums mounted back to back supply the bi-directional power needed in servo systems. Replacing the operator with a 2-phase induction servomotor permits using electronic or magnetic signal amplification. A rotating input avoids linear input and output of the simple windlass. Control and output ends of the multi-turn bands are both connected to gears mounted concentrically with the drum axis.

When servomotor rotates the control gear it locks the band-drum combination, forcing output gear to rotate with it. Clockwise rotation of the servomotor produces CW power output while second drum idles. Varying the servo speed, by changing servo voltage, varies output speed.

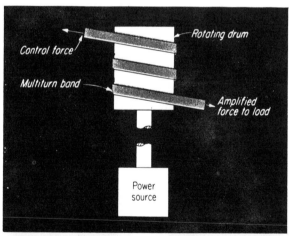

Control force
Rotating drum
Multiturn band
Amplified force to load
Power source

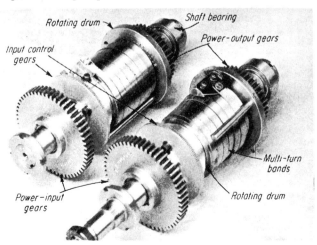

Rotating drum
Shaft bearing
Power-output gears
Input control gears
Multi-turn bands
Power-input gears
Rotating drum

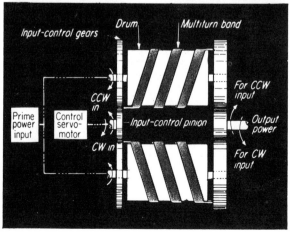

Input-control gears
Drum
Multiturn band
CCW in
Input-control pinion
For CCW input
Prime power input
Control servo-motor
Output power
CW in
For CW input

Guide to variable-speed belt and chain drives   D. Z. DVORAK

VARIABLE-SPEED drives provide an infinite number of speed ratios within a specific range. They differ from the stepped-pulley drives in that the stepped drives offer only a discrete number of velocity ratios.

A second article in this series (see the article on p. 244) covers mechanical "all-metal" drives. For the most part, these employ friction or pre-loaded cones, disks, rings, and spheres, which do involve a certain amount of slippage. Belt drives, on the other hand, have little slippage or frictional losses, and chain has none—it maintains a fixed phase relationship between the input and output shafts.

Belt drives

Belt drives have high efficiency and are relatively low in price. Most use V-belts, reinforced by steel wires, of up to 3-in. width.

Speed adjustment in belt drives is obtained through one of the four basic arrangements shown below.

Variable-distance system (Fig 1).

A variable-pitch sheave on the input shaft opposes a solid (fixed-pitch) sheave on the output shaft. To vary the speed, the center distance is varied, usually by means of an adjustable base motor of the tilting or sliding types (Fig 6).

Speed variations up to 4:1 are easily achieved, but torque and horsepower characteristics depend on the location of the variable-diameter sheave.

Fixed-distance system (Fig 2). Variable-pitch sheaves on both input and output shafts maintain a constant center distance between shafts. The sheaves are controlled by linkage. Either the pitch diameter of one sheave is positively controlled and the disks of the other sheave, being under spring tension, adjust automatically, or the pitch diameters of both sheaves are positively controlled by the linkage system (Fig 5). Pratt & Whitney has applied the system in Fig 5 to the spindle drive of numerically controlled machines.

Speed variations up to 11:1 are obtained, which means that with a 1200-rpm motor, the maximum output speed will be $1200 (11)^{\frac{1}{2}} = 3984$ rpm, and the minimum output speed $= 3984/11 = 362$ rpm.

Double-reduction system (Fig 3). Solid sheaves are on both the input and output shafts; but both sheaves on the intermediate shaft are of variable-pitch type. Center distance between input and output is constant.

Coaxial shaft system (Fig 4). The intermediate shaft in this arrangement permits the output shaft to be coaxial with that of the input shaft. To maintain a fixed center distance, all four sheaves must be of the variable-pitch type and controlled by linkage, similar to the system in Fig 6. Speed variation up to 16:1 is available.

Packaged belt units (Fig 7). These combine the motor and variable-pitch transmissions as an integral unit. The belts are usually ribbed, and speed ratios can be dialed by a handle.

text continued, next page

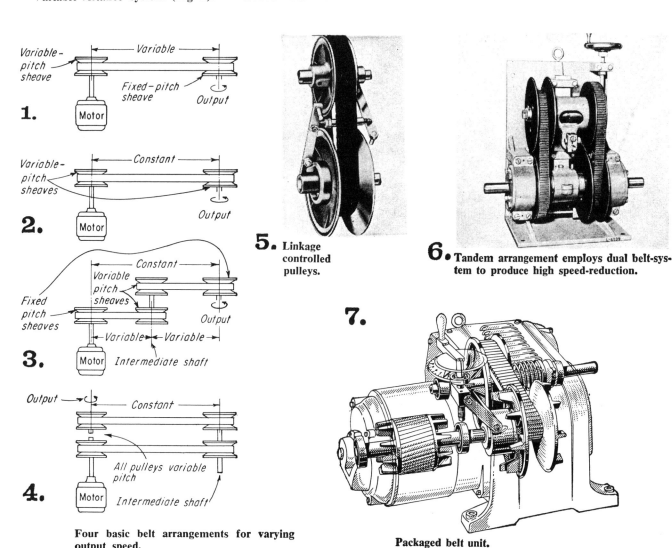

1. Variable-pitch sheave / Variable / Fixed-pitch sheave / Output / Motor

2. Variable-pitch sheaves / Constant / Output / Motor

3. Fixed pitch sheaves / Constant / Variable pitch sheaves / Output / Variable — Variable / Motor / Intermediate shaft

4. Output / Constant / All pulleys variable pitch / Motor / Intermediate shaft

5. Linkage controlled pulleys.

6. Tandem arrangement employs dual belt-system to produce high speed-reduction.

7.

Four basic belt arrangements for varying output speed.

Packaged belt unit.

Sheave designs

The axial shifting of variable-pitch sheaves is controlled by one of four methods:

Linkage actuation. The sheave assemblies in Fig 5 are directly controlled by linkages which, in turn, are manually adjusted.

Spring pressure. The cones of the sheaves in Fig 2 and 4 are axially loaded by spring force. A typical pulley of this type is illustrated in Fig 8. Such pulleys are used in conjunction with directly controlled sheaves, or with variable center-distance arrangements.

Cam-controlled sheave. The cones of this sheave (Fig 9) are mounted on a floating sheave free to rotate on the pulley spindle. Belt force rotates the cones whose surfaces are cammed by the incline plane of the spring. The camming action wedges the cones against the belt, thus providing sufficient pressure to prevent slippage at the higher speeds, as shown in the curve.

Centrifugal-force actuator. In this unique sheave arrangement (Fig 10) the pitch diameter of driving sheave is controlled by the centrifugal force of steel balls. Another variable-pitch pulley mounted on the driven shaft is responsive to the torque. As the drive speed increases, the centrifugal force of balls forces the sides of the driving sheave together. With a change in load, the movable flange of the driven sheave rotates in relation to the fixed flange. The differential rotation of the sheave flanges cams them together and forces the V-belt to the outer edge of the driven sheave, which is a lower transmission ratio. The driving sheave is also shifted as the load rises with decreasing speed. With a stall load, it is moved to the idling position. When the torque responsive sheave is the driving member, any increase in drive speed closes its flanges and opens the flanges of the centrifugal member, thus maintaining a constant output speed.

The drive has been employed in transmissions ranging from 2 to 12 hp.

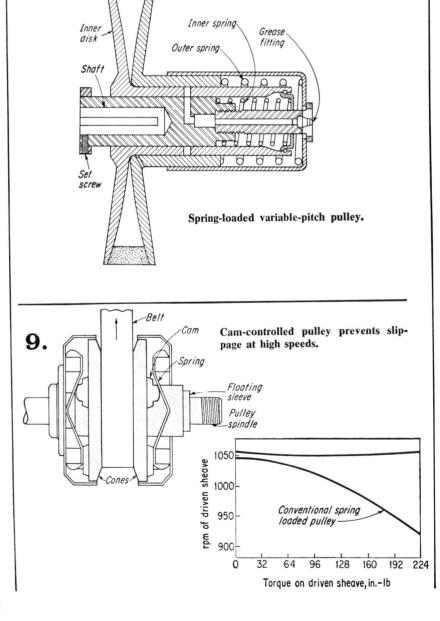

8.

Spring-loaded variable-pitch pulley.

9. Cam-controlled pulley prevents slippage at high speeds.

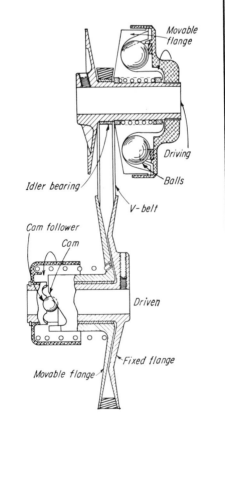

10.

Ball-controlled pulley has sides pressured by centrifugal force. Morse Chain Co.

Chain drives

PIV drive (Fig 11). This chain drive (positive, infinitely variable) eliminates any slippage between the self-forming laminated chain teeth and the chain sheaves. The individual laminations are free to slide laterally so as to take up the full width of the sheave. The chain runs in radially grooved faces of conical surface sheaves which are located on the input and output shafts. The faces are not straight cones, but have a slight convex curve to maintain proper chain tension at all positions. The pitch diameters of both sheaves are positively controlled by the linkage. Tooth action is positive throughout operating range. Capacity: to 25 hp, speed variation to 6:1.

Double-roller chain drive (Fig 12). This specially developed chain is built for capacities to 22 hp. The hardened rollers are wedged between the hardened conical sides of the variable-pitch sheaves. Radial rolling friction results in smooth chain engagement.

Single-roller chain drive (Fig 13). The double strand of this chain boosts the capacity to 50 hp. The scissor-lever control system maintains proper proportion of forces at each pair of sheave faces throughout the range.

11.

PIV drive chain grips radially grooved faces of variable-pitch sheave to prevent slippage.

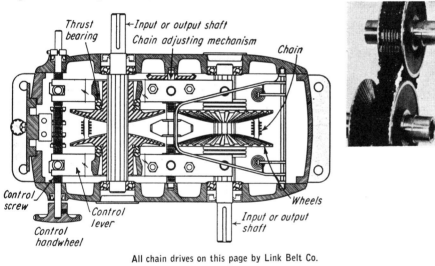

All chain drives on this page by Link Belt Co.

13.

RS drive for high horsepower applications.

12. RS-E drive combines strength with ease in changing speed.

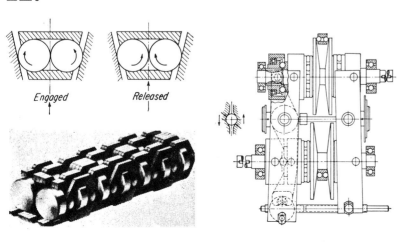

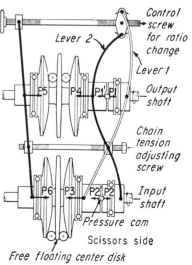

Timing belts shift smoothly to provide two-speed drive

Designed for possible use on moon buggy, its sprockets ride on a vertical turntable to new positions with the flick of a control cable

A novel system based on a series of sprockets and toothed belts is providing a simple way to obtain a two-speed drive without need for clutches, gears, or other conventional transmission components.

In the new drive developed by John L. Burch at the Marshall Space Flight Center, Huntsville, Ala., for possible use on Boeing's moon buggy, the speeds can be shifted by actuating a cable. Pulling on the cable, even while the drive is rotating under power, shifts speed smoothly from,

say, low speed to high. Pulling on the cable again shifts the speed back to low, and so on.

In the moon buggy, power is transmitted to each wheel by individual dc motors. With reversal of the motor, the drive produces one speed (high) in re verse.

For other uses, too. The simplicity of the Burch drive makes it a candidate for use on electric cars, golf carts, garden tractors, and other light-powered vehicles. Moreover, the unit is relatively inexpensive to produce and maintain, is light in weight, and has a long operating life without need for lubrication.

Available systems use gears and clutches and operate within lubricating

baths. Because no more than two teeth are engaged at a time, the loading on the gear teeth is heavy. In the Burch drive, at least four sprocket teeth on the toothed belt are engaged at any moment.

How it works. Key factor in the concept is a round rotatable plate, a sort of vertical turntable, that when released by a shift cable is kicked counterclockwise 180 deg—automatically—by the movement of the main drive belt.

The drive consists of two toothed belts (the main drive belt and the intermediate belt) and four sprockets. The input sprocket, sprocket 1, is double-lengthed and extends to the turntable where its shaft is supported by an angular-contact ball bearing.

The motor is mounted close to the sprockets. In the drawing, for clarity's sake, it is shown by means of the dashed lines as being pulled back. Sprocket 1 is mounted on the motor shaft, and the motor is bolted onto the turntable by means of the mounting sleeve.

In the slow-speed operation, sprocket 1 drives sprocket 2. Sprocket 3, mounted on the same shaft as sprocket 1 and the bolt that retains the motor, drives the output sprocket, 4. So far, the arrangement provides a straightforward stepdown in speed ratio of approximately 18.4 to 1.

Automatic shift. When the flexible cable is actuated by the operator of the vehicle, it pulls on the latching pawl. The pawl releases the turntable, which is rotated 180 deg by the clockwise rotation of the drive.

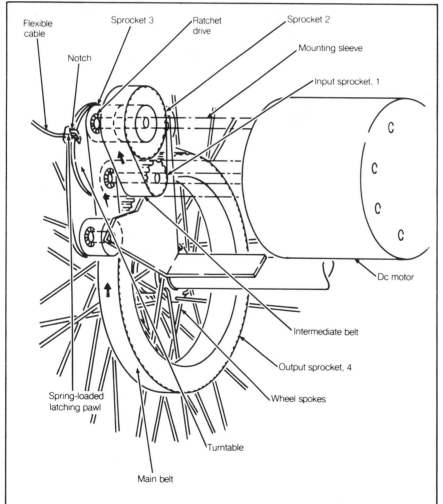

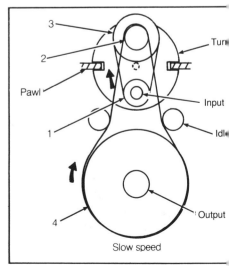

Slow speed

In timing-belt drive, release of latching pawl shifts smoothly from low speed to high, or vice versa. Sequence at right shows how input sprocket kicks over automatically

Boeing's Lunar Rover serves as test-bed to check design of two-speed drive

Because the turntable pivots around the centerline of sprockets *2* and *3*, for the first 30 deg or so, the impetus to the turntable comes from the motion of the intermediate belt. Soon, as the turntable continues its counterclockwise rotation, the extension of sprocket *1*, which can be designated as sprocket *1'*, comes in contact with the main belt and begins to drive the main belt directly, thus beginning to cut out the contribution of motion through sprocket *2*.

At this point, the output sprocket picks up speed because there is a smaller step-down ratio (about 6.5 to 1) between sprocket *1* and sprocket *4* than previously when the drive included the contribution of sprocket *2*. Moreover, during this transfer phase, both the slow-turning and fast-turning sprockets engage the main driving belt simultaneously. This phase of operation requires that the slower sprocket, *2*, be driven through a ratchet that allows its shaft to turn faster than the sprocket until the turntable completes its

rotation.

Sprocket *1*—and, incidentally, the motor and its mountings—continue their ride on the turntable until sprocket *1* is in position directly above sprockets *2* and *3*, at which point the spring-loaded pawl latches into the notch in the turntable (180 deg around from where the first notch is shown). The drive is now in a stable position to produce the high-speed output.

When the driver wishes to shift to low speed, all he needs to do is pull on the cable again. This again releases the turntable, which rotates counterclockwise to repeat the whole process.

Speed ratio equations. The equations for the high-speed ratio, R_H, and the low-speed ratio, R_L, are:

$$R_H = D_4/D_1$$

$$R_L = (D_2/D_1) \text{ x } (D_4/D_3)$$

Where symbol D denotes the pitch diameters of the sprocket, and the subscripts identify the sprockets. Also, D_1 is usually made equal to D_3.

Contributions. The moon-buggy specifications call for a four-wheel drive, which is why the vehicle employs a drive motor on each wheel. Engineers from General Electric cooperated with Marshall Space Flight Center to develop the special brushless dc motors for the application.

The timing belts are from Uniroyal Co., and Burch also ordered a cutting hob from Uniroyal so that the Space Flight Center could machine its own sprockets.

Burch's work has been sponsored by NASA, and inquiries about rights for commercial use of this invention should be addressed to NASA, Code GP, Washington, D.C. 20546. ∎

Tiny polyurethane belts time speeding shafts

As efficient, positive transmission of power, timing belts have an advantage over conventional belts in that they can precisely synchronize two or more shafts. This versatility is opening up many instrument applications.

Now, timing belts are available in "miniature" sizes—40 diametral pitch—which means that the tooth-to-tooth spacing is only 0.0816 in. Moreover, the belts are molded of polyurethane and ride on low-inertia aluminum pulleys.

The highly flexible polyurethane belts,

Miniature timing belts ride on low-inertia pulleys for fast starts, stops

reinforced with polyester cords, are stocked in ⅛- to 5/16-in. widths. Belt-tooth height is 0.018 in., and overall thickness is approximately 0.048 in. Pulleys are available from stock with as few as 10 teeth and 0.260-in. pitch dia.

Aluminum-alloy pulleys offer distinct advantages over molded or extruded pulleys in that they are unaffected by temperature changes and cleaning solvents, are non-hygroscopic, and can be easily modified by the user. ∎

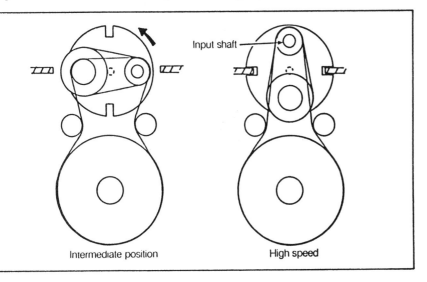

Intermediate position   High speed

Centrifugal clutch varies output of belt transmission

The gasoline version, developed by GM's Engineering Staff, Styling Staff, and Delco-Remy Div., was bolstered in the comparison tests by being equipped with a variable-ratio belt-type transmission. This belt drive, believed to be the first ever applied in an American car, helped the car achieve a remarkable fuel economy of 72 mpg at 30 mph.

Elegantly simple. The belt transmission turned out to be simple, inexpensive, and compatible with a low-horsepower gasoline engine. A key to its success is a variable-ratio mode that is tailored to use the available power more effectively.

Power is transmitted through a centrifugal clutch and a pair of variable-diameter pulleys, coupled by a V-belt (middle drawing, right). The driving pulley connects with the engine through the clutch; the driven pulley is connected with a gear-reduction unit coupled to the differential and the rear axles (photo, top right). The clutch, a pivotal-shoe type, is designed to engage at 1200 rpm engine speed.

The stationary member of each pulley is fixed to the shaft; the movable member is free to travel axially against the resistance of a Belleville spring (drawing, right). The driven pulley employs a compression coil spring in series with the Belleville spring, and the ratio of the transmission is varied by controlling the axial movement of the pulley.

As speed increases, centrifugally-actuated weights increase the effective pitch diameter of the driving pulley. A vacuum-actuated diaphragm opposes this centrifugal force, decreasing the pitch diameter as speed decreases and thus putting the drive into a low-range mode. This vacuum assist goes into action also during full-throttle operation and during braking. At such times, it responds to control through a valve on the carburetor linkage and, for the opposite response, through brake-line pressure.

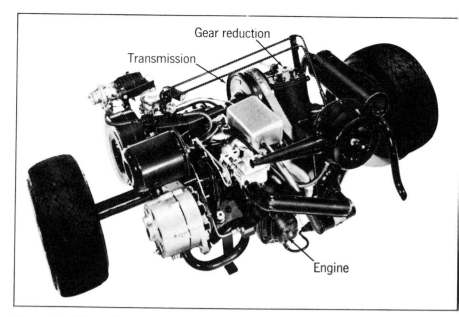

Compact, self-contained driveline was achieved by GM engineers in the gasoline-powered vehicle by mounting the major components solidly to the rear axle.

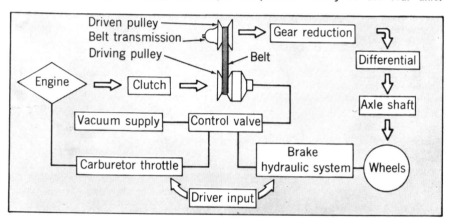

System-controlled belt transmission, together with clutch, gear unit, and differential, delivers engine power to the wheels in gasoline-driven experimental car.

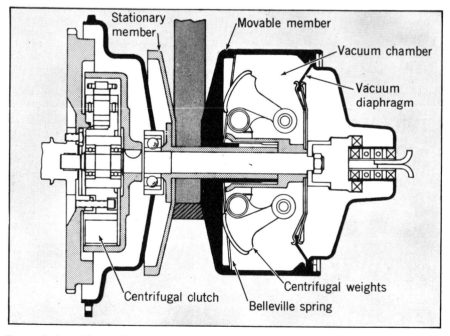

Transmission drive ratio is varied by applying centrifugal weights and vacuum diaphragm to the control of the axial movement of the driving pulley on a shaft.

Getting in step with hybrid belts

from *Design Engineering*

Imaginative fusions of belts, cables, gears and chains are expanding the horizons for light-duty synchronous drives

Doug McCormick

Researchers and manufacturers continually look for new ways of combining the best characteristics of belts with the load capacity and positive drive of gears and chains.

Familiar flat belts and V-belts, for example, are relatively light, quiet, inexpensive, and tolerant of alignment errors. Since they transmit power solely through frictional contacts, however, they function best at moderate speeds (4000-6000 fpm) under static loads. Their efficiencies drop slightly at low speeds, and centrifugal effects limit their capacities at high speeds. Moreover, they are inclined to slip under shock loads or when starting and braking. Even under constant rotation, standard belts tend to creep. Thus, these drives must be kept under tension to function properly, increasing loads on pulley shaft bearings.

Gears and chains, on the other hand, transmit power through bearing forces between positively engaged surfaces. They do not slip or creep, as measured by the relative motions of the driving and driven shafts. But the contacts themselves can slip a good deal as the chain rollers and gear teeth move in and out of mesh.

Positive drives are also very sensitive to the geometries of the mating surfaces. A gear's load is borne by one or two teeth, magnifying small tooth-to-tooth errors. A chain's load is more widely distributed, but chordal variations in the driving wheel's effective radius produce small oscillations in the chain's velocity.

To withstand these stresses, chains and gears must be carefully made from hard materials and must then be lubri-

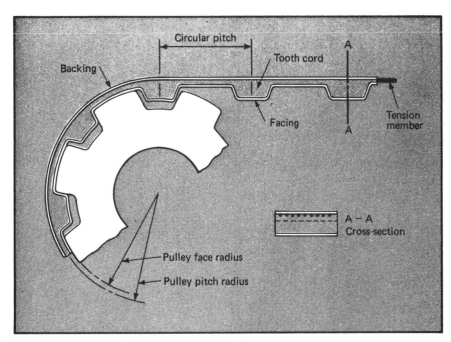

1. *Conventional timing belts have fiberglass or polyester tension members, bodies of neoprene or polyurethane, and trapezoidal tooth profiles*

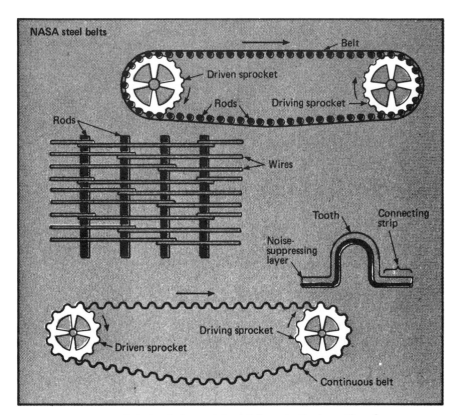

2. *New NASA metal timing belts exploit stainless steel's strength and flexibility, and are coated with sound- and friction-reducing plastic*

cated in operation. Even so, their operating noise betrays sharp impacts and friction between mating surfaces.

The cogged timing belt, with its trapezoidal teeth (Fig 1), is the best-known fusion of belt, gear and chain. Though these well-established timing belts can handle high powers (up to 800 hp), many of the newer ideas in synchronous belting have been incorporated into low and fractional horsepower drives for instruments and business machines.

Steel belts for reliability

Researchers at NASA's Goddard Space Flight Center (Greenbelt, MD) have turned to steel in the construction of long-lived toothed transmission belts for spacecraft instrument drives. The developers say the devices may have earthly uses in critical, hard-to-service applications.

The NASA engineers started looking for a belt design that would retain its strength and hold together for long periods of sustained or intermittent operation in hostile environments, including extremes of heat and cold.

Two steel designs emerged. In the more chain-like version (Fig 2a), wires running along the length of the belt are wrapped at intervals around heavier rods running across the belt. The rods do double duty, serving as link pins and as teeth that mesh with cylindrical recesses cut into the sprocket. The assembled belt is coated with plastic to reduce noise and wear.

In the second design (Fig 2b), a strip of steel is bent into a series of U-shaped teeth. The steel is supple enough to flex as it runs around the sprocket with its protruding transverse ridges, but the material resists stretching. This belt, too, is plastic-coated to reduce wear and noise.

The "V-belt" is best formed from a continuous strip of stainless steel "not much thicker than a razor blade," according to the agency, but a variation can be made by welding several segments together.

NASA has patented both belts, which are now available for commercial li-

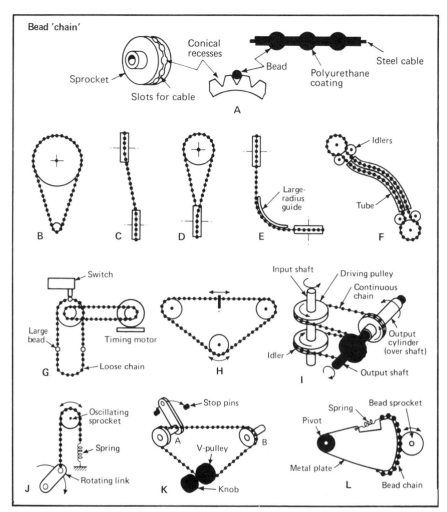

3. *Polyurethane-coated steel-cable 'chains'—both beaded and 4-pinned—can cope with conditions unsuitable for most conventional belts and chains*

censing. Researchers predict that they will be particularly useful in machines that must be dismantled to uncover the belt pulleys, in permanently encased machines, and in machines installed in remote places. In addition, stainless steel belts may find a place in high-

precision instrument drives because they neither stretch nor slip.

Though plastic-and-cable belts don't have the strength or durability of the NASA steel belts, they do offer versatility and production-line economy. One of the least expensive and most

I. Conventional timing belts

Type	Circular pitch, in.	Wkg. tension lb/in. width	Centr. loss const., K_c
Standard (Fig 1)			
MXL	0.080	32	10×10^{-9}
XL	0.200	41	27×10^{-9}
L	0.375	55	38×10^{-9}
H	0.500	140	53×10^{-9}
40DP	0.0816	13	—
High-torque (Fig 8)			
3 mm	0.1181	60	15×10^{-9}
5 mm	0.1968	100	21×10^{-9}
8 mm	0.3150	138	34×10^{-9}

Courtesy Stock Drive Products

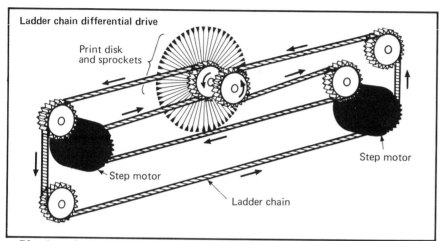

Ladder chain differential drive

Print disk and sprockets

Step motor

Step motor

Ladder chain

5. *Plastic-and-cable ladder chain in an impact-printer drive. In extreme conditions, such hybrids may serve many times longer than steel*

adaptable is the modern version of the bead chain, now common only in key chains and light-switch pull-cords.

The modern bead chain—if chain is the proper word—has no links. It has, instead, a continuous cable of stainless steel or aramid fiber which is covered with polyurethane. At controlled intervals, the plastic coating is molded into a bead (Fig 3a). The length of the pitches thus formed can be controlled to within 0.001 in.

In operation, the cable runs in a grooved pulley; the beads seat in conical recesses in the pulley face. The flexibility, axial symmetry, and positive drive of bead chain suit a number of applications, both common and uncommon:

■ An inexpensive, high-ratio drive that resists slipping and requires no lubrication (Fig 3b). As with other chains and belts, the bead chain's capacity is limited by its total tensile strength (typically 40 to 80 lb for a single-strand steel-cable chain), by the speed-change ratio, and by the radii of the sprockets or pulleys.

■ Connecting misaligned sprockets. If there is play in the sprockets, or if the sprockets are parallel but lie in different planes, the bead chain can compensate for up to 20 deg of misalignment (Fig 3c).

■ Skewed shafts, up to 90 deg out of phase (Fig 3d).

■ Right-angle and remote drives using guides or tubes (Fig 3e and 3f). These methods are suitable only for low-speed, low-torque applications. Otherwise, frictional losses between the guide and the chain are unacceptable.

■ Mechanical timing, using oversize beads at intervals to trip a microswitch (Fig 3g). The chain can be altered or exchanged to give different timing schemes.

■ Accurate rotary-to-linear motion conversion (Fig 3h).

■ Driving two counter-rotating outputs from a single input, using just a single belt (Fig 3i).

■ Rotary-to-oscillatory motion conversion (Fig 3j).

■ Clutched adjustment (Fig 3k). A regular V-belt pulley without recesses permits the chain to slip when it reaches a pre-set limit. At the same time, bead-pulleys keep the output shafts synchronized. Similarly, a pulley or sprocket with shallow recesses permits the chain to slip one bead at a time when overloaded.

■ Inexpensive "gears" or gear segments fashioned by wrapping a bead chain around the perimeter of a disk or solid arc of sheet metal (Fig 3l). The sprocket then acts as a pinion. (Other designs are better for gear fabrication.)

A more stable approach

Unfortunately, bead chains tend to cam out of deep sprocket recesses under high loads, so manufacturers like PIC Design (Ridgefield, CT) and Winfred M. Berg (East Rockaway, NY) began looking, and continue to look, for stronger and more stable bead shapes. In its first evolutionary step, the simple spherical bead grew limbs—two pins projecting at right angles to the cable axis (Fig 4). The pulley or sprocket looks like a spur gear grooved to accommodate the belt; in fact, the pulley can mesh with a conventional

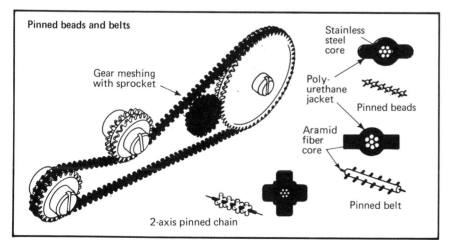

Pinned beads and belts

Gear meshing with sprocket

Stainless steel core

Poly-urethane jacket

Pinned beads

Aramid fiber core

Pinned belt

2-axis pinned chain

4. *Plastic pins eliminate the bead chain's tendency to cam out of pulley recesses, and permit greater precision in angular transmission*

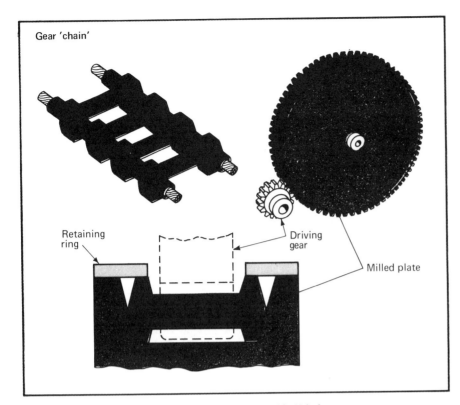

Gear 'chain'

Retaining
ring

Driving
gear

Milled plate

6. Gear chain can function as ladder chain, as a wide V-belt, or,
as here, as a gear surrogate meshing with a standard pinion

Parallel-cable drives

Another species of positive-drive belt uses parallel cables, sacrificing some flexibility for improved stability and greater strength. Here, the cables are connected by rungs molded into the plastic coating, givng the appearance of a ladder (Fig 6). This "ladder chain" also meshes with toothed pulleys, which need not be grooved.

Cable-and-plastic ladder chain is the basis for the differential drive system in a Hewlett-Packard impact printer

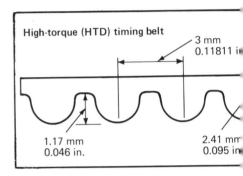

High-torque (HTD) timing belt

3 mm
0.11811 i

1.17 mm
0.046 in.

2.41 mm
0.095 in

7. Curved high-torque tooth profiles
(just introduced in 3-mm and 5-mm
pitches) increase load capacity of fine-
pitch neoprene belts

spur gear of proper pitch.

Versions of the belt are also available with two sets of pins, one projecting vertically and the other horizontally. This arrangement permits the device to drive a series of perpendicular shafts without twisting the cable, like a bead chain but without the bead chain's load

limitations. Reducing twist increases the transmission's lifetime and reliability.

These belt-cable-chain hybrids can be sized and connected in the field, using metal crimp-collars. Such non-factory splices generally reduce the cable's tensile strength by half.

(Fig 5). When the motors rotate in the same direction at the same speed, the carriage moves to the right or left. When they rotate in opposite directions, but at the same speed, the carriage remains stationary and the print-disk rotates. A differential motion of the motors produces a combined translation and rotation of the print-disk.

The hybrid ladder chain is also well suited to fabrication of large spur gears from metal plates or pulleys (Fig 6). Such a "gear" can run quietly in mesh with a pulley or a standard gear pinion of the proper pitch.

Another type of parallel-cable "chain," which mimics the standard roller chain, weighs just 1.2 oz/ft, requires no lubrication and runs almost silently. Such a chain drove the Kremer-prize-winning Gossamer Albatross.

A traditional note

A new high-capacity tooth profile is emerging on conventional cogged belts. These use standard cord and elastic body construction, but instead of the usual trapezoid, they use a curved tooth (Fig 7). The manufacturer, Uni-royal Power Transmission Products (Middlebury, CT), recently introduced 3-mm and 5-mm pitch versions.

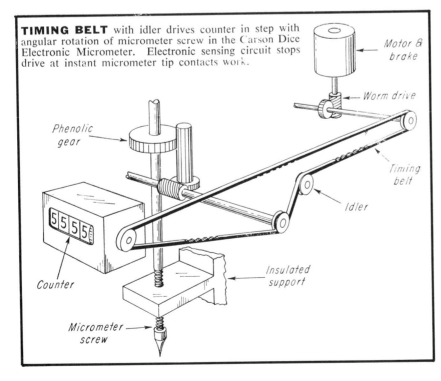

TIMING BELT with idler drives counter in step with angular rotation of micrometer screw in the Carson Dice Electronic Micrometer. Electronic sensing circuit stops drive at instant micrometer tip contacts work.

Motor &
brake

Worm drive

Phenolic
gear

Timing
belt

Idler

5 5 5 5

Counter

Insulated
support

Micrometer
screw

How to change center distance without affecting speed ratio

Increasing the gap between the roller and knife changes chain lengths from F to E. Since the idler moves with the roller sprocket, length G changes to H. The changes in chain length are similar in value but opposite in direction. Chain lengths E minus F closely approximate G minus H. Variations in required chain length occur because the chains do not run parallel. Sprocket offset is required to avoid interference. Slack produced is too minute to affect the drive because it is proportional to changes in the cosine of a small angle (2° to 5°). For the 72-in. chain, variation is 0.020 in.

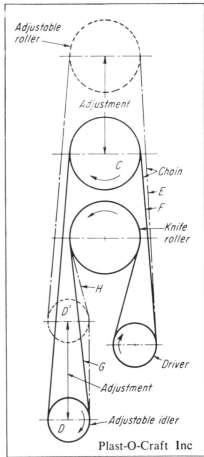

Plast-O-Craft Inc

Motor mount pivots for
Controlled tension

Belt tensioning proportional to load.

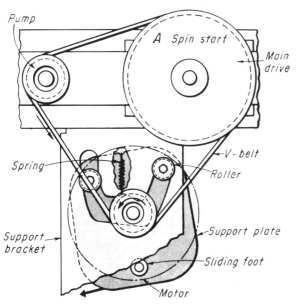

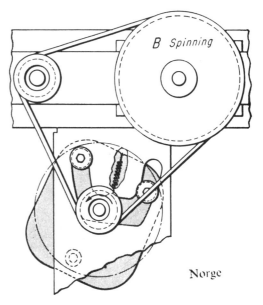

Norge

When the agitation cycle is completed the motor is momentarily idle with the right roller bottomed in the right-hand slot. When spin-dry starts (A) the starting torque produces a reaction at the stator pivoting the motor on the bottomed roller. The motor pivots until the opposite roller bottoms in the left-hand slot. The motor now swings out until restrained by the V-belt, which drives the pump and basket.

The motor, momentarily at zero rpm, develops maximum torque and begins to accelerate the load of basket, water and wash. The motor pivots (B) about the left roller increasing belt tension in proportion to the output torque. When the basket reaches maximum speed the load is reduced and belt tension relaxes. The agitation cycle produces an identical reaction in the reverse direction.

Mechanisms for Reducing

Pulsations in chain motion created by the chordal action of chain and sprockets can be minimized or avoided by introducing a compensating cyclic motion in driving sprocket.

EUGENE I. RADZIMOVSKY

Fig. 1—The large cast-tooth non-circular gear, mounted on the chain sprocket shaft, has wavy outline in which number of waves equals number of teeth on sprocket. Pinion has a corresponding noncircular shape. Although requiring special-shaped gears, drive completely equalizes chain pulsations.

Fig. 2—This drive has two eccentrically mounted spur pinions (1 and 2). Input power is through belt pulley keyed to same shaft as pinion 1. Pinion 3 (not shown), keyed to shaft of pinion 2, drives large gear and sprocket. However, mechanism does not completely equalize chain velocity unless the pitch lines of pinions 1 and 2 are noncircular instead of eccentric.

Fig. 3—Additional sprocket 2 drives noncircular sprocket 3 through fine-pitch chain 1. This imparts pulsating velocity to shaft 6 and to long-pitch conveyor sprocket 5 through pinion 7 and gear 4. Ratio of the gear pair is made same as number of teeth of sprocket 5. Spring-

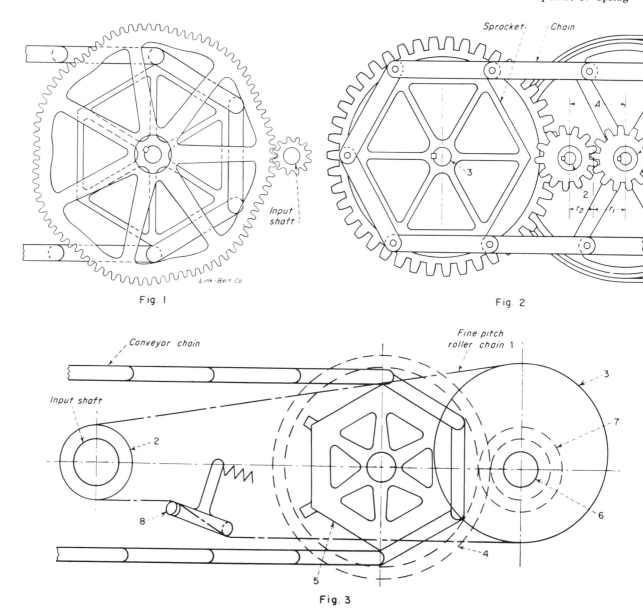

Fig. 1

Input shaft

Link-Belt Co

Fig. 2

Sprocket   Chain

Fig. 3

Conveyor chain

Input shaft

Fine pitch roller chain 1

Pulsations in Chain Drives

Mechanisms for reducing fluctuating dynamic loads in chain drives and the pulsations resulting therefrom include non-circular gears, eccentric gears, and cam activated intermediate shafts.

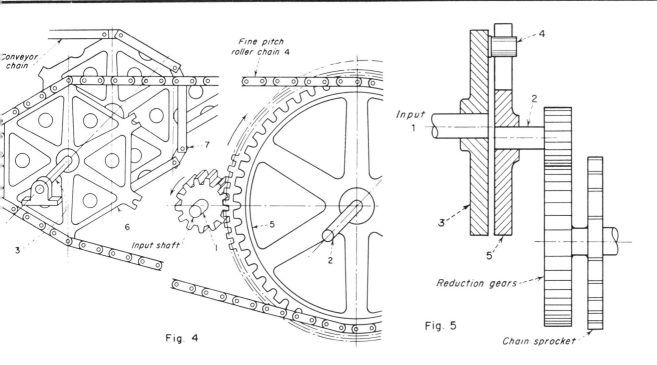

Fig. 4

Fig. 5

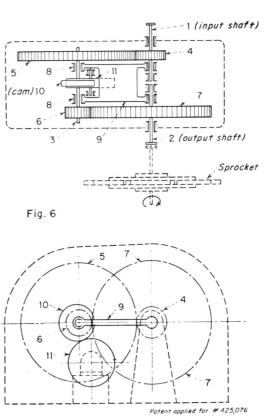

Fig. 6

actuated lever and rollers 8 take up slack. Conveyor motion is equalized but mechanism has limited power capacity because pitch of chain 1 must be kept small. Capacity can be increased by using multiple strands of fine-pitch chain.

Fig. 4—Power is transmitted from shaft 2 to sprocket 6 through chain 4, thus imparting a variable velocity to shaft 3, and through it, to the conveyor sprocket 7. Since chain 4 has small pitch and sprocket 5 is relatively large, velocity of 4 is almost constant which induces an almost constant conveyor velocity. Mechanism requires rollers to tighten slack side of chain and has limited power capacity.

Fig. 5—Variable motion to sprocket is produced by disk 3 which supports pin and roller 4, and disk 5 which has a radial slot and is eccentrically mounted on shaft 2. Ratio of rpm of shaft 2 to sprocket equals number of teeth in sprocket. Chain velocity is not completely equalized.

Fig. 6—Integrated "planetary gear" system (gears 4, 5, 6 and 7) is activated by cam 10 and transmits through shaft 2 a variable velocity to sprocket synchronized with chain pulsations thus completely equalizing chain velocity. The cam 10 rides on a circular idler roller 11; because of the equilibrium of the forces the cam maintains positive contact with the roller. Unit uses standard gears, acts simultaneously as a speed reducer, and can transmit high horsepower.

Smoother Drive Without Gears

The transmission in this motor scooter is torque-sensitive; motor speed controls continuously variable drive ratio. Operator merely works throttle and brake.

Variable-diameter V-belt pulleys . . .

connect motor and chain drive sprocket to give a wide range of speed reduction. The front pulley incorporates a 3-ball centrifugal clutch which forces the flanges together when the engine speeds up. At idle speed the belt rides on ballbearing between retracted flanges of the pulley. During starting and warmup a lockout prevents the forward clutch from operating.

Upon initial engagement the over-all drive ratio is approximately 18:1. As engine speed increases, belt rides higher up on the forward-pulley flanges until over-all drive ratio becomes approximately 6:1. Resulting variations in belt tension are absorbed by the spring-loaded flanges of the rear pulley. When clutch is in idle position, the V-belt is forced to the outer edge of the rear pulley by spring force. When the clutch engages, the floating half of the front pulley moves inward increasing its effective diameter and pulling the belt down between flanges of the rear pulley.

The transmission is torque-responsive. A sudden engine acceleration increases effective diameter of the rear pulley, lowering the drive ratio. It works this way. Increase in belt tension rotates the floating flange ahead in relation to the driving flange. The belt now slips slightly on its driver. It's at this time that nylon rollers on the floating flange engage cams on the driving flange, pulling the flanges together and increasing effective diameter of the pulley.

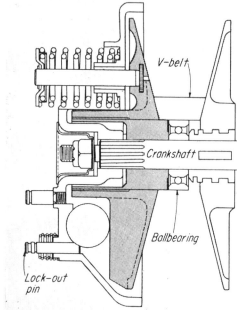

Flexible conveyor moves in waves

The trouble with most conventional conveyors used in tunneling and mining is that they can't negotiate curves, can't be powered at different points, are subject to malfunction because of slight misalignment, and require time-consuming adjustments to be lengthened or shortened.

Thomas E. Howard, U.S. Bureau of Mines, has invented a new type of conveyor belt—one that does not move forward—that may solve all of these problems. Howard has assigned patent rights to the government.

The conveyor, designed to move broken ore, rock, and coal in mines, moves material along a flexible belt. The belt is given a wave-like movement by the sequenced rising and dropping of supporting yokes beneath it.

The principle. The conveyor incorporates modules built in arcs and Y's in such a way that it can be easily joined with standardized sections to negotiate corners and either merge or separate streams of moving materials. It can be powered at any one point or at several points, and it incorporates automatic controls to actuate only those parts of the belt that are loaded, thereby reducing power consumption.

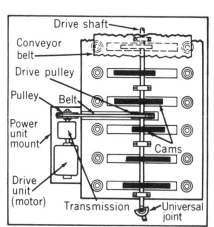

Two modules of undulatory conveyor are shown in partially cut-away view.

In tests at the bureau's Pittsburgh (Pa.) Mining Research Center, a simplified mechanical model of the device has moved rock at rates comparable to conventional belts.

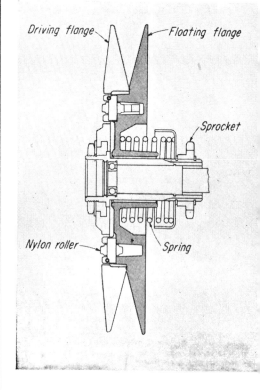

CHAPTER 6.

GEARED SYSTEMS AND VARIABLE-SPEED MECHANISMS

Traction drives move to higher powers

from *Design Engineering*

Thanks to better steels and a deeper knowledge of elastohydrodynamics, rolling-contact drives may be about to prove out commercially

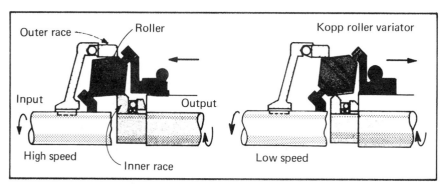

3. *The Kopp roller variator's conical rollers move back and forth to change path of contact with races, varying speed ratios*

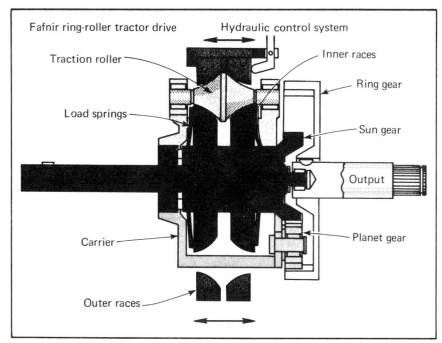

4. *In this 50 hp Fafnir ring-roller design, speed is adjusted by spreading and compressing the ring-race halves*

Doug McCormick

Every few years, it seems, there is a new flurry of enthusiasm for traction drives. Since the mid-1930s—when a few still-current industrial traction machines were introduced abroad and the Transitorq automotive transmission made its abortive first appearance here—the industry has endured several cycles of high expectations and commercial disappointment. Today, even the most ardent exponents of the technology tend to qualify their enthusiasm with "wait and see" caution: traction drives must prove themselves in the marketplace.

Meanwhile, the past decade has seen steady improvements in the metals and lubricants needed to overcome the stress problems that dogged earlier traction drives. The 1970s also witnessed the introductions of a good dozen new traction drives here—many of them capable of handling powers well above the 15 hp limit of most earlier models. The majority of these new designs have been imported from Europe and Japan, where they have been in regular use for many years.

Today, most traction drives are used in continuously variable transmissions (CVTs), which become increasingly popular as energy costs escalate and it grows more important to match the machine speed to the job without wasted motion. Traction CVTs often have marked advantages—in quietness, producibility, or efficiency—over current belt- and chain-drives, hydrostatic transmissions, and such electrical devices as eddy current clutches.

The best known traction drive is probably the Kopp variator. Swiss designer Jean Kopp's first CVT (introduced here

26 years ago) was a tilting-ball-and-disk machine, variations of which are still available from Cleveland Gear (Cleveland, OH, formerly a division of Eaton), the Winsmith Division of UMC Industries (Springville, NY), and the Koppers Company (Baltimore, MD). These units, with capacities of up to 17 hp and speed ranges of 12:1, cannot be called high-power, but a more recent Kopp design can.

Five years ago, Koppers brought out the Kopp double-cone roller variator (Fig 3), which carries up to 100 hp at up to a 12:1 speed range. The unit has a power density of 0.08 hp/lb for a 100

hp unit (lower for smaller machines) and efficiencies of 83% to 93%. It is interesting to note that Kopp variators did not begin using traction fluid to increase capacities until the late 1970s.

A mobile transmission

The grail of traction designers since the '30s has been a commercially viable mobile equipment transmission. In the next year or so, the Fafnir Bearing Division of Textron (New Britain, CT) hopes to begin marketing a new agricultural equipment transmission (Fig 4) that can carry up to 50 hp. Smaller powers may be varied over the full

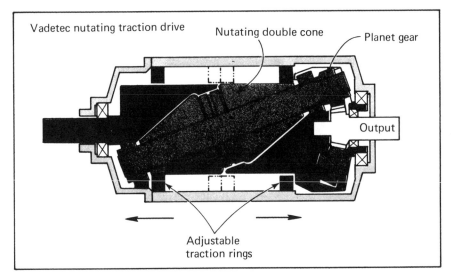

5. This Vadetec nutating design extends the cone-and-ring idea, adding interchangeable output planetary gear systems for greater flexibility

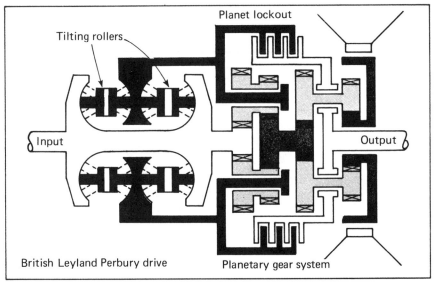

6. The planetary lockout on British Leyland's 100 hp Perbury drive permits a full input/output speed range from 1:0.4 reverse through 1:2 forward

range of speeds from reverse to forward. Depending on the accompanying planetary design, the drive could cover the range from 0.3 times input forward to 0.3 times input in reverse, or from 0 rpm to 0.5 times input forward.

The heart of the drive is a double-conical roller that drives a planetary gear cage. The roller rides between two two-part races—an inner race splined to the drive shaft, and a stationary outer race. The drive speed is controlled hydraulically by spreading or contracting the halves of the outer race, changing the roller rotational speed.

The unit was designed to compete in cost and performance with existing hydrostatic transmissions: cost projections run about $15/hp (far lower than the $150/hp common among industrial CVTs) and the power densities should be in the 0.6 hp/lb range. Efficiency took a backseat to cost, size and flexibility, so the unit's efficiency should peak at about 85% and be somewhat lower over some of its range, according to Fafnir engineers. Graham (Menomonee Falls, WI) and Mitsubishi also make CVTs on similar ring-roller principles.

Nutating traction drive
Vadetec Corp. (Troy, MI) has taken the long-established cone-and-ring traction principle and gone it one better, to design a compact and elegant nutating traction drive. The drive's input shaft is attached to a rotating cylinder, and set into the cylinder is a freely turning double-cone roller (Fig 5).

When the input shaft rotates, the cone roller is forced to roll along the traction ring. As the cone rotates, a planetary gear on the cone's output end drives one of several interchangeable output gear configurations. The cone's rotational speed—and hence output speed—is adjusted by moving the traction rings forward and backward to engage the cone over different radii.

Last summer, a Vadetec licensee completed "very successful" field tests of a 200 hp drive in off-road equipment. The company has also tested the nutating drive in a 400 hp centrifugal air-conditioner compressor operating at up to 10,000 rpm, which showed 96% efficiency, the company says.

Automotive entries
British Leyland (now known more properly as BL Technology Ltd, Solihull, England) has promised it will have an automotive traction CVT in production before the end to the decade, according to the company's chief transmission research engineer.

The BL design is a modification of the Perbury roller-and-toroid drive, with the addition of a planetary gear system (Fig 6). The rollers link one rotational radius (on the input side of a toroidal cavity) with another rotational radius on the output side. By tilting the plane of the rollers, one can alter the relationship between the radii and therefore the speed ratio.

When the planetary gear is engaged, it serves the same purpose as the output gear arrangements on the Fafnir and Vadetec machines: it produces what BL calls a "low regime," spanning the range from +0.2 times input (low speed forward) to −0.4 times input (low speed reverse). The gear system can also be locked out, however, to produce a "high regime," ranging from a 2.7:1 reduction to a 2:1 increase, for a total high speed range of a little over 5:1.

According to BL roadtests, the transmission typically operates at above 80% efficiency, rising to 91% as the vehicle climbs a slight grade at about 60 km/hr. The drive gives peak acceleration equivalent to that of a geared transmission driven by a 10% more powerful engine, while giving fuel savings of roughly 15% at highway cruising speeds.

New drive design: no teeth

from Design Engineering

When Advanced Energy Technologies (Boulder, CO) went to patent its "anti-friction drive," the U.S. Patent Office had to establish an entirely new category for this modular, two-stage eccentric trochoidal device (see figures).

According to one of the device's two inventors, engineer Bob Distin, the physical principles are exactly the same as those of an eccentric gear set, with one significant tribological exception: instead of transmitting forces through

as tooth-to-tooth contact through an oil film, he says the anti-friction drive transmits its power as trochoid-to-trochoid contact through ball bearings.

Thus, say the inventors, they have fully realized the dream that spawned modern gear technology—true rolling contact between metallic surfaces. (Even when established at the proper center distances, conventional gears give true rolling contact only at the pitch line.)

Eccentric drives are inherently "in-line" drives, but they all have one problem in common: re-converting the epicyclic motion to a constant concentric rotation. Designing the power take-offs for such drives often demands more ingenuity than the drive itself.

The anti-friction drive incorporates a "constant-velocity coupling" linking the two eccentric disks—what AET president Rory McFarland described as an "Oldham-like" device. McFarland has indicated that the coupling itself may find an independent market.

Oddly enough, the constant-velocity coupling may be excluded from many of the production drives. Instead, AET plans to add a pair of trochoidal races linking the two disks—producing what is essentially a three-stage reducer.

"Why bother to add stages when you can go from 1.1:1 to 10,000:1 in two

stages?" he asks. Ultimately, the company's plans call for a limited series of disk modules which can be speedily assembled to provide virtually any ratio—at almost constant cost. McFarland says that shortening the period of the trochoid (thus increasing the ratio) presents little problem. For higher power levels, the company plans a version in which rollers replace the balls.

Other advantages claimed for the anti-friction drive include:
- *Efficiency.* The anti-friction drive is markedly more efficient than the competition, and its advantages are more markets at higher reduction ratios.
- *Compactness.* Because about half of the drive ball bearings are loaded at all points of the cycle, driving stresses are lower than in geared drives. Designers can therefore use smaller drives to transmit more power.
- *Accuracy.* There is no built-in "backlash" in the anti-friction drive.
- *Silence.* The AET ball-bearing drive operates without the high-frequency impacts that generate gear noise.
- *Durability.* Early indications are that the anti-friction drive's service life conforms to standard bearing fatigue-life formulas.

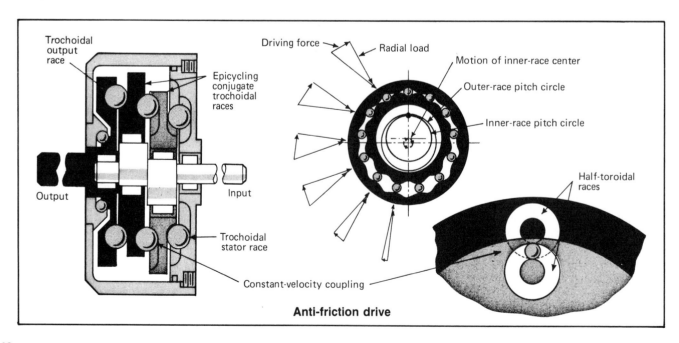

Anti-friction drive

242

Nutating-plate drive

Fig. 1 An exploded view of the Nutation Drive.

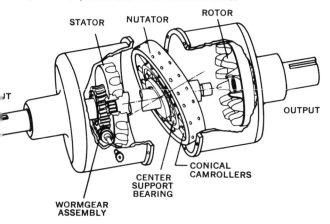

The Nutation Drive* is a mechanically positive, gearless power transmission with which high single-stage speed ratios can be obtained at high efficiencies. A nutating member carries camrollers on its periphery and causes differential rotation between the three major components of the drive: stator, nutator, and rotor. Correctly designed cams on the stator and rotor assure a low-noise engagement and mathematically pure rolling contact between camrollers and cams.

The drive is characterized by compactness, high speed ratio, and efficiency. Its unique design guarantees rolling contact between the power-transmitting members and even distribution of the load among a large number of these members. Both factors contribute to the drive's inherent low noise level and long, maintenance-free life. Vulnerability of the drive is limited by the small number of moving parts; furthermore, grease and solid lubrication seem adequate for many applications.

Kinetics of the Nutation Drive

Basic Components. The three basic components of the Nutation Drive are the stator, nutator, and rotor, as shown in Fig. 1. The nutator carries radially mounted conical camrollers that engage between cams on the rotor and stator. Cam surfaces and camrollers have a common vanishing point—the center of the nutator—and thereby line-contact rolling is assured between the rollers and the cams.

Nutation is imparted to the nutator through the center support bearing by the fixed angle of its mounting on the input shaft. One rotation of the input shaft causes one complete nutation of the nutator. Each nutation cycle forwards the rotor by an angle equivalent to the angular spacing of the rotor cams. During nutation the nutator is held from rotating by the stator, which transmits the reaction forces to the housing.

* Four U. S. patents (3,094,880, 3,139,771, 3,139,772, and 3,590,659) have been issued to A. M. Maroth.
From *Mechanical Engineering*
Based on a paper contributed by the ASME Design Engineering Division.

Cone drive needs no gears or pulleys

A new variable-speed-transmission cone drive operates without gears or pulleys. The drive unit has its own limited slip differential and clutch.

As the drawing shows, two cones made of brake lining material are mounted on a shaft directly connected to the engine. These drive

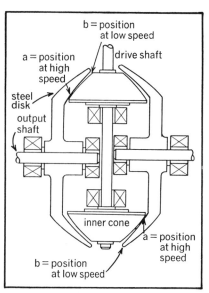

Cone drive operates without lubrication.

two larger steel conical disks mounted on the output shaft. The outer disks are mounted on pivoting frames that can be moved by a simple control rod.

To center the frames and to provide some resistance when the outer disks are moved, two torsion bars attached to the main frame connect and support the disk-support frames. By altering the position of the frames relative to the driving cones, the direction of rotation and speed can be varied.

The unit was invented by Marion H. Davis of V-Plex Corp, Indiana. □

Your guide to variable-speed
Mechanical drives

D. Z. DVORAK

Most of the mechanical types have a *limited* variable-speed range in that they cannot produce a zero or near-zero output speed. Those that include zero speed in their range are said to have an *infinite* variable-speed range. These terms are only trade jargon because all do produce an infinite variety of speeds.

Generally, the outputs are irreversible—if reversibility is required, the direction of input rotation must be changed. Because drive motors are directly coupled to the input shafts of the drive, split arrangements which permit the mechanical separation of the motor from the adjustable output speed unit are not possible.

Mechanical drives have been grouped in nine major classes, based on their operating principle:

1) Cone drives
2) Disk drives
3) Ring drives
4) Spherical drives
5) Multiple-disk drives
6) Impulse drives
7) Controlled-differential drives
8) Belt drives
9) Chain drives

Many of the drives are available as *commercial* units; they can be obtained in specific sizes and horsepowers from one or more manufacturers. Some drives are not available commercially and must be custom designed.

Table I includes the horsepower *capabilities* of the drive. The horsepower *requirements* of the application are determined by the basic equation

$$\text{hp} = \frac{T\,n}{63,000}$$

where T = torque, in.-lb; n = speed, rpm.

The drive must also be matched to the torque requirements of the application. The three basic types of horsepower-torque characteristics of variable-speed drives are illustrated in Fig 1:

Constant horsepower application—The torque decreases almost hyperbolically with increase in speed. This requirement is frequently found with machine tools, particularly on spindle drives. The critical condition here is at minimum speed when the torque and the stress of the mechanical parts is at a maximum.

Constant torque application—The horsepower requirements increase proportionally to speed, as is the case with many conveyors, reciprocating compressors, printing presses, and machine-tool feed drives, or where the load is almost a pure friction load. The selection of the drive should be based on the power requirement at the maximum speed.

Variable horsepower and speed—This requirement is common with propellers and centrifugal pumps. The critical condition is at maximum speed. The power delivered at low speed is usually more than is needed.

Only stepless drives are described here. Most of the drives are of medium capacity; however, some are limited to 5 hp or less. All are over 1 hp except the impulse drives.

All drives, with the exception of the PIV drive, are based on the friction principle. Therefore, a certain amount of slip, which increases with torque, can be expected. Slippage serves as a safety device to prevent damage from overload. But high slippage is undesirable because it decreases speed regulation, efficiency, and life of the system. All drives except the belt-driven units operate in an oil bath or mist.

1.

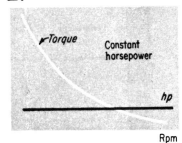

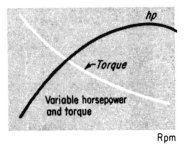

Table I

Type of Drive	Manufacturer
Cone drives	Graham Transmissions Inc. Menomonee Falls, Wis.
	Simpo Kogyo Co Ltd Karahashi, Minami-ku Kyoto, Japan
Disk drives	Sentinel (Shrewsbury) Ltd Shrewsbury, England
	Block and Vaupel Wuppertal, Germany
Ring drives	Master Electric Div of Reliance Electric Dayton 1, Ohio
	H Stroeter Dusseldorf, Germany
	Excelermatic Inc Rochester, N.Y.
Spherical drives	New Departure Div of General Motors Bristol, Conn.
	Perbury Engineering Ltd. England
	Cleveland Worm & Gear Div of Eaton Mfg Co Cleveland
	Excelecon Corp
	Friedr. Cavallo Berlin-Neukoelln Germany
Multiple-disk drives	Ligurtecnica Genoa, Italy
	Reeves Pulley Div of Reliance Electric Co Columbus, Ind
Impulse drives	Morse Chain Co Ithaca, NY
	Zero Max Co Minneapolis 8, Minn
Differential drives	Link-Belt Co
	Stratos Div of Fairchild Engine & Airplane Corp Babylon, NY
	Lombard Governor Corp Ashland, Mass

Commercial variable-speed drives

Trade name	Fig No.	Basic elements	Maximum horsepower capacity	Maximum speed variation	Horsepower-torque characteristics	Peak efficiency	Comments
Graham	4	Tapered rollers Fixed ring Planetary gears carrier	5 hp	Non-reversing type: 1/3 of input rpm to zero Reversing type: 1/5 of rpm in either direction	See curve in Fig 4 High torque at high speed also available	85% at full load High	For low speed and zero-speed applications
Ring cone RC	5	Tapered rollers Preloaded ring	10 hp	4:1	constant hp	High	Similar to ring drive. Max output speed for 10 hp units = 2400 rpm
Ring cone SC	6	Planetary cones Track ring	20 hp	From 4:1 to 24:1	Combination constant torque and constant hp	85%	Employs a planetary cone system in place of planetary gears, with movable track ring (in place of ring gear)
F.U.	8	Disks beveled roller	20 hp	6:1 (10:1)	constant hp	90%	Also made in France by La Filiere Unicum, Paris
—	9	Planetary friction disks	—	—	—	—	Friction wheels act as planets. Orbits are adjustable
Speed ranger	10	Steel ring Variable-pitch sheaves	3 hp	8:1 (16:1)	constant torque	90%	Available with output speeds from 2 to 4100 rpm. Made in Germany by Hans Heynay, Munich
—	11	Special-shape rings Inverted sheave wheels	—	10:1	constant torque	—	Similar in principle to the Speed Ranger (above), except that special shape rings are employed
—	12	Rings Beveled disks	5 hp	12:1	—	96%	Novel principle
Transitorq	14	Spherical disks Tilting rollers	20 hp	6:1 (10:1)	constant hp or constant torque	—	Not in production
—	15	Double spherical disks Tilting rollers	—	—	—	—	Reportedly still in development stage
Cleveland	16	Spherical beveled disks Axle-mounted balls	15 hp	9:1	constant hp constant torque over low range also available	To 90%	High torque at low speed. Made in Germany by Eisenwerk Wulfel, in Switzerland by: Aciera SA, Le Locle
Excelecon	17	Input concave disks Beveled rollers	15 hp	9:1	Constant hp .	90%	High torque at low speed
Cavallo	18	Axially-free ball Cone disks	—	—	—	—	Very simple and compact
—	19	Multiple-disks Balls	33 hp	5:1	—	To 95%	Multiple-disks permit high horsepower
Beier	20	Input tapered disks Output rimmed disks	60 hp	4:1	Combination constant torque and constant hp, see Fig 33	85%	Units up to several hundreds hp built in England by Beier Infinitely Variable Gear Co, London SE1
Morse	22	Gear-linkage system One way clutch	1.5 hp	4½:1 to 120:1 (180 rpm max)	constant torque 175 ft-lb	over 95%	Slight pulsating motion of output is of distinct advantage to feeders and mixer
Zero-Max	23	Linkage system One way clutch	¾ hp	0 to ¼ input rpm (2000 rpm max)	constant torque	—	
—		Differential gears Speed variator	25 hp				
—		Differential gears Speed variator	75 hp	A great variety of torque-rpm characteristics can be achieved depending on arrangement.		Varies	For very accurate control applications. Expensive
—		Differential gears Speed variator	15 hp				

Table includes only drives in the 1 to 100 hp range. Maximum hp capacity column gives maximum hp of standard units only.
Units are frequently available in smaller sizes, sometime with fraction hp output.

Cone drives

The simpler types in this group have a cone or tapered roller in combination with a wheel or belt (Fig 2). They have evolved from the stepped-pulley system. Even the more sophisticated designs are capable of only a limited (although infinite) speed range and generally must be spring-loaded to reduce slippage.

Adjustable cone drive (Fig 2A). This is perhaps the oldest variable-speed friction system and is usually custom built. Power from the motor-driven cone is transferred to the output shaft by the friction wheel, which is adjustable along the cone side to change the output speed. The speed depends upon the ratio of diameters at point of contact.

Two-cone drive (Fig 2B). The adjustable wheel is the power transfer element, but this drive is difficult to preload because both input and output shafts would have to be spring loaded. The second cone, however, doubles the speed reduction range.

Cone-belt drives (Fig 2C and D). In (C) the belt envelops both cones; in (D) a long-loop endless belt runs between the cones. Stepless speed adjustment is obtained by shifting the belt along the cones. The cross section of the belt must be large enough to transmit the rated force, but the width must be kept to a minimum to avoid a large speed differential over the belt width.

Electrically coupled cones (Fig 3). This patented device (US Patent 3,048,046) is composed of thin laminates of paramagnetic material. The laminates are separated with semidielectric materials which also localize the effect of the inductive field. There is a field generating device within the driving cone. Adjacent to the cone is a positioning motor for the field generating device. The field created in a

2.

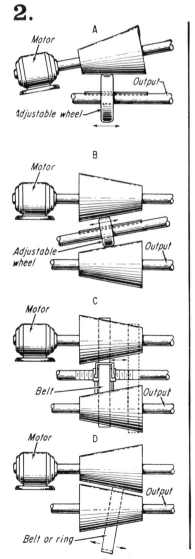

3.

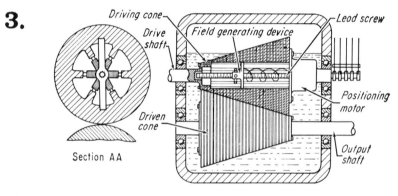

Section AA

4.

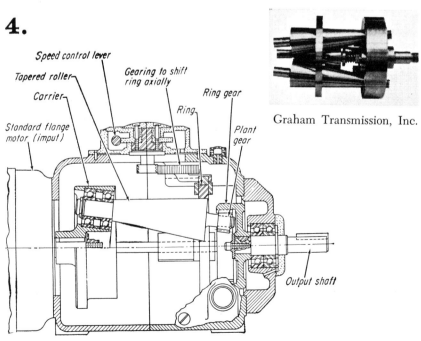

Graham Transmission, Inc.

particular section of the driving cone induces a magnetic effect in the surrounding lamination. This causes the laminate—and its opposing lamination—to couple and rotate with the drive shaft. The ratio of diameters of the cones, at the point selected by positioning the field generating component, determines the speed ratio.

Graham drive (Fig 4). This commercial unit combines a planetary-gear set and three tapered rollers (only one of which is shown). The ring is positioned axially by a cam and gear arrangement. The drive shaft rotates the carrier with the tapered rollers, which are inclined at an angle equal to their taper so that their outer edge is parallel to the centerline of the assembly. Traction pressure between rollers and ring is created by centrifugal force, or spring loading of the rollers. At the end of each roller a pinion meshes with a ring

gear. The ring gear is part of the planetary gear system and is coupled to the output shaft.

The speed ratio depends on the ratio of the diameter of the fixed ring to the effective diameter of the roller at the point of contact, and is set by the axial position of the ring. The output speed, even at its maximum, is always reduced to about one-third of input speed because of the differential feature. When the angular speed of the driving motor equals the angular speed of the centers of the tapered rollers around their mutual centerline (which is set by the axial position of the nonrotating friction ring), the output speed is zero. This drive is manufactured up to 3 hp; efficiency is to 85%.

Cone-and-ring drive (Fig 5). Here, two cones are encircled by a preloaded ring. Shifting the ring axially varies the output speed. This prin-

ciple is similar to that of the cone-and-belt drive (Fig 2C). In this case, however, the contact pressure between ring and cones increases with load to limit slippage.

Planetary-cone drive (Fig 6). This is basically a planetary gear system but with cones in place of gears. The planet cones are rotated by the sun cone which, in turn, is driven by the motor. The planet cones are pressed between an outer nonrotating ring and the planet holder. Axial adjustment of the ring varies the rotational speed of the cones around their mutual axis. This varies the speed of the planet holder and the output shaft. Thus the mechanism resembles that of the Graham drive (Fig 4) pictured on the facing page.

Speed adjustment range of the unit illustrated is from 4:1 to 24:1. The system is built in Japan in capacities of 2 hp and under.

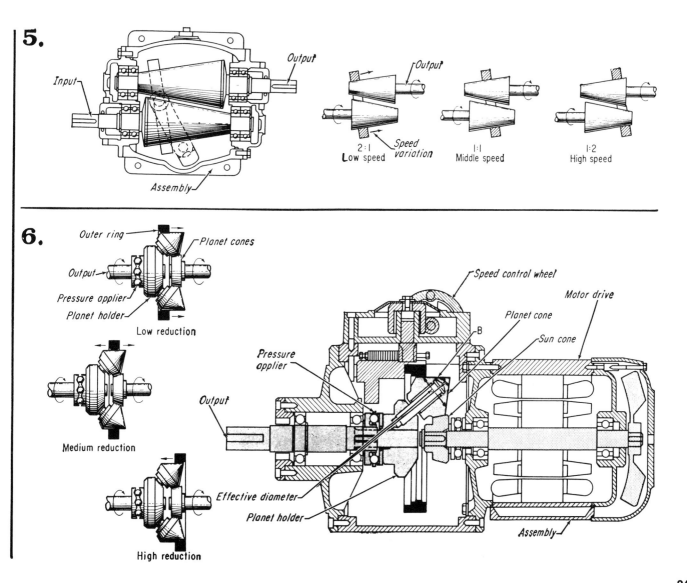

5.
Input — Output — Assembly
2:1 Low speed — Speed variation — 1:1 Middle speed — 1:2 High speed

6.
Outer ring — Planet cones — Output — Pressure applier — Planet holder
Low reduction
Medium reduction
High reduction
Pressure applier — Output — Effective diameter — Planet holder — Speed control wheel — Motor drive — Planet cone — Sun cone — B — Assembly

Disk drives

Adjustable disk drives (Fig 7A and B). The output shaft in (A) is perpendicular to the input shaft. If the driving power, the friction force, and the efficiency stay constant, then the output torque decreases in proportion to increasing output speed. The wheel is made of friction material, the disk of steel. Because of relatively high slippage, only small torques can be transmitted. The wheel can move over the center of the disk as this system has infinite speed adjustment.

To increase the speed range, a second disk can be added. This arrangement (Fig 7B) also makes input and output shafts parallel.

Spring-loaded disk drive (Fig 8). To reduce slippage, the contact force between the rolls and disks in this commercial drive is increased by means of the spring assembly in the output shaft. Speed adjustments are made by rotating the leadscrew to shift the cone roller in the vertical direction. The unit illustrated is built to 4-hp capacity. With a double assembly of rollers, units are available to 20 hp. Units operate to 92% efficiency. Standard speed range is 6:1, but units of 10:1 have been built. The power transferring components, which are made from hardened steel, operate in an oil mist, thus minimizing wear.

Planetary disk drive (Fig 9). Four planet disks replace planet gears in this friction drive. Planets are mounted on levers which control radial position and therefore control the orbit. Ring and sun disks are spring loaded.

7.

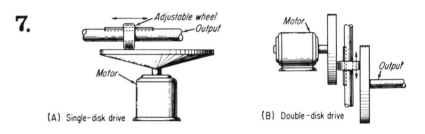

(A) Single-disk drive   (B) Double-disk drive

10a.

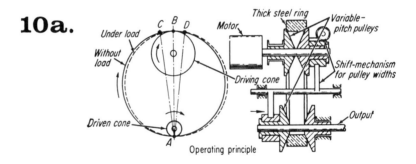

Operating principle

Ring drives

Ring-and-pulley drive (Fig 10). A thick steel ring in this type drive encircles two variable pitch (actually *variable width*) pulleys. A novel gear-and-linkage system simultaneously changes the width of both pulleys (see Fig 10 B). For example, when the top pulley opens, the sides of the bottom pulley close up. This reduces the effective pitch diameter of the top pulley and increases that of the bottom pulley, thus varying the output speed.

Normally, the ring engages the pulleys at point A and B. However, under load, the driven pulley resists rotation and the contact point moves from B to D because of the very small elastic deformation of the ring. The original circular shape of the ring is changed to a slightly oval form and the distance between points of contact decreases. This wedges the ring between the pulley cones and increases the contact pressure between ring and pulleys in proportion to the load applied, so that constant horsepower at all speeds is obtained. The unit is available to 3-hp capacity; speed variations to 16:1, with practical range about 8:1.

Some manufacturers employ rings with special-shaped cross section, (Fig 11) by inverting one of the sets of sheaves.

Double-ring drive (Fig 12). Power transmission is through two steel traction rings that engage two sets of disks mounted on separate shafts. Such a drive requires that the outer disks be under a compression load by a spring system (not illlustrated). The rings are hardened and convex-ground to reduce wear. Speed is changed by tilting the ring support cage, forcing the rings to move the desired position.

8.

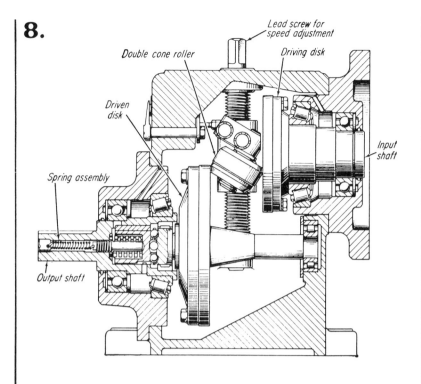

Lead screw for speed adjustment

Double cone roller

Driving disk

Driven disk

Input shaft

Spring assembly

Output shaft

9

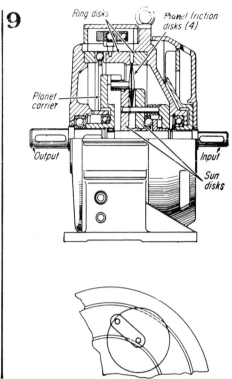

Ring disks

Planet friction disks (4)

Planet carrier

Output

Input

Sun disks

10b.

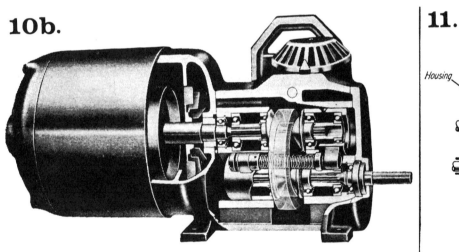

11.

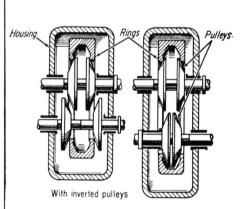

Housing

Rings

Pulleys

With inverted pulleys

12.

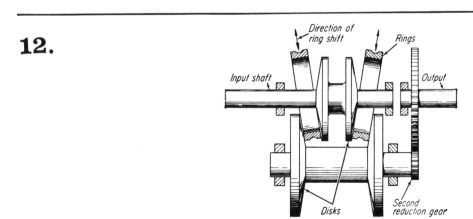

Direction of ring shift

Rings

Input shaft

Output

Disks

Second reduction gear

Spherical drives

Sphere-and-disk drives (Fig 13 and 14). Speed variations in Fig 13 are obtained by changing the angle that the rollers make in contacting spherical disks. As illustrated, the left spherical disk is keyed to the driving shaft, the right contains the output gear. The sheaves are loaded together by a helical spring.

One commercial unit, Fig 14, is a coaxial input and output shaft version of the previous arrangement. The rollers are free to rotate on bearings and can be adjusted to any speed between the limits of 6:1 to 10:1. An automatic device regulates the contact pressure of the rollers, maintaining the pressure exactly in proportion to the imposed torque load.

Double-sphere drive (Fig 15). Higher speed reductions are obtained by grouping a second set of spherical disks and rollers. This also reduces operating stresses and wear. The input shaft runs through the unit and carries two opposing spherical disks. The disks drive the double-sided output disk through two sets of three rollers. To change the ratio, the angle of the rollers is varied. The disks are axially loaded by hydraulic pressure.

Tilting-ball drive (Fig 16). Power is transmitted between disks by steel balls whose rotational axes can be tilted to change the relative lengths of the two contact paths around the balls, and hence the output speed. The ball axes can be tilted uniformly in either direction; the effective rolling radii of balls and disks produce speed variations up to 3:1 increase, or 1:3 decrease, with the total up to 9:1 variation in output speed.

Tilt is controlled by a cam plate through which all ball axes project. To prevent slippage under starting or shock load, torque responsive mechanisms are located on the input and output sides of the drive. The axial pressure created is proportional to the applied torque. A worm drive positions the plate. The drives are manufactured to 15-hp capacity. Efficiency characteristics are shown in the chart.

Sphere-and-roller drive (Fig 17). The roller, with spherical end surfaces, is eccentrically mounted between the coaxial input and output spherical disks. Changes in speed ratio are

13.

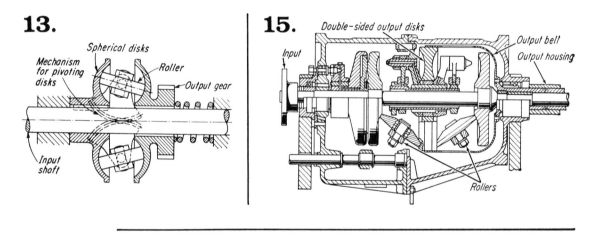

15.

14.

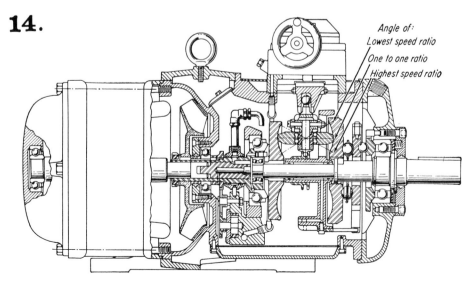

made by changing the angular position of the roller.

The output disk rotates at the same speed as the input when the roller centerline is parallel to the disk centerline, as in (A). When the contact point is nearer the centerline on the output disk and further from the centerline on the input disk, as in (B), the output speed exceeds that of the input. Conversely, when the roller contacts the output disk at a large radius, as in (C), the output speed is reduced.

A loading cam maintains the necessary contact force between the disks and power roller. Speed range is to 9:1; efficiency close to 90%.

Ball-and-cone drive (Fig 18). This is a remarkably simple drive. Input and output shafts are offset. Two opposing cones with 90-deg internal vertex angles are fixed to each shaft. The shafts are preloaded against each other. Speed variation is obtained by positioning the ball which contacts the cones. Note that the ball can shift laterally in relation to the ball plate. The conical cavities, as well as the ball, have hardened surfaces and the drive operates in an oil bath.

17.

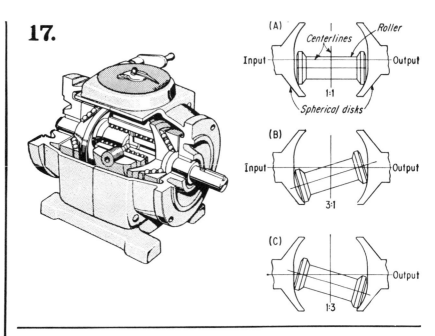

18.

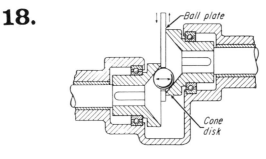

16.

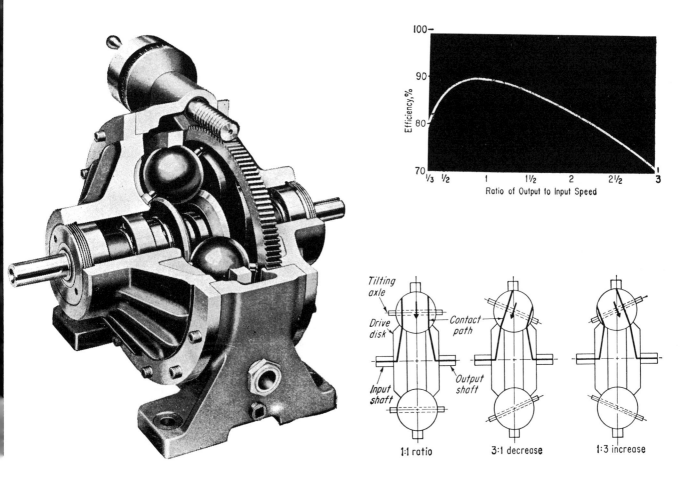

Multiple disk drives

Ball-and-disk drive (Fig 19). Friction disks are mounted on splined shafts to allow axial movement. Steel balls are carried by swing arms, rotate on guide rollers, and are in contact with driving and driven disks. Belleville springs provide the loading force between the balls and the disks. Position of balls controls the ratio of contact radii, and thus the speed.

Only one pair of disks is required to provide the desired speed ratio—purpose of multiple disks is to increase the torque capacity. And, if the load changes, a centrifugal loading device increases or decreases the axial pressure in proportion to the speed. The helical gears permit the output shaft to be coaxial with the input shaft. Speed ratios are from 1:1

19.

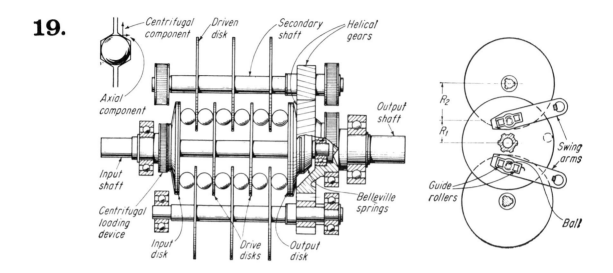

21.

22.

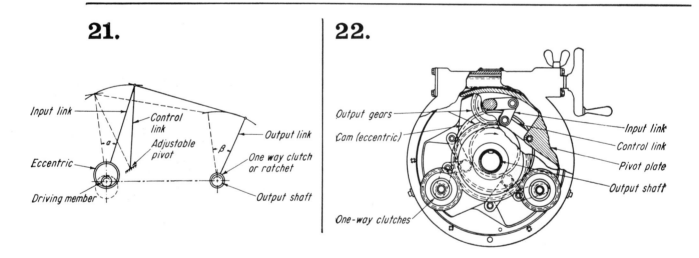

Impulse drives

Variable-stroke drive (Fig 21). This is a combination of a four-bar linkage with a one-way clutch or ratchet. The driving member rotates the eccentric which, through the linkage, causes the output link to rotate a fixed amount. On the return stroke, the output link overrides the output shaft. Thus a pulsating motion is transmitted to the output shaft which in many applications such as feeders and mixers is a distinct advantage. Shifting the adjustable pivot varies the speed ratio. By adding eccentrics, cranks, and clutches in the system, the fre-

quency of pulsations per revolution can be increased to produce a smoother drive.

Morse drive (Fig 22). The oscillating motion of the eccentric on the output shaft imparts motion to the input link, which in turn rotates the output gears. Travel of the input link is regulated by the control link which oscillates around its pivot and carries the roller, which rides in the eccentric cam track. Usually, three linkage systems and gear assemblies overlap the motions, two linkages on return, while the third is driving. Turning the hanto 1:5, efficiency to 92%. Small units to 9 hp; large units to 38 hp.

Oil-coated disks (Fig 20). Power is transmitted without metal-to-metal contact at 85% efficiency. The interleaved disk sets are coated with oil during operation. At the points of contact axial pressure applied by the rim disks compresses the oil film, increasing its viscosity. The cone disks transmit motion to the rim disks by shearing the molecules of the high-viscosity oil film.

Three stacks of cone disks (only one stack shown) surround the central rim stack. Speed changes are produced by moving the cone disks radially toward (output speed increases) or away (decreases) from the rim disks. A spring and cam on the output shaft maintain pressure on the disks at all times.

The drive is available to 60 hp, but much larger units have been built. For small units, air cooling is provided; for big units, water cooling is required.

Under normal conditions the drive has the capacity to transmit its rated power with negligible slip—1% at high speed, 3% at low speeds.

20.

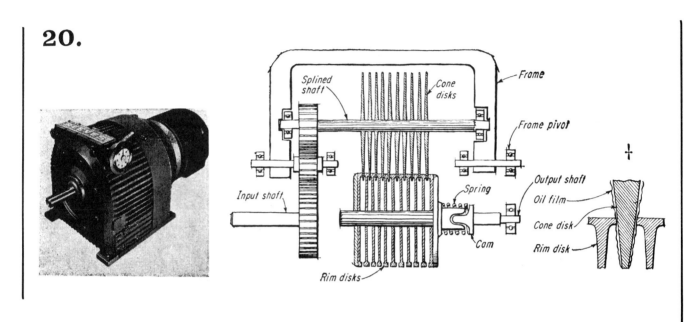

23.

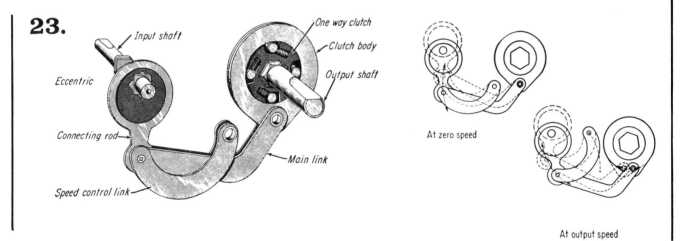

At zero speed

At output speed

dle repositions the control link and changes the oscillation angles of the input link, intermediate gear, and input gear. This drive is specified as constant-torque device with limited range. Maximum torque output is 175 ft-lb at maximum input speed of 180 rpm. The speed variation is between 4½:1 to 120:1.

Zero-Max drive (Fig 23). This drive, which is also based on the variable-stroke principle, delivers with 1800-rpm input, 7200 or more impulses per min to the output shaft at all speed settings above zero. The pulsating nature of this drive is again damped by the number of parallel working sets of mechanisms between the input and output shaft. The illustration above shows only one of these working sets.

At zero output speed, the eccentric on the input shaft moves the connecting rod up and down through an arc. There is no back and forth motion to the main link. To set the output speed, the pivot is moved (upwards in the illustration), thereby changing the direction of the connecting rod motion and imparting a back and forth motion to the main link. The one-way clutch mounted on the output shaft provides the ratchet action. Reversing the input shaft rotation will not reverse the output. However, the reversing can be accomplished in two ways—by a special reversible clutch, or by a bell-crank mechanism in gearhead models.

This drive belongs to the infinite speed range class, because its output speed passes through zero. Maximum input speed: 2000 rpm. Speed range: from zero to ¼ of input speed. Maximum capacity: to ¾ hp.

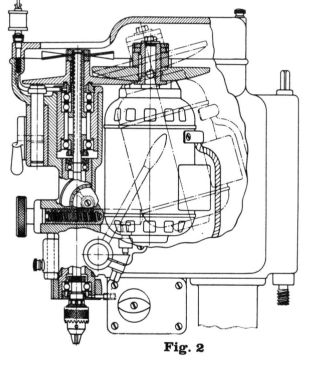

Fig. 1

Variable-Speed Drives –
Additional Variations

Fig. 1 –The well-known Sellers' disks consist of a device for transmitting power between fixed parallel shafts. Convex disks mounted freely on a rocker arm and pressing firmly against the flanges of the shaft wheels by a coiled spring form the intermediate sheave. Speed ratio changed by moving rocker lever. No reverse possible, but driven shaft may rotate above or below driver speed. Convex disk must be mounted on self aligning bearings to ensure good contact at all positions.

Fig. 2 –A curved disk device made possible by motorization. Motor is swung on its pivot in such a manner as to change the effective diameters of the contact circles. A compact drive for a small drill press.

Fig. 3 –Another motorized modification of an older device. Principle similar to Fig. **3**, but with only two shafts. Ratio changed by sliding motor in Vee guides.

Fig. 4 –Two cones mounted close together and making actual contact through a squeezed belt. Speed ratio changed by shifting belt longitudinally. Taper on cones must be moderate in order to avoid excessive wear on sides of belt.

Fig. 5 –These speed cones are mounted at any convenient distance apart and connected by a Condor Whipcord belt, whose outside edges consist of an envelope of tough, flexible, wear-resisting rubberized fabric built to withstand the wear caused by the belt edge travelling at a slightly different velocity to the part of the cone in actual contact. Speed ratio changed by sliding belt longitudinally.

Fig. 2

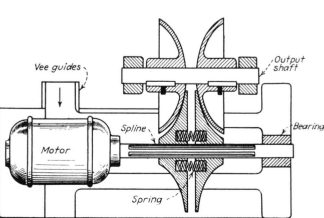

Fig. 3

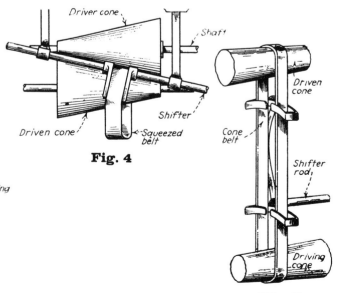

Fig. 4

Fig. 5

Fig. 6 —Another device to avoid belt "creep" and wear in speed cone transmissions. The inner bands are tapered on the inside and present a flat or crowned surface to the belts in all positions. Speed ratio changed by moving inner bands rather than main belts.

Fig. 7 —Another device for avoiding belt wear when using speed cones. Creeping acting of belt not entirely eliminated, and universal joints present problem of cost and maintenance.

Fig. 8 —An extension of the principle used in Fig. 7 whereby a roller is substituted for the belt, giving more compactness.

Fig. 9 —The main component of this drive is a hollow cone driven by a conical roller. Speed ratio changed by sliding driving unit in Vee guides. Note that when roller is brought to the center of the hollow cone, the two run at identical speed with the same characteristics as a cone clutch. This feature makes system very attractive where heavy torque at motor speed is required in combination with lower speeds for light preliminary operations.

Fig. 10 —In this transmission, the cones are mounted in line and supported by the same shaft. One cone is keyed to main shaft and the other is mounted on a sleeve. Power transmitted by series of rocking shafts and rollers. Pivoting rocking shafts and allowing them to slide changes speed ratio.

Fig. 11 —This J. F. S. transmission uses curved surfaces on its planetary rollers and races. The cone-shaped inner races revolve with the drive shaft, but are free to slide longitudinally on sliding keys. Strong compression springs keep these races in firm contact with the three planetary rollers.

Fig. 12 —Featuring simplicity with only five major parts, this Graham transmission employs three tapered rollers carried by a spider fastened to the drive shaft. Each roller has a pinion meshing with a ring gear connected to the output shaft. The speed of the rollers, and in turn, the speed of the output shaft is varied by moving contact ring longitudinally, thus changing the ratio of the diameters in contact.

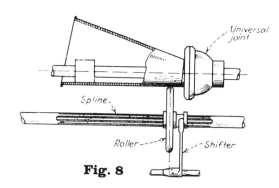

Fig. 8

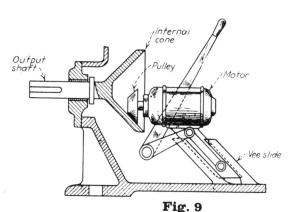

Fig. 9

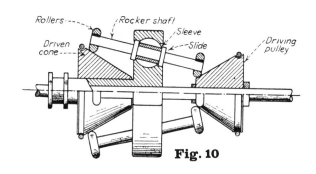

Fig. 10

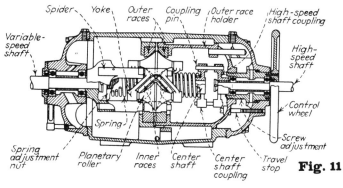

Fig. 11

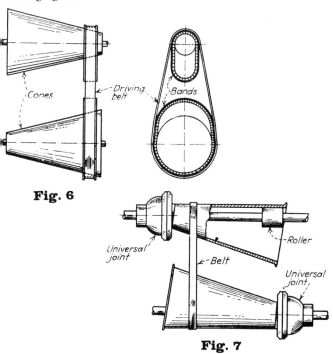

Fig. 6

Fig. 7

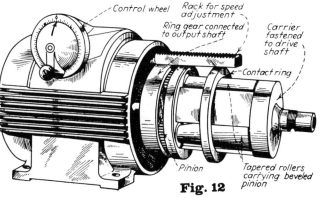

Fig. 12

Ratchet and Inertia Variable-Speed Drives

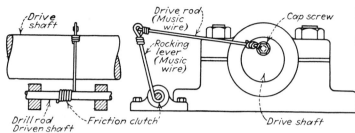

Fig. 1

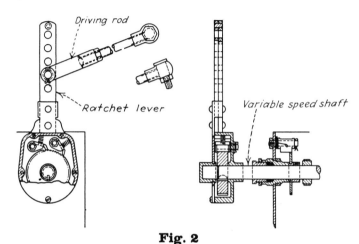

Fig. 2

Fig. 3

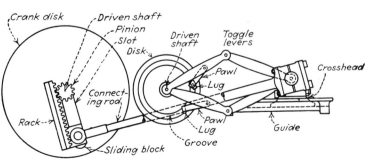

Fig. 4

Ratchet and inertia type mechanisms for variable-speed driving of heavy or light loads

CYRIL DONALDSON

Fig. 1 —A handy variable speed device suitable only for very light drives in laboratory or experimental work. Drive rod receives motion from drive shaft and rocks lever. Friction clutch made by winding wire around drill-rod in lathe with diameter slightly smaller than diameter of driven shaft. Speed ratio changed when unit is stationary, by varying length of rods, or throw of eccentric.

Fig. 2 —This Torrington lubricator drive illustrates the general principles of ratchet transmission devices. Reciprocating motion from a convenient sliding part, or from an eccentric, rocks the ratchet lever, which gives the variable speed shaft an intermittent uni-directional motion. Speed ratio can be changed only when unit is stationary. Placing fork of driving rod in different hole varies the throw of the ratchet lever.

Fig. 3 —An extension of the principle illustrated above, this Lenney transmission replaces the ratchet with an over-running clutch. Speed of the driven shaft can be varied while the unit is in motion by changing the position of the connecting lever fulcrum.

Fig. 4 —Another transmission employing the principle shown above. Crank disk imparts motion to connecting rod. Crosshead moves toggle levers which in turn give uni-directional motion to clutch wheel by means of friction pawls engaging in groove. Speed ratio changed by varying throw of crank with the aid of the rack and pinion.

Fig. 5 —A variable-speed transmission used on gasoline railroad section cars. The connecting rod from the crank, mounted on the constant speed shaft, rocks the oscillating lever and actuates the over-running clutch, thus giving intermittent but uni-directional motion to the variable speed shaft. The toggle link keeps the oscillating lever within a prescribed path. Speed ratio changed by swinging bell crank towards position shown in dotted lines, around pivot attached to frame. This varies the movement of the over-running clutch. Several units out of phase with each other are necessary for continuous motion of the shaft.

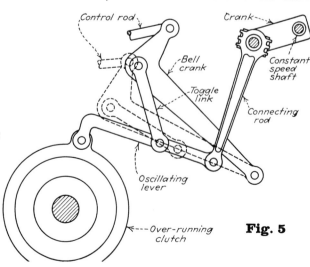

Fig. 5

Fig. 6 —This Thomas transmission is an integral part of an automobile engine in which piston motion is transferred by means of the conventional connecting rod to long arm of a bell-crank lever oscillating about a fixed fulcrum. Attached to the short arm of the bell crank lever is a horizontal connecting rod which in turn rotates the crankshaft. Crankshaft motion is rendered continuous and steady by means of a flywheel, but no power other than that required to operate auxiliaries is taken from this shaft. The main power output is transferred from the bell-crank lever to the over-running clutch by means of a third connecting rod. Speed ratio is changed by sliding the top end of the third connecting rod within the bell-crank lever by means of a cross-head and guide mechanism. Highest ratio is obtained when the crosshead is farthest from the fulcrum and movement of the cross-head toward the fulcrum reduces the ratio until a "neutral" position is reached when the center line of the connecting rod coincides with the fulcrum.

Fig. 7 —Another automobile transmission system built as an integral part of the engine, this Constantinesco torque converter features an inherently automatic change of speed ratio according to the speed and load on the engine. The constant speed shaft rotates a crank which in turn operates two oscillating levers having inertia weights at one end, while the other ends are attached by links to the rocking levers. Incorporated in these rocker levers are over-running clutches. Since at low engine speeds the inertia weights oscillate through a wide angle at low speed, the reaction of the inertia force on the other end of the lever is very slight, and the link imparts no motion to the rocker lever. Speed increase of engine causes inertia weight reaction to increase thereby rocking small end of oscillating lever as the crank rotates. Consequent motion rocks rocking lever by means of link and the variable shaft is driven in one direction.

Fig. 8 —Featuring a differential gear with an adjustable escapement, this transmission by-passes a variable proportion of the drive shaft revolutions. Constant speed shaft rotates freely mounted worm wheel carrying two pinion shafts. The firmly fixed pinions on these shafts in turn rotate the sun gear which meshes with other planetary gears, rotating the small worm gear attached to the variable speed output shaft.

Fig. 9 —In this Morse transmission, an eccentric cam, integral with the constant speed input shaft, rocks three ratchet clutches by means of a series of linkage systems containing three rollers running in a circular groove cut in the cam face. Uni-directional motion is conveyed to the output shaft from the clutches by planetary gearing. Speed ratio is changed by rotating anchor ring containing fulcrum of links, thus varying the stroke of the levers.

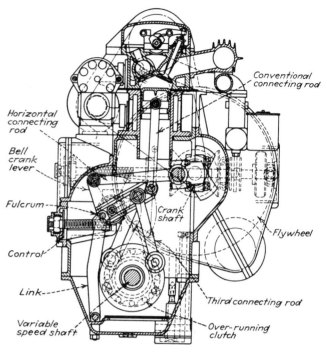

Fig. 6

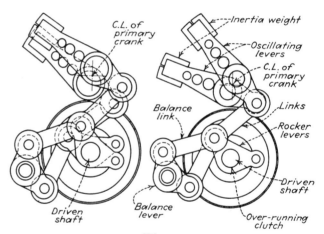

Fig. 7

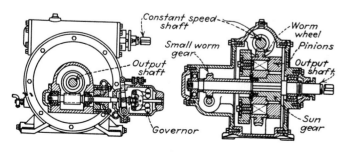

Fig. 8

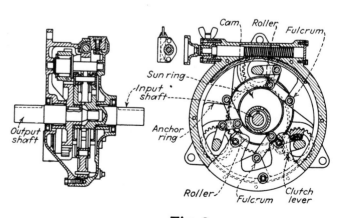

Fig. 9

Precision ball bearings replace gears in tiny speed reducers

Miniature bearings are taking over the role of gears in speed reducers where a very high speed change, either a speed reduction or speed increase, is desired in a tiny space. Such ball bearing reducers employing ingenious arrangements, as those designed by MPB Corp., Keene, N.H. (photo and drawings), are already providing speed ratios as high as 300-to-1 in a space ½-in. dia by ½-in. long.

And at the same time the designs give quiet operation, with both the input and output shafts rotating on the same line.

Much of the surge in interest for the ball bearing reducers, says Philip Dusini, engineering manager at MPB, is due to the pressure on mechanical engineers to compact their designs to meet the miniaturization gains in the electronic fields. "This need for compactness exists not only in space-borne instrumentation but in consumer products as well," says Dusini. Capstan drives of portable recorder and office dictating equipment as well as high-torque dental drills are now beginning to use them."

The advantages of the bearing-reducer concept lie in its simplicity. A conventional precision ball bearing is used as an epicyclic or planetary gearing device. The bearing inner ring, outer ring, and ball complement become, in a sense, the sun gear, internal gear and planet pinions.

Power transmission functions occur with either a single bearing or with two or more in tandem. Contact friction or traction between the bearing components transmits the torque. To prevent slippage, the bearings are preloaded just the right amount to achieve balance between transmitted torque and operating life.

Input and output functions always rotate in the same direction, irrespective of the number of bearings, and different results can be achieved by slight alterations in bearing characteristics. All these factors lead to specific advantages:

■ **Space saving.** The outside diameter, bore, and width of the bearings set the envelope dimensions of the unit. The housing need be only large enough to hold the bearings. In most cases the speed-reducer bearings can be built into the total system, conserving more space.

■ **Quiet operation.** The traction drive is between nearly perfect concentric circles with component roundness and concentricity, controlled to precise tolerances of 0.00005 in. or better. Moreover, operation is not dependent in any way on con-

ventional gear teeth. Thus quiet operation is inherent.

■ **High speed ratios.** Through design ingenuity and use of special bearing races, virtually any speed-reducing or speed-increasing ratio can be achieved. MPB studies show that speed ratios of 100,000-to-1 are theoretically possible with only two bearings.

■ **Low backlash.** Backlash is restricted mainly to the clearance between backs and ball retainer. Because the balls are preloaded, backlash is almost completely eliminated.

"On the other hand," says Dusini, "ball bearing reducers are limited as to the amount of torque that can be transmitted."

The three MPB units (below) illustrate the variety of designs possible:

■ **Torque increaser** (top drawing). This simple design, used to boost the output torque in an air-driven dental handpiece, provides a 2½-to-1 speed reduction. The speed reduces as the bearing's

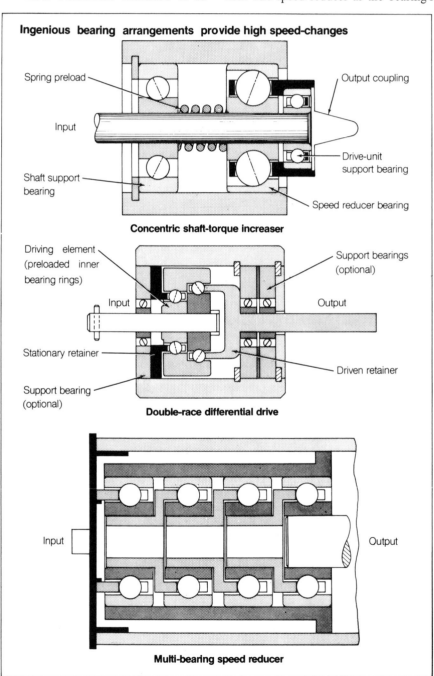

Ingenious bearing arrangements provide high speed-changes

Spring preload · Input · Shaft support bearing · Output coupling · Drive-unit support bearing · Speed reducer bearing

Concentric shaft-torque increaser

Driving element (preloaded inner bearing rings) · Input · Stationary retainer · Support bearing (optional) · Support bearings (optional) · Output · Driven retainer

Double-race differential drive

Input · Output

Multi-bearing speed reducer

outer ring is kept from rotating while the inner ring is driven; the output is taken from a coupling that is integral with the ball retainer.

The exact speed ratio depends on the bearing's pitch diameter, ball diameter, or contact angle. By stiffening the spring, the amount of torque transmitted increases, thereby increasing the force across the ball's normal line of contact.

■ **Differential drive** (middle drawing). This experimental reduction drive, now under test, uses the inner rings of a preloaded pair of bearings as the driving ele-

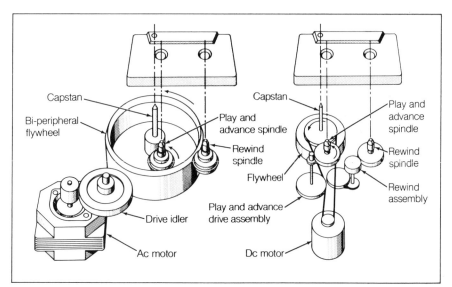

Capstan
Bi-peripheral flywheel
Play and advance spindle
Rewind spindle
Drive idler
Play and advance drive assembly
Ac motor

Capstan
Play and advance spindle
Rewind spindle
Flywheel
Rewind assembly
Play and advance drive assembly
Dc motor

Lightweight flywheel in new design (left) has higher inertia than conventional design (right). Its dual peripheries serve as drives for friction rollers.

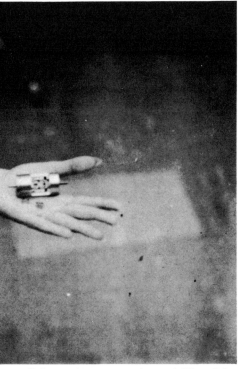

Cutaway shows compactness of differential drive, with instrument-type bearings

ment. The ball retainer of one bearing is the stationary element, and the opposing ball retainer is the driven element. The common outer ring is free to rotate. Keeping the differences between the two bearings small provides extremely high speed reductions. A typical test model has a speed reduction ratio of 200-to-1 and transmits 1 in.-oz of torque.

■ **Multi-bearing reducer** (bottom). This stack of four precision bearings achieves a 26-to-1 speed reduction to drive the recording tape of a dictating machine.

Its 1-in. by 3-in. compactness houses both the drive motor and reduction unit completely within the drive capstan. The balls are preloaded by assembling each bearing with a controlled interference or negative radial play. ■

Multifunction flywheel smoothes friction drive in tape cassette

A cup-shaped flywheel is performing a dual function in tape recorders by acting as a central drive for friction rollers as well as a high inertia wheel.

The flywheel is the heart of a drive train to be used in a new line of Wollensak cassette audio-visual tape recorders.

The models include record-playback and playback-only portables and decks. Slide synchronization and remote control models will be available in April and audio-active model next fall.

Fixed parameters. The cassette concept has several fixed parameters—the size of the tape container (4 x 2½ in.), the distance between the hubs onto which the tape is wound, and the operating speed. The speed, standardized at 1⅞ ips, made it possible to enclose enough tape in the container for lengthy recordings. Cassettes are available for recording on one side for 30, 45, or 60 min.

The new designs include a motor comparable in size and power to those used in standard reel-to-reel recorders, and a large bi-peripheral flywheel and sturdy capstan that reduces wow and flutter and drives the tape. A patent application has been filed for the flywheel design.

The motor drives the flywheel and capstan assemblies. The flywheel moderates or overcomes variations in speed that cause wow and flutter. The accuracy of the tape drive is directly related to the inertia of the flywheel and the accuracy of the flywheel and capstan. The greater the

inertia the more uniform is the tape drive, and the less pronounced the wow and flutter.

The flywheel is nearly twice as large as the flywheel of most portable cassette recorders, which average less than 2 in. dia. Also, a drive idler is used on the Wollensak models while thin rubber bands and pulleys are employed in conventional portable recorders.

Take-up and rewind. In the new tape drive system, the flywheel drives the take-up and rewind spindle. In play or fast-advance mode, the take-up spindle makes contact with the inner surface of the counterclockwise moving flywheel, moving the spindle counterclockwise and winding the tape onto the hub. In the rewind mode, the rewind spindle is brought into contact with the outer periphery of the flywheel, driving it clockwise and winding the tape onto the hub.

According to Wollensak engineers, the larger ac motor has a service life five times that of a dc motor.

Basic performance for all of the new models are identical: frequency response of 50 to 8000 Hz; wow and flutter of less than 0.25%; signal-to-noise ratio of more than 46 db and a 10-w amplifier.

All the models also have identical operating controls. One simple lever controls fast forward or reverse tape travel. A three-digit, pushbutton resettable counter permits the rapid location of specific portions of recorded programs. ■

Controlled differential drives

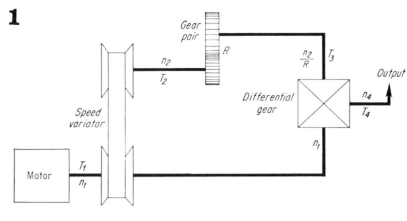

1

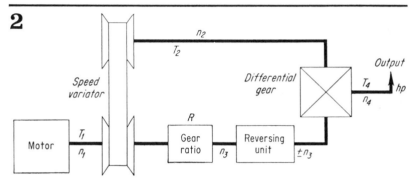

2

By coupling a differential gear assembly to one of the variable speed drives you can increase the horsepower capacity—at the expense of the speed range—or you can increase the speed range—at the expense of the horsepower range. Numerous combinations of the variables are possible.

The type of differential depends on the manufacturer. Some systems have bevel gears, others have planetary gears. Both single and double differentials are employed. Variable-speed drives with differential gears are commercially available up to 30 hp.

Horsepower - increase differential (Fig 1). The differential is coupled so that the output of the motor is fed into one side and the output of the speed variator into the other side. An additional gear pair is employed as illustrated.

Output speed $\quad n_4 = \frac{1}{2}\left(n_1 + \frac{n_2}{R}\right)$

Output torque

$$T_4 = 2T_3 = 2RT_2$$

Output hp

$$\text{hp} = \left(\frac{Rn_1 + n_2}{63,025}\right)T_2$$

hp increase

$$\Delta\text{hp} = \left(\frac{Rn_1}{63,025}\right)T_2$$

Speed variation

$$n_{4\,\text{max}} - n_{4\,\text{min}}$$

$$= \frac{1}{2R}\left(n_{2\,\text{max}} - n_{2\,\text{min}}\right)$$

Speed range increase differential (Fig 2). This arrangement achieves a wide range of speed with the low limit at zero or in reverse direction.

3

Bruning Model 50 Whiteprinter

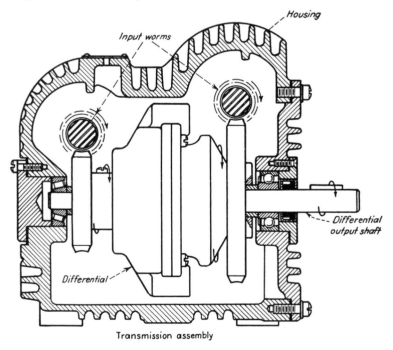

Transmission assembly

VARIABLE SPEED TRANSMISSION consists of two sets of worm gearing feeding into a differential mechanism. Output shaft speed depends on difference in rpm between the two input worms. When worm speeds are equal, output is zero. Each worm shaft carries a cone-shaped pulley. These pulleys are mounted so that their tapers are in opposite directions. Shifting the position of the drive belt on these pulleys has a compound effect on output speed.

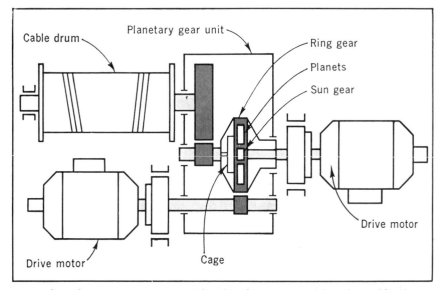

Power flow from two motors combine in planetary to drive the cable drum.

Twin-motor planetary gears provide safety plus dual-speed

A major fear that many operators and owners of hoists and cranes have is the possible catastrophic damage that can occur if the driving motor of a unit should fail for any reason. An interesting solution to this problem is to feed the power of two motors of equal capacity into a planetary gear drive.

Power supply. Each of the motors is designed to supply half the required output power to the hoisting gear (diagram above). One motor drives the ring gear, which has both external and internal teeth. The second motor feeds directly to the sun gear.

Both the ring gear and sun gear rotate in the same direction. If both gears rotate at the same speed, the planetary cage, which is coupled to the ouput, will also revolve at the same speed (and in the same

direction). It would be as if the entire inner works of the planetary were fused together. There would be no relative motion. Then, if one motor fails, the cage will revolve at half its original speed, and the other motor can still lift with undiminished capacity. The same principle holds true when the ring gear rotates more slowly than the sun gear.

No need to shift gears. Another advantage is that two working speeds are available by a simple switching arrangement. This makes it unnecessary to shift gears to obtain either speed.

The diagram shows a unit for a steel mill crane. The gear units are manufactured by Simmering-Graz-Pauker of Vienna, Austria. □

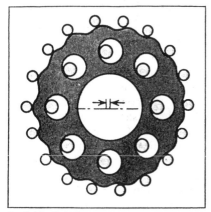

Cycloid disk is eccentrically driven to contact the pins on input shaft.

than there are pins (17 to 18 for the drive in the drawing), the disk during each shaft rotation loses one position in the opposite direction for a speed-reduction ratio of $(18 - 17)/17 = 1/17$. Double-reduction cycloids are being built at ratios up to 7500:1.

Phase adjustment. The problem of how to change the register or the phase orientation between two elements such as two coaxial shafts or a gear and a sprocket is neatly solved with an adjusting ring that contains an "inside-out" Harmonic Drive unit (drawing), designed by the Harmonic Drive Div. of United Shoe Machinery.

The external member acts as the adjusting knob. The internal spline

Cycloid drives take on improved forms

Return of the cycloids. Several companies in the past have hit upon the idea of substituting cycloid teeth and pins (drawing above) for the gear teeth in two-gear planetary systems. Now a Japanese-based firm, the Sumitomo Machinery Corporation of America, has brought the cycloid drives back, but with an

ingenious arrangement that combines two cycloid disks back-to-back, 180 degrees out of phase, to balance the centrifugal loads.

In the drive, a cycloid disk is forced to contact the pins at consecutive points by an eccentric mounted on the input shaft. Because there is one tooth less on the disk

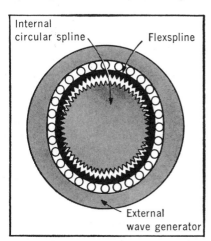

Stepless adjustment without backlash is provided by "inside-out" drive unit

is attached to one shaft, the Flexspline to the other shaft. Rotating the external member one degree changes the relationship between the two shafts approximately 1/40th of a degree, thus permitting stepless, no-backlash adjustment.

Mechanisms simplify timing between shafts

A new concept in machinery design has emerged, almost accidentally, through some recent developments in timing mechanisms. In brief, the concept, as defined by R. V. Hendershot, president of Candy Mfg. Co. (Chicago), involves "dynamic timing control"—the use of mechanisms with the ability to change the timing relationships between multiple machine movements.

For years, engineers have been trying to overcome the problems of achieving accurate timing relationships by mechanical means.

The only solution so far has been to select mechanical actions in an assembly and make them adjustable, so they may be appropriately synchronized to the timing of primary motions. This approach has not proved adequate.

Nonstop correction. Adjustable timing, according to Hendershot, has the disadvantage that it requires that a machine be stopped for timing correction. Once the machine has been stopped, it is difficult for a mechanic to determine how much adjustment is required. A series of trial-and-error adjustments then often results in only "good enough" settings, not the optimum setting. Moreover, adjustable mechanisms usually employ set screws, friction clamps, or slots that can slip.

One of the few techniques for dynamic timing known to designers is the chain take-up method (top drawing). According to Hendershot, whose company made the sketches in the Leonardo da Vinci style, the chain method is a low-cost way of precisely timing or changing angular position between two shafts while they are running.

Control is infinitely adjustable within a prescribed range, and timing is positive—the shafts cannot slip. Adjusting the control knob varies the length of chain between the input and output shafts of the timing unit, so it is possible to change the phase between the two shafts up to a full 360 deg.

More elaborate. The running differential drive (middle drawing) is a more sophisticated means of controlling timing or registration. By altering the position of the spider gears through use of a worm-gear

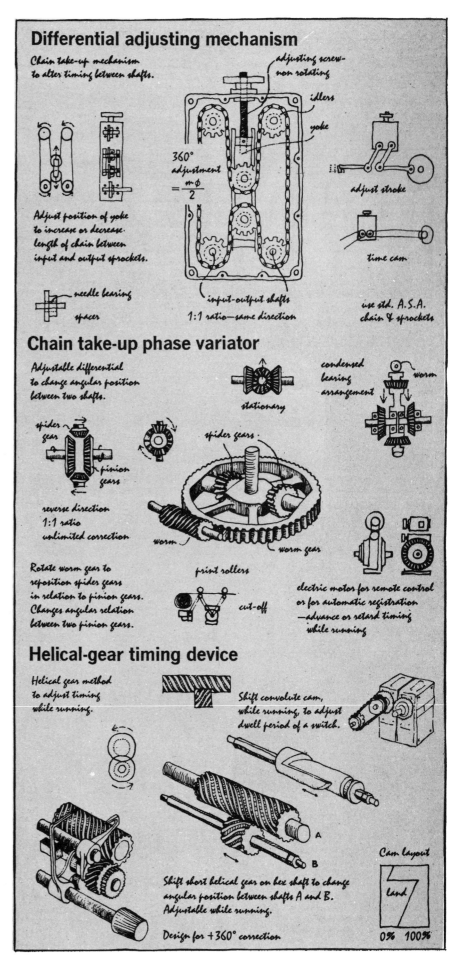

Differential adjusting mechanism

Chain take-up mechanism to alter timing between shafts.

Adjust position of yoke to increase or decrease length of chain between input and output sprockets.

needle bearing spacer

$360°$ adjustment $= \dfrac{\pi \phi}{2}$

adjusting screw— non rotating

idlers

yoke

input-output shafts

1:1 ratio—same direction

adjust stroke

time cam

use std. A.S.A. chain & sprockets

Chain take-up phase variator

Adjustable differential to change angular position between two shafts.

spider gear

pinion gears

reverse direction 1:1 ratio unlimited correction

Rotate worm gear to reposition spider gears in relation to pinion gears. Changes angular relation between two pinion gears.

stationary

spider gears

worm

worm gear

print rollers

cut-off

condensed bearing arrangement

worm

electric motor for remote control or for automatic registration —advance or retard timing while running

Helical-gear timing device

Helical gear method to adjust timing while running.

Shift convolute cam, while running, to adjust dwell period of a switch.

A

B

Shift short helical gear on hex shaft to change angular position between shafts A and B. Adjustable while running.

Design for +360° correction

Cam layout

land

0% 100%

262

control, the timing can be advanced or retarded continuously.

This type of control is useful when there is a cumulative error, such as in web processing equipment, and where 360-deg. adjustment is inadequate. The running differential type is also well suited to an electric motor drive on the control worm gear for remote or automated applications.

Newest method. The helical-gear method is the newest means of positively adjusting the timing while running. Shifting one helical gear parallel to another imparts a rotary motion between the two. By designing the travel of one to equal the length of a complete revolution of one tooth of the other, a 360-deg. phase adjustment can be realized (bottom drawing, facing page).

This principle has been known to designers, but in this device it is coupled with a spiral cam to create an adjustable-while-running cam switch for the timing control of electrically operated machine elements. Both timing and dwell can be altered while the machine is running.

Hendershot cites an application in which the timing-switch device is driven by a parent machine at a one-to-one ratio.

Switching operation. An engineer may want a solenoid, motor, hydraulic cylinder, clutch-brake, or other machine device turned on for a certain period of dwell, say 90-deg. or a quarter-revolution during each full revolution of the machine cycle. Adjustment of the dwell period is accomplished by rotating the lower knob, which changes the percentage of on-off time per cycle.

It is just as important, however, in such switching operations to establish precisely the trip-on point within the machine cycle. This is where the pair of parallel shaft helical gears comes in. Turning the upper adjusting knob (even while the machine is running) advances or retards the trip-on point.

All these operations previously required the design of an accurate cam with the desired rise-and-dwell period, plus the assembly and pinning of this cam on its shaft at a prescribed angular position so the proper trip-on point is obtained.

The three timing mechanisms are being produced by Candy Mfg. Co.

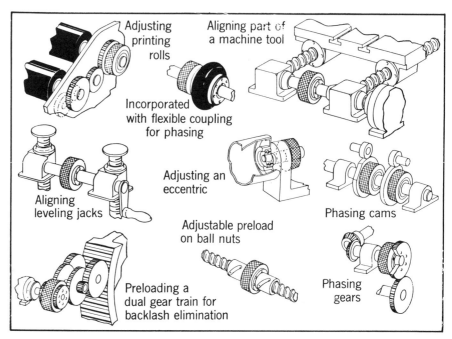

Coupling serves in a variety of applications when the phase relationship between two shafts, rollers, or other operating components must be changed.

Coupling design offers precise angular shaft adjustments

The problem of how to change the orientation between two revolving components, such as shafts (drawings above), is neatly solved with the new improved version of the phase-adjusting coupling developed by USM Corp's Gear Systems Div., Beverly, Mass. The new design of the "Infinit Indexer" allows manual angular shaft adjustments to be made easily by merely loosening a set screw and rotating a knurled outer ring.

The device resembles a coupling but contains a version of the well-known harmonic drive. The drive has a solid spline that is attached by the user to one shaft of the application. A second flexible spline, which meshes with the solid spline, is attached to the second shaft. The flexible spline has two more teeth than the solid spline. Both splines are contained within a knurled "wave generator," setting up a differential relationship.

Easy adjustment. Normally the harmonic drive elements in the coupling are "locked up," thereby providing the characteristics of a one-to-one coupling. Adjustment in phase is made by turning the knurled wave generator with relation to the driven shaft, each full turn resulting in 3.6 deg. of travel of the drive element. Thus, it is extremely simple to make adjustments down to just one minute of arc.

The new design is self-locking, has no backlash, and needs only two sizes to encompass a torque range from 0 to 8000 in.-lb. USM Corp. predicts many applications as a phase-shifting coupling to precisely adjust or correct register of in-line elements of conveyor drives, packaging machinery, machine tools, textile machinery, printing presses, and similar types of equipment.

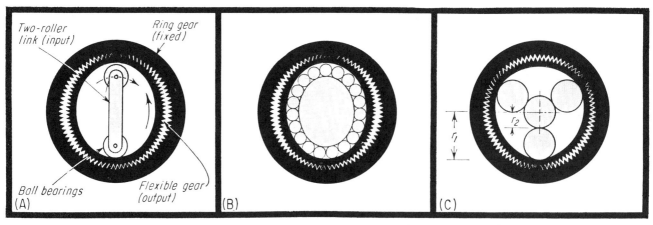

THREE VERSIONS OF DRIVE. Flexible gear is deflected in (A) by two-roller link; (B) by elliptical cam rolling within ballbearings; (C) by planetary-gear system for still-higher speed change.

How the Drive Works

The rotary version has a ring gear with internal teeth mating with a flexible gear with external teeth—see (A) in previous illustration. These teeth are straight-sided, and both gears have the same circular pitch, hence the areas of engagement are in full mesh. But the flexible inner gear has fewer teeth than the outside gear and therefore its pitch circle is smaller. Third member of the drive is a link with two rollers which rotates within the flexible gear, causing it to mesh with the ring gear progressively at diametrically opposite points. This propagates a traveling strain, or deflection wave in the flexible gear—hence, United Shoe's tradename, Harmonic Drive. If motion of the center link or "wave generator" is clockwise, and the ring gear is held fixed, the flexible gear will rotate (or "roll") counterclockwise at a slower rate, with constant angular velocity.

Teeth are stationary where in mesh, thus acting as splines in full contact. Movement of the flexible, driven member is confined to that area where teeth are disengaged. Each rotation of the center link moves the flexible gear a distance equal to the tooth differential between the two gears. Thus, speed ratio between center link (input) and flexible gear (output) is

$$\frac{V_o}{V_i} = \frac{N_f - N_r}{N_f}$$

where V_o = output velocity, rpm; V_i = input velocity, rpm; N_r = number of teeth (or pitch dia) of the ring gear; N_f = number of teeth (or pitch dia if permitted to take its full circular form) of the flexible gear.

For a drive with, say, 180 teeth in the ring gear and 178 teeth in the flexible gear (for the drive illustrated, difference in the number of teeth must be an even integer), the speed ratio will be

$$\frac{V_o}{V_i} = \frac{178 - 180}{178} = -\frac{1}{89}$$

Negative sign indicates that the input and output move in opposite directions. Actually, any one of the three basic parts can be held fixed and the other two used interchangeably as input and output.

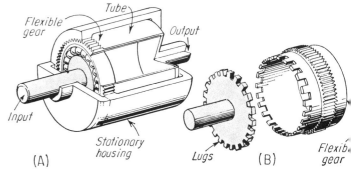

TWO METHODS OF COUPLING
flexible gear to shafts:
(A) by means of tubing; (B) with lugs.

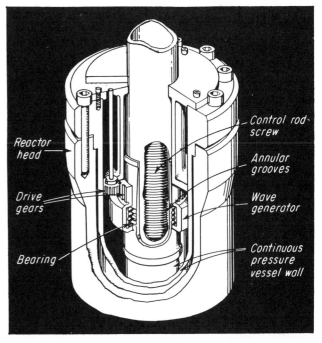

ROTARY-TO-LINEAR VERSION moves the control-rod linearly in this nuclear reactor head without need for mechanical contact through the sealed inner tube.

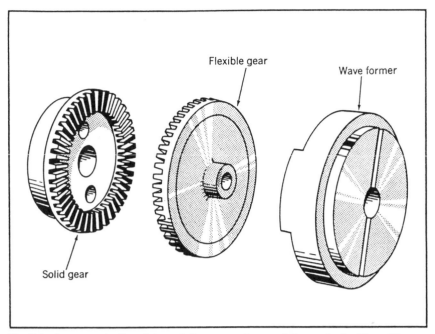

Flexible gear

Wave former

Solid gear

Flexible face gear is flexed by rotating wave former into contact with solid gear at point of mesh. The two gears have slightly different numbers of teeth.

Flexible face-gears make efficient high-reduction drives

A new system of flexible face-gearing is providing designers with a means of obtaining high-ratio speed reductions in compact trains with concentric input and output shafts.

With the new approach, reduction ratios range from 10:1 to 200:1 for single-stage reducers, whereas ratios of millions to one are possible for multi-stage trains. Patents on the flexible face-gear reducers are held by Clarence Slaughter of Power Tronics Co. (Grand Rapids, Mich.).

Building blocks. Single-stage gear reducers consist of three basic parts: a flexible face-gear made of plastic or thin metal; a solid, non-flexing face-gear; and a wave form-er with one or more sliders and rollers to force the flexible gear into mesh with the solid gear at points where the teeth are in phase.

The high-speed input to the system usually drives the wave former. Low-speed output can be derived from either the flexible or the solid face-gear; the gear not connected to the output is fixed to the housing.

Teeth make the difference. Motion between the two gears depends on a slight difference in their number of teeth. The difference is usually one or two teeth, but drives with gears that have up to a difference of 10 teeth have been devised.

On each revolution of the wave former there is a relative motion between the two gears that equals the difference in their numbers of teeth. The reduction ratio equals the number of teeth in the output gear divided by the difference in their numbers of teeth.

Two-stage and four-stage gear reducers are obtained by combining flexible and solid gears with multiple rows of teeth and driving the flexible gears with a common wave former.

There are also many special applications. Hermetic sealing is accomplished by making the flexible gear serve as a full seal and by taking-off output rotation from the solid gear.

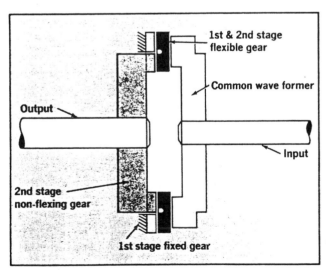

1st & 2nd stage flexible gear

Common wave former

Output

Input

2nd stage non-flexing gear

1st stage fixed gear

Two-stage speed reducer is driven by a common wave former operating against an integral flexible gear for both stages.

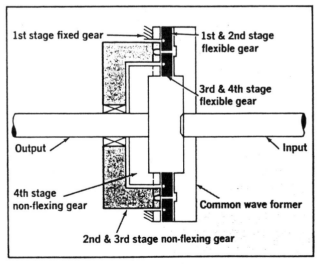

1st stage fixed gear

1st & 2nd stage flexible gear

3rd & 4th stage flexible gear

Output

Input

4th stage non-flexing gear

Common wave former

2nd & 3rd stage non-flexing gear

Four-stage speed reducer can, theoretically, attain reductions of millions to one. The train is both compact and simple.

Ball-gear speed variator

All sorts of products from computers to heavy-duty power drives may benefit from the smoothness and compactness of an infinitely variable, positive, in-line drive developed by Jack J. Gilbert, president of J. J. Gilbert Co., Spring Valley, N.Y. Automotive applications are an early possibility.

Gilbert has applied for patents on the drive and on a key feature of it: a right-angle ball-gear train for the control input to the planetary gear carrier (drawing). The ball teeth of the gear, held in sockets or fixed retainers, are free to roll as they mesh with a spiral groove in a face gear rotating at right angles. Thus, sliding friction is minimized.

Operating efficiency of the ball-gear train at least equals that of spur gears, says Gilbert, and may approach that of ball bearings.

Low-power control. The infinitely variable drive consists of a planetary gear train with input to the sun gear and output taken from the internal ring gear. A low-power secondary input to the planet-gear carrier controls speed. That's where the ball-gear train comes in.

With constant-speed input to the drive, output can be infinitely and steplessly varied by changing speed of the control input. As the control speed is increased, output speed decreases gradually to a stop, then accelerates in the opposite direction. The control input rotates only in one direction.

Speed at which the output becomes stationary is determined by the combination of drive input speed and the ratio between the ball-gear and the planetary train. Control input may be driven separately by a small electric motor, or it can take off from the drive input through a friction-wheel drive. At low power, slippage would be negligible.

Solving a problem. Gilbert devised the ball-tooth gear mechanism to solve the problem of connecting the control input to the planet gear carrier. Worm gearing proved unsatisfactory. The worm wheels tended to override the worms, making speed control difficult. The ball gear showed no tendency to override

Small motor with spiral-grooved face gear delivers control input to novel "ball" gear in setup by inventor Jack Gilbert, right. Drawings below show the whole drive system, with detail at far right showing how balls can be retained by two-piece split race, press-fitted on hub after assembly.

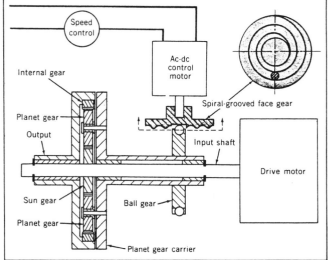

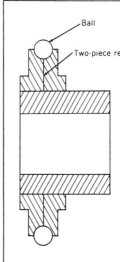

or back-drive the spiral-grooved face gear, and the friction loss was much less than in worm drives.

Tests by Gilbert indicate that point contact is achieved between ball and groove. The grooves, with a radial cross section, are cut in a constant-pitch spiral, and the contact point with the groove has been proven to follow an orbital path on the surface of the ball.

Broad choice. An almost limitless variety of gear ratios is obtainable with the ball-gear train by changing the number of teeth on the ball gear or by using multiple-lead spiral grooves on the face gear. To increase the arc of contact between ball and groove, the face gear can be machined into a dish shape.

When a large face gear is used, location of the point of mesh with the ball gear is not critical. It can occur anywhere along a radius of the face gear. Several ball gears, each operating at a different ratio, can work with the same face gear.

The ball-gear train is hailed as most effective in applications that require high efficiency at high gear ratios. It can compete favorably with worm drives, which are notably low in efficiency.

Balls are those used in standard ball bearings. They can be mounted in open sockets, with fixed retainers to hold them, or the race members can be split and then pressed together on a hub after assembly of the balls.

Compact Rotary Sequencer

Two coaxial rotations, one clockwise and one counterclockwise, are derived from a single clockwise rotation.

A proposed rotary sequencer is assembled from a conventional planetary differential gearset and a latching mechanism. Its single input and two rotary outputs (one clockwise and one counterclockwise) are coaxial, and the output torque is constant over the entire cyle. Housed in a lightweight, compact, cylindrical package, the sequencer requires no bulky ratchets, friction clutches, or cam-and-track followers. Among its possible applications are sequencing in automated production-line equipment, in home appliances, and in vehicles.

The sequencer is shown in Figure 1. A Sun gear connects with four planetary gears that engage a ring gear. With the ring gear held stationary, clockwise rotation of the Sun gear causes the entire planetary-gear carrier also to rotate clockwise. If the planetary-gear carrier is held fixed, the ring gear will rotate counterclockwise when the Sun gear rotates clockwise.

Figure 2 shows the latch. It consists of a hook (the carrier hook) that is rigidly attached to the planetary-gear carrier, a ring that is rigidly attached to the ring gear, and a latch pivot arm with a pair of latch rollers attached to one end. The other end of the pivot arm rotates about a short shaft that extends from the fixed wall of the housing.

The sequencer cycle starts with the ring latch roller resting in a slot in the ring. This locks the ring and causes the planetary-gear carrier to rotate clockwise with the input shaft (Figure 2a). When the carrier hook has rotated approximately three-quarters of a complete cycle, it begins to engage the planet-carrier latch roller (Figure 2b), causing the latch pivot arm to rotate and the ring latch roller to slip out of its slot (Figure 2c). This frees the ring and ring gear for counterclockwise motion, while locking the carrier. After a short interval of concurrent motion, the planetary-gear output shaft ceases its clockwise motion, and the ring-gear output shaft continues its clockwise motion.

When the ring reaches the position in Figure 2d, the cycle is complete, and the input shaft is stopped. If required, the input can then be rotated counterclockwise, and the sequence will be reversed until the starting position (Figure 2a) is reached again.

In a modified version of the sequencer, the latch pivot arm is shortened until its length equals the radii of the rollers. This does away with the short overlap of output rotations when both are in motion. For this design, the carrier motion ceases before the ring begins its rotation.

This work was done by Walter T. Appleberry of Rockwell International Corp. for **Johnson Space Center,** *Houston, Texas. For commercial use, contact the center; refer to MSC-19514.*

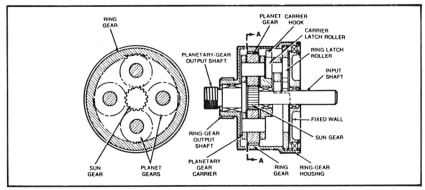

Figure 1. The **Rotary Sequencer** has a ring-gear output (in color) coaxial with a planetary-gear output (in gray). Clockwise rotation of the input is converted to clockwise rotation of the planetary-gear output followed by counterclockwise rotation of the ring-gear output. The sequence is controlled by the latch action described in Figure 2.

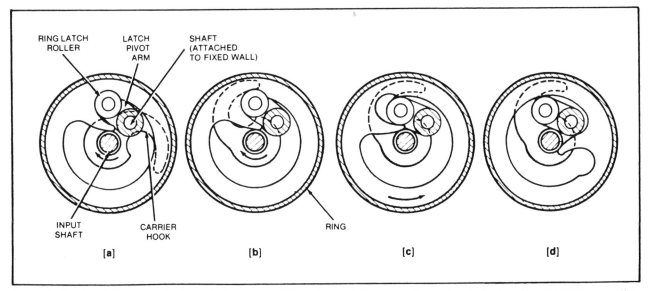

Figure 2. The **Latch Sequence** is shown in four steps: (a) The input shaft rotates the carrier clockwise while the ring latch roller holds the ring gear stationary; (b) the carrier hook begins to engage the carrier latch roller; (c) the ring latch roller begins to move out of its slot, and carrier motion ceases while the ring begins to move; and (d) the sequence has ended with the ring in its final position.

Planetary gear systems

Designers keep finding new and useful planetaries.
Forty-eight popular types are given here with their
speed-ratio equations

JOHN H. GLOVER

Minuteman cover drive (American Electric Co.)

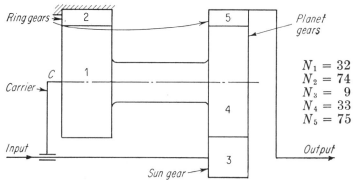

Ring gears
Carrier
Input
Planet gears
Output
Sun gear

$$N_1 = 32$$
$$N_2 = 74$$
$$N_3 = 9$$
$$N_4 = 33$$
$$N_5 = 75$$

Ring gear 2 fixed; ring gear 5 output

Speed-ratio equation

$$R = \frac{1 + \dfrac{N_4 N_2}{N_3 N_1}}{1 - \dfrac{N_4 N_2}{N_5 N_1}} = \frac{1 + \dfrac{(33)(74)}{(9)(32)}}{1 - \dfrac{(33)(74)}{(75)(32)}} = -541\tfrac{2}{3}$$

Planetary gear for pulling 95-ton blast-resistant
lid to cover and uncover underground Minute-
man missiles. Schematic at left. Author's equations lead directly to the speed-ratio equation
for the system, boxed at left.

Symbols

C = carrier (also called "spider")—a non-gear
member of a gear train whose rotation
affects gear ratio

N = number of teeth
R = overall speed reduction ratio
1, 2, 3, etc = gears in a train (corresponding to
labels on schematic diagram)

Double-eccentric drives

Input is through double-throw crank (carrier). Gear 1

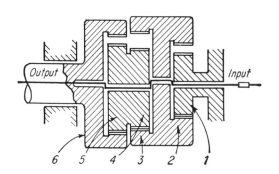

Output
Input
6 5 4 3 2 1

$$R = \frac{1}{1 - \dfrac{N_5 N_3 N_1}{N_6 N_4 N_2}}$$

When $N_1 = 103$, $N_2 = 110$, $N_3 = 109$,
$N_4 = 100$, $N_5 = 94$, $N_6 = 96$

$$R = \frac{1}{1 - \dfrac{(94)(109)(103)}{(96)(100)(110)}} = 1505$$

Coupled planetary drives (Ref. 1)

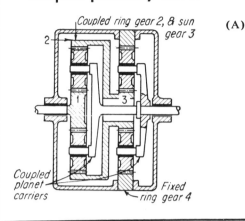

(A)

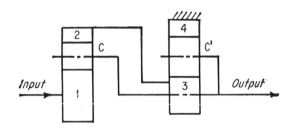

Coupled ring gear 2, & sun gear 3

2

Coupled planet carriers

Fixed ring gear 4

$$R = 1 - \frac{N_2 N_4}{N_1 N_3}$$

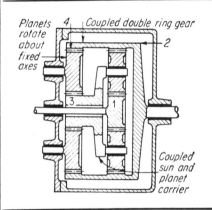

(B)

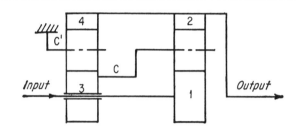

Planets rotate about fixed axes

4   Coupled double ring gear

2

3   1

Coupled sun and planet carrier

$$R = \left(1 + \frac{N_2}{N_1}\right)\left(-\frac{N_4}{N_3}\right) - \frac{N_2}{N_1}$$

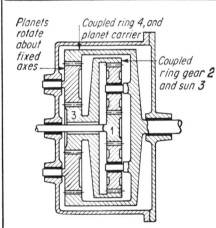

(C)

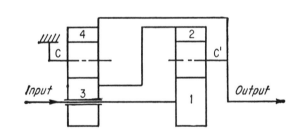

Planets rotate about fixed axes

Coupled ring 4, and planet carrier

Coupled ring gear 2 and sun 3

3   1

$$R = 1 + \frac{N_2}{N_1}\left(1 + \frac{N_4}{N_3}\right)$$

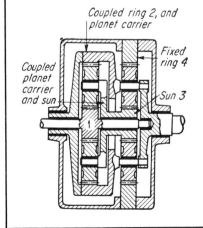

(D)

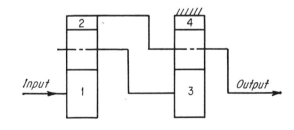

Coupled ring 2, and planet carrier

Fixed ring 4

Coupled planet carrier and sun

Sun 3

1

$$R = 1 + \frac{N_4}{N_3}\left(1 + \frac{N_2}{N_1}\right)$$

Fixed-differential drives

Output is difference between speeds of two parts leading to high reduction ratios

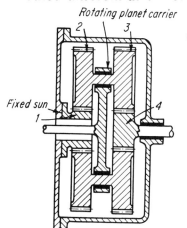

Rotating planet carrier

Fixed sun

(A)

$N_1 = 20$
$N_2 = 31$
$N_3 = 32$
$N_4 = 19$

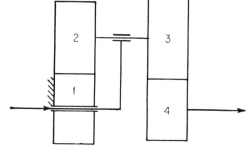

$$R = \frac{1}{1 - \dfrac{N_3 N_1}{N_4 N_2}} = \frac{1}{1 - \dfrac{(32)(20)}{(19)(31)}} = -11.549$$

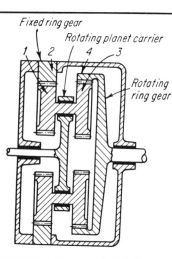

Fixed ring gear
Rotating planet carrier
Rotating ring gear

(B)

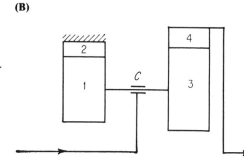

$$R = \frac{1}{1 - \dfrac{N_3 N_2}{N_4 N_1}}$$

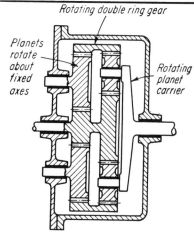

Rotating double ring gear

Planets rotate about fixed axes

Rotating planet carrier

(C)

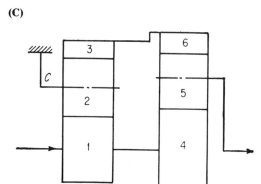

$$R = \frac{1 + (N_4/N_6)}{(N_4/N_6) - (N_1/N_3)}$$

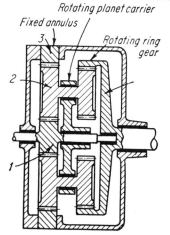

Rotating planet carrier
Fixed annulus
Rotating ring gear

(D)

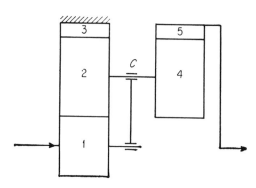

$$R = \frac{1 + (N_3/N_1)}{1 - \dfrac{N_4 N_3}{N_5 N_2}}$$

Simple planetaries and inversions

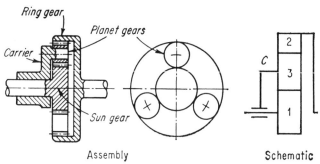

Ring gear
Planet gears
Carrier
Sun gear
Assembly
Schematic

Input member	Fixed member	Output member	Speed-ratio equation
1	C	2	$R = -N_2/N_1$
2	C	1	$R = -N_1/N_2$
1	2	C	$R = 1 + (N_2/N_1)$
2	1	C	$R = 1 + (N_1/N_2)$
C	2	1	$R = \dfrac{1}{1 + (N_2/N_1)}$
C	1	2	$R = \dfrac{1}{1 + (N_1/N_2)}$

Input member	Fixed member	Output member	Speed-ratio equation
1	C	3	$R = \dfrac{N_2 N_3}{N_1 N_4}$
1	3	C	$R = 1 - \dfrac{N_2 N_3}{N_1 N_4}$
3	1	C	$R = 1 - \dfrac{N_1 N_4}{N_2 N_3}$
3	C	1	$R = \dfrac{N_4 N_1}{N_3 N_2}$
C	1	3	$R = 1 \Big/ \left(1 - \dfrac{N_1 N_4}{N_2 N_3}\right)$
C	3	1	$R = 1 \Big/ \left(1 - \dfrac{N_2 N_3}{N_1 N_4}\right)$

Continued on next page

Humpage's bevel gears

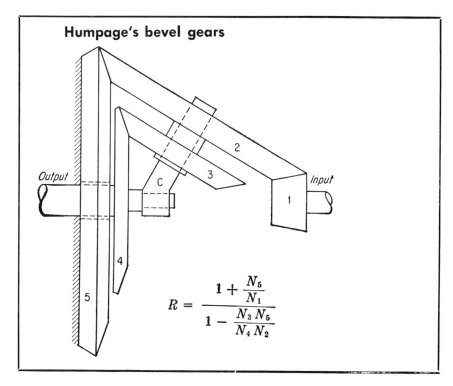

$$R = \frac{1 + \dfrac{N_5}{N_1}}{1 - \dfrac{N_3 N_5}{N_4 N_2}}$$

References:

1. D. W. Dudley, ed, *Gear Handbook*, pp 3-19 to 3-25, McGraw-Hill.

Two-speed Fordomatic (Ford Motor Co.)

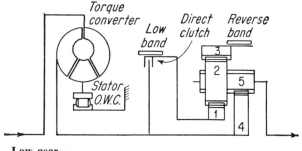

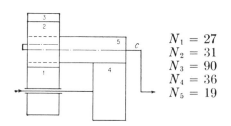

$$N_1 = 27$$
$$N_2 = 31$$
$$N_3 = 90$$
$$N_4 = 36$$
$$N_5 = 19$$

Low gear—
gear 1 fixed

$$R = 1 + \frac{N_1}{N_4} = 1.75$$

Reverse gear—
gear 3 fixed

$$R = 1 - \frac{N_3}{N_4} = -1.50$$

Note: Power-Glide Transmission is similar to above, but with $N_1 = 23$, $N_2 = 28$, $N_3 = 79$, $N_4 = 28$, $N_5 = 18$. This produces identical ratios in low and reverse.

$$R = 1 + \frac{23}{28} = 1.82 \qquad R = 1 - \frac{79}{28} = -1.82$$

Cruise-O-Matic 3-speed transmission (Ford Motor Co.)

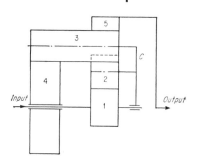

Long planet, $N_3 = 18$
Short planet, $N_2 = 18$
Sun gears, $N_4 = 36$, $N_1 = 30$
Ring gears, $N_5 = 72$

Low gear—Input to 1 C fixed

$$R = \frac{N_5}{N_1} = 2.4$$

Intermediate gear—
Input to 1, gear 4 fixed

$$R = \frac{1 + \dfrac{N_4}{N_1}}{1 + \dfrac{N_4}{N_5}} = 1.467$$

Reverse gear—
Input to 4, C fixed

$$R = \frac{N_5}{N_1} = -2.0$$

Hydramatic 3-speed transmission (General Motors)

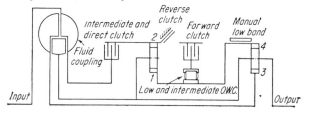

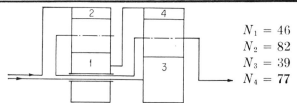

$$N_1 = 46$$
$$N_2 = 82$$
$$N_3 = 39$$
$$N_4 = 77$$

Low gear—
Input to 3, 4 fixed

$$R = 1 + \frac{N_4}{N_3} = 2.97$$

Intermediate gear— Input to 2, 1 fixed

$$= 1 + \frac{N_1}{N_2} = 1.56$$

Reverse gear—Input to 3, 2 fixed

$$R = 1 - \frac{N_4 N_2}{N_3 N_1} = -2.52$$

Triple planetary drives

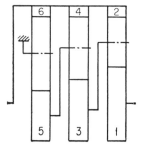

(Ref. V.Ya. Sukhina, U.S.S.R.)

Input to gear 1, output from gear 6

$$R = \left(1 + \frac{N_2}{N_1}\right)\left[\left(1 + \frac{N_4}{N_3}\right)\left(-\frac{N_6}{N_5}\right) - \frac{N_4}{N_3}\right] - \frac{N_2}{N_1}$$

(B)

(C)

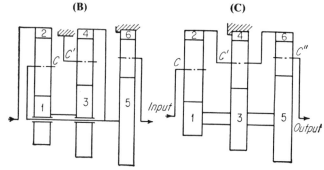

(B) $$R = \left[1 + \frac{N_1}{N_2}\left(1 + \frac{N_4}{N_3}\right)\right]\left(1 + \frac{N_6}{N_5}\right)$$

(C) $$R = \left[1 + \frac{N_4/N_3}{1 + (N_2/N_1)}\right]\Big/\left[1 + \frac{N_4/N_3}{1 + (N_6/N_5)}\right]$$

Ford tractor drives Ring gear 3 coupled to sun gear 1; split output.

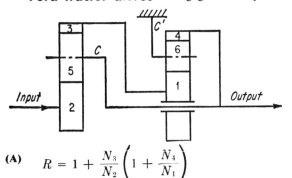

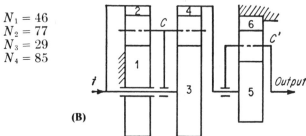

$$N_1 = 46$$
$$N_2 = 77$$
$$N_3 = 29$$
$$N_4 = 85$$

(A) $\qquad R = 1 + \dfrac{N_3}{N_2}\left(1 + \dfrac{N_4}{N_1}\right)$

$$R = \left(\dfrac{1 + (N_1/N_2)}{1 - \dfrac{N_1 N_3}{N_2 N_4}}\right)\left(1 + \dfrac{N_6}{N_5}\right)$$

(B)

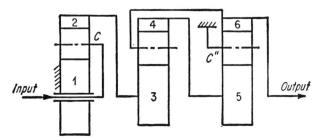

(C) $\qquad R = \dfrac{1}{1 + \dfrac{N_1}{N_2}}\left[1 + \dfrac{N_4}{N_3}\left(1 + \dfrac{N_6}{N_5}\right)\right]$

(D) $\qquad R = \dfrac{N_3}{N_4}\left(1 - \dfrac{N_4}{N_3} + \dfrac{N_2}{N_1}\right)$

Lycoming turbine drive

$$R = \left(1 + \dfrac{N_3}{N_2}\right)$$
$$\times\left(1 + \dfrac{N_4}{N_1}\right)$$

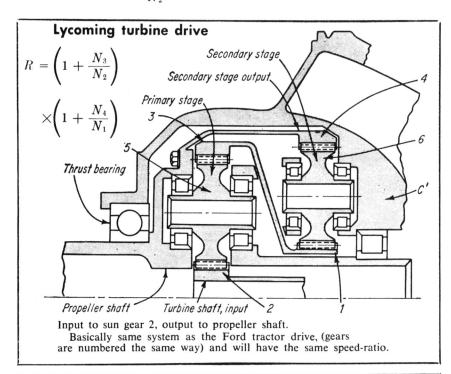

Input to sun gear 2, output to propeller shaft.
Basically same system as the Ford tractor drive, (gears are numbered the same way) and will have the same speed-ratio.

Compound spur-bevel gear drive

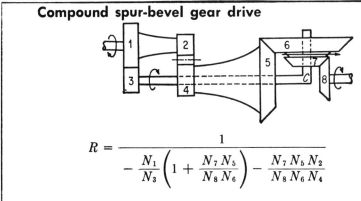

$$R = \dfrac{1}{-\dfrac{N_1}{N_3}\left(1 + \dfrac{N_7 N_5}{N_8 N_6}\right) - \dfrac{N_7 N_5 N_2}{N_8 N_6 N_4}}$$

Harmonic drive
(United Shoe Machinery Corp.)

Input:
Wave generator
Output:
Flexspline
Circular spline, fixed

High ratio, negative

Input:
Wave generator
Output:
Circular spline
Flexspline:
Fixed

High ratio, positive

Circular spline
Flexspline
Wave generator:
Fixed

Near unity, positive

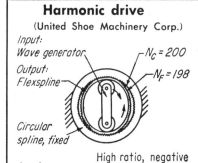

$N_C = 200$
$N_F = 198$

Top to bottom:

$$R = \dfrac{1}{1 - \dfrac{N_c}{N_F}} = -99$$

$$R = R_i = \dfrac{1}{1 - \dfrac{N_F}{N_c}} = 100$$

$$R = N_C/N_F = 100/99$$

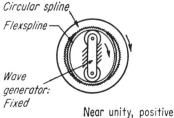

Two-gear planetary drives

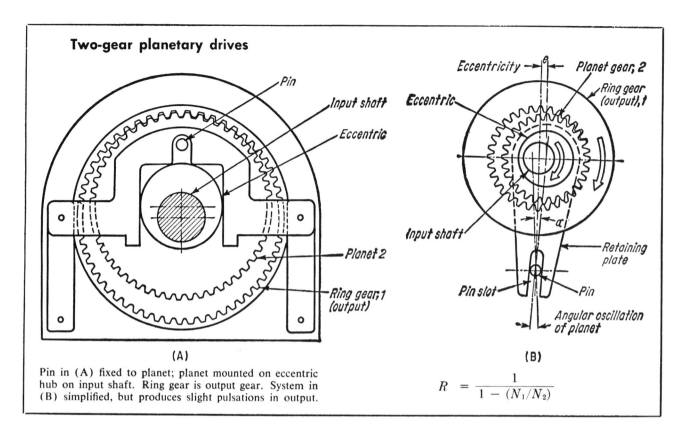

(A)

(B)

Pin in (A) fixed to planet; planet mounted on eccentric hub on input shaft. Ring gear is output gear. System in (B) simplified, but produces slight pulsations in output.

$$R = \frac{1}{1 - (N_1/N_2)}$$

Planocentric drive   (General Electric Co.)

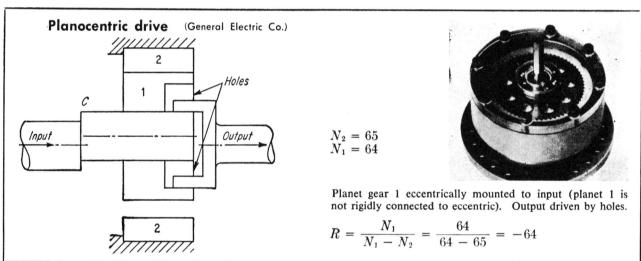

$N_2 = 65$
$N_1 = 64$

Planet gear 1 eccentrically mounted to input (planet 1 is not rigidly connected to eccentric). Output driven by holes.

$$R = \frac{N_1}{N_1 - N_2} = \frac{64}{64 - 65} = -64$$

Wobble-gear drive

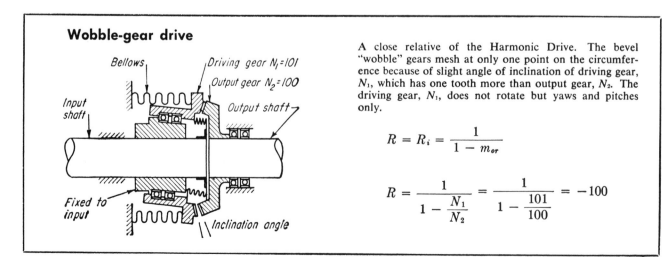

A close relative of the Harmonic Drive. The bevel "wobble" gears mesh at only one point on the circumference because of slight angle of inclination of driving gear, N_1, which has one tooth more than output gear, N_2. The driving gear, N_1, does not rotate but yaws and pitches only.

$$R = R_i = \frac{1}{1 - m_{or}}$$

$$R = \frac{1}{1 - \dfrac{N_1}{N_2}} = \frac{1}{1 - \dfrac{101}{100}} = -100$$

Types of Noncircular Gears

Noncircular gears generally cost more than competitive devices such as linkages and cams. But with the development of modern production methods, such as the tape-controlled gear shaper, cost has gone down considerably. Also, in comparison with linkages, noncircular gears are more compact and balanced—and can be more easily balanced. These are important considerations in high-speed machinery. Further, the gears can produce continuous, unidirectional cyclic motion—a point in their favor when compared with cams. The disadvantage of cams is that they offer only reciprocating motion.

Applications can be classed into two groups:

• Where only an over-all change in angular velocity of the driven member is required: quick-return drives, intermittent mechanisms as in printing presses, planers, shears, winding machines, automatic-feed machines.

• Where precise, nonlinear functions must be generated, as in computing machines for extracting roots of numbers, raising numbers to any power, generating trigonometric and logarithmic functions.

TYPES OF NONCIRCULAR GEARS

It is always possible to design a special-shaped gear to roll and mesh properly with a gear of any shape—sole requirement is that distance between the two axes must be constant. However, the pitch line of the mating gear may turn out to be an open curve, and the gears can be rotated only for a portion of a revolution—as with two logarithmic-spiral gears (illustrated in Fig 1).

True elliptical gears can only be made to mesh properly if they are twins, and if they are rotated about their focal points. However, gears resembling ellipses can be generated from a basic ellipse. These "higher-order" ellipses (see Fig 2) can be meshed in various interesting combinations to rotate about centers A, B, C or D. For example, two 2nd-order elliptical gears can be meshed to rotate about their geometric center; however, they will produce two complete speed cycles per revolution. Difference in contour between a basic ellipse and a 2nd-order ellipse is usually very slight. Note also that the 4th-order "ellipses"

continued, next page

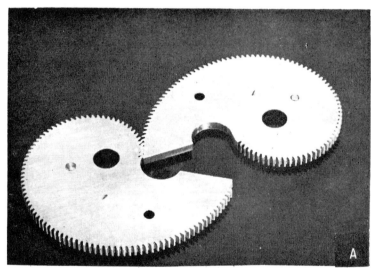

1 LOGARITHMIC SPIRAL GEARS in (A) are open curved, usually employed in computing devices. Elliptical-shape gears (B) are closed curved, frequently found in automatic machinery. Special-shape gears (C) offer wider range of velocity and acceleration characteristics.

resemble square gears (this explains why the square gears, sometimes found as ornaments on tie clasps, illustrated in Fig 3, actually work).

A circular gear, mounted eccentrically, can roll properly only with specially derived curves (shown in Fig 4). One of the curves, however, closely resembles an ellipse. For proper mesh, it must have twice as many teeth as the eccentric gear. When the radiis r, and eccentricity, e, are known, the major semi-axis of the elliptical-shape gear becomes $2r + e$, and the minor $2r - e$. Note also that one of the gears in this group must have internal teeth to roll with the eccentric gear. Actually, it is possible to generate internal-tooth shapes to rotate with noncircular gears of any shape (but, again, the curves may be of the open type).

Noncircular gears can also be designed to roll with special-shaped racks (shown in Fig 5). Combinations include: an elliptical gear and a sinusoid-like rack (a 3rd order ellipse is illustrated but any of the elliptical rolling curves can be used in its place—main advantage is that when the ellipse rolls, its axis of rotation moves along a straight line); and a logarithmic spiral and straight rack (the rack, however, must be inclined to its direction of motion by the angle of the spiral).

DESIGN EQUATIONS

Equations for noncircular gears are shown here in functional form for three common design requirements. They are valid for any noncircular gear pair. Symbols are defined in the box on the next page:

CASE I—Polar equation of one curve and center distance

2 Basic and High-order Elliptical Gear Combinations

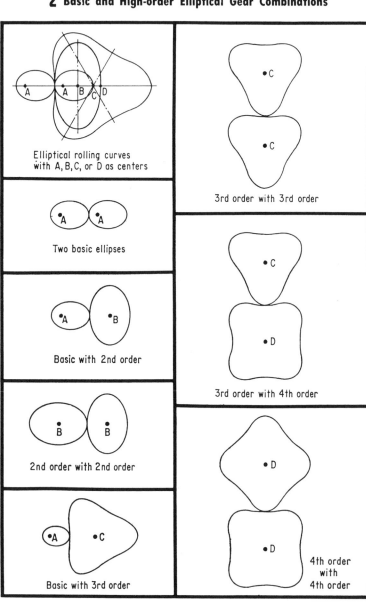

Elliptical rolling curves with A, B, C, or D as centers

Two basic ellipses

Basic with 2nd order

2nd order with 2nd order

Basic with 3rd order

3rd order with 3rd order

3rd order with 4th order

4th order with 4th order

3 SQUARE GEARS ON TIE CLASP seem to defy basic kinematic laws, are actually a takeoff on a pair of 4th order ellipses.

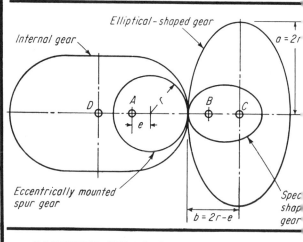

4 ECCENTRIC SPUR GEAR rotating about point A, will mesh properly with any of the three gears shown with centers at points B, C and D.

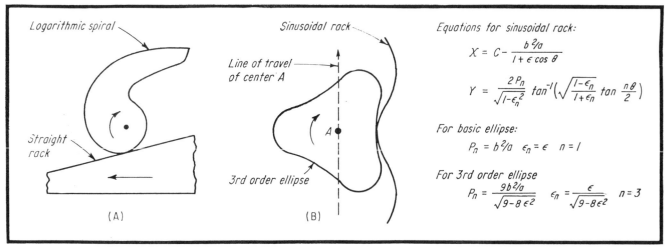

Equations for sinusoidal rack:

$$X = C - \frac{b^2/a}{1 + \epsilon \cos \theta}$$

$$Y = \frac{2 P_n}{\sqrt{1 - \epsilon_n^2}} \tan^{-1}\left(\sqrt{\frac{1 - \epsilon_n}{1 + \epsilon_n}} \tan \frac{n\theta}{2}\right)$$

For basic ellipse:

$$P_n = b^2/a \quad \epsilon_n = \epsilon \quad n = 1$$

For 3rd order ellipse

$$P_n = \frac{9b^2/a}{\sqrt{9 - 8\epsilon^2}} \quad \epsilon_n = \frac{\epsilon}{\sqrt{9 - 8\epsilon^2}} \quad n = 3$$

5 RACK AND GEAR COMBINATIONS are possible with noncircular gears. Straight rack for logarithmic spiral (A) must move obliquely; center of 3rd order ellipse (B) follows straight line.

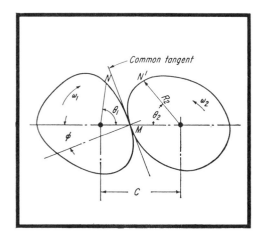

Symbols

a = semi-major axis of ellipse
b = semi-minor axis of ellipse
C = center distance (see above sketch)
ϵ = eccentricity of an ellipse = $\sqrt{1 - (b/a)^2}$
e = eccentricity of an eccentrically mounted
    spur gear
N = number of teeth
P = diametral pitch
r_c = radius of curvature
R = active pitch radius
S = length of periphery of pitch circle
X, Y = rectangular coordinates
θ = polar angle to R
ϕ = angle of obliquity
ω = angular velocity
$f(\theta), F(\theta), G(\theta)$ = various functions of θ
$f'(\theta), F'(\theta), G'(\theta)$ = first derivatives of functions of θ

are known; to find the polar equation of the mating gear:

$$R_1 = f(\theta_1)$$
$$R_2 = C - f(\theta_1)$$
$$\theta_2 = -\theta_1 + C \int \frac{d\theta_1}{C - f(\theta_1)}$$

CASE II—Relationship between angular rotation of two members and center distance are known; to find polar equations of both members:

$$\theta_2 = F(\theta_1)$$
$$R_1 = \frac{C F'(\theta_1)}{1 + F'(\theta_1)}$$
$$R_2 = C - R_1 = \frac{C}{1 + F'(\theta_1)}$$

CASE III—Relationship between angular velocities of two members and center distance are known; to find polar equations of both members:

$$\omega_2 = \omega_1 G(\theta_1)$$
$$R_1 = \frac{C G(\theta_1)}{1 + G(\theta_1)}$$
$$R_2 = C - R_1$$
$$\theta_2 = \int G(\theta_1) d\theta_1$$

Velocity equations and characteristics of five types of noncircular gears are listed in the table on the next page.

CHECKING FOR CLOSED CURVES

Gears can be quickly analyzed to determine whether their pitch curves are open or closed by means of the following equations:

In Case I, if $R = f(\theta) = f(\theta + 2N\pi)$, the pitch curve is closed.

In Case II, if $\theta_1 = F(\theta_2)$ and $F(\theta_0) = 0$, the curve is closed when the equation $F(\theta_0 + 2\pi/N_1) = 2\pi/N_2$ can be satisfied by substituting integers or rational fractions for N_1 and N_2. If fractions must be used to solve this

Characteristics of Five Noncircular Gear Systems

Type	Comments	Basic equations	Velocity equations ω_1 = constant
 Two ellipses rotating about foci	Gears are identical. Comparatively easy to manufacture. Used for quick-return mechanisms, printing presses, automatic machinery	$R = \dfrac{b^2}{a[1 + \epsilon \cos\theta]}$ ϵ = eccentricity $= \sqrt{1 - \left(\dfrac{b}{a}\right)^2}$ $a = \frac{1}{2}$ major axis $b = \frac{1}{2}$ minor axis	$\omega_2 = \omega_1 \left[\dfrac{r^2 + 1 + (r^2 - 1)\cos\theta_2}{2r}\right]$ where $r = \dfrac{R\ max}{R\ min}$
 2nd Order elliptical gears rotating about their geometric centers	Gears are identical. Geometric properties well known. Better balanced than true elliptical gears. Used where two complete speed cycles are required for one revolution	$R = \dfrac{2ab}{(a + b) - (a - b)\cos 2\theta}$ $C = a + b$ a = maximum radius b = minimum radius	$\omega_2 = \omega_1 \left[\dfrac{r^2 + 1 - (r^2 - 1)\cos 2\theta_2}{2r}\right]$ where $r = \dfrac{a}{b}$
 Eccentric circular gear rotating with its conjugate	Standard spur gear can be employed as the eccentric. Mating gear has special shape	$R_1 = \sqrt{a^2 + e^2 + 2ae \cos\theta_1}$ $\theta_2 = \theta_1 + C\displaystyle\int \dfrac{d\theta_1}{C - R_1}$ $C = R_1 + R_2$	$\dfrac{\omega_2}{\omega_1} = \dfrac{\sqrt{a^2 + e^2 + 2ae \cos\theta_1}}{C - \sqrt{a^2 + e^2 + 2ae \cos\theta_1}}$
 Logarithmic spiral gears	Gears can be identical although can be used in combinations to give variety of functions. Must be open gears	$R_1 = Ae^{k\theta_1}$ $R_2 = C - R_1$ $\quad = Ae^{k\theta_2}$ $\theta_2 = \dfrac{1}{k}\log(C - Ae)^{k\theta_1}$ e = natural log base	$\dfrac{\omega_2}{\omega_1} = \dfrac{Ae^{k\theta_1}}{C - Ae^{k\theta_1}}$
 Sine-function gears	For producing angular displacement proportional to sine of input angle. Must be open gears	$\theta_2 = \sin^{-1}(k\theta_1)$ $R_2 = \dfrac{C}{1 + k\cos\theta_1}$ $R_1 = C - R_2$ $\quad = \dfrac{Ck\cos\theta_1}{1 + k\cos\theta_1}$	$\dfrac{\omega_2}{\omega_1} = k\cos\theta_1$

equation, the curve will have double points (intersect itself), which is, of course, an undesirable condition.

In Case III, if $\theta_2 = \int G(\theta_1)\ d\theta_1$, let $G(\theta_1)\ d\theta_1 = F(\theta_1)$, and use the same method as for Case II, with the subscripts reversed.

With some gear sets, the mating gear will be a closed curve only if the correct center distance is employed. This distance can be found from the equation:

$$4\pi = \int_0^{2\pi} \frac{d\theta_1}{C - f(\theta_1)}$$

ow to Prevent Reverse Rotation

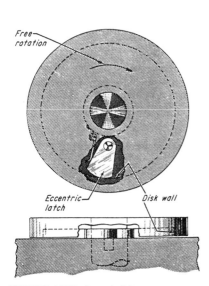

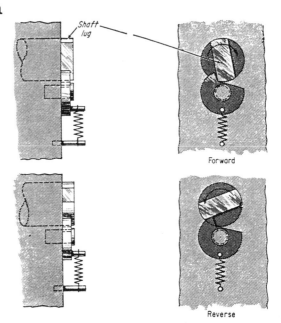

Forward

Reverse

ECCENTRIC LATCH allows shaft to rotate in one direction; attempted reversal immediately causes latch to wedge against disk wall.

LUG ON SHAFT pushes the notched disk free during normal rotation. Disk periphery stops lug to prevent reverse rotation.

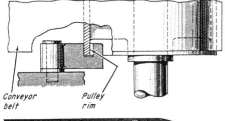

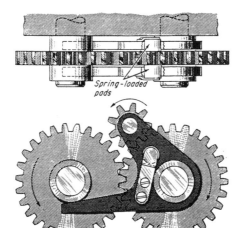

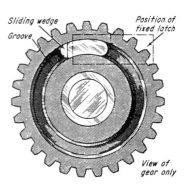

View of gear only

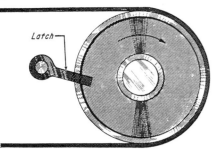

LATCH ON RIM of pulley is free only when rotation is in direction shown. This arrangement is ideal for conveyor-belt pulleys.

SPRING-LOADED FRICTION PADS contact the right gear. Idler meshes and locks gear set when rotation is reversed.

FIXED WEDGE AND SLIDING WEDGE tend to disengage when the gear is turning clockwise. The wedges jam in reverse direction.

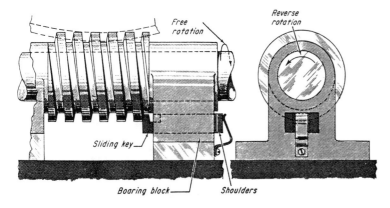

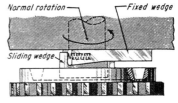

SLIDING KEY has tooth which engages the worm threads. In reverse rotation key is pulled in until its shoulders contact block.

Gear-Shift Arrangements

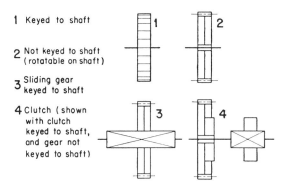

1 Keyed to shaft

2 Not keyed to shaft (rotatable on shaft)

3 Sliding gear keyed to shaft

4 Clutch (shown with clutch keyed to shaft, and gear not keyed to shaft)

Fig. 1. Schematic symbols used in the illustrations to represent gears and clutches.

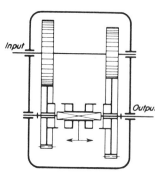

Fig. 2. Double-clutch drive. Two pairs of gears permanently in mesh. Pair I or II transmits motion to output shaft depending on position of coupling; other pair idles. Coupling shown in neutral position with both gear pairs idle. Herring-bone gears recommended for quiet running.

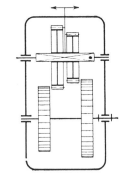

Fig. 3. Sliding-change drive. Gears meshed by lateral sliding. Up to three gears can be mounted on sliding sleeve. Only one pair in mesh in any operating position. Drive simpler, cheaper and more extensively used than drive of Fig. 2. Chamfering side of teeth facilitates engagement.

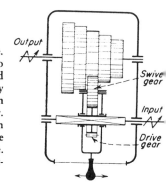

Fig. 4. Swivel-gear drive. Output gears are fastened to shaft. Handle is pushed down, then shifted laterally to obtain transmission through any output gear. Not suitable for transmission of large torques because swivel gear tends to vibrate. Over-all ratio should not exceed 1:3.

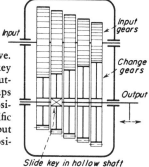

Fig. 5. Slide-key drive. Spring-loaded slide key rides inside hollow output shaft. Slide key snaps out of shaft when in position to lock a specific change gear to output shaft. No central position is shown.

Slide key in hollow shaft

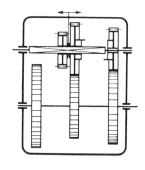

Fig. 6. Combination coupling and slide gears. Three ratios: direct mesh for ratios I and II; third ratio transmitted through gears II and III which couple together.

Fig. 7. Double-shift drive. One shift must always be in a neutral position which may require both levers to be shifted when making a change. However only two shafts are used to achieve four ratios.

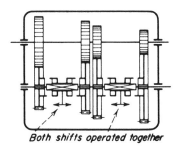

Both shifts operated together

13 ways of arranging gears and clutches to obtain changes in speed ratios.

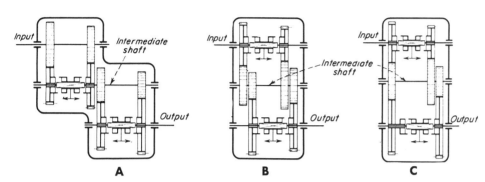

Fig. 8. (A) Triple shaft drive gives four ratios. Output of first drive serves as input for second. Presence of intermediate shaft obviates necessity for always insuring that one shift is in neutral position. Wrong shift lever position can not cause damage. (B) Space-saving modification. Coupling is on shaft *A* instead of intermediate shaft. (C) Still more space saved if one gear replaces a pair on intermediate shaft. Ratios can be calculated to allow this.

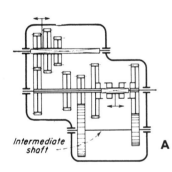

Fig. 9. Six ratios available with two couplings and (A) ten gears, (B) eight gears. Up to six gears in permanent mesh. It is not necessary to insure that one shift is in neutral.

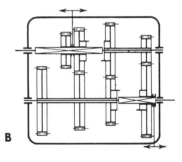

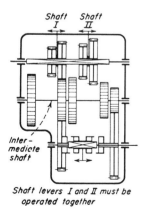

Fig. 10. Eight-ratio drive uses two slide gears and a coupling. This arrangement reduces number of parts and meshes. Position of shifts I and II are interdependent. One shift must be in neutral if other *is* in mesh.

Shaft levers *I* and *II* must be operated together

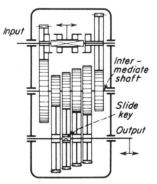

Fig. 11. Eight ratios; coupled gear drive and slide-key drive in series. Comparatively low strength of slide key limits drive to small torque.

Data based on material and sketches in AWF und VDMA Getrieblaetter, published by Ausschuss fuer Getriebe heim Ausschuss fuer Wirtschaftiche Fertigung, Leipzig, Germany.

SHIFTING MECHANISMS FOR GEARS AND CLUTCHES

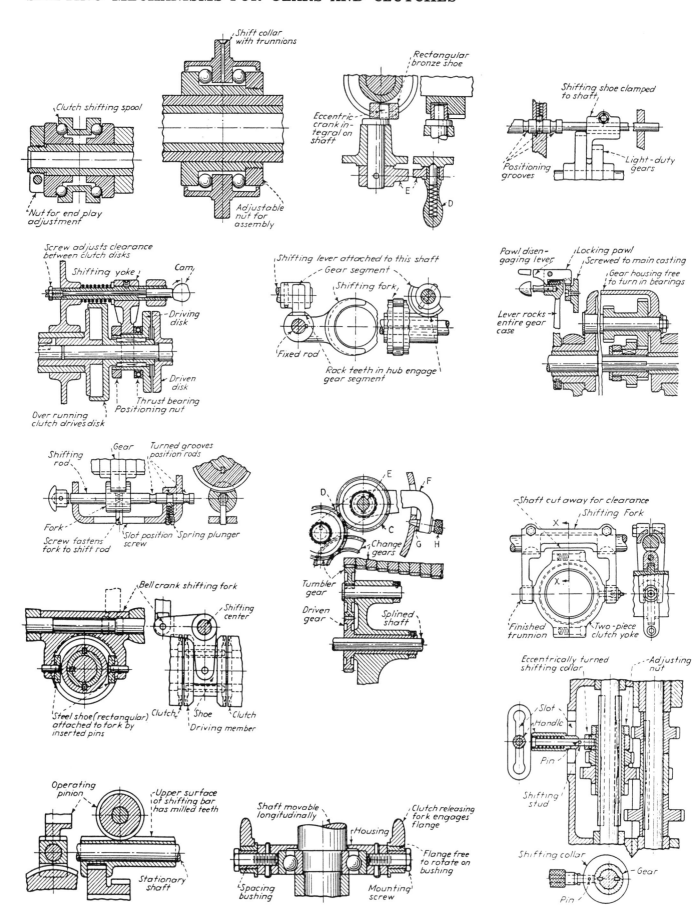

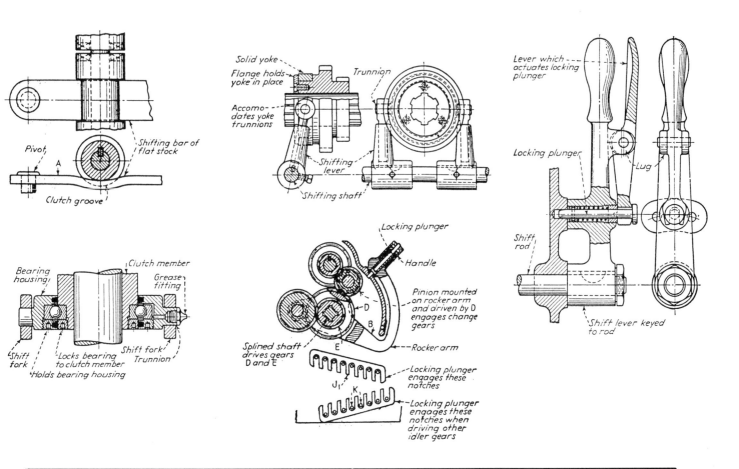

Spiderless differential

If you've ever failed to drive out of a ditch because one wheel spun uselessly while the other sat torqueless and immobile, you'll thank the inventors (Seliger and Hegar) of the limited-slip differential pictured below.

In straight running, it performs as a drive axle, driven by the driveshaft pinion through the ring gear. The differential action comes about only when one wheel loses traction, travels along a different arc, or otherwise attempts to turn at a speed different from that of the other. Then the wedge-type two-way over-running clutch (second sketch) disengages, freeing the wheel to spin without drag.

Variations. Each clutch has three positions: forward drive, idle, and reverse drive. Thus there are numerous combinations of drive-idle, depending on road conditions and turn direction. US Patent 3,124,972 describes a few:

• For left turns, the left wheel is driving, and the right wheel is forced to turn faster—thus over-running and disengaging the clutch. A friction ring built into each clutch assembly does the shifting. Wear is negligible.

• If power should be removed from the driveshaft during the left turn, the friction rings will shift each clutch and cause the left wheel to run free and

the right wheel to drag in full coupling with the driveshaft.

• If on the straightaway, under power, one wheel is lifted out of contact with the road, the other immediately transmits full torque. (The conventional spider differential does just the opposite.)

On or off. Note one limitation, however: There is no gradual division of power. A wheel is either clutched in and turning at exactly the same speed as its opposite, or it is clutched out. It is not the same sort of mechanism as the conventional spider differential, which divides the driving load variably at any ratio of speeds.

Hi-Torq Products Corp.

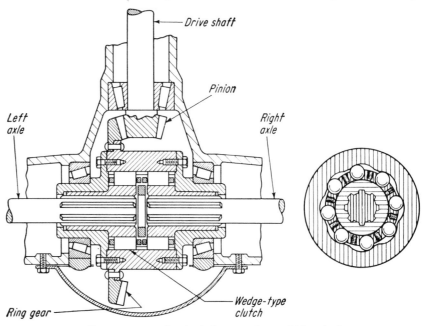

Two-way over-running clutch disengages the non-driving wheel

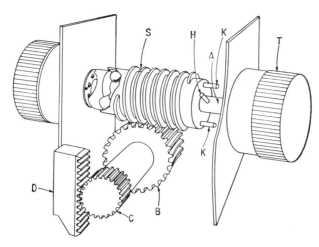

use, or for rapid observations. It is most suitable in high magnifications, less so for low-power work where a greater range of fine adjustment is necessary. The mechanism runs in ball bearings, totally enclosed so lubrication is unnecessary.

Turning the knob continuously in one direction provides the coarse adjustment. When direction is reversed, the fine adjustment is automatically engaged for about ⅓ turn of the knob. Turning beyond this amount at either end shifts back to coarse adjustment.

Worm **S** is loosely mounted on shaft **A**, along which it can move a short distance. Drive knob **T** is rigidly attached to the shaft. As soon as drive pin **H** on shaft **A** engages one of the stop pins **K** on the worm, the latter is rotated directly. It in turn rotates worm wheel **B** and pinion **C**, which in turn drives rack **D** on the table lift. This is the coarse adjustment. But a reversal of the knob disengages the coarse feed and moves the worm gear (**S**) along the shaft a very short distance through a mechanism consisting of an inclined plane and ball. This causes very slight rotation of the worm wheel (**B**) and pinion (**C**), so movement of rack **D** is correspondingly limited. This fine feed can be continued, or reversed, within the limits of stops (**K**).

One-knob control—New and controversial in microscopy is this single knob control for coarse and fine adjustment, available in three series of Leitz microscopes, particularly those for classroom and student

---

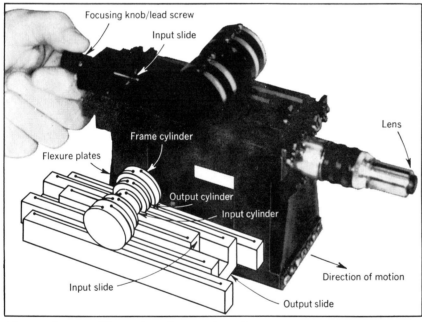

Linear micron-positioner, based on series of connected cylinders of unequal diameters, can adjust a camera lens to within 10 millionths of an inch.

Fine-focus adjustments

Achieving fine focus control on high-resolution cameras usually means that an expensive and intricate gear system must be built. Now, John J. Dalton of IBM's Research Laboratory, Kingston, NY, has designed a simple, low-cost mechanism that can adjust a camera lens to within 10 millionths of an inch.

The ingenious system, called a linear micron-positioner, is based on the differential circumference of connected concentric cylinders of unequal diameters. When flexible bands, in this case shim stock, are fixed to these cylinders, the difference in takeup between the bands which occurs when the cylinders are rotated—is proportional to the differential circumference. If the bands connected to one cylinder are referenced to a fixed frame, and the bands connected to the other cylinder are referenced to a movable member, rotation of the cylinder on the fixed frame will result in relative motion of the movable member equal to the difference in band takeup.

Three cylinders. Dalton's mechanism consists of three interconnected, different-sized cylinders to provide reduced displacement. A small input cylinder and the focusing knob are attached to a lead screw.

When the lead screw is turned, the movement between the sliding and fixed frames is very small. In fact when Dalton was demonstrating his mechanism he had to use cylinders whose diameter difference was large enough to show the relative forward motion.

Temperature effects are negligible, because all the basic elements, except the lead screw, have opposing forces. The only element having friction, and subject to wear in the device, is the lead screw; the friction here provides a holding force.

Theoretically there is no limit to the reduction ratio using the principle of different sized cylinders. Dalton has explored the possibility of making a linear positioner with a 10,000:1 ratio.

Applications. In addition to its use in optical systems, the same mechanical principle could be used to obtain precise adjustments to the axis of an X-Y measurement table; or to position miniature components during manufacture. □

atchet-tooth speed-change drive

in-line shaft drive, with reduction ratios of 1:1 and 1:16 or 1:28, combined in a le element, has been designed by Telefunken of West Germany. It consists cally of friction wheels which grip each other elastically.

rown wheels with a gear ratio of 1:1 used for the coarse adjustment; and ion spur gearing, with a ratio of or 1:28 for the fine or vernier stment.

spring (see diagram) applies pres- to the fine-adjustment pinion, pre- ting backlash while the coarse ad- ment is in use; and uncouples the se adjustment when the vernier is ght into play by forward move- t of the front shaft. The spring makes sure that the front shaft is ays in gear.

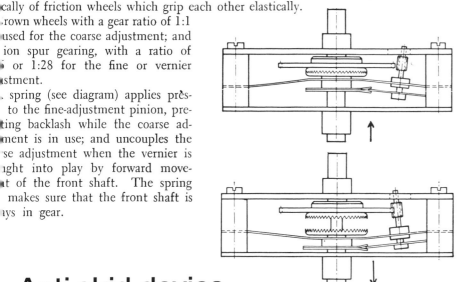

Anti-skid device

To prevent the skids that jack-knife trailer trucks, vehicles or ground-loop airplanes, a British-made device senses the buildup of skid forces and automatically compensates for them in the braking system.

These devices, called Maxaret units, are made by Dunlop Co., Ltd., Coventry, England, use the inertia of a flywheel (drawing below) to actuate a linkage system that regulates oil supply in the brake system.

Interacting forces. As the drawing of the aircraft Maxaret shows, the assembly's linkage operates two valves in the hydraulic lines. The inside of the wheel hub is fitted with a drive spring anchored at one

end. As this spring tries to expand, it makes contact with the inner surface of a drum. A flywheel, mounted on bearings, surrounds this drum, and the two components are kept in contact by drum segments that fit into the web of the flywheel. A mainspring, in trying to unwind, keeps the drum and flywheel firmly together.

In the center of the flywheel web, two balls lie at the apex of a cam profile. These balls are covered by a spring-loaded thrust rod connected with a linkage system. The entire Maxaret unit is fitted on the torque plate of the brake.

How it works. In normal braking, the drive spring transmits wheel motion to the drum and flywheel. The

thrust balls remain at the apex of the cam profile, and the thrust rod holds the hydraulic valve open, allowing pressure on the brake.

When a skid is imminent, the wheel motion starts to slow down, but the inertia keeps the Maxaret's flywheel spinning at its original speed. Three things now happen: As the flywheel turns faster than the slowing drum, the mainspring winds up until, at 60 deg. of rotation, the web of the flywheel makes contact with the drum segments. During this shift in relation between flywheel and drum, the two balls move up the cam profile, pushing the thrust rod back, axially. When the rod is pushed back, the linkage moves to close the hydraulic inlet valve and open the exhaust valve, releasing the brake pressure.

Finally, the drive spring tends to collapse while the flywheel is driving the drum, because the spring is being wound up. Thus, the drum-flywheel unit can rotate freely for a few seconds to hold off on brake pressure. In a plane, this effect ensures that brakes are not applied while a wheel is off the ground in a bumpy landing. If a wheel locks while clear of the ground, the inertia of the flywheel is absorbed.

When the plane or truck wheel starts to pick up speed again, ending the incipient skid condition, the process is reversed. Brake pressure is automatically restored when the drum reaches the same speed as the fly-wheel and the mainspring releases its energy, putting the two elements again into firm contact. □

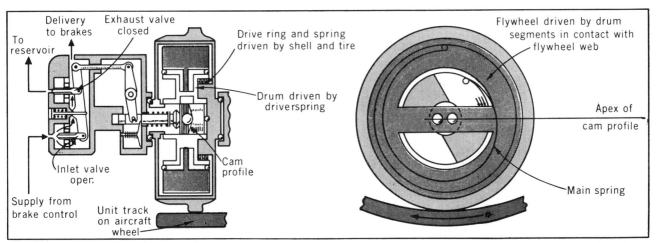

When wheel is rotating during normal braking, two balls within the flywheel web rest at the apex of a cam profile (left). But, when skid is sensed the balls are forced up the profile by the relative motion of flywheel and drum.

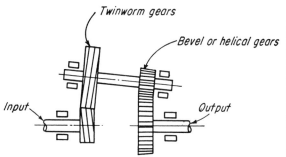

Twinworm gears

Bevel or helical gears

Input

Output

APPLICATIONS FOR TWINWORM

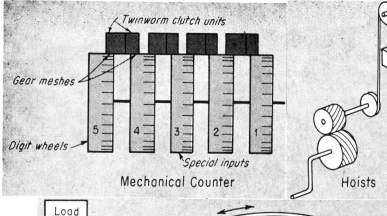

Twinworm clutch units

Gear meshes

Digit wheels

5   4   3   2   1

Special inputs

Mechanical Counter   **Hoists**

Load

Rack

Worm

Motor or handle

Racks

Worm

Motor

Ring (be ante

Large Rings

NICHOLAS CHIRONIS

The term "self-locking" as applied to gear systems denotes a drive which gives the input gear freedom to rotate the output gear in either direction—but the output gear locks with the input when an outside torque attempts to rotate the output in either direction. This characteristic is often sought by designers who want to be sure that loads on the output side of the system cannot affect position of the gears. Worm gears are one of the few gear systems that can be made self-locking, but at the expense of efficiency—they seldom exceed 40%, when self-locking.

An Israeli engineer displayed a simple dual-worm gear system that not only provided self-locking with over 90% efficiency, but exhibited a new phenomenon which the inventor calls "deceleration-locking."

A point in favor of the inventor—B. Popper, an engineer with the Scientific Department of the Israel Ministry of Defense in Tel Aviv—is that his "Twinworm" drive has been employed in Israel-designed counters and computers for several years and with marked success.

The Twinworm drive is quite simply constructed. Two threaded rods, or "worm" screws, are meshed together. Each worm is wound in a different direction and has a different pitch angle. For proper mesh, the worm axes are not parallel, but slightly skewed. (If both worms had the same pitch angle, a normal, reversible drive would result—similar to helical gears.) But by selecting proper, and different, pitch angles, the drive will exhibit either self-locking, or a combination of self-locking and deceleration-locking characteristics, as desired. Deceleration-locking is a completely new property best described in this way.

When the input gear decelerates (for example, when the power source is shut off, or when an outside force is applied to the output gear in a direction which tends to help the output gear) the entire transmission immediately locks up and comes to an abrupt stop moderated only by any elastic "stretch" in the system.

Almost any type of thread will work with the new drive—standard, 60° screw threads, Acme threads, or any arbitrary shallow-profile thread. Hence, the worms can be produced on standard machine-shop equipment.

JOBS FOR THE NEW DRIVE

Applications for Twinworm can be divided into two groups:

(1) Those employing self-locking characteristics to prevent the load from affecting the system.

(2) Those employing deceleration-locking characteristics to brake the system to an abrupt stop if the input decelerates.

Self-locking occurs as soon as tan ϕ_1 is equal to or smaller than μ, or when

$$\tan \phi_1 = \frac{\mu}{S_1}$$

Angles ϕ_1 and ϕ_2 represent the respective pitch angles of the two worms, and $\phi_2 - \phi_1$ is the angle between the two worm shafts (angle of misalignment). Angle ϕ_1 is quite small (usually in the order of 2 to 5°).

Here, S_1 represents a "safety factor" (selected by the designer) which must be somewhat greater than one to make sure that self-locking is maintained even if μ should fall below an assumed value. Neither ϕ_2 nor the angle ($\phi_2 - \phi_1$) affects self-locking.

Deceleration-locking occurs as soon as tan ϕ_2 is also equal to or smaller than μ; or, if a second safety factor S_2 is employed (where $S_2 > 1$), when

$$\tan \phi_2 = \frac{\mu}{S_2}$$

For the equations to hold true, ϕ_2 must always be made greater than ϕ_1. Also, μ refers to the idealized case where the worm threads are square. If the threads are inclined (as with Acme-threads or V-threads) then a modified value of μ must be employed, where

$$\mu_{modified} = \frac{\mu_{true}}{\cos \theta}$$

Relationship between input and output forces during rotation is:

$$\frac{P_1}{P_2} = \frac{\sin \phi_1 + \mu \cos \phi_1}{\sin \phi_2 + \mu \cos \phi_2}$$

Efficiency

$$\eta = \frac{1 + \mu/\tan \phi_2}{1 + \mu/\tan \phi_1}$$

CHAPTER 7.

COUPLING, CLUTCHING AND BRAKING DEVICES

Novel linkage that can couple offset shafts . . .

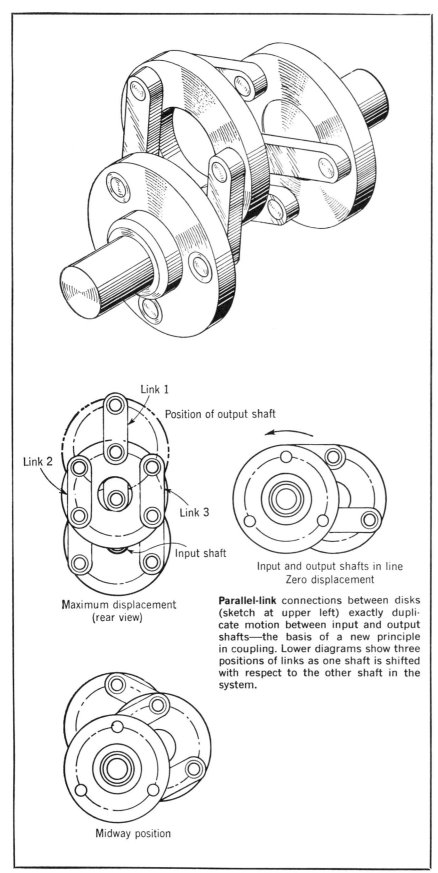

Link 1

Position of output shaft

Link 2

Link 3

Input shaft

Maximum displacement
(rear view)

Input and output shafts in line
Zero displacement

Parallel-link connections between disks (sketch at upper left) exactly duplicate motion between input and output shafts—the basis of a new principle in coupling. Lower diagrams show three positions of links as one shaft is shifted with respect to the other shaft in the system.

Midway position

An unorthodox yet remarkably simple arrangement of links and disks forms the basis of a versatile type of parallel-shaft coupling. This type of coupling—essentially three disks rotating in unison and interconnected in series by six links (drawing, left)—can adapt to wide variations in axial displacement while running under load.

Changes in radial displacement do not affect the constant-velocity relationship between input and output shafts, nor do they initial radial reaction forces that might cause imbalance in the system. These features open up unusual applications in automotive, marine, machine-tool, and rolling-mill machinery (drawings, facing page).

How it works. The inventor of the coupling, Richard Schmidt of Schmidt Couplings, Inc., Madison, Ala., notes that a similar link arrangement has been known to some German engineers for years. But these engineers were discouraged from applying the theory because they erroneously assumed that the center disk had to be retained by its own bearing. Actually, Schmidt found, the center disk is free to assume its own center of rotation. In operation, all three disks rotate with equal velocity.

The bearing-mounted connections of links to disks are equally spaced at 120 deg. on pitch circles of the same diameter. The distance between shafts can be varied steplessly between zero (when the shafts are in line) and a maximum that is twice the length of the links (drawings, left). There is no phase shift between shafts while the coupling is undulating. □

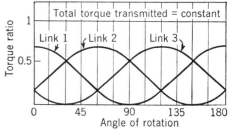

Torque transmitted by three links in group adds up to a constant value regardless of the angle of rotation.

... simplifies the design of a variety of products

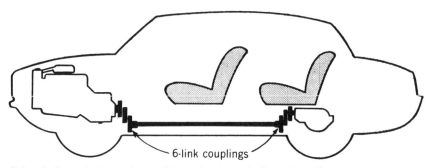

6-link couplings

Drive shaft can be lowered to avoid causing hump in floor of car. Same arrangement can be applied to other applications to bypass an object.

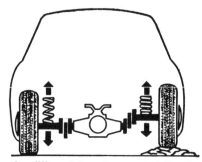

Car differential can be mounted directly to frame, while coupling transmits driving torque and permits wheels to bounce up and down. Arrangement also keeps wheels vertical during shock motion.

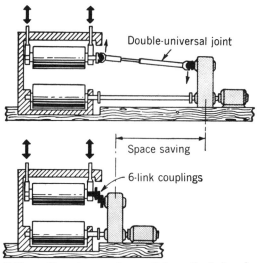

Double-universal joint

Space saving

6-link couplings

Rolling mill needs a way to permit top roller to be adjusted vertically. Double universal joint, normally used, causes radial forces at the joints and requires more lateral space than the 6-link coupling.

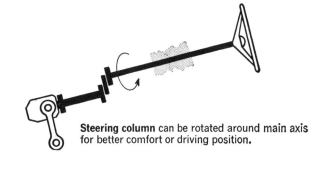

Steering column can be rotated around main axis for better comfort or driving position.

Machine for pounding road beds uses unbalanced shaft to induce large-amplitude vibration. Coupling prevents vibrations from passing on to transmission and frame.

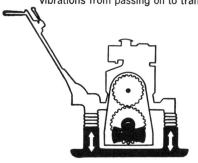

shaft

springs

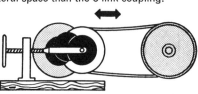

Belt drive can be adjusted for proper tension without need for moving entire base.

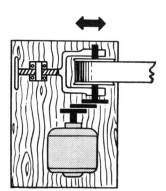

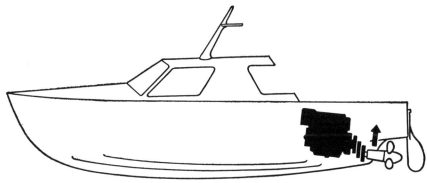

Inboard motor is segregated from propeller shock and vibration and can be mounted higher.

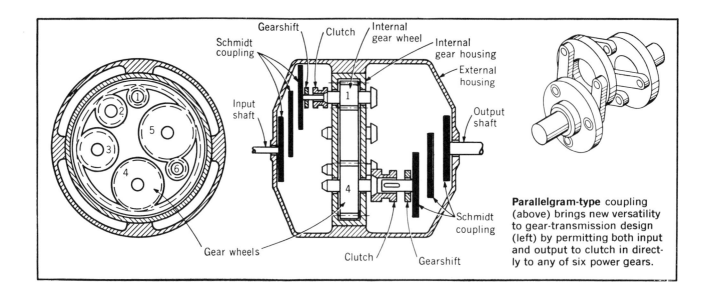

Parallelgram-type coupling (above) brings new versatility to gear-transmission design (left) by permitting both input and output to clutch in directly to any of six power gears.

Novel coupling shifts shafts to simplify transmission design

A unique disk-and-link coupling that can handle large axial displacement between shafts, while the shafts are running under load, is opening up new approaches to transmission design. It was developed by Schmidt Couplings, Inc., Madison, Ala.

The coupling (drawing, upper right) maintains a constant transmission ratio between input and output shafts while the shafts undergo axial shifts in their relative positions. This permits gear-and-belt transmissions to be designed that need fewer gears and pulleys.

Half as many gears. In the internal-gear transmission above, a Schmidt coupling on the input side permits the input to be "plugged-in" directly to any one of six gears, all of which are in mesh with the internal gear wheel.

On the output side, after the power flows through the gear wheel, a second Schmidt coupling permits a direct power takeoff from any of the same six gears. Thus, any one of 6 x 6 minus 5 or 31 different speed ratios can be selected while the unit is running. A more orthodox design would require almost twice as many gears.

Powerful pump. In the worm-type pump (bottom left), as the input shaft rotates clockwise, the worm rotor is forced to roll around the inside of the gear housing, which has a helical groove running from end to end. Thus, the rotor centerline will rotate counterclockwise to produce a powerful pumping action for moving heavy media.

In the belt drive (bottom right), the Schmidt coupling permits the belt to be shifted to a different bottom pulley while remaining on the same top pulley. Normally, because of the constant belt length, the top pulley would have to be shifted, too, to provide a choice of only three output speeds. With the new arrangement, nine different output speeds can be obtained. □

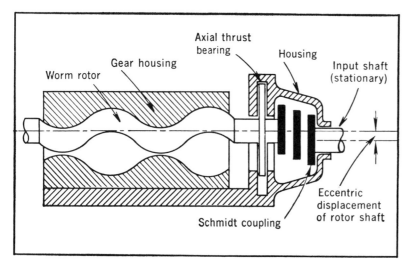

Coupling allows helical-shape rotor to wiggle for pumping purposes.

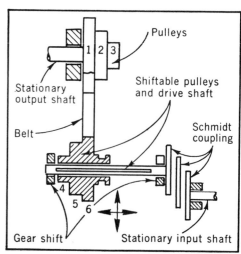

Coupling takes up slack when bottom shifts.

New constant-velocity universal joint transmits motion with true fidelity from input to output shafts, even when angle between shafts varies more than 90 deg.

Intertwining links produce true constant-motion universal

A new way of arranging linkages to form a universal joint (photo, above) helps maintain a true constancy of motion while transmitting power between intersecting shafts. Moreover, the device, called the Uni-Tru universal joint, can perform this task while the angle between the shafts is 90 deg. or even more.

The developers of Uni-Tru—Southwestern Industries, Inc., Los Angeles—claim their product is the world's first universal joint consisting of pin-connecting links that can provide such constant velocity.

Defining terms. The term "constant-velocity joint" means a coupling that can transmit motion with true fidelity from one shaft to another, whether the input motion be constant, whether it has cycling overtones, or whether it is oscillatory.

A conventional pin-connected universal joint accelerates and decelerates the output shaft undesirably during each revolution of the input, if the shafts are at an angle. Such a joint usually is employed in pairs to neutralize this effect.

Uni-Tru is based on a geometric concept that says constant-velocity transmission is possible when connecting linkages form a plane of symmetry midway between the intersecting shafts (drawings, right) and when the center lines of the shafts intersect within that plane. The linkage arrangement in the Uni-Tru can maintain the desired symmetry at all angles of transmission.

Under test. Southwestern Industries proved this concept mathematically before building the first prototype. The device is still in its developmental stage: Several configurations and sizes, ranging from miniature instruments to transmission drives up to 800 hp, are being tested.

The testing program includes installations on drive systems ranging from large diesel marine drives through servo-system controls. Applications are expected in instruments or devices where angular drives are required, including automobiles, trucks, tractors, construction equipment, textile machinery, conveyors, and metalworking machines.

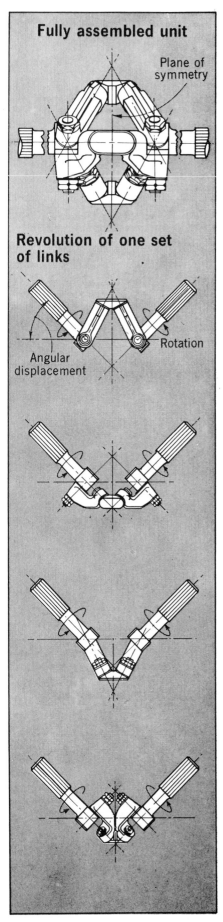

Connecting linkages form symmetrical plane that stays in place as shafts rotate.

Interlocking space-frames flex as they transmit shaft torque

New coupling tolerates unusually high degrees of misalignment, with no variation in the high torque that's being taken from the shaft

A new concept in flexible drive-shaft couplings permits unusually large degrees of misalignment and axial motion during the transmission of high amounts of torque. Moreover, the rotational velocity of the driven member remains constant during transmission at angular misalignments; in other words, cyclic pulsations are not induced as they would be if,

Inventor Bossler shows how rectangular plates improve bolt clearance. Version for parallel misalignment is in foreground

say, a universal coupling or a Hooke's joint were employed.

The new coupling consists essentially of a series of square space-frames, each bent to provide offsets at the diagonals and each bolted to adjacent members at alternate diagonals. The concept is the brainchild of Robert B. Bossler, Jr., chief of Mechanical Systems Research, Kaman Aerospace Corp., Bloomfield, Conn.

Proving an idea. Bossler performed most of the original development in his home on his own. Kaman Aerospace became interested in the device and is continuing its development. The company has already received both NASA and Naval Research funds. Kacarb Products, also of Bloomfield and a subsidiary of

Kaman Aerospace, will produce the coupling under the trade name, Kaflex, in a range of sizes for both aerospace and industrial use.

Because there are no moving parts, the new concept has high potential for applications where lubrication is difficult or environmental conditions such as moisture and dust are detrimental to existing high-power couplings. As a typical industrial example, the coupling has been installed in a power drive system at the Kaman plant and has been in daily operation since last February.

Moreover, Kaman is in the process of testing a unit for helicopter operation and will be flying this unit in a Navy UH-2 helicopter toward the end of the year.

Essence of a coupling. The Bossler coupling at first glance seems rather a straightforward solution to the problem of providing flexibility in joining two rotating shafts. But Bossler arrived at it only after a series of innovations, starting with a long, hard look at a drive shaft and ending with the government's granting of a patent (U.S. Patent No. 3,177,684).

Couplings accommodate the inevitable misalignments between rotating shafts in a drive train. These misalignments are caused by imperfect parts, dimensional variations, temperature changes, and deflections of the supporting structures. The couplings accommodate by either moving-contact or flexing.

Most couplings, however, have parts with moving contact that require lubrication and maintenance. The rubbing parts also absorb power. Moreover, the lubricant and the seals limit coupling environment and coupling life. Parts wear out, and the coupling may develop a large resistance to movement as the parts deteriorate. Then, too, in many designs, the coupling does not provide true constant velocity.

For flexibility. "I studied the various types of couplings on the market and at first developed a new one with moving contact," says Bossler. "After exhaustive tests, I became convinced that if there were to be the improvements I wanted, I had to come up with a coupling that flexed without any sliding or rubbing."

Flexible-coupling behavior, however, is not without design problems. Any flexible coupling can be proportioned with strong, thick, stiff members that easily transmit a design torque and provide the

stiffness to operate at a design speed. However, misalignment requires flexing of these members. The flexing produces alternating stresses that can limit coupling life. The greater the strength and stiffness of a member, the higher the alternating stress from a given misalignment. Therefore, strength and stiffness provisions to transmit torque at speed will be detrimental to misalignment capability.

"The problem of design is to proportion the flexible coupling to accomplish torque transmission and misalignment for the lowest system cost," says Bossler. "So I took a look at a drive shaft—after all, it is a good example of power transmission—and I wondered how I would go about converting it into one with flexibility.

"I began to evolve it by following basic principles. How does a drive shaft transmit torque? Well, by tension and compression. I began paring it down to the important struts that could transmit torque and found that they are curved beams. But a curved beam in tension and compression is not as strong as a straight beam, so I wound up with the beams straight in a square space-frame with what might be called a double helix arrangement. One helix contains elements in compression; the other helix, elements in tension."

Flattening the helix. The total number of plates should be an even number to obtain constant velocity characteristics during misalignment. But even with an odd number, the cyclic speed variations are minute, not nearly the magnitude of those in a Hooke's joint.

Although the analysis and resulting equations developed by Bossler are based on a square-shaped unit, he has since concluded that the perfect square is not the ideal for the coupling, because of the position of the mounting holes. The flatter the helix—in other words the smaller the distance S—the more misalignment the coupling will tolerate.

Hence, Bossler is now making the space frames slightly rectangular (similar to the ones he is holding in the photo) instead of square. In this design, the boltheads that fasten the plates together are offset from adjoining pairs, providing enough clearance for design of a "flatter" helix. The difference in stresses between a coupling using square-shaped plates

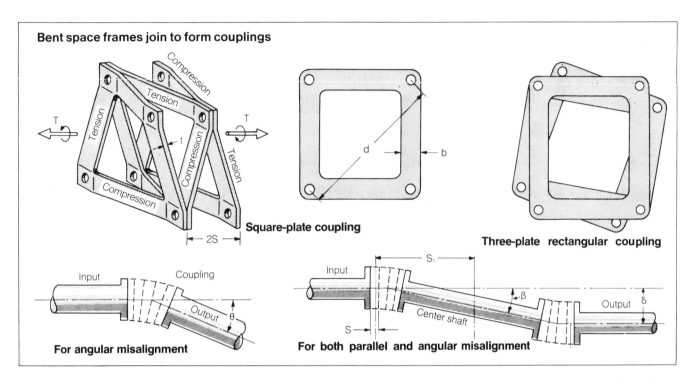

Bent space frames join to form couplings

Square-plate coupling

Three-plate rectangular coupling

For angular misalignment

For both parallel and angular misalignment

Design equations for the Bossler coupling

Ultimate torque capacity

(1) $\quad T = 11.62 \, \dfrac{Ebt^3}{dn^{0.9}}$

Maximum stress per degree of misalignment.

(2) $\quad \sigma_{max} = 0.0276 \, Et/L$

Minimum thickness to meet required torque strength

(3) $\quad t = 0.4415 \left(\dfrac{dT}{bE} \right)^{1/3} n^{0.3}$

Weight of coupling with minimum-thickness plates

(4) $\quad W = 1.249w \left(\dfrac{T}{E} \right)^{1/3} d^{4/3} \, b^{2/3} \, n^{1.3}$

Maximum permissible misalignment

(5) $\quad \theta_{max} = 54.7 \left[\dfrac{bd^2}{TE^2} \right]^{1/3} \sigma_c \, n^{0.7}$

Maximum permissible misalignment (simplified)

(6) $\quad \theta/d = 10.9 \, \dfrac{n^{0.7}}{T^{1/3}}$

Maximum permissible offset-angle

(7) $\quad \beta = 54.7 \left[\dfrac{bd^2}{TE^2} \right]^{1/3} \dfrac{\sigma_c \, C}{n^{0.3}}$

where: $C = \displaystyle\sum_{x=1}^{x=n} \left[1-(x-1) \, \dfrac{S}{S_1} \right]^2$

Maximum permissible offset-angle (simplified)

(8) $\quad \beta/d = \dfrac{10.9 \, C}{T^{1/3} \, n^{0.3}}$

Critical speed frequency

(9) $\quad f = \dfrac{60}{2\pi} \left(\dfrac{k}{M} \right)^{1/2}$

where: $k = \dfrac{24 \, (EI)_c}{(nS)^3}$ and $(EI)_c = 0.886 Ebt^3 S/L$

List of symbols

b = Width of an element
d = Diameter at the bolt circle
E = Modulus of elasticity
f = First critical speed, rpm
I = Flatwise moment of inertia of an element $= bt^3/12$
k = Spring constant for single degree of freedom
L = Effective length of an element. This concept is required because joint details tend to stiffen the ends of the elements. $L = 0.667 \, d$ is recommended
M = Mass of center shaft plus mass of one coupling with fasteners
n = Number of plates in each coupling
S = Offset distance by which a plate is out of plane
t = Thickness of an element
T = Torque applied to coupling, useful ultimate, usually taken as lowest critical buckling torque
w = Weight per unit volume
W = Total weight of plates in a coupling
$(EI)_c$ = Flexural stiffness, the moment that causes one radian of flexural angle change per unit length of coupling
β = Equivalent angle change at each coupling during parallel offset misalignment, deg
ϑ = Total angular misalignment, deg
σ_c = Characteristic that limits stress for the material: yield stress for static performance, endurance limit stress for fatigue performance

and one with slightly rectangular plates is so insignificant that the square-shape equations can be employed with confidence.

Design equations. By making a few key assumptions and approximations, Bossler boiled the complex analytical relationships down to a series of straightforward design equations and charts. The derivation of the equations and the resulting verification from tests are given in the NASA report *The Bossler Coupling,* CR-1241, priced at $3 from the Clearinghouse for Federal Scientific & Technical Information, Springfield, Va. 22151.

Torque capacity. The ultimate torque capacity of the coupling before buckling that may occur in one of the space-frame struts under compression is given by Eq.1. The designer usually knows or establishes the maximum continuous torque that the coupling must transmit. Then he should allow for possible shock loads and overloads. Thus, the clutch should be designed to have an ultimate torque capacity that is at least twice as much, and perhaps three times as much,

Simplicity of the two-plate unit is evident in this power-drive coupling for industrial service, already in use

as the expected continuous torque, according to Bossler.

Induced stress. At first glance, Eq.1 seems to allow much leeway in selecting the clutch size. The torque capacity is easily boosted, for example, by picking a smaller bolt-circle diameter, *d,* which makes the clutch smaller, or by making the plates thicker. But either solution would make the clutch also stiffer, hence would restrict the misalignment permitted before the clutch becomes overstressed. The stress-misalignment relationship is given in Eq.2, which shows the maximum flatwise bending stress produced when a plate is misaligned 1 deg

and is then rotated to transmit torque.

Plate thickness. For optimum misalignment capability, the plates should be selected with the least thickness that will provide the required torque strength. To determine the minimum thickness, Bossler found it expedient to rearrange Eq.1 into the form shown in Eq.3. The weight of any coupling designed in accordance to the minimum-thickness equation can be determined from Eq.4.

Maximum misalignment. Angular misalignment occurs when the centerlines of the input and output shafts intersect at some angle—the angle of misalignment. When the characteristic limiting stress is known for the material selected—and for the coupling's dimensions—the maximum allowable angle of misalignment can be computed from Eq.5.

If this allowance is not satisfactory, the designer may have to juggle the size factors, by, say, adding more plates to the unit. To simplify Eq.5, Bossler made some shrewd assumptions in the ratio of endurance limit to modulus and in the ratio of *dsb* to obtain Eq.6.

Parallel offset. This condition exists when the input and output shafts remain parallel but are displaced laterally. As with Eq.6, Eq.7 is a performance equation and can be reduced to design curves. Bossler obtained Eq.8 by making the same assumptions as in the previous case.

Critical speed. Because of the non-circular configurations of the coupling, it is important that the operating speed of the unit be higher than its critical speed. It should not only be higher but also should avoid an integer relationship.

Bossler worked out a handy relationship for critical speed (Eq.9) that employs a somewhat idealized value for the spring constant, *k.*

"Tests have proved that all these relationships work quite well," says Bossler.

Bossler also makes other recommendations where weight reduction is vital:
■ Size of plates. Use the largest *d* consistent with envelope and centrifugal force loading. Usually, centrifugal force will not be a problem below 300 ft/sec tip speed.
■ Number of plates. Pick the least *n* consistent with the required performance.
■ Thickness of plates. Select smallest *t* consistent with the required ultimate torque.
■ Joint details. Be conservative; use high-strength tension fasteners with high preload. Provide fretting protection. Make element centerlines and bolt centerlines intersect at a point.
■ Offset distance. Use the smallest *S* consistent with clearance.
Nicholas P. Chironis

Off-center pins cancel misalignment of shafts

Two Hungarian engineers have developed an all-metal coupling (drawing below) for connecting shafts where alignment is not exact —that is, where the degree of misalignment does not exceed the magnitude of the shaft radius.

The coupling is applied to shafts that are being connected for either high-torque or high-speed operation and that must operate at maximum efficiency. Knuckle joints are too costly, and they have had too much play; elastic joints are too vulnerable to the influences of high loads and vibrations.

How it's made. In essence, the coupling consists of two disks, each keyed to a splined shaft. One disk bears four fixed-mounted steel studs

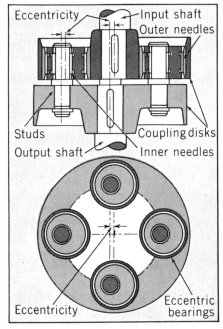

Eccentrically bored bearings rotate to make up for misalignment between shafts.

at equal spacing; the other disk has large-diameter holes drilled at points facing the studs.

Each large hole is fitted with a bearing that rotates freely inside it by rollers or needles. The bore of the bearings, however, is off-center. The amount of eccentricity of the bearing bore is identical to the deviation of the two shaft center lines.

In operation, input and output shafts can be misaligned, yet they still rotate with the same angle of relationship they would have if perfectly aligned.

Hinged links, torsion bushings give drives a soft start

Centrifugal force automatically draws up the linkage legs, while torsional resistance of bushings opposes the deflection forces

A spidery linkage system combined with a rubber torsion bushing system have formed a new type of power-transmission coupling. Developed by a British company, Twiflex Couplings Ltd., Twickenham, Middlesex, England, the device (photo and drawing below) provides ultra-soft starting characteristics. In addition to the torsion system, it also uses centrifugal force to draw up the linkage legs automatically, thus providing additional soft coupling at high speeds to absorb and isolate any torsional vibrations arising from the prime mover.

The new TL coupling has found jobs in coupling marine main propulsions to gearbox-propeller systems. Here the coupling reduces the propeller vibration modes to negligible proportions even at high critical speeds. Other applications are also foreseen, including diesel drives, heavy machine tools, and off-the-road construction equipment. The coupling's range is from 100 hp at 4000 rpm to 20,000 hp at 400 rpm.

Articulating links. The key factor in the TL coupling, an improvement over an earlier Twiflex design, is the circular grouping of hinged linkages connecting the driving and driven coupling flanges. The forked or tangential links have resilient precompressed bonded-rubber bushings at the outer flange attachments, while the other pivots ride on bearings.

When torque is applied to the coupling, the linkages deflect in a positive or negative direction from the neutral position (drawings, below). Deflection is opposed by the torsional resistance of the rubber bushings at the outer pins. When the coupling is rotating, the masses of the linkages give rise to centrifugal forces that further oppose coupling deflection. Therefore, the working position of the linkages depends both on the applied torque and on the speed of rotation.

Tests of the coupling's torque-/deflection characteristics under load have shown that the torsional stiffness of the coupling increases progressively with speed and with torque when deflected in the positive direction. Although the geometry of the coupling is asymmetrical, the torsional characteristics are similar for either direction of drive in the normal working range. Either half of the coupling can be used as the driver for either hand of rotation.

The linkage configuration permits the coupling to be tailored to meet the exact stiffness requirements of individual systems or to provide ultra-low torsional stiffness at values substantially softer than other positive-drive couplings.

These characteristics enable the Twiflex coupling to perform several tasks:

- It detunes the fundamental mode of torsional vibration in a power-transmission system. The coupling is especially soft at low speeds, which permits complete detuning of the system.

- It decouples the driven machinery from engine-excited torsional vibration. In a typical geared system, major machine modes driven by gearboxes are not excited if the ratio of coupling stiffness to transmitted torque is less than about 7:1— a ratio easily provided by the Twiflex coupling.

- It protects the prime mover from impulsive torques generated by driven machinery. Generator short circuits and other impulsive torques are frequently of sufficient duration to cause high response torques in the main shafting.

Using the example of the TL 2307G coupling design—which is suitable for 10,000 hp at 525 rpm—the torsional stiffness at working points is largely determined by coupling geometry and is, therefore, affected to a minor extent by variations in the properties of the rubber bushings. Moreover, the coupling can provide torsional-stiffness values that are accurate within 5.0%.

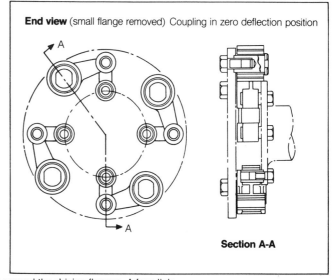

End view (small flange removed) Coupling in zero deflection position

A

A

Section A-A

Articulating links of new coupling (left) are arranged circumferentially around the driving flanges. A four-link design (right) can easily handle torques from a 100-hp prime mover driving at 4000 rpm.

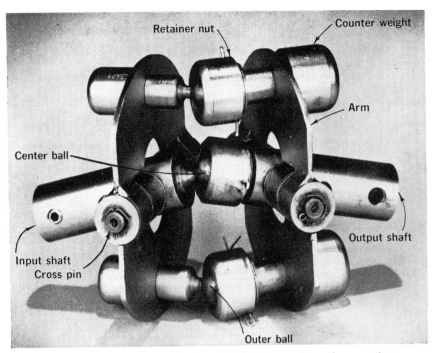

Novel arrangement of pivots and ball-socket joints transmits uniform motion.

Universal joint relays power 45 deg. at constant speeds

During the long Minnesota nights last winter, Malton Miller, a farmer by day, turned engineer and designed a universal joint that transmits power at constant speeds through angles up to 45 deg. James Brodie, a mechanical engineer who owns Brodie Engineering Corp., St. Paul, saw the novel mechanism and liked it. He obtained the manufacturing rights and has now developed and tested models that can transmit up to 20 hp.

It has not been possible until now to transmit power at constant speeds with only one universal joint. Engineers have had to specify an intermediate shaft between two Hooke's joints or use a Rzappa-type joint to get the desired effect.

Ball-and-socket. Basically, the True-Speed joint is a system of ball-and-socket connections with large contact area (low unit pressure) to transmit torsional forces across the joint. This arrangement minimizes problems when high bearing pressures build up against running surfaces. The low-friction bearings also increase efficiency. The joint is balanced to keep vibration at high speeds to a minimum.

The joint consists of driving and driven halves. Each half has a coupling sleeve to its end of the drive-shaft, a pair of driving arms opposite each other and pivoted on a cross pin that extends through the coupling sleeve, and a ball-and-socket coupling at the end of each driving arm.

As the joint rotates, angular flexure in one plane of the joint is accommodated by the swiveling of the ball-and-socket couplings and, in the 90-deg. plane, by the oscillation of the driving arms about the transverse pin. As rotation occurs, torsion is transmitted from one half of the joint to the other half through the swiveling ball-and-socket couplings and the oscillating driving arms.

Balancing. Each half of the joint, in effect, rotates about its own center shaft, so each half is considered separate for balancing. The center ball-and-socket coupling serves only to align and secure the intersection point of the two shafts. It does not transmit any forces to the drive unit as a whole.

Balancing for rotation is achieved by equalizing the weight of the two driving arms of each half of the joint. Balancing the acceleration forces due to the oscillation of the ball-and-socket couplings, which are offset from their swiveling axes, is achieved by the use of counterweights extending from the opposite side of each driving arm.

The outer ball-and-socket couplings work in two planes of motion, swiveling widely in the plane perpendicular to the main shaft and swiveling slightly about the transverse pin in the plane parallel to the main shaft. In this coupling configuration, the angular displacement of the driving shaft is perfectly duplicated in the driven shaft, providing constant rotational velocity and constant torque at all shaft intersection angles.

Bearings. The only bearing parts are the ball-and-socket couplings and the driving arms on the transverse pins. Needle bearings support the driving arms on the tranverse pin, which is hardened and ground. A high-pressure grease lubricant coats the bearing surfaces of the ball-and-socket couplings. Under maximum rated loadings of 600 psi on the ball-and-socket surfaces, there is no appreciable heating or power loss due to friction.

Capabilities. Units have been laboratory-tested at all rated angles of drive under dynamometer loadings. Although the presently available units are for smaller capacities, a unit designed for 20 hp. at 550 rpm, suitable for tractor power take-off drive, has been tested by Brodie Engineering Corp.

Similar couplings have been designed, for instance the pump coupling shown in the diagram on page 327. But the True-Speed drive differs in that the speed and transfer elements are positive. With the pump coupling, on the other hand, the speed might fluctuate because of spring bounce. □

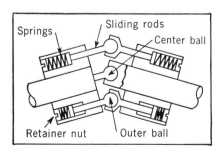

Earlier version for angled shafts required spring-loaded sliding rods.

Flexible shafts take on more jobs

Extensive use of flexible shafts in a wide range of
machine drives and industrial applications is making
this component one of the designer's favorite choices

Even after decades of successful rise, the flexible shaft is still fighting to become accepted as a basic element for power and motion transmission, just as gears, belts and chains are. Recently, a technical committee of the American Gear Manufacturers Association (AGMA), describing to AGMA members its work to update standards for flexible couplings, brushed aside questions from the floor as to why flexible shafts were not among the components under committee study.

Yet the flexible shaft is probably the simplest way to transmit rotary motion between two points on a machine.

The major drawback to a flexible shaft —its relatively high torsional windup when transmitting large torques—can be kept to a minimum by suitable core construction and design. Some heavy-duty types can handle 20 hp; some smaller diameter types can operate to 30,000 rpm.

Versatility in design. The benefits derived from use of flexible shafts are impressive:

■ Simplicity of design—Greater freedom in the location of components. A pump, for example, can be located at the most expeditious position on a truck.

■ Unaffected by relative motion—Ideal for driving a shifting or portable tool. The operator of the tool is relieved from having to carry a heavy motor or engine.

■ Absorbing vibration—Automatically dampen shock and vibration and are trouble-free in high-vibration applications, such as concrete vibrators and vibration-testing tables.

■ Simplicity in installation—Accuracy of alignment, so important in conventional drives, is not required for smooth, even power transmission when using flexible shafts. Time saved in installation.

■ High efficiency—The efficiency ranges from 80 to 90%.

■ Multiple power outlets—Flexible shafts have been successfully designed to provide for multiple power outlets from a single source.

■ Flexible coupling—Transmits power inexpensively between closely coupled parts inside a piece of equipment, usually less than 18 in. apart. Such shafts generally do not require casings.

Much of a flexible shaft's characteristics depends on how it is made. Basically, a unit consists of a core (the flexible shaft), a flexible casing or housing, and a core end-fitting or casing ferrule. The words core, cable, and flexible shaft sometimes are used synonymously.

The rotating core is the crucial component of the flexible shaft. It's usually built upon a single wire on which one or more layers of wire (each layer consisting of two or three wires) are wound. Layers are wound in opposite directions. In most types, the wires not only increase in size but the number of wires increases in each layer (Fig below).

The direction of rotation of a flexible shaft is the direction which tends to tighten the last layer of wire on the core shaft. The core casing or housing usually is marked with the proper rotation. A CCW-wound core should always be driven clockwise, and vice versa. If driven in the opposite direction, the torque capacity of the system drops about 40%.

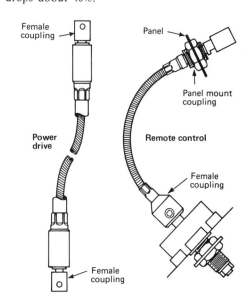

Remote control flexible cables differ from power drive flexible cables in minor but important ways (S.S. White)

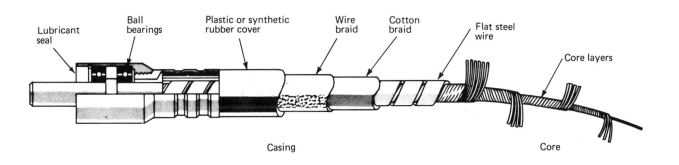

The flexible core rotates within the flat steel wire guide in this typical assembly. The casing with its various layers is supported
separately from the rotating core itself, and firmly contains and guides the core

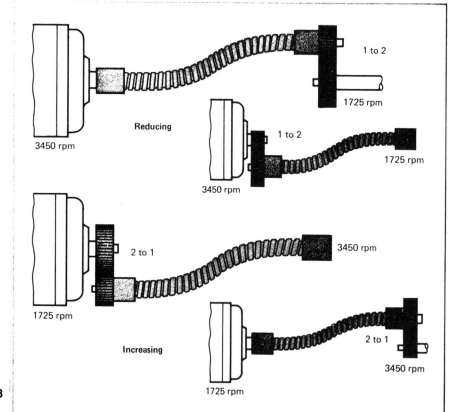

Differences between power-drive and remote-control cores

Both cores ½ in. diameter (stow)	Power drive core	Remote control core
Construction:		
No. of layers	6	8
No. of wires	27	80
Diameter of wires on outside layer	0.063	0.041
No. of wires on outside layer	6	16
Deflection at 80 lb-in. of torque/ft of length, degrees		
In direction of windup of outside layer	18	3
In direction of unwind of outside layer	81	5

Defining equations for flexible shafts are given in the accompanying text and are

Because of variations in construction, a flexible shaft core of the same diameter can have its characteristics changed significantly by varying the number of wires per layer, the number of layers of wire, the diameter of the wires, the material of the wires and the spacing between the wires. Manufacturers tend to restrict the variations to two or three types of cores. The table, for example, shows how the construction and deflection (twist) varies between the so-called power-drive cores and remote-control cores.

The casing acts as a support to prevent the rotating core from looping during operation under torsional load. It also serves as a bearing for the core and protects it from moisture, dust and abrasion, and retains any lubricant surrounding the core.

Metallic casings consist of interlocking windings of wire (drawing), generally of two-wire or four-wire construction. This facilitates bending, resulting in casings that are strong, durable and moderate in cost. They are sufficiently grease-tight for most purposes, but not oil-tight, and are typically made of galvanized steel, stainless steel or phosphor bronze wire.

Covered casings are for medium-duty and heavy-duty applications and are fabricated with various liners, some metallic and others either plastic or cotton. Metallic liners include flat steel spiral wire for heavy-duty work, and multi-wire construction where greater flexibility is needed. These casings often are reinforced with layers of braid. The outer wrap is either plastic or rubber for oil- and water-tight protection.

Couplings and fittings. Many styles of core fittings and casing couplings are available for coupling the flexible shafts to power sources and driven loads. Because only the core is rotating, the core and casing must be connected independently of each other.

Core fittings include both male and female types, typically secured in position by threads, set screws, splines, or keyways. Casing couplings are attached to the casing by swaging, crimping, or soldering. Today there are quick-disconnect versions for hand tools. For high power drives, one of the core fittings should be free to slide back and forth in the rotating part to take care of the relative motion and avoid strain.

The basic factors for selecting a flexible shaft are (1) rpm of the shaft, (2) horsepower to be transmitted, or the actual amount of torque to be transmitted at a

Flexible gear reducers and increasers are readily built with flexible shaft elements combined with conventional gearing. These examples are by Elliott

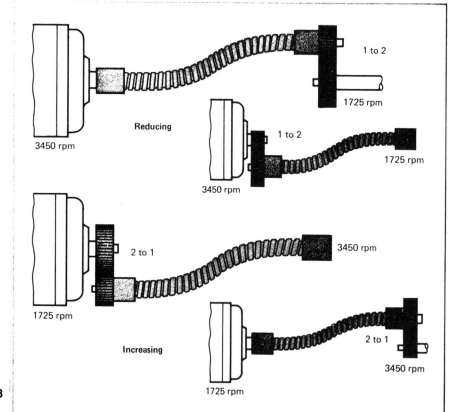

Reducing
3450 rpm
1 to 2
1725 rpm

1 to 2
3450 rpm
1725 rpm

2 to 1
1725 rpm
3450 rpm

Increasing
1725 rpm
2 to 1
3450 rpm

Do's and don'ts. For trouble-free service determine torque requirements before selecting a flexible shaft. Use the maximum figure encountered during start-up and possible overloads. Design the drive to operate at the highest possible speeds. For transmission of a given amount of horespower, the higher the speed, the lower the torque on the shaft.

Operate the shaft in the largest possible radius. Don't bend a core into a radius smaller than recommended, even during shipment or installation. Make sure that the flexible shaft is installed properly. Use a floating-shaft fitting if possible, because the shaft may contract or elongate under loading or flexing. End fittings should protrude far enough to engage mating elements properly. Avoid lengths greater than 10 ft. Use solid shafts for long, straight runs and couple them at one or both ends to a section of flexible shaft.

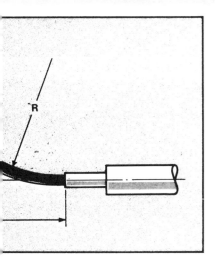

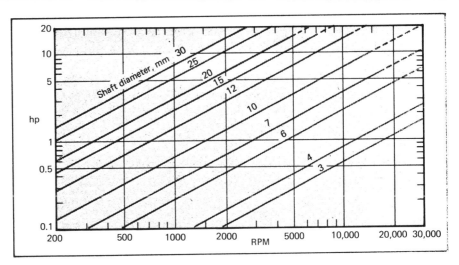

based on this typical configuration of input, flexible shaft, and output

Load carrying capacity of flexible shafts is determined by a combination of diameter and speed. This graph is Suhner's, but each manufacturer has its own

given rpm, (3) minimum bend or operating radius of the flexible shaft in the actual operating position (if multiple bends are involved, use the radius of the smallest bend), and (4) length of the flexible shaft.

The torque, T, to be transmitted may be determined by knowing the power load and the rotation speed N in revolutions per minute:

$$T \text{ (in-lb)} = 63,000 \text{ hp}/N \text{ (inches)}$$
$$T \text{ (cm-kg)} = 71,620 \text{ hp}/N \text{ (metric)}$$

Knowing either the torque or horsepower and speed helps determine the proper core diameter (and from this also the casing size). One manufacturer has a graph (above) that relates rotational speed and horsepower to the flexible shaft diameters necessary to carry the loading.

The radius of bend should be kept as large as possible. In most applications, the bend radius can be approximated from the layout of the installation, but in some offset conditions this approach becomes inaccurate. The best way to calculate the radius of curvature R and length L of a flexible shaft is with the following formulas (also see drawings.):
Parallel offset:

$$R = \frac{x^2 + y^2}{4y}$$

$$L = \frac{\pi R}{90} \arcsin \frac{x}{2R}$$

$$x = \sqrt{y(4R - y)}$$

$$y = 2R - \sqrt{4R^2 - x^2}$$

If the minimum radius of curvature for the particular application is not larger than the minimum operating radius of the core, then some adjustments in dimensions x and y should be made.

Successful applications. Typically there are power-drive and remote-control situations. In power-drive applications, the flexible shaft operates most efficiently in the windup direction (which tends to

tighten the outer layer of wires). Manual remote control or intermittent power-driven operation may be in both directions.

Multiple-spindle drilling and tapping costs are reduced considerably with pneumatic feed cylinders that are flexible-shaft driven and powered by an electric motor, particularly for multi-drilling or tapping from different directions. It's six times cheaper than with individual pneumatic motors at each spindle.

Flexible tool shafts available from some manufacturers are a core and case assembly with a motor connection on one end and handpiece on the other end. The handpiece has a threaded wheel arbor to attach tools such as grinding wheels. Also

available are 18-in. slip joints used, for example, to operate pumps on tractor trailers from the power-transmission shaft.

These manufacturers supplied information:
S. S. White Industrial Products, Piscataway, N.J. 08854
Stow Manufacturing Co., Binghamton, N.Y.
Suhner Industrial Products, Rome Ga. 30161
B.W. Elliott Manufacturing Co., Binghamton, N.Y. 13902

This typical application for flexible cables is for multi-drilling equipment. The cables are driven from one central power source in the machine base (Suhner)

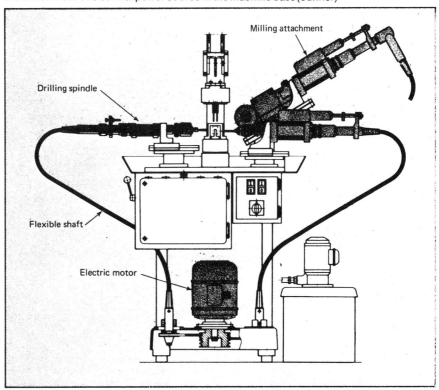

SPRING-WRAPPED SLIP CLUTCHES

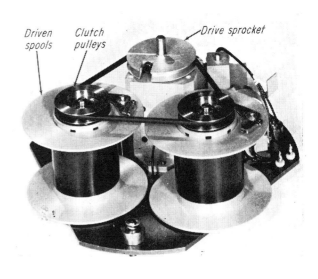

Driven spools · Clutch pulleys · Drive sprocket

The simple spring clutch becomes even more useful when designed to slip at a predetermined torque. Unaffected by temperature extremes or variations in friction, these clutches are simple—can even be "homemade." Here are two of the author's designs, with information published on the dual-spring slip-type for the first time. Two dual-spring clutches are incorporated in the tape drive pictured above

J KAPLAN

Spring clutches are handy devices for driving a load in one direction and uncoupling it when the output is overdriven or direction of the input rotation is reversed. A few years ago, the author successfully modified a spring clutch to give a predetermined slip in either direction—hence the designation of this type as a "slip clutch." A stepped helical spring was employed to accomplish that modification. It has since been developed further by introducing an intermediate clutch member between two helical springs. This dual-spring innovation is preferred where more accuracy of the output torque is required.

Most designs employ either a friction-disk clutch or a shoe clutch to obtain a predetermined slip (in which the input drives output without slippage until a certain torque level is reached—then a drag-slippage occurs). But the torque capacity (or slip torque) for friction-disk clutches is the same for both directions of rotation.

By contrast, the stepped-spring slip clutch, pictured on next page, can be designed to have either the same or different torque capacities for each direction of rotation. Torque levels where slippage occurs are independent of each other, thus providing wide latitude of design.

The element producing slip is the stepped spring. Outside diameter of the large step of the spring is assembled tightly in the bore of the output gear. Inside diameter of the smaller step fits tightly over the shaft. Rotation of the shaft in one direction causes the coils in contact with the shaft to grip tightly, and the coils inside the bore to contract and produce slip. Rotation in the opposite direction reverses the action of the spring parts, and slip is effected on the shaft.

DUAL-SPRING SLIP CLUTCH

This innovation also permits bidirectional slip and independent torque capacities for the two directions of rotation. It requires two springs, one right-handed and one left-handed, for coupling the input, intermediate

and output members. These members are coaxial, with the intermediate and input free to rotate on the output shaft. Rotation of input in one direction causes the spring, which couples the input and intermediate members, to grip tightly. The second spring, which couples the intermediate and output members, being oppositely wound, tends to expand and slip. Rotation in the opposite direction reverses action of the two springs so that the spring between input and intermediate members provides the slip. Because this design permits greater independence in the juggling of dimensions, it is preferred where more accurate slip-torque values are required.

REPEATABLE PERFORMANCE

Spring-wrap slip clutches and brakes have remarkably repeatable slip-torque characteristics which do not change with service temperature. Torque capacity remains constant with or without lubrication, and is unaffected by variations in the coefficient of friction. Thus, breakaway torque capacity is equal to the sliding torque capacity. This stability obviates overdesign of slip members to obtain reliable operation. Such advantages are absent in more commonly used slip clutches.

BRAKE AND CLUTCH COMBINATIONS

An interesting example of how slip brakes and clutches worked together to maintain proper tension in a tape drive, in either direction of operation, is pictured above and shown schematically on opposite page. A brake here is merely a slip clutch with one side fastened to the frame of the unit. Stepped-spring clutches and brakes are shown for simplicity although, in the actual drive, dual-spring units were employed.

The sprocket wheel drives both the tape and belt. This allows linear speed of the tape to be constant (one of the requirements). The angular speed of the spools, however, will vary as they wind or unwind. Problem here

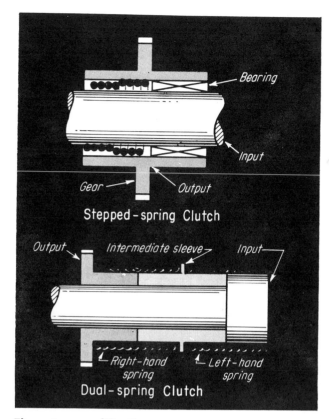

Stepped-spring Clutch

Dual-spring Clutch

These two modifications . . .
of spring clutches obtain independent slip characteristics in either direction of rotation.

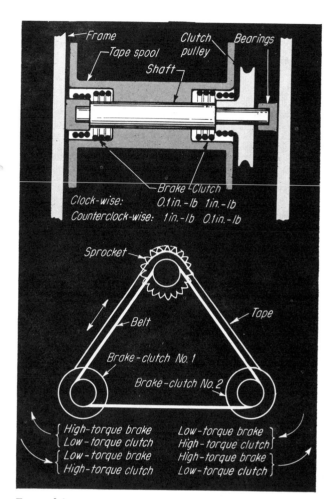

Clock-wise: 0.1in.-lb   1in.-lb
Counterclock-wise: 1in.-lb   0.1in.-lb

High-torque brake	Low-torque brake
Low-torque clutch	High-torque clutch
Low-torque brake	High-torque brake
High-torque clutch	Low-torque clutch

Tape drive . . .
requires two slip clutches and two brakes to assure proper tension for bi-directional rotation. Detail of spool (above) shows clutch and brake unit.

is to maintain proper tension in the tape at all times and in either direction. This is done by employing a brake-clutch combination. In a counterclockwise direction, for example, the brake may be a "low-torque brake" in which it resists with a 0.1 in.-lb torque. The clutch in this direction is a "high-torque clutch"—it will provide a 1-in.-lb torque. Thus the clutch overrides the brake with a net torque of 0.9 in.-lb.

When the drive is reversed, the same brake may now act as a high-torque brake, resisting with a 1 in.-lb torque, while the clutch acts as a low-torque clutch, resisting with 0.1 in.-lb. Thus in the first direction the clutch drives the spool; in the other, the brake overcomes the clutch and provides a steady resisting force to provide tension in the tape. Of course, the clutch also permits the pulley driven by the belt to overdrive.

Two brake-clutch units are required. The second unit will provide opposing torque values—as listed in the diagram above. The drive necessary to advance the tape only in a clockwise direction would be the slip clutch in unit 2 and the brake in unit 1. Advancing the tape in the other direction calls for use of the clutch in unit 1 and the brake in unit 2.

For all practical purposes, the low torque values in the brakes and clutches can be made negligible by specifying minimum interference between spring and the bore or shaft. The low torque is amplified in the spring clutch to the level necessary to drive the tensioning torques of the brake and slip clutches.

Action thus produced by the simple arrangement of directional slip clutches and brakes cannot otherwise be duplicated without resorting to more complex designs.

Torque capacities of spring-wrap slip clutches and brakes using round, rectangular and square wire are, respectively:

$$T = \frac{\pi E d^4 \delta}{32 D^2}; \; T = \frac{E b t^3 \delta}{6 D^2}; \; T = \frac{E t^4 \delta}{6 D^2}$$

where E = modulus of elasticity, psi; d = wire dia, in.; D = dia of shaft or bore, in.; δ = diametral interference between spring and shaft, or spring and bore, in.; t = wire thickness, in.; b = width of rectangular wire, in.; and T = slip torque capacity, lb-in.

Minimum interference moment (on the spring gripping lightly) required to drive the slipping spring is:

$$M = \frac{T}{e^{\mu \theta} - 1}$$

where e = natural logarithmic base (e = 2.716); θ = angle of wrap of spring per shaft, radians, μ = coefficient of friction, M = interference moment between spring and shaft, lb-in.

DESIGN EXAMPLE

Required: to design a tape drive similar to the one shown above. Torque requirements for the slip clutches and brakes for the two directions of rotation in are:

(1) Slip clutch in normal takeup capacity (active function) is 0.5 to 0.8 in.-lb.

(2) Slip clutch in override direction (passive function) is 0.1 in.-lb (max).

(3) Brake in normal supply capacity (active function) is 0.7 to 1.0 in.-lb.

(4) Brake in override direction (passive function) is 0.1 in.-lb (max).

Assume that the dual-spring design shown previously is to be used with 0.750-in. drum diameters. Also available is an axial length for each spring, equivalent to 12 coils which are divided equally between the bridged shafts. Using round wire, calculate the wire diameter of the springs if 0.025 in. is max diametral interference desired for the active functions. For the passive functions use round wire which produces a spring index not more than 25.

Slip clutch, active spring:

$$d = \sqrt[4]{\frac{32D^2T}{\pi E \delta}} = \sqrt[4]{\frac{(32)(0.750)^2(0.8)}{\pi(30 \times 10^6)(.025)}} = 0.050 \text{ in.}$$

Min diametral interference is $(0.025)(0.5)/0.8 = 0.016$ in. Consequently, ID of the spring will vary from 0.725 to 0.734 in.

Slip clutch, passive spring:

$$\text{Wire dia} = \frac{\text{drum dia}}{\text{spring index}} = \frac{0.750}{25} = 0.030 \text{ in.}$$

Diametral interference:

$$\delta = \frac{32D^2T}{\pi Ed^4} = \frac{(32)(0.750^2)(0.1)}{\pi(30 \times 10^6)(0.030)^4} = 0.023 \text{ in.}$$

Assuming a min coefficient of friction of 0.1, determine min diametral interference for a spring clutch sufficient to drive the max slip clutch torque of 0.8 lb-in.:

$$M = \frac{T}{e^{\mu\theta} - 1} = \frac{0.8}{e^{(0.1\pi)(6)} - 1}$$

Min diametral interference:

$$\text{min} = 0.023 \times \frac{0.019}{0.1} = 0.0044 \text{ in.}$$

ID of the spring is therefore 0.727 to 0.745 in.

Brake springs

By similar computations wire dia of the active brake spring is 0.053 in., with an ID that varies from 0.725 and .733 in.; wire dia of the passive brake spring is 0.030 in., with its ID varying from 0.727 to 0.744 in.

Controlled-slip concept adds new uses for spring clutches

A remarkably simple change in spring clutches is solving a persistent problem in tape and film drives—how to keep drag tension on the tape constant, as its spool winds or unwinds. Shaft torque has to be varied directly with the tape diameter and many designers resort to using electrical control systems, but that involves additional components, and an extra motor, making it an expensive solution. The self-adjusting spring brake (Fig 1) developed by Joseph Kaplan and his Machine Components Corp, Farmingdale, NY, gives a constant drag torque ("slip" torque) that is

easily and automatically varied by a simple lever arrangement actuated by the tape spool diameter (Fig 2). The new brake is also being employed to test the output of motors and solenoids by providing levels of accurate slip torque.

Kaplan is using his "controlled-slip" concept in two other new products. In the controlled-torque screwdriver (Fig 3) a stepped spring provides a 1¼-in.-lb slip when turned in either direction. It avoids overtightening machine screws in delicate instrument assemblies. A stepped spring is also the basis for the go no-go torque gage

that permits production inspection of output torques to within +1%.

Interfering spring. The three products are the latest in a series of slip clutches, drag brakes, and slip couplings developed by Kaplan for instrument brake drives. All are actually outgrowths of the spring clutch. The spring in such a clutch is normally prevented from gripping the shaft by means of a detent. Upon release of the detent the spring will grip the shaft. If the shaft is turning in the proper direction, it is self-energizing. In the other direction, the spring merely overrides. Thus the spring clutch is a "one-way" clutch.

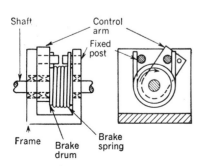

1. Variable-torque drag brake ...

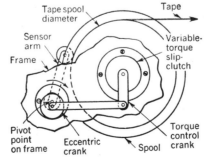

2. ... holds tension constant on tape

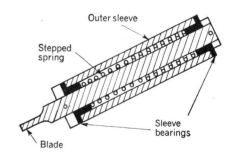

3. Constant-torque screwdriver

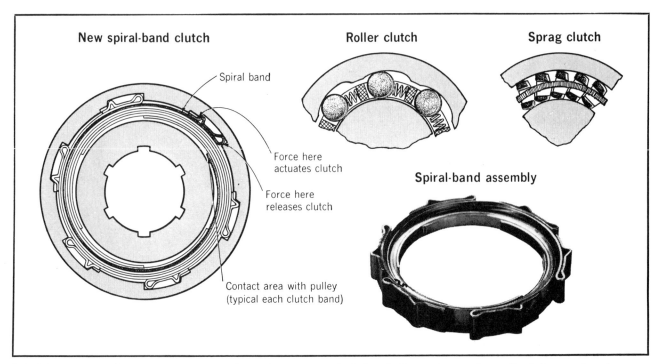

Spring bands grip tightly to drive overrunning clutch

A new type of overrunning clutch that takes up only half the usual space employs a series of spiral-wound bands instead of the conventional rollers or sprags to transmit high torques. The new design (drawings, above) also simplifies the assembly, cutting costs as much as 40% by eliminating more than half the parts in conventional clutches.

The key to the savings in cost and bulk is the new design's freedom from the need for a hardened outer race. Roller and sprag types must have hardened races because they transmit power by a wedging action between the inner and outer races.

Role of spring bands. Overrunning clutches, including the spiral-band type, slip and overrun when reversed—in drawing above, when outer member is rotated clockwise and inner ring is the driven member.

The new clutch, developed by National Standard Co., Niles, Mich., contains a set of high-carbon spring-steel bands (six in the design illustrated) that grip the inner member when the clutch is driving. The outer member merely serves to retain the spring anchors and to play a part in actuating the clutch. Since it isn't subject to wedging action, it can be made of almost any material, and this accounts for much of the cost saving. For example, in the automotive torque converter in the drawing at right, the bands fit into the aluminum die-cast reactor.

Reduced wear. The bands are spring-loaded over the inner member of the clutch, but they are held and rotated by the outer member. The centrifugal force on the bands thus releases much of the force on the inner member and considerably decreases the overrunning torque. Wear is, therefore, greatly reduced.

The inner portion of the bands fits into a V-groove in the inner member. When the outer member is reversed, the bands wrap, creating a wedging action in this V-groove. This action is similar to that of a spring clutch with a helical-coil spring, but the spiral-band type has very little unwind before it overruns, compared with the coil type. Thus it responds faster.

Edges of the clutch bands carry the entire load, and there is also a compound action of one band upon

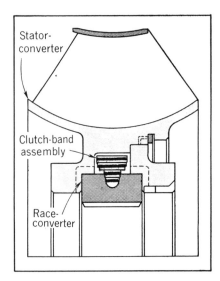

Spiral clutch bands can be bought separately to fit in user's assembly.

another. As the torque builds up, each band pushes down on the band beneath it, so each tip is forced more firmly into the V-groove.

National Standard plans to sell the bands as separate components, without the inner and outer clutch members (which the user customarily builds as part of his product). The bands are rated for torque capacities from 85 to 400 ft.-lb. Applications include auto transmissions and starters and industrial machinery. □

Slip clutch, bidirectional type combine to control torque

A new type of torque-limiting knob uses a dual set of miniature clutches—a detent slip clutch in series with a novel bidirectional locking clutch—to prevent the driven member from backturning the knob. The bidirectional clutch in the knob locks the shaft from backlash torque emanating from within the panel, and the slip clutch limits the torque transmitted from outside the panel. The new device is the brainchild of Ted Chanoux, vice-president of Beier-Chanoux Corp., Medford, N.Y., and inventor of several mechanical devices.

The new design (drawing below) is the result of an attempt to solve a problem that often plagues design engineers. A mechanism behind the panel—a display counter, for example—a precision potentiometer, and switches must be operated by a shaft that protrudes from the panel. The mechanism, though, must not be able to turn the shaft. Only the operator in front of the knob can turn the shaft, and he must be limited in the amount of torque he is able to apply.

Solving design problem. "This problem occurred in the design of a navigational system for aircraft," says Chanoux. "The counter gives a longitudinal or latitudinal readout. When the aircraft is ready to take off, the navigator or pilot sets the counter to some nominal figure, depending on the location of his starting point, and he energizes the system. The

computer then takes the directional information from the gyro, the air speed from instruments in the wings, plus other data and feeds a readout at the counter.

"The entire mechanism was subjected to vibration, acceleration and deceleration, shock, and other high-torque loads, all of which could feed back through the system and could sometimes move the counter. The new knob device positively locks the mechanism shaft against the vibration, shock loads, and accidental turning and limits the input torque to the system at a preset value."

Operation. To turn the shaft, the operator depresses the knob 1/16 in. and turns it in the desired direction. When it is released, the knob retracts, and the shaft immediately and automatically locks to the panel or frame with zero backlash. Should the shaft torque exceed the preset value, such as hitting a mechanical stop after several turns, or should the knob turn in the retracted position, the knob will slip to protect the system mechanism.

Internally, pushing in the knob turns both the detent clutch and the bidirectional-clutch release cage via the keyway. The fingers of the cage extend between the clutch rollers so rotation of the cage cams out the rollers, which are usually kept jammed between the clutch cam and the outer race by means of the roller springs. This action permits rotation of

the cam and instrument shaft both clockwise and counterclockwise, but it locks the shaft securely against inside torque up to 30 oz.-in.

Applications. The detent clutch can be adjusted to limit the input torque to the desired values without removing the knob from the shaft. Outside diameter of the shaft is only 0.900 in., and the total length is 0.940 in. The exterior material of the knob is anodized aluminum, black or gray, and all other parts are stainless steel. The device is designed to meet the military requirements of MIL-E-5400 class 3 and MIL-K-3926 specifications.

Chanoux says potential applications include counter and reset switches and controls for machines and machine tools, radar systems, and precision potentiometers.

Eight-joint coupler

A novel linkage combines two parallel linkage systems in a three-dimensional arrangement to provide wide angular and lateral off-set movements in pipe joints. By including a bellows between the connecting pipes, the connector can join high-pressure and high-temperature piping such as found in refineries, steam plants, and stationary power plants.

Key components in the device are four pivot levers (drawing) mounted

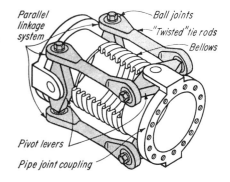

in two planes. Each pivot lever has provisions for a ball joint at each end. "Twisted" tie rods, with holes in different planes, connect the pivot levers to complete the system. The arrangement permits each pipe face to twist through an appreciable arc and also to shift orthogonally with respect to the other.

Longer tie rods can be used by joining several bellows together by means of center tubes.

The device was developed by Ralph Kuhm Jr of North American Aviation, El Segundo, Calif.

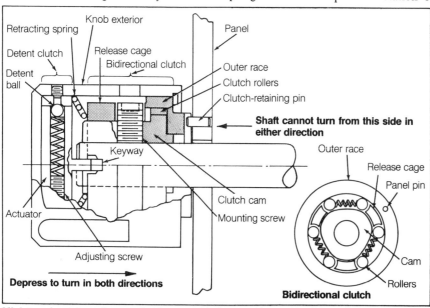

Miniature knob is easily operated from outside the panel by pushing it in and turning it in desired direction. When released, the bidirectional automatically locks the shaft against all conditions of shock and vibration.

Simple clutch/brake system makes tape recorder foolproof

Often the way to make something foolproof and troublefree is to make it simple. And that's exactly what the 3M Co. did for the design of its new Wollensak reel-to-reel tape recorder.

The key innovation in the recorder's tape transport system is a direction-sensing clutch/brake that is a model of simplicity. It consists of five parts that replace a whole array of complex levers, braking arms, and drive pulleys. The parts are assembled on the shaft of a dc motor that drives the tape in fast-forward, rewind, and takeup.

A driving helix is the only part that is secured to the motor shaft. Two pulleys are mounted on the shaft, one on either side of the helix. Between them, a shuttle clutch disk with an internal helix is assembled on the drive helix. Friction material is carried on both faces of the shuttle disk, and a retaining ring is snapped in place above the upper pulley to hold the whole assembly on the motor shaft.

Sliding by. The shuttle slides freely on the drive helix, and the pulleys are free to rotate on the shaft unless their motion is impeded or otherwise influenced by contact with friction material on the shuttle disk faces.

When the shuttle is centered between the pulleys, there is only a few thousandths of an inch clearance. But the friction material is sponge-like and highly compressible, so the shuttle has considerably greater side motion than its narrow clearance would indicate. Maximum side motion is 1/8 in.

When the motor is turned to play/record or fast-forward, it runs counterclockwise. At start-up, the inertia of the shuttle and the action of the helix forces the shuttle down the helix, bringing the friction material into contact with the lower pulley to smoothly accelerate the take-up reel through a belt drive.

When rewinding the tape, the motor runs clockwise. The shuttle rises on the helix and similarly engages the upper pulley to drive the rewind reel.

Putting a stop to it. If the dc motor is stopped, the inertia of the clutch causes it to disengage from the pulley it was driving and to move up or down the helix to engage the opposite pulley. This action smoothly brakes the unwinding reel, putting tension on the tape to stop both it and the reel driven by the pulley that was disengaged by the shuttle. The shuttle always senses and applies braking to the unwinding reel.

The dc motor is stopped quickly, but without shock, by dynamic braking, a magnetic action applied to the armature by short-circuiting the windings. It causes a counter emf to develop a braking force that disappears as the motor slows. There is nothing that will wear out in this system.

The clutch/brake not only eliminates a multiplicity of levers in the braking system, but it also eliminates the need to adjust and readjust levers to compensate for wear. Wear of the friction material has little or no effect on operation.

Two motors better than one. Another innovation in the system is the use of two motors—the small, 65% efficient dc motor drives the tape transport and a small, necessarily inefficient ac motor provides torque for the capstan tape drive. In previous tape recorders, both functions were powered by a larger ac motor.

Only 1/2-in.-oz. torque is required to drive the capstan at uniform tape speed. The tape-transport system needs 8-in.-oz. torque for a 2-min. rewind.

The 2-in.-oz. motor runs cool, because it is much smaller, thus eliminating the fan and fan noise. It also draws less current and radiates less hum. There is less physical vibration, so mounts are not needed and the design center is fixed.

Separating the take-up reel physically from the capstan drive prevents variations in take-up action from taking energy out of the tape drive. It also decreases the overall wow and flutter created by the basic drive system.

Simple, direction-sensing clutch/brake automatically accelerates and decelerates the tape reels in a system that can move 1200 ft. of tape in only 1½ min.

Walking pressure plate delivers constant torque

The new design of an automatic clutch causes the driving plate to move around the surface of the driven plate, to prevent the clutch plates from overheating if the load gets too high. This "walking" action enables the clutch to transmit full engine torque for hours, without serious damage to the clutch plates or the engine.

The automatic centrifugal clutch, manufactured by K-M Clutch Co., Van Nuys, Calif., combines the principles of a governor and a wedge to transmit torque from the engine to the drive shaft (drawing).

How it works. As the engine builds-up speed, the weights attached to the levers have a tendency to move towards the rim of the clutch plate, but are stopped by retaining springs. When the shaft speed reaches 1600 rpm, however, centrifugal force overcomes the resistance of the springs, and the weights move outward. Simultaneously, the tapered end of the lever wedges itself in a slot in pin E, which is attached to the driving clutch plate. The wedging action forces both the pin and the clutch plate to move into contact with the driven plate.

A pulse of energy is transmitted to the clutch each time a cylinder fires. With every pulse, the lever arm moves outward, and there is an increase in pressure between the faces of the clutch. Before the next cylinder fires, both the lever arm and the driving plate return to their original positions. This pressure fluctuation between the two faces is repeated throughout the firing sequence of the engine.

Plate walks. If the load torque exceeds the engine torque, the clutch immediately slips, but full torque transfer is maintained without serious overheating. What happens is that the pressure plate momentarily disengages from the driven plate. However, as the plate rotates and builds up torque, it again comes in contact with the driven plate. In effect, the pressure

Centrifugal clutch combines governor and wedge principles to transmit torque.

plate "walks" around the contact surface of the driven plate, enabling the clutch to continuously transmit full engine torque.

Applications. The clutch has undergone hundreds of hours of development testing on 4-stroke engines that ranged from 5 to 9 hp. According to Walter Humphrey of K-M Clutch Co., the clutch enables designers to use smaller motors than they previously could because of its no-load starting characteristics.

The clutch also acts as a brake to hold engine speeds within safe limits. For example, if the throttle accidentally opens when the driving wheels or driven mechanisms are locked, the clutch will stop.

The clutch can be fitted with sprockets, sheaves, or a stub shaft. It operates in any position, and can be used in either direction. In fact, K-M is thinking of using the clutch in a marine application in which the applied torque would come from the direction of the driven plate.

The pressure plate is made of cast iron, and the driven-plate casting of magnesium. To prevent too much wear the steel fly weights and fly levers are pre-hardened. □

When centrifugal force overcomes resistance of spring force, the lever action forces the plates together.

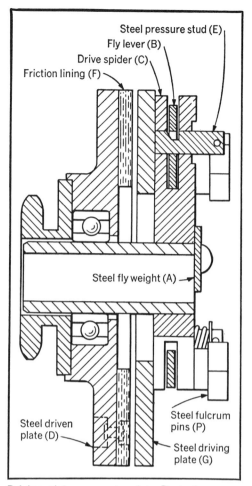

Driving plate moves to plate D closing the gap, when speed reaches 1600 rpm.

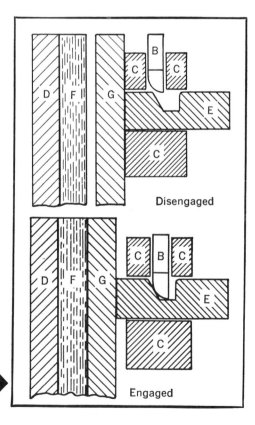

Conical-rotor motor provides instant clutching or braking

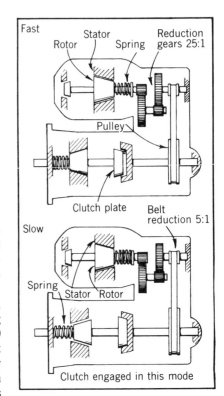

Fast

Rotor Stator Spring Reduction gears 25:1

Pulley

Clutch plate Belt reduction 5:1

Slow

Spring Stator Rotor

Clutch engaged in this mode

By reshaping the rotor of an ac electric motor, engineers at Demag Brake Motors, Wyandotte, Mich., have found that the axial component of the magnetic forces can be used to act on a clutch, or a brake. Moreover the motors can be arranged in tandem to obtain fast or slow speeds with instant clutching or braking.

As a result, this type motor is being used in many applications where instant braking is essential—for example in an elevator when the power supply fails. The principle can also be used to obtain a vernier effect, which is useful in machine-tool operations.

Operating principles. The Demag brake motor operates on a sliding-rotor principle. When no power is being applied, the rotor is pushed slightly away from the stator in an axial direction by a spring. However, with power the axial vector of the magnetic forces overcomes the spring pressure and causes the rotor to slide forward almost fully into the stator. The maximum distance in an axial direction is 0.18 in. This effect is used so that a combined fan and clutch, mounted on the rotor shaft, engages with a brake drum when power is stopped, and disengages when power is applied.

In Europe, the conical-rotored motor is used where rapid braking is essential to overcome time consuming overruns, or where accurate braking and precise angular positioning are critical—such as in packaging machines.

New arrangement. For instance, if two motors are used, one running at 900 rpm and the other at 3600 rpm, the unit can be used so that at a precise moment the travel may be reduced from a fast speed to an inching speed. This is achieved as follows: When the main motor (running at 3600 rpm, and driving a conveyer table at fast speed) is stopped, the rotor slides back, and the clutch plate engages with the other rotating clutch plate, which is being driven through a reduction gear system by the slower running motor. Because the second motor is running at 900 rpm and the reduction through the gear and belts is 125:1, the speed is greatly reduced.

Fast-reversal reel drive...

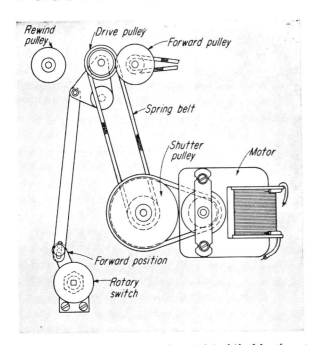

Rewind pulley Drive pulley Forward pulley

Spring belt

Shutter pulley Motor

Forward position

Rotary switch

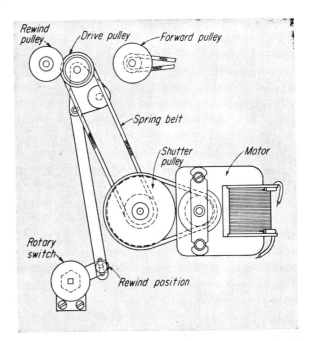

Rewind pulley Drive pulley Forward pulley

Spring belt

Shutter pulley Motor

Rotary switch

Rewind position

for both forward movement and rewind is shifted by the rotary switch; it also controls lamp and drive motor. A short lever on the switch shaft is linked to an over-center mechanism on which the drive wheel is mounted. During the shift from forward to rewind, the drive pulley crosses its pivot point so that spring tension of the drive belt maintains pressure on the driven wheel. Drive from shutter pulley is 1:1 by spring belt to drive pulley and through a reduction when the forward pulley is engaged. When rewind is engaged, the reduction is eliminated and film rewinds at several times forward speed.

Zooming Is Motorized

Push one button and camera "moves in";
push another and it "backs up."

Bell and Howell

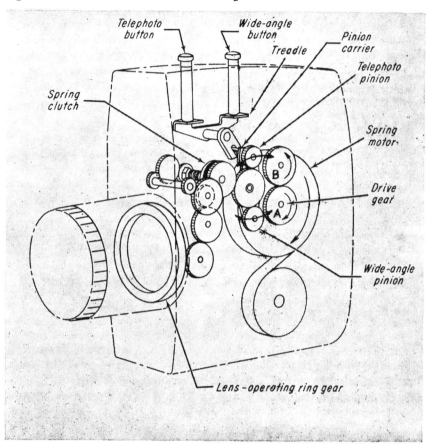

MOVABLE ELEMENTS of lens are controlled by two pushbuttons located at top of camera. The forward button, marked "TELE," adjusts the lens towards the forward or telephoto position; the rear button, marked "WIDE," adjusts it back to the wide-angle position. The buttons actuate a treadle that shifts the drive-pinion carrier. Gear A is driven by the spring motor and meshed with gear B. Both rotate, in opposite directions, whenever the spring motor is operating. Depressing the TELE button pivots the carrier so that the telephoto pinion engages gear B which shifts lens elements forward; pressing the WIDE button engages the wide-angle pinion with gear A, which shifts lens back.

Mounting both pinions on common pinion carrier prevents jams or stoppages. Spring clutch prevents damage when the lens reaches the limits of its travel. When zooming at regular film speed (16 frames per sec) the lens switches from telephoto to wide angle in six seconds. At slow-motion speed (48 frames per sec) the cycle is completed in two sec. The result is that zoom scenes appear the same on the screen whether taken at normal or slow motion.

Dual Clutches Speed Response of Missile Controls

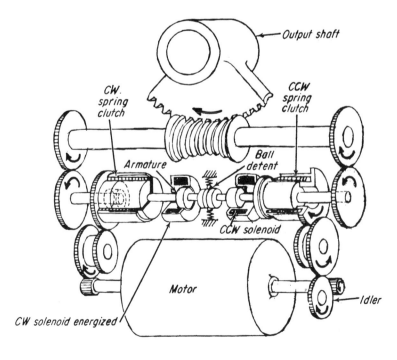

DRIVE MOTOR rotates continuously, so response is fast enough for control applications. Solenoid-operated clutches respond to signals and connect output to the end of motor that will give the desired direction of movement. An idler in one of the gear trains reverses direction of drive to one clutch. Solenoid armatures are mechanically interconnected, so only one clutch can be engaged at a time. When both clutches are deenergized, spring-loaded ball detent centers solenoids, and locks the non-reversible worm-and-gear segment.

. . . Electromechanical actuator is designed for the Bumblebee missile project, a joint effort of the Applied Physics Lab of Johns Hopkins Univ and Convair. The unit has a response time of 8 millisec from start of command signal to full torque.

7 Low-cost Designs for ...

OVERRUNNING CLUTCHES

All are simple devices that can be constructed inexpensively in the laboratory workshop.

JAMES F MACHEN

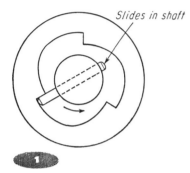

1 Lawnmower type

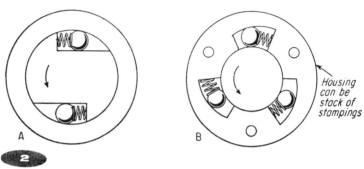

2 Wedging balls or rollers: internal (A); external (B)

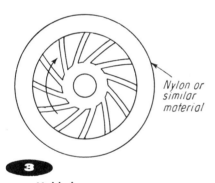

3 Molded sprags (for light duty)

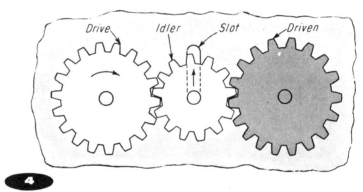

4 Disengaging idler rises in slot when drive direction is reversed

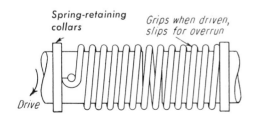

5 Slip-spring coupling

6 Internal ratchet and spring-loaded pawls

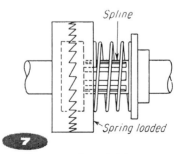

7 One-way dog clutch

Spring-loaded pins aid sprags in one-way clutch

PARIS—Sprags combined with cylindrical rollers in a bearing assembly may be a simple, low-cost way to meet the torque and bearing requirements of most machine applications. Designed and built by Est. Nicot here, this new unit gives one-direction-only torque transmission in an overrunning clutch. In addition, it also serves as a roller bearing.

Torque rating of the clutch depends on the number of sprags. A minimum of three, equally spaced around the circumference of the races, is generally necessary to get acceptable distribution of tangential forces on the races.

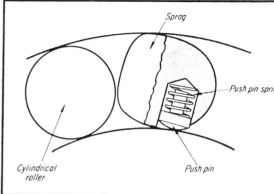

RACES ARE CONCENTRIC; locking ramp is provided by the sprag profile, which is composed of two nonconcentric curves of different radius. Spring-loaded pin holds the sprag in the locked position until torque is applied in the running direction. Stock roller bearing can not be verted, as the hard-steel races of bearing are too britt handle the locking impact of the sprags. Sprags and ro can be mixed to give any desired torque value.

Roller-type clutch, designed in England . . .

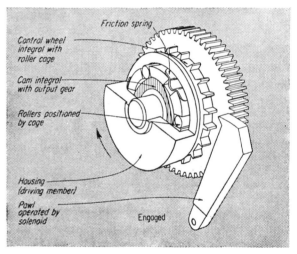

. . . is adaptable to either electrical or mechanical activation, and will control ½ hp at 1500 rpm with only 7 watts of power in the solenoid. The rollers are positioned by a cage (integral with the toothed control wheel—see diagram) between the ID of the driving housing and the cammed hub (integral with the output gear).

When the pawl is disengaged, the drag of the housing on the friction spring rotates the cage and wedges the rollers into engagement. This permits the housing to drive the gear, through the cam.

When the pawl engages the control wheel while the housing is rotating, the friction spring slips inside the housing and the rollers are kicked back, out of engagement. Power is therefore interrupted.

According to the designer, Tiltman Langley Ltd, Redhill Aerodrome, Surrey, the unit will operate over the full temperature range of −40 to 200 F.

Positive drive is provided by this new British roller clutch design. Company says it's unusually small and compact.

Two-speed operation provided by new cam clutch

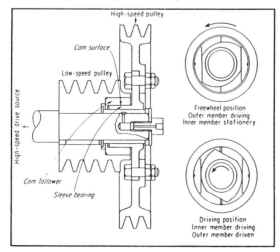

The clutch consists of two rotary members (see diagrams), so arranged that the outer (follower) member acts on its pulley only when the inner member is driving. When the outer member is driving, the inner member idles. First application is in a dry-cleaning machine where the clutch is used as an intermediate between an ordinary and a high-speed motor to provide two output speeds that are used alternately.

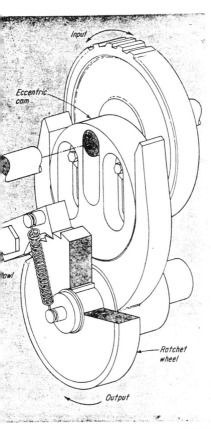

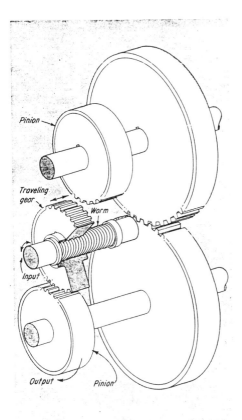

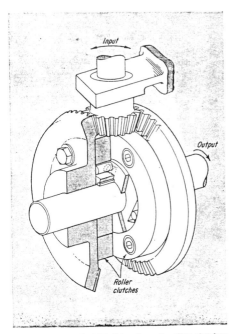

TWO BEVEL GEARS drive through roller clutches. One clutch catches in one direction; the other catches in the opposite direction. There is negligible interruption of smooth output rotation when input direction changes.

ECCENTRIC CAM adjusts over a range of high-reduction ratios, but unbalance limits it to low speeds. When direction of input changes, there is no lag in output rotation. Output shaft moves in steps because of ratchet drive through pawl which is attached to U-follower.

TRAVELING GEAR moves along worm and transfers drive to other pinion when input rotation changes direction. To ease engagement, gear teeth are tapered at ends. Output rotation is smooth, but there is a lag after direction changes as gear shifts. Gear cannot be wider than axial offset between pinions, or there will be interference.

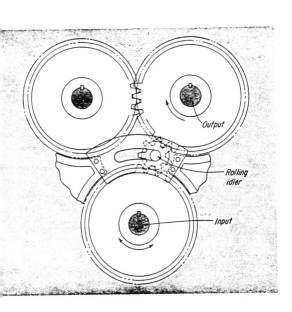

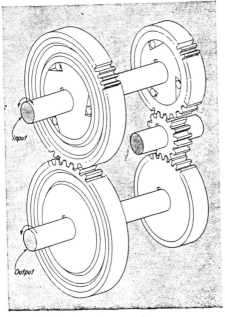

ROLLING IDLER also gives smooth output and slight lag after input direction changes. Small drag on idler is necessary, so that it will transfer into engagement with other gear and not sit spinning in between.

ROLLER CLUTCHES are on input gears in this drive, again giving smooth output speed and little output lag as input direction changes.

Springs, shuttle-gear, gliding-ball in

Novel one-way drives

These three devices change
oscillating motion into one-way rotation
for feeding operations and counters

R H MEIER

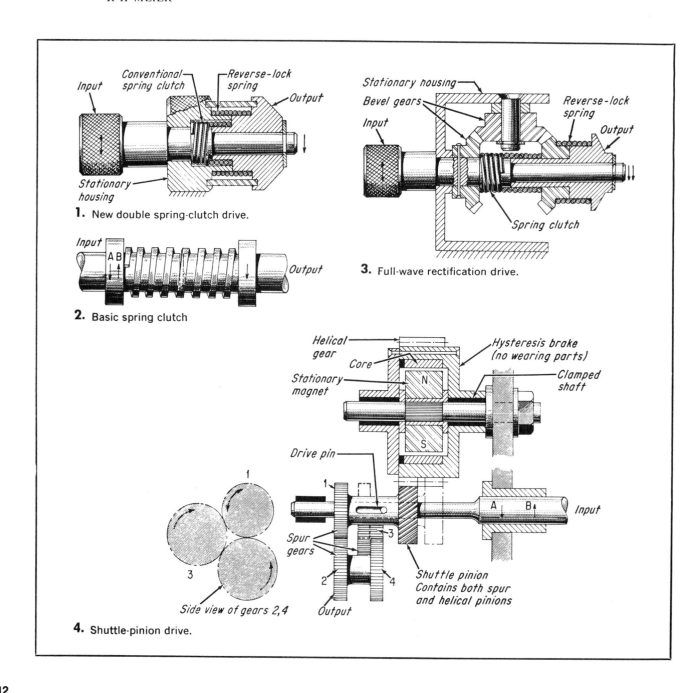

1. New double spring-clutch drive.

2. Basic spring clutch

3. Full-wave rectification drive.

4. Shuttle-pinion drive.

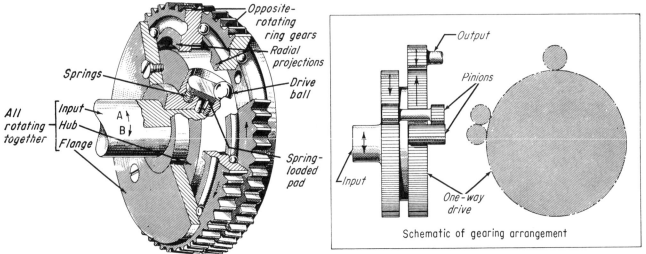

5. Reciprocating-ball drive.

WHILE engaged in the design of a money order imprinter for the Canada Post Office, I devised a one-way drive, Fig 1, which appears to have novel features.

The problem was to convert the oscillating motion of the input crank (20 deg in this case) into a one-way motion to advance the inking ribbon. Basically I used one of the simplest of known devices for obtaining a one-way drive—a spring clutch which is a helical spring joining two co-linear butting shafts (Fig 2). The spring is usually made of square or rectangular cross-section wire.

Such a clutch transmits torque in one direction only—it overrides when reversed. The helical spring, which bridges both shafts, need not be fastened at either end; a slight interference fit will do. Rotating the input shaft in the direction tending to wind the spring (direction *A* in Fig 2) will cause the spring to grip both shafts and thus transmit motion from the input to the output shaft. Reversing the input unwinds the spring, and it overrides the output shaft with a drag—but this drag, slight as it was, caused a problem in operation.

Double-clutch drive

The trouble with the simple approach of Fig 2 was that there was absolutely no friction in the tape drive which would have allowed the spring clutch to slip on the shafts on the return stroke. Thus the output moved in sympathy with the input, and the desired one-way drive was not obtained.

At first, we attempted to add friction artificially to the output, but this resulted in an awkward design. Finally the problem was elegantly solved,

Fig 1, by using a second helical spring, slightly larger than the first and serving exactly the same purpose, viz to transmit motion in one direction only. This spring, however, joined the output shaft and a stationary cylinder. In this way, with the two springs of the same hand, the undesirable return motion of the ribbon drive was immediately arrested, and a positive one-way drive simply obtained.

This compact drive can be considered to be a *half wave rectifier* in that it transmits motion in one direction only, and suppresses motion in the reverse direction.

Full-wave rectifier

The principles described above will also produce a *full wave rectifier* by introducing some reversing gears, Fig 3. In this application the input drive in one direction is directly transmitted to the output as before, but on the reverse stroke the input is passed through reversing gears so that the output appears in the opposite sense, in other words, the original sense of the output is maintained. Thus the output moves forward twice for each back-and-forth movement of the input.

Shuttle-gear drive

A few years ago, I developed a one-way drive which harnesses the axial thrust of a pair of helical gears to shift a pinion, Fig 4. Although at first glance it may look somewhat complicated, the drive is inexpensive to make and has been operating successfully with little wear.

When the input rotates in direction *A,* it drives the output through spur gears *1* and *2*. The shuttle pinion is also driving the helical gear whose rotation is resisted by the magnetic

flux built up between the stationary permanent magnet and the rotating core. This magnet-core arrangement is actually a hysteresis brake and its constant resisting torque produces an axial thrust in mesh of the helical pinion acting to the left. Reversing the input reverses the direction of thrust which shifts the shuttle pinion to the right. The drive then operates through gears *1, 3* and *4* which nullifies the reversion to produce the same direction output.

Reciprocating-ball drive

When the input rotates in direction *A,* Fig 5, the drive ball trails to the right and its upper half engages one of the radial projections in the right ring gear to drive it in the direction as the input. The slot for the ball is milled at 45 deg to the shaft axes and extends to the flanges on each side.

When the input is reversed, the ball extends to the flanges on each side. trails to the left and deflects to permit the ball to ride over to the left ring gear, and engage its radial projection to drive the gear in the direction of the input.

Each gear, however, is constantly in mesh with a pinion, which in turn is in mesh with the other. Thus, no matter which direction the input is turned, the ball positions itself under one or another ring gear, and the gears will maintain their respective sense of rotation (the rotation shown in Fig 5). Hence, an output gear in mesh with one of the ring gears will rotate in one direction only.

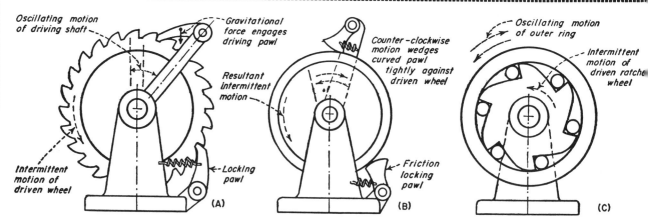

1 Elementary over-riding clutches: (A) Ratchet and Pawl mechanism is used to convert reciprocating or oscillating movement to intermittent rotary motion. This motion is positive but limited to a multiple of the tooth pitch. (B) Friction-type is quieter but requires a spring device to keep eccentric pawl in constant engagement. (C) Balls or rollers replace the pawls in this device. Motion of the outer race wedges rollers against the inclined surfaces of the ratchet wheel.

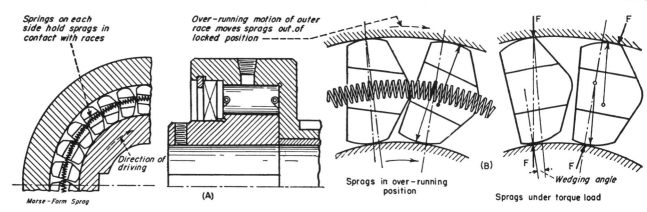

4 With cylindrical inner and outer races, sprags are used to transmit torque. Energizing springs serves as a cage to hold the sprags. (A) Compared to rollers, shape of sprag permits a greater number within a limited space; thus higher torque loads are possible. Not requiring special cam surfaces, this type can be installed inside gear or wheel hubs. (B) Rolling action wedges sprags tightly between driving and driven members. Relatively large wedging angle insures positive engagement.

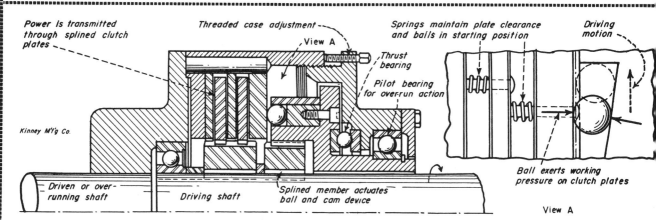

6 Multi-disk clutch is driven by means of several sintered-bronze friction surfaces. Pressure is exerted by a cam actuating device which forces a series of balls against a disk plate. Since a small part of the transmitted torque is carried by the actuating member, capacity is not limited by the localized deformation of the contacting balls. Slip of the friction surfaces determine the capacity and prevent rapid, shock loads. Slight pressure of disk springs insure uniform engagement.

Over-Riding Clutches

A. DeFEO

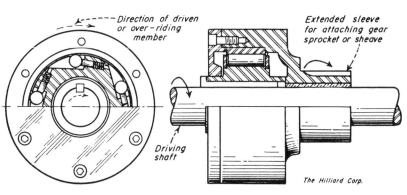

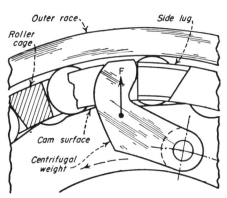

2 Commercial over-riding clutch has springs which hold rollers in continuous contact between cam surfaces and outer race; thus there is no backlash or lost motion. This simple design is positive and quiet. For operation in the opposite direction, the roller mechanism can easily be reversed in the housing.

3 Centrifugal force can be used to hold rollers in contact with cam and outer race. Force is exerted on lugs of the cage which controls the position of the rollers.

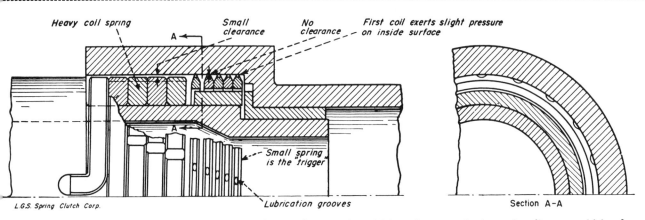

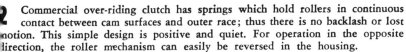

5 Engaging device consists of a helical spring which is made up of two sections: a light trigger spring and a heavy coil spring. It is attached to and driven by the inner shaft. Relative motion of outer member rubbing on trigger causes this spring to wind-up. This action expands the spring diameter which takes up the small clearance and exerts pressure against the inside surface until the entire spring is tightly engaged. Helix angle of spring can be changed to reverse the over-riding direction.

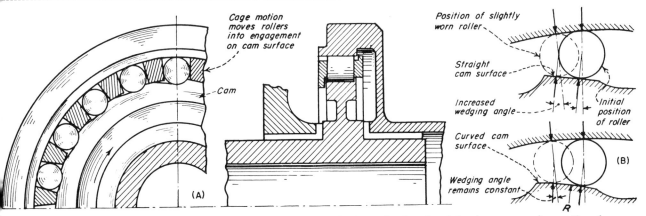

7 Free-wheeling clutch widely used in power transmission has a series of straight-sided cam surfaces. An engaging angle of about 3 deg is used; smaller angles tend to become locked and are difficult to disengage while larger ones are not as effective. (A) Inertia of floating cage wedges rollers between cam and outer race. (B) Continual operation causes wear of surfaces; 0.001 in. wear alters angle to 8.5 deg. on straight-sided cams. Curved cam surfaces maintain constant angle.

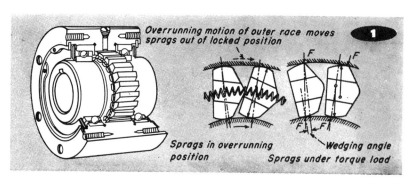

Overrunning motion of outer race moves sprags out of locked position

Sprags in overrunning position

Wedging angle

Sprags under torque load

1

Precision Sprags . . .

act as wedges and are made of hardened alloy steel. In the Formsprag clutch, torque is transmitted from one race to another by wedging action of sprags between the races in one direction; in other direction the clutch freewheels.

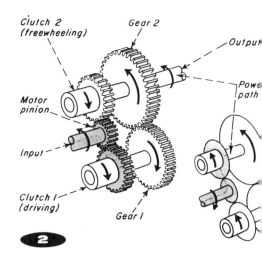

Clutch 2 (freewheeling) Gear 2 Output

Motor pinion Powe path

Input

Clutch I (driving) Gear I

2

2-Speed Drive — I . . .

requires input rotation to be reversible. Counterc wise input as shown in the diagram drives gear 1 thr clutch 1; output is counterclockwise; clutch 2 over- Clockwise input (schematic) drives gear 2 through c 2; output is still counterclockwise; clutch 1 over-

3

2-Speed Drive — II . . .

for grinding wheel can be simple, in-line design if over-running clutch couples two motors. Outer race of clutch is driven by gearmotor; inner race is keyed to grinding-wheel shaft. When gearmotor drives, clutch is engaged; when larger motor drives, inner race over-runs.

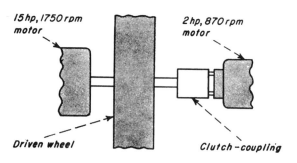

15 hp, 1750 rpm motor

2 hp, 870 rpm motor

Driven wheel

Clutch-coupling

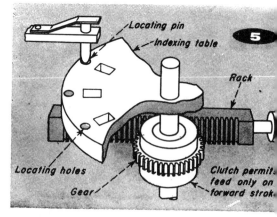

Locating pin

Indexing table

5

Rack

Locating holes

Gear

Clutch permit. feed only on forward stroke

Indexing Table . . .

is keyed to clutch shaft. Table is rotated by forward st of rack, power being transmitted through clutch b outer-ring gear only during this forward stroke. Inde is slightly short of position required. Exact position is located by spring-loaded pin, which draws table forw to final positioning. Pin now holds table until next po stroke of hydraulic cylinder

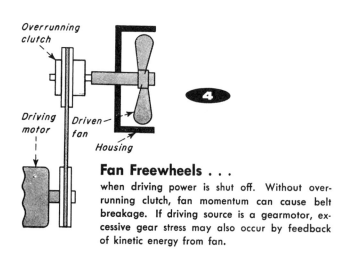

Overrunning clutch

Driving motor

Driven fan

Housing

4

Fan Freewheels . . .

when driving power is shut off. Without over-running clutch, fan momentum can cause belt breakage. If driving source is a gearmotor, excessive gear stress may also occur by feedback of kinetic energy from fan.

CLUTCHES

problems. Here are some clutch setups.

W. EDGAR MULHOLLAND
JOHN L. KING, JR.

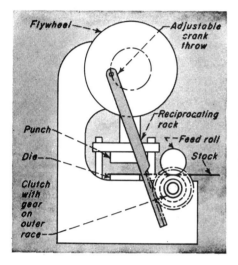

6

Punch Press Feed ..

is so arranged that strip is stationary on down-stroke of punch (clutch freewheels); feed occurs during upstroke when clutch transmits torque. Feed mechanism can easily be adjusted to vary feed amount.

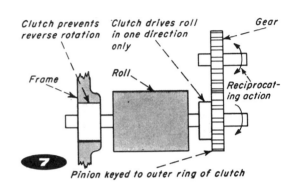

7

Indexing and Backstopping . . .

is done with two clutches so arranged that one drives while the other freewheels. Application here is for capsuling machine; gelatin is fed by the roll and stopped intermittently so blade can precisely shear material to form capsules.

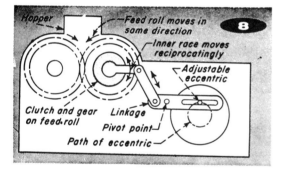

8

Intermittent Motion . . .

of candy machine is adjustable; function of clutch is to ratchet the feed rolls around. This keeps the material in the hopper agitated.

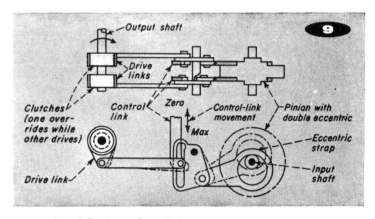

9

Double-impulse Drive . . .

employs double eccentrics and drive clutches. Each clutch is indexed 180° out of phase with the other. One revolution of eccentric produces two drive strokes. Stroke length, and thus the output rotation, can be adjusted from zero to max by the control link.

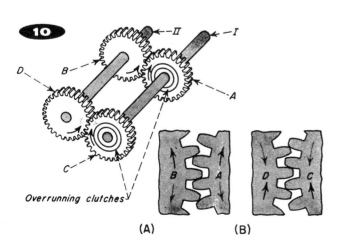

10

(A)   (B)

Anti-backlash Device . . .

uses over-running clutches to insure that no backlash is left in the unit. Gear A drives B and shaft II with the gear mesh and backlash as shown in (A). The over-running clutch in gear C permits gear D (driven by shaft II) to drive gear C and results in the mesh and backlash shown in (B). The over-running clutches never actually over-run. They provide flexible connections (something like split and sprung gears) between shaft I and gears A, C to allow absorption of all backlash.

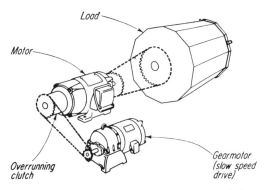

Fig. 1—Overrunning permits torque transmission in one direction; and free wheels or over-runs in the opposite direction. For example, gear motor drives the load by transmitting torque through the overrunning clutch and the high speed shaft. Energizing the high speed motor, causes the inner member to rotate at the rpm of the high speed motor, the gear motor continues to drive the inner member, but the clutch is freewheeling.

Fig. 2—Backstopping permits rotation in one direction only—clutch serves as a counter rotation holding device. An example is a clutch mounted on the headshaft of a conveyor, with the outer race restrained by means of torque-arming to the stationary frame of the conveyor. If for any reason power to the conveyor is interrupted, the backstopping clutch will prevent the buckets from running backwards and dumping the load.

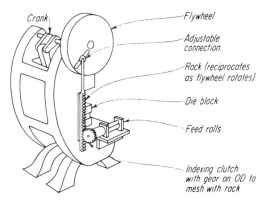

Fig. 3—Indexing is the transmission of intermittent rotary motion in one direction; an example, on the roll feeds of a punch press. On each stroke of the press crankshaft, a feed stroke on the roll feed is accomplished by means of the rack and pinion system. Rack and pinion system feeds the material into the dies of punch press.

Applications of sprag-type clutches

W. T. CHERRY

Sprag clutches of the overrunning type transmit torque in one direction and reduce speed, rest, hold, or free wheel in the reverse direction. Applications include overrunning, backstopping and indexing. Selection—similar to other mechanical devices—requires a review of torque to be transmitted, overrunning speed, type of lubrication, mounting characteristics, environmental conditions and shock conditions that may be met.

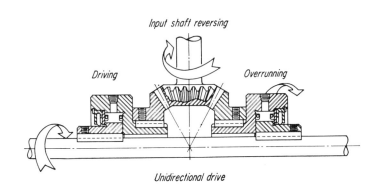

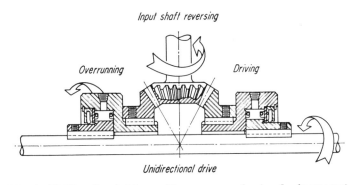

Fig. 4—Unidirectional drives with reverse mechanism by incorporating two overrunning clutches into the gears, sheaves or sprockets. Here, a 1:1 ratio right-angle drive is depicted having a reversing input shaft. The output shaft rotates clockwise regardless of input shaft direction. By changing gear sizes, combinations of continuous or intermittent unidirectional output relative to input is possible.

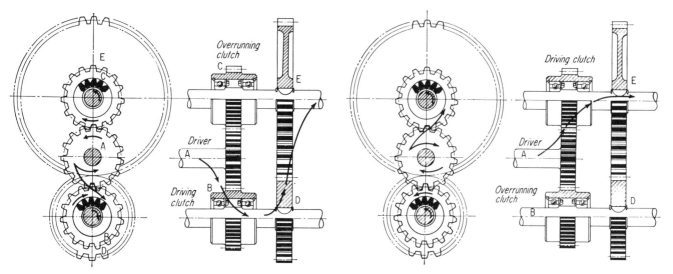

Fig. 5—Two speed unidirectional output is possible by using spur gears and reversing the direction of the input shaft. Rotation of shaft A power gears B, D, and E to output. Counterclockwise rotation engages lower clutch, freewheeling upper clutch as gear C is traveling at a faster rate than the shaft caused by reduction between gears B and E. Clockwise rotation of A engages upper clutch, while lower clutch freewheels because of the speed increase between gears D and E.

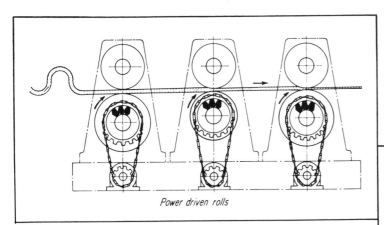

Power driven rolls

Fig. 6 (Left)—Speed-differential or compensation is required where a different speed range for a function is desired while retaining the same basic speed for all other functions. A series of individually power driven rolls may have different surface speeds because of drive or diameter variations of the rolls. An overrunning clutch permits the rolls of slower peripheral speed to overspeed and adjust to the material speed.

Fig. 7—Speed differential application permits operation of engine accessories within a narrow speed range while the engine operates over a wide range. Pulley No. 2 contains the overrunning clutch. When the friction or electric clutch is disengaged, driver pulley drives pulley No. 2 through the overrunning clutch, rotating the driven shaft. Engagement of the friction or electric clutch causes high speed driven shaft rotation causing an overrun condition in the clutch at pulley No. 2.

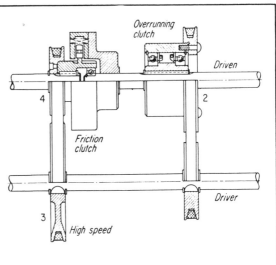

Fig. 8—High inertia dissipation is desirable to avoid driving back through a power system, or in machines having high resistances, to prevent power train damage. If the engine is shut down and the generator was under a "no-load" condition, there would be a tendency to twist off the generator shaft. Overrunning clutch allows generator deceleration at a slower rate than the engine deceleration.

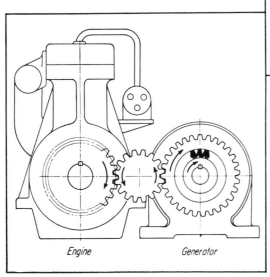

Engine   Generator

Small Mechanical Clutches for

Clutches used in calculating machines must have: (1) Quick response—lightweight moving parts; (2) Flexibility—permit multiple members to

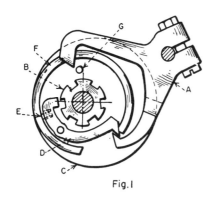

Fig. 1

PAWL AND RATCHET SINGLE CYCLE CLUTCH (Fig. 1). Known as Dennis Clutch, parts *B*, *C* and *D*, are primary components, *B*, being the driving ratchet, *C*, the driven cam plate and, *D*, the connecting pawl carried by the cam plate. Normally the pawl is held disengaged by the lower portion of clutch arm *A*. When activated, arm *A* rocks counter-clockwise until it is out of the path of rim *F* on cam plate *C* and permits pawl *D* under the effect of spring *E* to engage with ratchet *B*. Cam plate *C* then turns clockwise until, near the end of one cycle, pin *G* on the plate strikes the upper part of arm *A* camming it clockwise back to its normal position. The lower part of *A* then performs two functions: (1) cams pawl *D* out of engagement with the driving ratchet *B* and (2) blocks further motion of rim *F* and the cam plate.

PAWL AND RATCHET SINGLE CYCLE DUAL CONTROL CLUTCH—(Fig. 2). Principal parts are: driving ratchet, *B*, directly connected to the motor and rotating freely on rod *A*; driven crank, *C*, directly connected to the main shaft of the machine and also free on *A*; and spring loaded ratchet pawl, *D*, which is carried by crank, *C*, and is normally held disengaged by latch *E*. To activate the clutch, arm *F* is raised, permitting latch *E* to trip and pawl *D* to engage with ratchet *B*. The left arm of clutch latch *G*, which is in the path of the lug on pawl *D*, is normally permitted to move out of interference by the rotation of the camming edge of crank *C*. For certain operations block *H* is temporarily lowered, preventing motion of latch *G*, resulting in disengagement of the clutch after part of the cycle until subsequent raising of block *H* permits motion of latch *G* and resumption of the cycle.

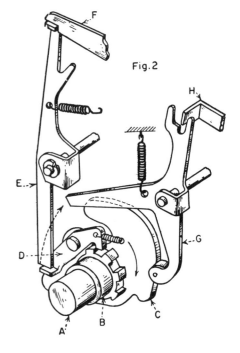

Fig. 2

PLANETARY TRANSMISSION CLUTCH (Fig. 3). A positive clutch with external control, two gear trains to provide bi-directional drive to a calculator for cycling the machine and shifting the carriage. Gear *A* is the driver, gear *L* the driven member is directly connected to planet carrier *F*. The planet consists of integral gears *B* and *C*; *B* meshing with sun gear *A* and free-wheeling ring gear *G*, and *C* meshing with free-wheeling gear *D*. Gears *D* and *G* carry projecting lugs, *E* and *H* respectively, which can contact formings on arms *J* and *K* of the control yoke. When the machine is at rest, the yoke is centrally positioned so that the arms *J* and *K* are out of the path of the projecting lugs permitting both *D* and *G* to free-wheel. To engage the drive, the yoke rocks clockwise as shown, until the forming on arm *K* engages lug *H* blocking further motion of ring gear *G*. A solid gear train is thereby established driving *F* and *L* in the same direction as the drive *A* and at the same time altering the speed of *D* as it continues counter-clockwise. A reversing signal rotates the yoke counter-clockwise until arm *J* encounters lug *E* blocking further motion of *D*. This actuates the other gear train of the same ratio.

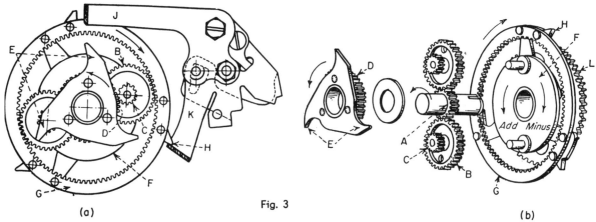

(a)   Fig. 3   (b)

Precise Service

MARVIN TAYLOR

control operation; (3) Compactness—for equivalent capacity positive clutches are smaller than friction; (4) Dependability; and (5) Durability.

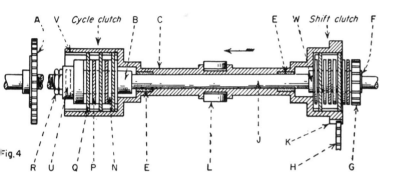

Fig.4

MULTIPLE DISK FRICTION CLUTCH (Fig. 4). Two multiple disk friction clutches are combined in a single two-position unit which is shown shifted to the left. A stepped cylindrical housing C enclosing both clutches is carried by self-lubricated bearing E on shaft J and is driven by the transmission gear H meshing with the housing gear teeth K. At either end, the housing carries multiple metal disks Q that engage keyways V and can make frictional contact with formica disks N which, in turn, can contact a set of metal disks P which have slotted openings for coupling with flats on sleeves B and W. In the position shown, pressure is exerted through rollers L forcing the housing to the left making the left clutch compact against adjusting nuts R, thereby driving gear A via sleeve B which is connected to jack shaft J by pin U. When the carriage is to be shifted, rollers L force the housing to the right, first relieving the pressure between the adjoining disks on the left clutch then passing through a neutral position in which both clutches are disengaged and finally making the right clutch compact against thrust bearing F, thereby driving gear G through sleeve W which rotates freely on the jack shaft.

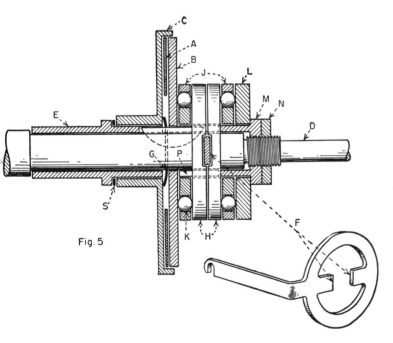

Fig. 5

SINGLE PLATE FRICTION CLUTCH (Fig. 5). The basic clutch elements, formica disk A, steel plate B and drum C, are normally kept separated by spring washer G. To engage the drive, the left end of a control arm is raised, causing ears F, which sit in slots in plates H, to rock clockwise spreading the plates axially along sleeve P. Sleeves E and P and plate B are keyed to the drive shaft; all other members can rotate freely. The axial motion loads the assembly to the right through the thrust ball bearings K against plate L and adjusting nut M, and to the left through friction surfaces on A, B and C to thrust washer S, sleeve E and against a shoulder on shaft D, thus enabling plate A to drive the drum C.

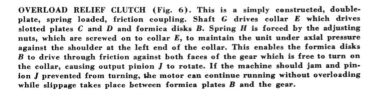

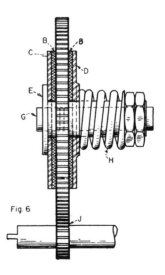

Fig. 6

OVERLOAD RELIEF CLUTCH (Fig. 6). This is a simply constructed, double-plate, spring loaded, friction coupling. Shaft G drives collar E which drives slotted plates C and D and formica disks B. Spring H is forced by the adjusting nuts, which are screwed on to collar E, to maintain the unit under axial pressure against the shoulder at the left end of the collar. This enables the formica disks B to drive through friction against both faces of the gear which is free to turn on the collar, causing output pinion J to rotate. If the machine should jam and pinion J prevented from turning, the motor can continue running without overloading while slippage takes place between formica plates B and the gear.

Mechanisms for STATION CLUTCHES

HERMANN HILL

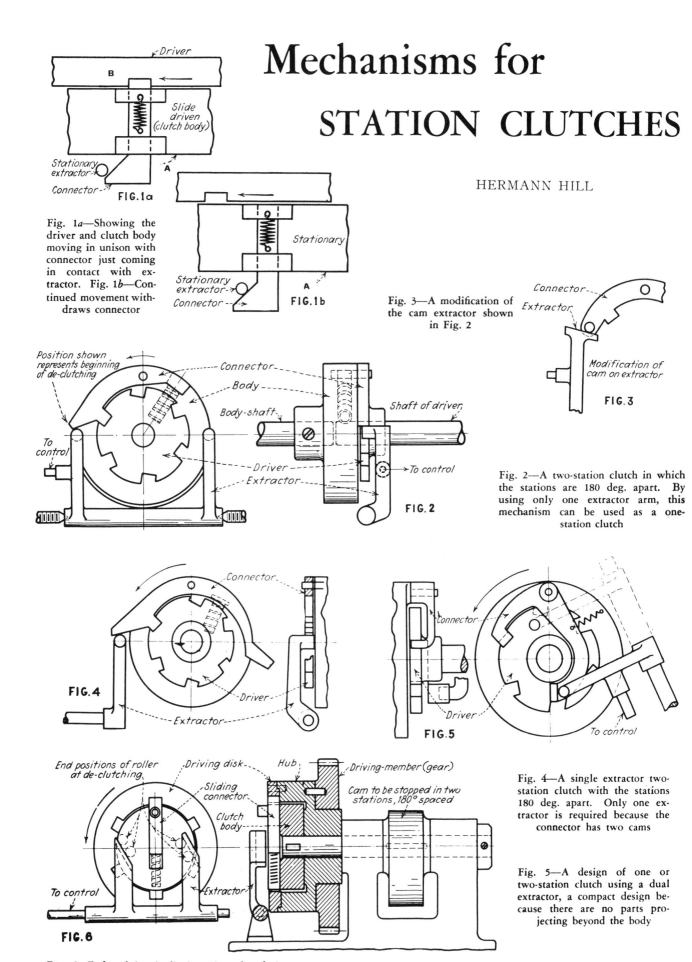

Fig. 1a—Showing the driver and clutch body moving in unison with connector just coming in contact with extractor. Fig. 1b—Continued movement withdraws connector

Fig. 1a

Fig. 1b

Fig. 3—A modification of the cam extractor shown in Fig. 2

Modification of cam on extractor

FIG.3

Fig. 2—A two-station clutch in which the stations are 180 deg. apart. By using only one extractor arm, this mechanism can be used as a one-station clutch

FIG. 2

FIG.4

FIG.5

Fig. 4—A single extractor two-station clutch with the stations 180 deg. apart. Only one extractor is required because the connector has two cams

Fig. 5—A design of one or two-station clutch using a dual extractor, a compact design because there are no parts projecting beyond the body

FIG.6

Fig. 6—End and longitudinal section of a design of a station clutch using internal driving recesses

Innumerable variations of these station clutches may be designed for starting and stopping machines at selected points in their cycle of operation

Fig. 8—Another design of one or two-station clutch, using a single or dual extractor with stations spaced 180 deg. apart

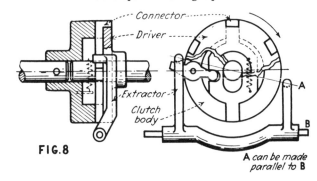

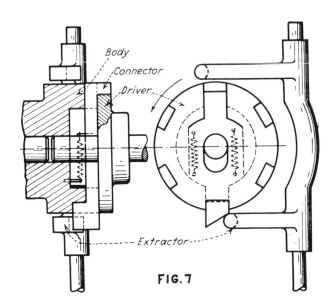

FIG.7

Fig. 7—A one or two-station clutch, depending on the use of a single or a dual extractor, the stations being spaced 180 deg. apart

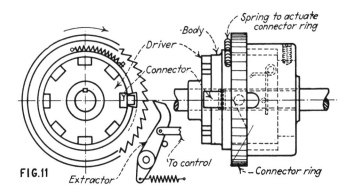

FIG.8

A can be made parallel to B

Fig. 11—A typical design of multi-station clutch of the non-selective type for instantaneous stopping at any position

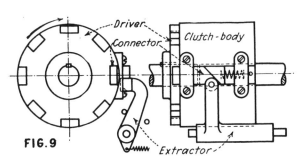

FIG.9

Fig. 9—A design of one-station clutch of the axail conector type

FIG.11

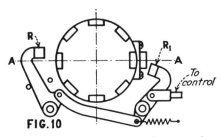

FIG.10

Fig. 10—In this design of two-station clutch the roller R and R₁ of the extractor may also be arranged on the center line A-A

Fig. 12—A design of multi-station clutch with remote control. The extractor pins are actuated by solenoids which either hold the extractor pin in position against spring pressure, or release the pin

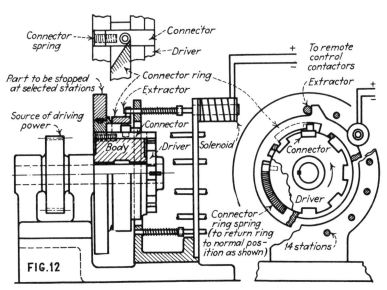

FIG.12

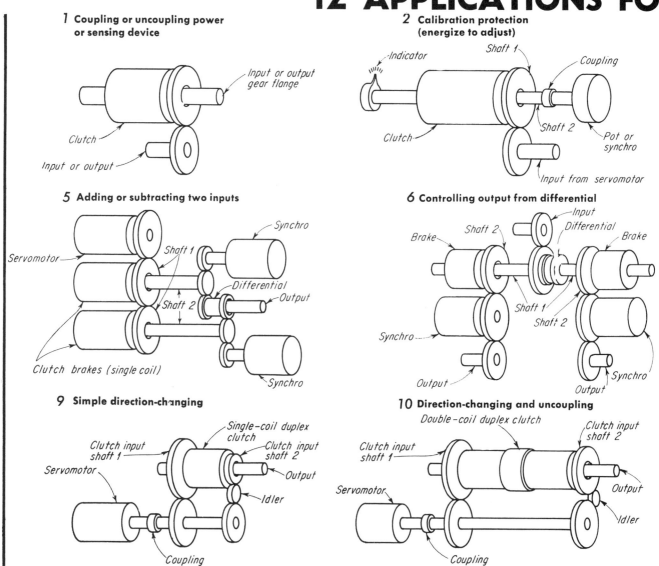

1 Coupling or uncoupling power or sensing device

Input or output gear flange

Clutch

Input or output

2 Calibration protection (energize to adjust)

Indicator

Shaft 1

Coupling

Clutch

Shaft 2

Pot or synchro

Input from servomotor

5 Adding or subtracting two inputs

Servomotor

Shaft 1

Synchro

Differential

Output

Shaft 2

Clutch brakes (single coil)

Synchro

6 Controlling output from differential

Brake

Shaft 2

Input

Differential

Brake

Synchro

Shaft 1

Shaft 2

Output

Output

Synchro

9 Simple direction-changing

Single-coil duplex clutch

Clutch input shaft 1

Clutch input shaft 2

Servomotor

Output

Idler

Coupling

10 Direction-changing and uncoupling

Double-coil duplex clutch

Clutch input shaft 1

Clutch input shaft 2

Servomotor

Output

Idler

Coupling

MAGNETIC FRICTION CLUTCHES

The most common type of electromagnetic control clutch—the simplest and most adaptable—is the magnetic friction clutch. It works on the same principle as a simple solenoid-operated electric relay with a spring return to normal. Like the relay, it is a straightforward automatic switch for controlling the flow of power (in this case, torque) through a circuit.

Rotating or fixed field?

This is a question primarily of magnetic design. Rotating-field clutches employ a rotating coil, energized through brushes and slip rings. Fixed-field units have a stationary coil. Rotating-field units are still the more common, but there has been marked trend toward the fixed-field design.

Generally speaking, a rotating-field clutch is a two-member unit, with the coil carried in the driving (input) member. It can be mounted directly on a motor or speed-reducer shaft without loading down the driving motor. In the smaller sizes, it offers a better ratio of size to rated output than the fixed-field type, although the rotating coil increases inertia in the larger models.

A fixed-field clutch, on the other hand, is a three-member unit, with rotating input and output members and a stationary coil housing. It eliminates the need for brushes and slip rings, but it demands additional bearing supports, and may require close tolerances in mounting.

PURELY MAGNETIC CLUTCHES

Probably less familiar than the friction types are hysteresis and eddy-current clutches, which operate on straight magnetic principles and do not depend upon mechanical contact between their members. The two styles are almost identical in construction, but the magnetic segments of the hysteresis clutch are electrically isolated and those of the eddy-current type are interconnected. The magnetic analogy of both styles is similar in that the flux path is passed between the two clutch members.

Hysteresis clutches

The hysteresis clutch is a proportional-torque control device which—as its name implies—exploits the hysteresis effect in a permanent-magnet rotor ring to produce a substantially constant torque that is almost completely independent of speed (except for slight, unavoidable secondary

ELECTROMAGNETIC CLUTCHES AND BRAKES

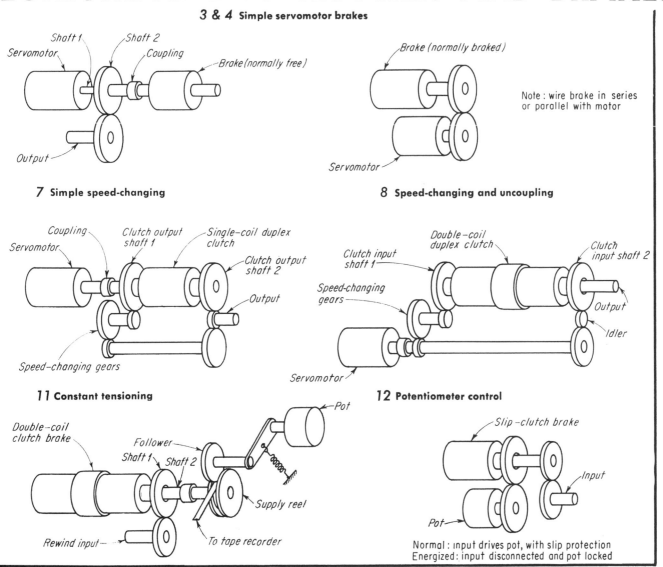

3 & 4 Simple servomotor brakes

Shaft 1 — Shaft 2 — Coupling
Servomotor — Brake (normally free)
Output

Brake (normally braked)
Note: wire brake in series or parallel with motor
Servomotor

7 Simple speed-changing

Coupling — Clutch output shaft 1 — Single-coil duplex clutch
Servomotor — Clutch output shaft 2 — Output
Speed-changing gears

8 Speed-changing and uncoupling

Double-coil duplex clutch — Clutch input shaft 2
Clutch input shaft 1 — Output
Speed-changing gears — Idler
Servomotor

11 Constant tensioning

Double-coil clutch brake — Pot
Follower
Shaft 1 — Shaft 2
Supply reel
Rewind input — To tape recorder

12 Potentiometer control

Slip-clutch brake
Input
Pot
Normal: input drives pot, with slip protection
Energized: input disconnected and pot locked

---

eddy-current torques—which do not seriously reduce performance). It is capable of synchronous driving or continuous slip, with almost no torque variation at any slip differential for a given control current. Control-power requirement is small enough for vacuum-tube drive. Typical applications include wire or tape tensioning, servo-control actuation, and torque control in dynamometers.

Eddy-current clutches

Eddy-current clutches, on the other hand, are inherently speed-sensitive devices. They exhibit virtually no hysteresis, and develop torque by dissipating eddy currents through the electrical resistance of the rotor ring. This torque is almost a linear function of slip speed. These clutches are best used in speed-control applications, and as oscillation dampers.

PARTICLE AND FLUID MAGNETIC CLUTCHES

There is no real difference between magnetic-particle and magnetic-fluid clutches, except that the magnetic medium in the first is a dry powder and in the second it is a similar powder—but suspended in oil. In either type, the ferromagnetic medium is introduced into the airgap between the input and output faces, which do not actually contact one another. When the clutch coil is energized, the particles are excited in the magnetic field between the faces; as they shear against each other, they produce a drag torque between the clutch members.

Theoretically, these clutches can approach the proportional control characteristics of a hysteresis clutch within the small weight and size limits of a comparably rated miniature friction clutch. But in practice the service life of miniature magnetic-particle clutches has so far been too short for industrial usage.

OTHER MAGNETIC CLUTCHES

Two sophisticated concepts—neither of them yet developed to the point of practical application—may prove of great academic interest to anyone researching this field.

Electrostatic clutches, which use high voltages instead of magnetic field to create force-producing suspensions.

Magnetostrictive clutches using a magnetic force to change the dimensions of a crystal or metal bar poised between two extremely precise faces.

FEDERICO STRASSER

HOOKE'S JOINTS

The commonest form of a universal coupling is Hooke's joint. This can transmit torque efficiently up to a maximum shaft-alignment angle of about 36°. At slow speeds, on hand-operated mechanisms, the permissible angle can reach 45° Simplest arrangement for a Hooke's joint is two forked shaft-ends coupled by a cross-shaped piece. There are many variations and a few of them are included here.

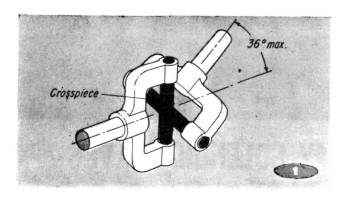

36° max.

Crosspiece

Basic design . . .

of Hooke's joint can transmit heavy loads. Anti-friction bearings are a refinement often used.

Pinned sphere . . .

replaces crosspiece in this design. Result is a more compact joint.

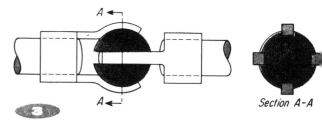

Section A-A

Grooved sphere . . .

is modification of pinned sphere. Tongues on fastening sleeves are bent over sphere on assembly. Greater sliding contact of tongues in grooves makes ample lubrication essential at high torques and alignment angles.

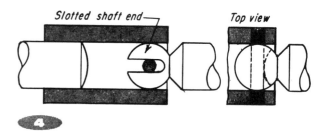

Slotted shaft end

Top view

Pinned sleeve . . .

fastened to one shaft engages forked, spherical end on other shaft to provide joint that also allows for axial shaft-movement. In this example, however, angle between shafts can only be small. Also, joint is only suitable for low torques.

CONSTANT-VELOCITY COUPLINGS

The disadvantage of a single Hooke's joint is that velocity of driven shaft varies. Its maximum velocity can be found by multiplying driving-shaft speed by secant of the shaft angle;

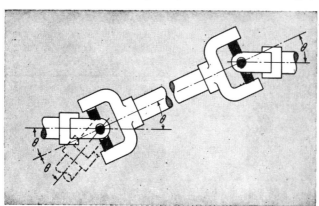

for minimum speed multiply by the cosine. An example of speed variation: Driving shaft rotates at 100 rpm; angle between shafts is 20°. Min output is 100 x 0.9397, which equals 93.97 rpm; max output is 1.0642 x 100, or 106.4 rpm Thus the difference is 12.43 rpm. When output speed is high, output torque is low and vice versa. This is an objectionable feature in some mechanisms. However, two universal joints connected by an intermediate shaft solve the problem.

Constant-velocity . . .

joint made by coupling two Hooke's joints must have input and output angles equal for correct action. Also, the forks must be assembled so that they will always be in the same plane. Shaft-alignment angle may be double that for a single joint.

SHAFT-COUPLINGS

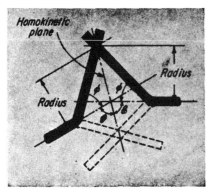

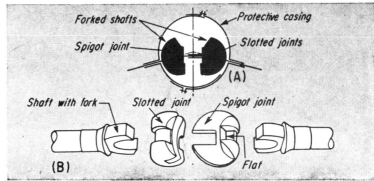

Single constant-velocity . . .

coupling is based on principle (6) that contact point of the two members must always lie on the homokinetic plane. Their rotation speed will then always be equal because radius to contact point of each member will always be equal. Such simple couplings are ideal for toy, instrument and other light-duty mechanisms. For heavy duty, such as front-wheel drive of military vehicles, a more

complex coupling, shown diagramatically (7A) has two joints close coupled with a sliding member between them. Exploded view (7B) shows these members. There are other designs for heavy-duty universal couplings; one, known as the Rzeppa, consists of a cage that keeps six balls in the homokinetic plane at all times. Another constant-velocity joint, the Bendix-Weiss, incorporates balls also.

MISCELLANEOUS COUPLINGS

Triple-strand spring

Flexible shaft . . .

allows any shaft angle. Such shafts, if long, should be supported to prevent backlash and coiling.

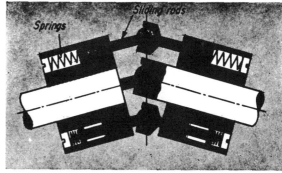

Pump-type coupling . . .

is so called because reciprocating action of sliding rods can be used to drive pistons in cylinders.

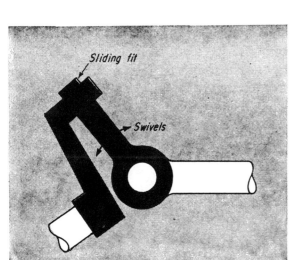

Light-duty coupling . . .

is ideal for many economical mechanisms. Sliding swivel-rod must be kept well lubricated.

TYPICAL METHODS OF COUPLING

Methods of coupling rotating shafts vary from simple bolted flange constructions to complex spring and synthetic rubber mechanisms. Some types incorporating chain belts, splines, bands, and rollers are described and illustrated below.

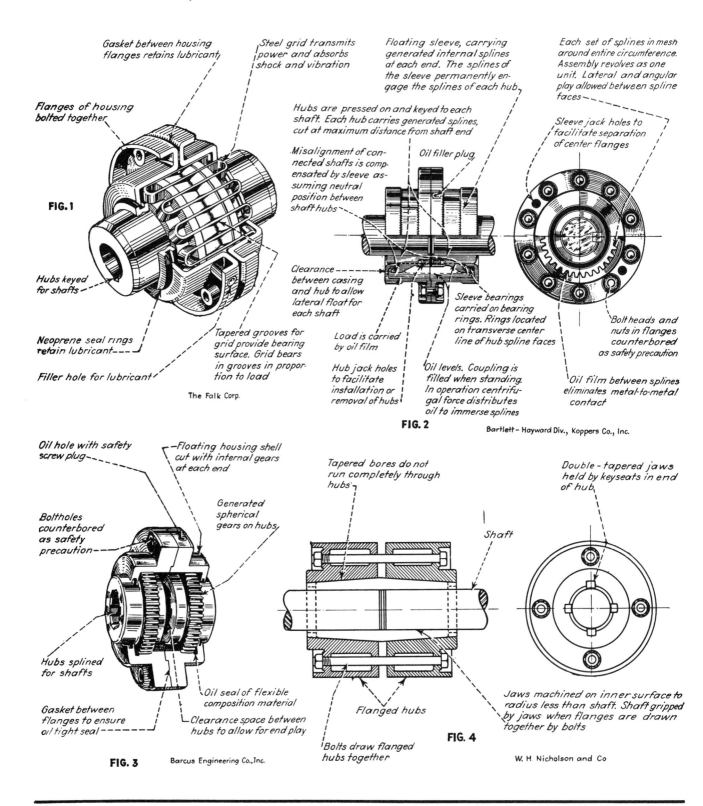

Gasket between housing flanges retains lubricant

Steel grid transmits power and absorbs shock and vibration

Floating sleeve, carrying generated internal splines at each end. The splines of the sleeve permanently engage the splines of each hub

Each set of splines in mesh around entire circumference. Assembly revolves as one unit. Lateral and angular play allowed between spline faces

Flanges of housing bolted together

Hubs are pressed on and keyed to each shaft. Each hub carries generated splines, cut at maximum distance from shaft end

Sleeve jack holes to facilitate separation of center flanges

Misalignment of connected shafts is compensated by sleeve assuming neutral position between shaft hubs

Oil filler plug

FIG. 1

Hubs keyed for shafts

Clearance between casing and hub to allow lateral float for each shaft

Sleeve bearings carried on bearing rings. Rings located on transverse center line of hub spline faces

Bolt heads and nuts in flanges counterbored as safety precaution

Neoprene seal rings retain lubricant

Tapered grooves for grid provide bearing surface. Grid bears in grooves in proportion to load

Load is carried by oil film

Oil film between splines eliminates metal-to-metal contact

Filler hole for lubricant

The Falk Corp.

Hub jack holes to facilitate installation or removal of hubs

Oil levels. Coupling is filled when standing. In operation centrifugal force distributes oil to immerse splines

FIG. 2

Bartlett – Hayward Div., Koppers Co., Inc.

Oil hole with safety screw plug

Floating housing shell cut with internal gears at each end

Tapered bores do not run completely through hubs

Double-tapered jaws held by keyseats in end of hub

Boltholes counterbored as safety precaution

Generated spherical gears on hubs

Shaft

Hubs splined for shafts

Oil seal of flexible composition material

Flanged hubs

Jaws machined on inner surface to radius less than shaft. Shaft gripped by jaws when flanges are drawn together by bolts

Gasket between flanges to ensure oil tight seal

Clearance space between hubs to allow for end play

Bolts draw flanged hubs together

FIG. 4

FIG. 3   Barcus Engineering Co., Inc.

W. H. Nicholson and Co

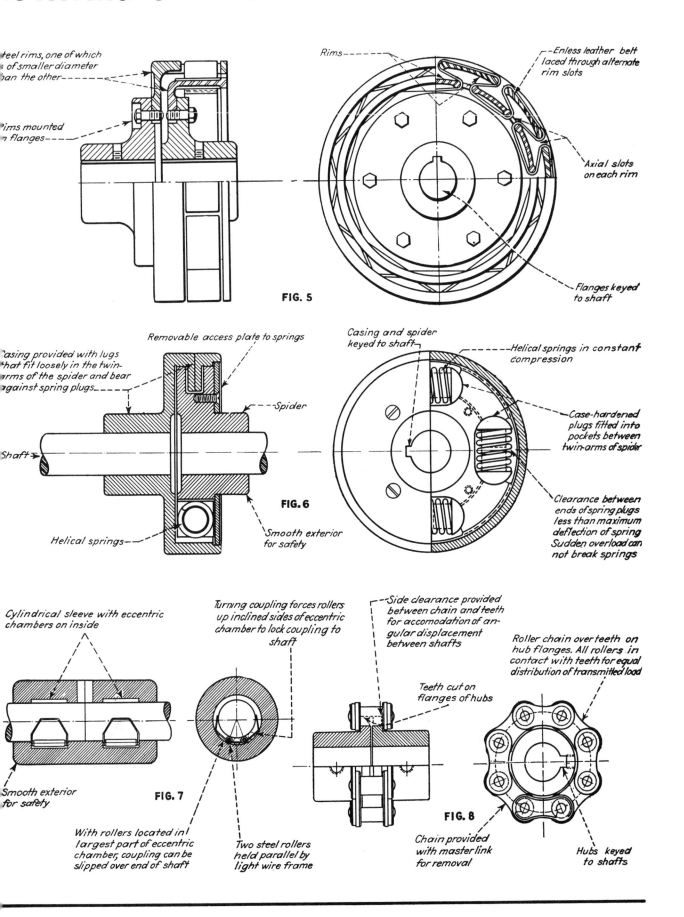

Steel rims, one of which is of smaller diameter than the other

Rims mounted on flanges

FIG. 5

Rims

Enless leather belt laced through alternate rim slots

Axial slots on each rim

Flanges keyed to shaft

Casing provided with lugs that fit loosely in the twin-arms of the spider and bear against spring plugs

Removable access plate to springs

Spider

Shaft

Helical springs

FIG. 6

Smooth exterior for safety

Casing and spider keyed to shaft

Helical springs in constant compression

Case-hardened plugs fitted into pockets between twin-arms of spider

Clearance between ends of spring plugs less than maximum deflection of spring Sudden overload can not break springs

Cylindrical sleeve with eccentric chambers on inside

Turning coupling forces rollers up inclined sides of eccentric chamber to lock coupling to shaft

Side clearance provided between chain and teeth for accomodation of angular displacement between shafts

Roller chain over teeth on hub flanges. All rollers in contact with teeth for equal distribution of transmitted load

Teeth cut on flanges of hubs

Smooth exterior for safety

FIG. 7

With rollers located in largest part of eccentric chamber, coupling can be slipped over end of shaft

Two steel rollers held parallel by light wire frame

FIG. 8

Chain provided with master link for removal

Hubs keyed to shafts

Couplings, continued

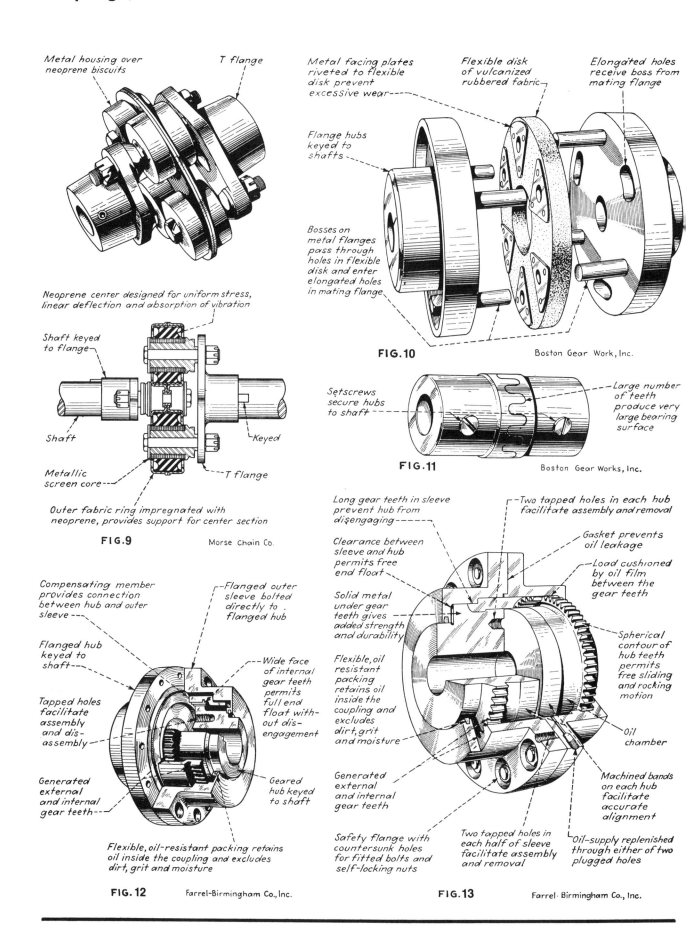

Metal housing over neoprene biscuits

T flange

Neoprene center designed for uniform stress, linear deflection and absorption of vibration

Shaft keyed to flange

Shaft

Metallic screen core

Outer fabric ring impregnated with neoprene, provides support for center section

Keyed

T flange

FIG. 9   Morse Chain Co.

Metal facing plates riveted to flexible disk prevent excessive wear

Flexible disk of vulcanized rubbered fabric

Elongated holes receive boss from mating flange

Flange hubs keyed to shafts

Bosses on metal flanges pass through holes in flexible disk and enter elongated holes in mating flange

FIG. 10   Boston Gear Work, Inc.

Setscrews secure hubs to shaft

Large number of teeth produce very large bearing surface

FIG. 11   Boston Gear Works, Inc.

Compensating member provides connection between hub and outer sleeve

Flanged hub keyed to shaft

Tapped holes facilitate assembly and disassembly

Generated external and internal gear teeth

Flanged outer sleeve bolted directly to flanged hub

Wide face of internal gear teeth permits full end float without disengagement

Geared hub keyed to shaft

Flexible, oil-resistant packing retains oil inside the coupling and excludes dirt, grit and moisture

FIG. 12   Farrel-Birmingham Co., Inc.

Long gear teeth in sleeve prevent hub from disengaging

Clearance between sleeve and hub permits free end float

Solid metal under gear teeth gives added strength and durability

Flexible, oil resistant packing retains oil inside the coupling and excludes dirt, grit and moisture

Generated external and internal gear teeth

Safety flange with countersunk holes for fitted bolts and self-locking nuts

Two tapped holes in each hub facilitate assembly and removal

Gasket prevents oil leakage

Load cushioned by oil film between the gear teeth

Spherical contour of hub teeth permits free sliding and rocking motion

Oil chamber

Machined bands on each hub facilitate accurate alignment

Oil-supply replenished through either of two plugged holes

Two tapped holes in each half of sleeve facilitate assembly and removal

FIG. 13   Farrel-Birmingham Co., Inc.

Shaft couplings that utilize internal and external gears, balls, pins, and non-metallic parts to transmit torque are shown herewith.

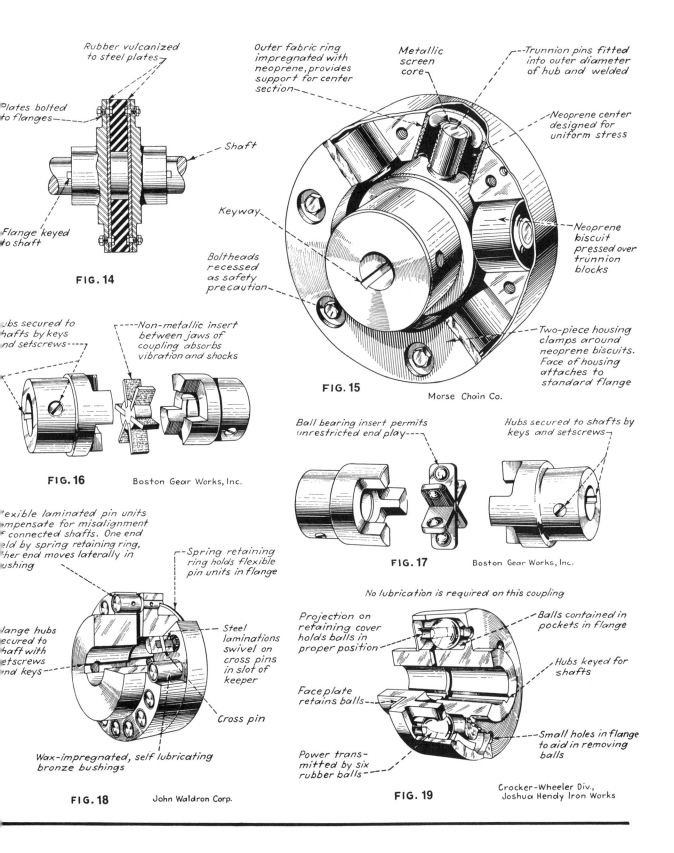

Rubber vulcanized to steel plates

Plates bolted to flanges

Shaft

Flange keyed to shaft

FIG. 14

Outer fabric ring impregnated with neoprene, provides support for center section

Metallic screen core

Trunnion pins fitted into outer diameter of hub and welded

Neoprene center designed for uniform stress

Keyway

Boltheads recessed as safety precaution

Neoprene biscuit pressed over trunnion blocks

Two-piece housing clamps around neoprene biscuits. Face of housing attaches to standard flange

FIG. 15   Morse Chain Co.

Hubs secured to shafts by keys and setscrews

Non-metallic insert between jaws of coupling absorbs vibration and shocks

FIG. 16   Boston Gear Works, Inc.

Ball bearing insert permits unrestricted end play

Hubs secured to shafts by keys and setscrews

FIG. 17   Boston Gear Works, Inc.

Flexible laminated pin units compensate for misalignment of connected shafts. One end held by spring retaining ring, other end moves laterally in bushing

Spring retaining ring holds flexible pin units in flange

Flange hubs secured to shaft with setscrews and keys

Steel laminations swivel on cross pins in slot of keeper

Cross pin

Wax-impregnated, self lubricating bronze bushings

FIG. 18   John Waldron Corp.

No lubrication is required on this coupling

Projection on retaining cover holds balls in proper position

Balls contained in pockets in flange

Hubs keyed for shafts

Faceplate retains balls

Small holes in flange to aid in removing balls

Power transmitted by six rubber balls

FIG. 19

Crocker-Wheeler Div., Joshua Hendy Iron Works

Linkages for Band Clutches
and Brakes

A. C. RASMUSSEN

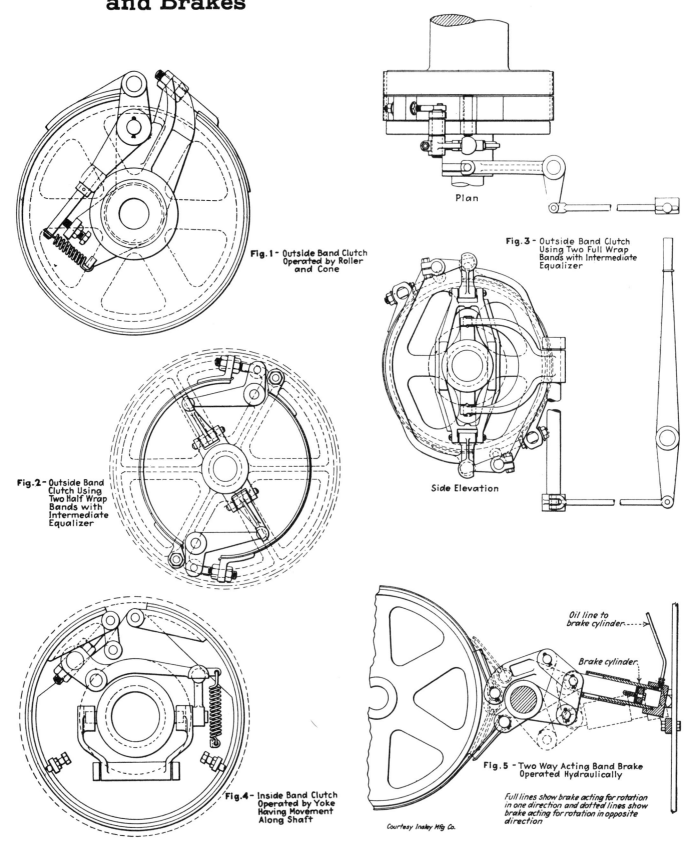

Fig. 1 - Outside Band Clutch Operated by Roller and Cone

Fig. 2 - Outside Band Clutch Using Two Half Wrap Bands with Intermediate Equalizer

Plan

Fig. 3 - Outside Band Clutch Using Two Full Wrap Bands with Intermediate Equalizer

Side Elevation

Fig. 4 - Inside Band Clutch Operated by Yoke Having Movement Along Shaft

Oil line to brake cylinder

Brake cylinder

Fig. 5 - Two Way Acting Band Brake Operated Hydraulically

Full lines show brake acting for rotation in one direction and dotted lines show brake acting for rotation in opposite direction

Courtesy Insley Mfg Co.

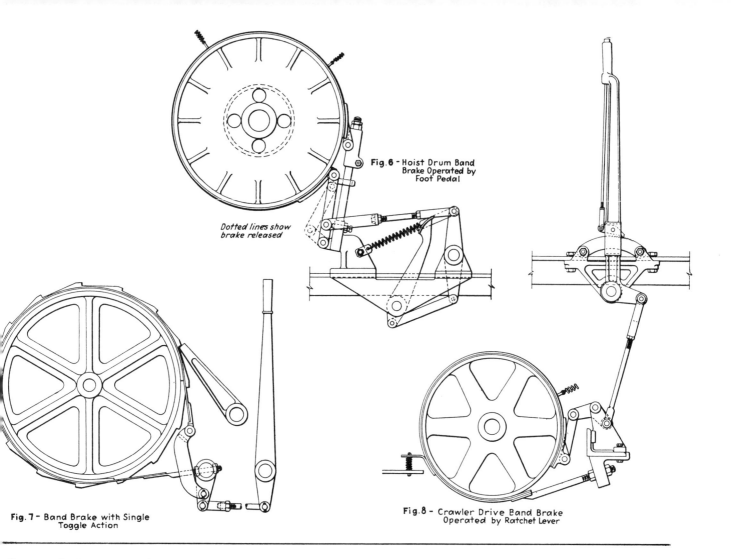

Fig. 6 - Hoist Drum Band Brake Operated by Foot Pedal

Dotted lines show brake released

Fig. 7 - Band Brake with Single Toggle Action

Fig. 8 - Crawler Drive Band Brake Operated by Ratchet Lever

Special coupling mechanisms

PREBEN W. JENSEN

Parallel-link coupling

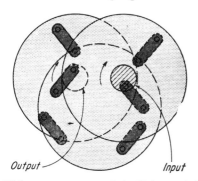

Output   Input

This arrangement of six links and three disks can synchronize the motion between adjacent, parallel shafts.

Here's another arrangement to synchronize shafts, but without need for links.

Six-disk coupling

Input   Output

Bent-pin coupling

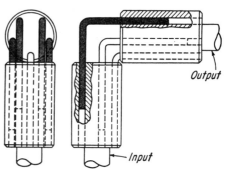

Output

Input

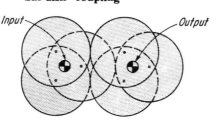

As the input rotates, the five bent pins will move in and out of the drilled holes to impart a constant velocity rotation to the right angle output shaft. The device can transmit constant velocity at angles other than 90 degrees, as shown.

SPECIAL LINK
COUPLING MECHANISMS

H. G. CONWAY
Cheltenham, England

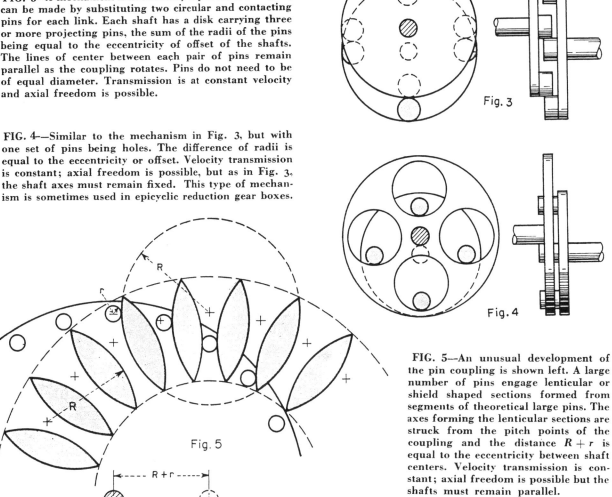

FIG. 1—If the velocity does not have to be constant a pin and slot coupling can be used. Velocity transmission is irregular as the effective radius of operation is continually changing, the shafts must remain parallel unless a ball joint is used between the slot and pin. Axial freedom is possible but any change in the shaft offset will further affect the fluctuation of velocity transmission.

FIG. 2—The parallel-crank mechanism is sometimes used to drive the overhead camshaft on engines. Each shaft has at least two cranks connected by links and with full symmetry for constant velocity action and to avoid dead points. By using ball joints at the ends of the links, displacement between the crank assemblies is possible.

FIG. 3—A mechanism kinematically equivalent to Fig. 2, can be made by substituting two circular and contacting pins for each link. Each shaft has a disk carrying three or more projecting pins, the sum of the radii of the pins being equal to the eccentricity of offset of the shafts. The lines of center between each pair of pins remain parallel as the coupling rotates. Pins do not need to be of equal diameter. Transmission is at constant velocity and axial freedom is possible.

FIG. 4—Similar to the mechanism in Fig. 3, but with one set of pins being holes. The difference of radii is equal to the eccentricity or offset. Velocity transmission is constant; axial freedom is possible, but as in Fig. 3, the shaft axes must remain fixed. This type of mechanism is sometimes used in epicyclic reduction gear boxes.

FIG. 5—An unusual development of the pin coupling is shown left. A large number of pins engage lenticular or shield shaped sections formed from segments of theoretical large pins. The axes forming the lenticular sections are struck from the pitch points of the coupling and the distance $R + r$ is equal to the eccentricity between shaft centers. Velocity transmission is constant; axial freedom is possible but the shafts must remain parallel.

CHAPTER 8

TORQUE-LIMITING, TENSIONING AND GOVERNING DEVICES

Clutch-servo system winds and regulates tension in tapes

A spring-wound mechanical clutch that provides a uniform but adjustable drag torque, and a photoelectric servo system that sets up an optical boundary to keep tape from wandering, have solved the problem of automatically guiding and pre-tensioning the winding of tape.

The tape, made of special transparent porous material, must be wound precisely in successive overlays from supply reels that are wound randomly on extra wide reels as they are manufactured. The delicate nature of the tape precludes use of any mechanical guidance implements. At the same time, the tape must be wound so its tension is uniform when it passes under a high-power microscope. (The tapes are employed for biological studies.)

Combining optics, mechanisms. The unique solution (sketch), which is expected to have application in other types of tape drives, has been worked out by Vincent Spitaleri and his colleagues at the Cutler-Hammer Div. of Airborne Instruments Laboratory, Deer Park, N. Y. The final unit shown in the photo is being used at Space General Corp., El Monte, Calif.

A simple optical system provides a collimated light source. The light passes at either side of the tape at the proper conjugate distance. Two boundary photo-sensitive cells sense both edge positions as the tape wanders transversely during the transporting mode. Interruption of any light beam by a tape edge will unbalance the photocell bridge circuit, and will result in an electrical difference between the interrupted cell and the non-interrupted cell.

Signals from the photocells then are fed into a differential amplifier (schematic) that determines the correct electrical polarity to be amplified by the power amplifier. A dc servo motor with a very high gear reduction ratio is fed by the power amplifier.

The motor rotates either clockwise or counterclockwise in relation to the electrical polarity. Rotation of the motor turns a cam that shifts

Mechanical and electrical systems combine to regulate tape winding

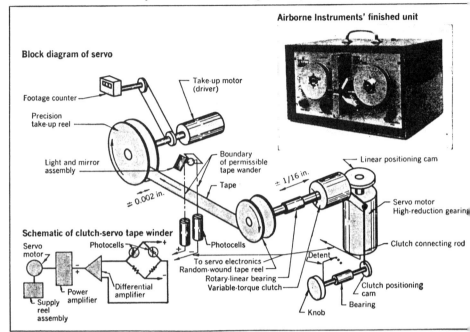

Block diagram of servo

Airborne Instruments' finished unit

Footage counter • Take-up motor (driver) • Precision take-up reel • Light and mirror assembly • Boundary of permissible tape wander • ± 1/16 in. • Linear positioning cam • Tape • ± 0.002 in. • Servo motor High-reduction gearing • Clutch connecting rod

Schematic of clutch-servo tape winder

Servo motor • Photocells • Photocells • To servo electronics • Random-wound tape reel • Rotary-linear bearing • Variable-torque clutch • Detent • Clutch positioning cam • Differential amplifier • Power amplifier • Supply reel assembly • Knob • Bearing

the supply reel linearly (the supply reel is mounted on the variable torque brake shaft, which in turn is mounted in a linear-rotary bearing assembly). This linear translation corrects the transverse tape position between the supply reel and the take-up reel.

When the tape is positioned in the ideal path, both edge-sensing photocells receive equal illumination, producing a nulling condition that causes the servo system to become inoperative.

The servo system transversely corrects randomly wound tape from ± ⅛ in. on the supply reel, to ±0.002 in. on the precision reel at a speed of 5 in./sec.

Clutch controls tension. To provide uniform tension, Spitaleri chose a spring-wound clutch, which produces a constant slip (or "drag") torque that is varied by changing the angular position of its control arm. Such clutches, manufactured by Machine Components Corp., Farmingdale, N.Y., use an interference fit between the spring and clutch shaft to obtain the torque. Tests have shown that variations in the friction coefficient between spring and shaft produce only negligible

changes in torque, so the clutches are remarkably reliable in producing a specific drag.

Because of the thinness of the tape, the variation in tension from beginning to end of the reel-winding is less than 9%. Spitaleri, therefore, did not feel the need for a sensor arm to ride on the tape spool liameter and to adjust the torque minutely as the tape unreeled. Instead he turned to a simpler way—a knob and calibrated detent arrangement that permits easy setting of desired winding tension.

Tape samples bacteria. The new tapes are expected to be widely applied in scientific projects of various types, in space probes, and in biological bacteria-monitoring systems. The plastic tapes are made porous by bombarding them with neutrons, and then can be used for continuous air or space sampling of bacteria and other minute solids.

According to Airborne Instruments and General Electric Co., which supplies the tape from its Pleasanton (Calif.) facilities, the tape is extremely difficult to handle, because of its high inherent static charge from radiation.

Caliper brakes help maintain proper tension in press feed

A simple cam-and-linkage arrangement (drawing above) works in team with two caliper disk brakes to provide automatic tension control for paper feeds on a web press.

The problem in such feed systems is to maintain controlled tension on paper that's being drawn off at 1200 fpm from a roll up to 42 in. wide and 36 in. dia. Such rolls, when full, weigh 2000 lb. Also required is almost instantaneous emergency stopping of the press.

Friction-disk brakes are subject to lining wear, but they can make millions of stops before relining.

In the new system, two pneumatic disk brakes made by Tol-O-Matic, Inc., Minneapolis, are used on each roll, gripping two separate 12-in. disks that provide maximum heat dissipation. To provide a desired constant drag tension on the rolls, the brakes are always under air pressure. A dancer roll riding on the paper web can, however, override the brakes at any time by operating a cam that adjusts a pressure regulator controlling brake effort.

If the web should break or the paper run out on the roll, the dancer roll will allow maximum braking. The press can be stopped in less than one revolution.

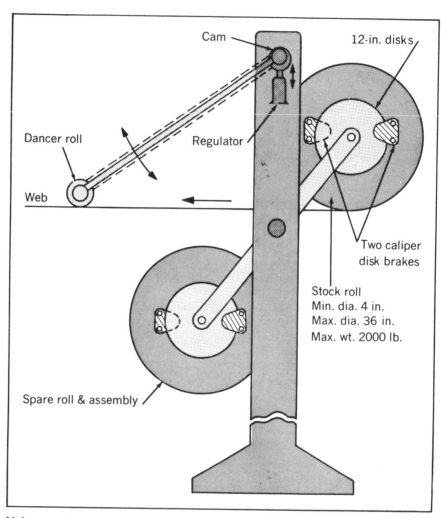

Linkage system works in combination with pressure regulator and caliper disk brakes to stop a press fast from a high speed if the web should break.

Sensors aid clutch/brakes

Two clutch/brake systems, teamed with magnetic pickup sensors, cut paper sheets to exact lengths. One magnetic pickup senses the teeth on a rotating sprocket. The resulting pulses, which are related to the paper length, are counted and a cutter wheel is actuated by the second clutch/brake system. The flywheel on the second system enhances the cutting force.

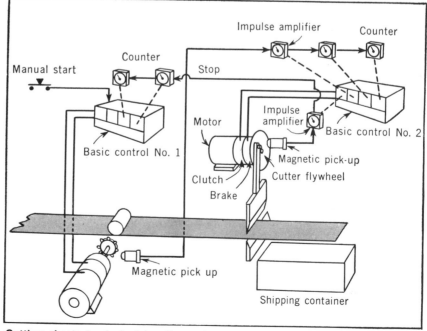

Cutting sheets to desired lengths and counting how many cuts are made is a simpler process with control system.

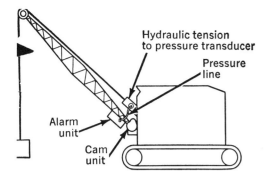

Warning device prevents overloading of boom

Cranes can now be protected against unsafe loading by a device in which movable electrical contacts are shifted by a combination of fluidic power and a cam-and-gear arrangement (drawing, below).

The device takes into consideration the two key factors in safe loading of a crane boom: the boom angle (low angles create a greater overturning torque than high angles) and the compression load on the boom, which is greatest at high boom angles. Both factors are translated into inputs that are integrated to actuate the electrical warning system that alerts the crane operator that a load is unsafe to lift.

How it works. In a prototype built for Thew-Lorain Inc by US Gauge, Sellersville, Pa, a tension-to-pressure transducer (drawing, above) senses the load on the cable and converts it into a hydraulic pressure that is proportional to the tension. This pressure is applied to a Bourdon-tube pressure gage with a rotating pointer that carries a small permanent magnet (details in drawing below). Two miniature magnetic reed switches are carried by another arm that moves on the same center as the pointer.

This arm is positioned by a gear and rack controlled by a cam, with a sinusoidal profile, that is attached to the cab. As the boom is raised or lowered, the cam shifts the position of the reed switches so they will come into close proximity of the magnet on the pointer and sooner or later make contact. The timing of this contact depends partly on the movement of the pointer that carries the magnet. On an independent path, the hydraulic pressure representing cable tension is shifting the pointer to the right or left on the dial.

When the magnet contacts the reed switches, the alarm circuit is closed and remains closed during a continuing pressure increase without retarding the movement of the point. In the unit built for Thew-Lorain, the switches are arranged in two stages, the first to trigger an amber warning light and the second to light a red bulb and also sound an alarm bell.

Over-the-side or over-the-rear loading requires a different setting of the Bourdon pressure-gage unit than does over-the-front loading. A cam built into the cab pivot post actuates a selector switch.

Constant watch on cable tension

A simple lever system is solving the problem of how to keep track of varying tension loads on a cable as it is wound on its drum.

Thomas Grubbs of NASA's Manned Spacecraft Center in Houston devised the system, built around two pulleys mounted on a pivoted lever. The cable is passed between the pulleys (drawing) so an increase in cable tension causes the lever to pivot. This in turn pulls linearly on a flat metal tongue to which a strain gage has been cemented. Load on the lower pulley is proportional to tension on the cable. Stretch of the strain gage changes an electrical current that gives a continuous, direct reading of the cable tension.

The two pulleys on the pivoting lever are free to translate on their axes of rotation to allow proper positioning of the cable as it traverses the take-up drum.

A third pulley might be added to the two-pulley assembly to give some degree of adjustment to strain-gage sensitivity. Located in the plane of the other two pulleys, it would be positioned to reduce the strain on the tongue (for heavy loads) or increase the strain (for light loads). □

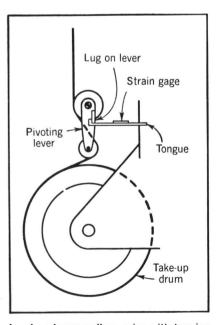

Load on lower pulley varies with tension on cable, and pivoting of lever gives direct reading via strain gage.

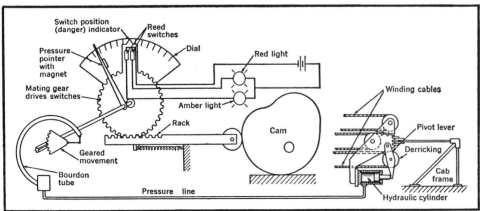

Cam on the cab positions an arm, bearing reed switches, according to boom angle; pressure pointer reacts to cable tension.

TORQUE-LIMITERS PROTECT
LIGHT-DUTY DRIVES

In such drives the light parts break easily when overloaded. These eight devices disconnect them from dangerous torque surges.

L KASPER

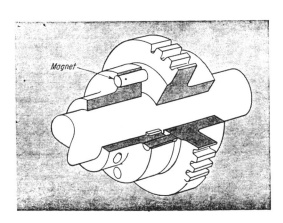

MAGNETS transmit torque according to their number and size. In-place control is limited to lowering torque capacity by removing magnets.

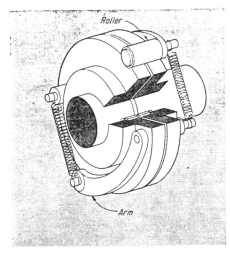

ARMS hold rollers in slots which are cut across disks mounted on ends of butting shafts. Springs keep rollers in slots; over-torque forces them out.

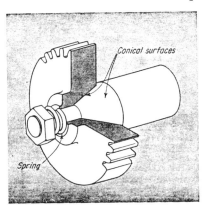

CONE CLUTCH is formed by mating taper on shaft to beveled hole through gear. Tightening down on nut increases torque capacity.

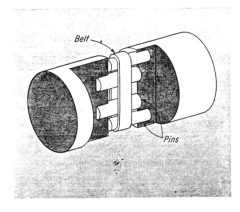

FLEXIBLE BELT wrapped around four pins transmits only lightest loads. Outer pins are smaller than inner pins to ensure contact.

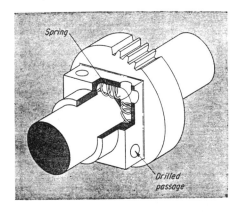

SPRINGS inside drilled block grip the shaft because they distort during mounting of gear.

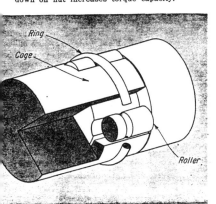

RING fights natural tendency of rollers to jump out of grooves cut in reduced end of one shaft. Slotted end of hollow shaft is like a cage.

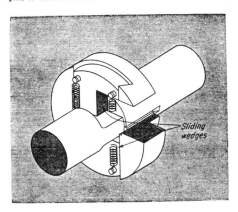

SLIDING WEDGES clamp down on flattened end of shaft; spread apart when torque gets too high. Strength of springs which hold wedges together sets torque limit.

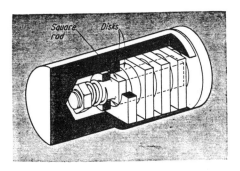

FRICTION DISKS are compressed by adjustable spring. Square disks lock into square hole in left shaft; round ones lock onto square rod on right shaft.

ways to
PREVENT OVERLOADING

These "safety valves" give way if machinery jams, thus preventing serious damage.

PETER C NOY

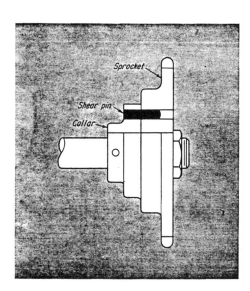

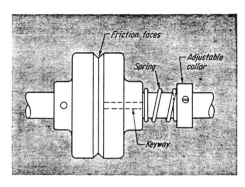

SHEAR PIN is simple to design and reliable in service. However, after an overload, replacing the pin takes a relatively long time; and new pins aren't always available.

FRICTION CLUTCH. Adjustable spring tension that holds the two friction surfaces together sets overload limit. As soon as overload is removed the clutch reengages. One drawback is that a slipping clutch can destroy itself if unnoticed.

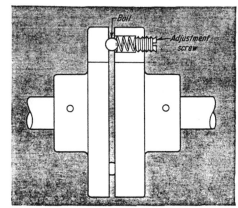

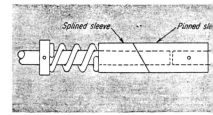

MECHANICAL KEYS. Spring holds ball in dimple in opposite face until overload forces the ball out. Once slip begins, wear is rapid, so device is poor when overload is common.

ANGLE-CUT CYLINDER. With just one tooth, this plified version of the jaw clutch. Spring tension sets

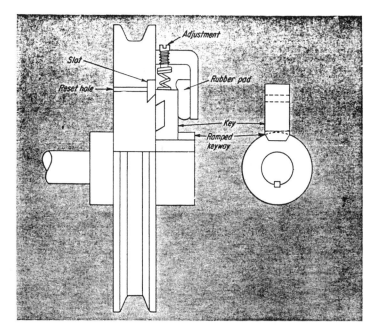

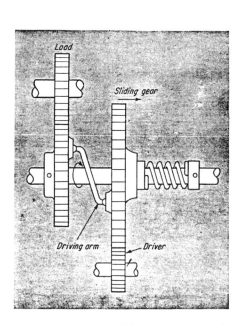

RETRACTING KEY. Ramped sides of keyway force key outward against adjustabe spring. As key moves outward, a rubber pad—or another spring—forces the key into a slot in the sheave. This holds the key out of engagement and prevents wear. To reset, push key out of slot by using hole in sheave.

DISENGAGING GEARS. Axial forces of spring and driving arm balance. Overload overcomes spring force to slide gears out of engagement. Gears can strip once overloading is removed, unless a stop holds gears out of engagement.

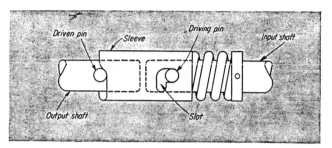

CAMMED SLEEVE connects input and output shafts. Driven pin pushes sleeve to right against spring. When overload occurs, driving pin drops into slot to keep shaft disengaged. Turning shaft backwards resets.

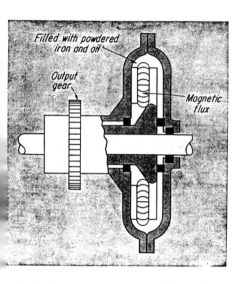

MAGNETIC FLUID COUPLING is filled with slurry made of iron or nickel powder in oil. Controlled magnetic flux that passes through fluid varies slurry viscosity, and thus maximum load over a wide range. Slip ring carries field current to vanes.

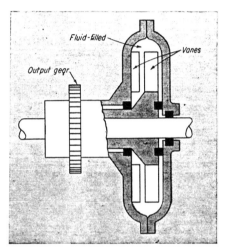

FLUID COUPLING. Maximum load can be closely controlled by varying viscosity and level of fluid. Other advantages are smooth transmission and low heat rise during slip.

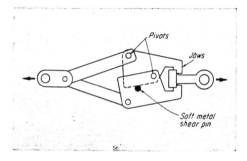

TENSION RELEASE. When toggle-operated blade shears soft pin, jaws open to release eye. A spring that opposes the spreading jaws can replace the shear pin.

NG PLUNGER is for reciprocating motion with le overload only when rod is moving left. Spring esses under overload.

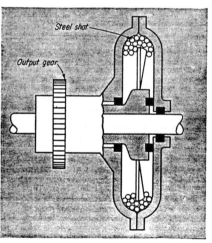
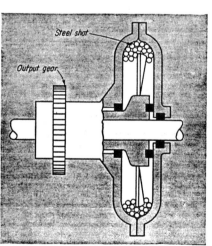

STEEL-SHOT COUPLING transmits more torque as speed increases. Centrifugal force compresses steel shot against case, increasing resistance to slip. Adding more steel shot also increases resistance to slip.

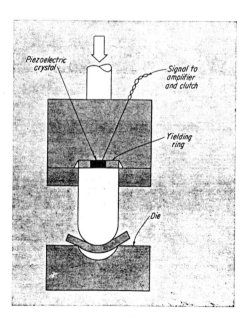

PIEZOELECTRIC CRYSTAL sends output signal that varies with pressure. Clutch at receiving end of signal disengages when pressure on the crystal reaches preset limit. Yielding ring controls compression of crystal.

7 ways to LIMIT SHAFT

Traveling nuts, clutch plates, gear fingers, and pinned members are the bases of these ingenious mechanisms.

Mechanical stops are often required in automatic machinery and servomechanisms to limit shaft rotation to a given number of turns. Two problems t guard against, however, are: Excessive forces caused by abrupt stops; larg torque requirements when rotation is reversed after being stopped.

I M ABELES

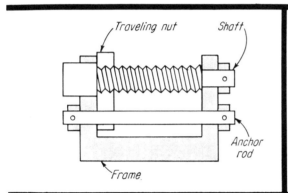

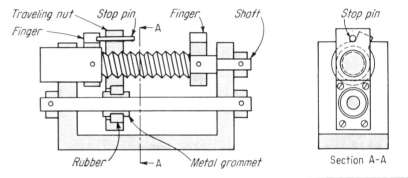

TRAVELING NUT moves (1) along threaded shaft until frame prevents further rotation. A simple device, but nut jams so tight that a large torque is required to move the shaft from its

stopped position. This fault is overcome at the expense of increased length by providing a stop pin in the traveling nut (2). Engagement between pin and rotating finger must be shorter

than the thread pitch so pin can cl finger on the first reverse-turn. rubber ring and grommet lessen pact, provide a sliding surface. grommet can be oil-impregnated met

CLUTCH PLATES tighten and stop rotation as the rotating shaft moves the nut against the washer. When rotation is reversed, the clutch plates can turn with the shaft from **A** to **B.** During this movement comparatively low torque is required to free the nut from the clutch plates. Thereafter, subsequent movement is free of clutch friction until the action is repeated at other end of the shaft. Device is recommended for large torques because clutch plates absorb energy well.

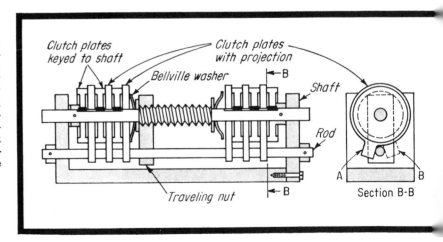

ROTATION

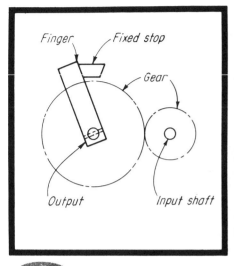

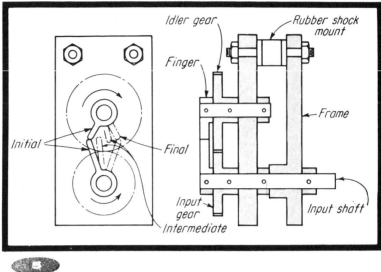

4

SHAFT FINGER on output shaft hits resilient stop after making less than one revolution. Force on stop depends upon gear ratio. Device is, therefore, limited to low ratios and few turns unless a worm-gear setup is used.

5

TWO FINGERS butt together at initial and final positions, prevent rotation beyond these limits. Rubber shock-mount absorbs impact load. Gear ratio of almost 1:1 ensures that fingers will be out of phase with one another until they meet on the final turn. Example: Gears with 30 to 32 teeth limit shaft rotation to 25 turns. Space is saved here but gears are costly.

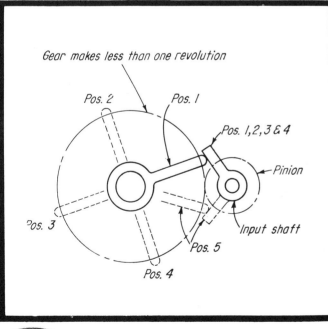

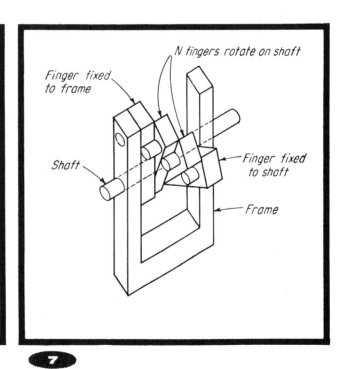

6

LARGE GEAR RATIO limits idler gear to less than one turn. Sometimes stop fingers can be added to already existing gears in a train, making this design simplest of all. Input gear, however, is limited to a maximum of about 5 turns.

7

PINNED FINGERS limit shaft turns to approximately $N+1$ revolutions in either direction. Resilient pin-bushings would help reduce impact force.

343

Mechanical Systems for Controlling Tension and Speed

J. H. GEPFERT

THE KEY TO THE SUCCESSFUL OPERA-TION of any continuous processing system that is linked together by the material being processed is positive speed synchronization of the individual driving mechanisms. Typical ex-amples of such a system are steel strip lines, textile equipment, paper ma-chines, rubber and plastic processers and printing presses. In each of these cases, the material will become wrinkled, marred, stretched or other-wise damaged if precise control is not maintained.

The automatic control for such a system contains three basic elements: The signal device or indicator, which senses the error to be corrected; the controller, which interprets the indi-cator signal and amplifies it, if neces-sary, to initiate control action; and the transmission, which operates from the controller to change the speed of the driving mechanism to correct the error.

Signal indicators for continuous sys-tems fall into two general classifica-

TABLE I — PRIMARY INDICATORS

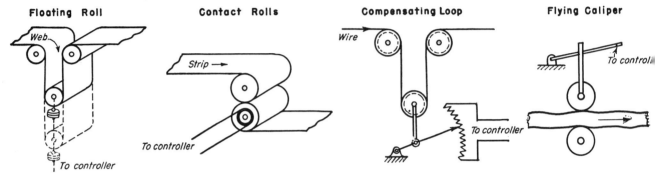

USE: Constant tension winding
Registry control; Section synchronizing.

USE: Velocity control;
Cutter feed control.

USE: Line tension control;
Section synchronizing.

USE: Thickness control;
Diameter control.

TABLE II — SECONDARY INDICATORS

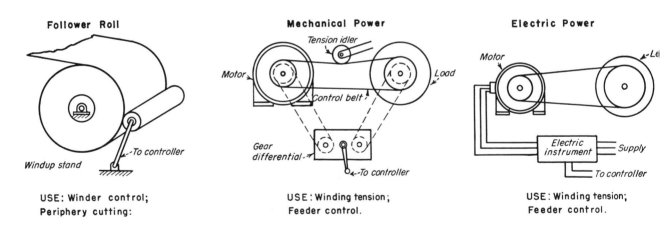

USE: Winder control;
Periphery cutting:

USE: Winding tension;
Feeder control.

USE: Winding tension;
Feeder control.

TABLE III — CONTROLLERS AND ACTUATORS

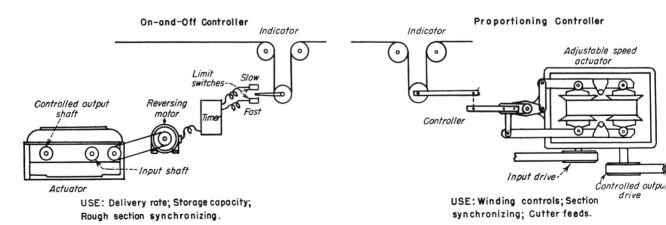

USE: Delivery rate; Storage capacity;
Rough section synchronizing.

USE: Winding controls; Section
synchronizing; Cutter feeds.

tions: Primary indicators that measure the change in speed or tension of the material by direct contact with the material; and secondary indicators, which measure a change in the material from some reaction in the system that is proportional to the change.

The primary type is inherently more accurate because of its direct contact with the material. These indicators take the form of contact rolls, float-ing or compensating rolls, resistance bridges and flying calipers, as illustrated in Table I. In each case, any change in the tension, velocity or pressure of the material is indicated directly and immediately by a displacement or change in position of the indicator element. The primary indicator, therefore, shows deviation from an established norm, regardless of the factors that have caused the change.

Secondary indicators, shown in Table II, are used in systems where the material cannot be in direct contact with the indicator or when the space limitations of a particular application make their use undesirable. This type indicator introduces into the control system a basic inaccuracy which is the result of measuring an error in the material from a reaction that is not exactly proportional to the error. The control follows the summation of the errors in the material and the indicator itself.

The controlling devices, which are operated by the indicators, determine the degree of speed change required to correct the error, the rate at which the correction must be made, and the stopping point of the control action after the error has been corrected. The manner in which the corrective action of the controller is stopped determines both the accuracy of the control system and the type of control equipment required.

Three general types of control action are illustrated in Table III. Their selection for any individual application is based on the degree of control action required, the amount of power available for initiating the control, that is: the torque amplification required, and the space limitations of the equipment.

The on-and-off control with timing action is the simplest of the three types. It functions on the basis that, when the indicator is displaced, the timer contact energizes the control in the proper direction for correcting the error. The control action continues until the timer stops the action. After a short interval, the timer again energizes the control system and, if the error still exists, control action is continued in the same direction. Thus, the control process is a step-by-step action to make the correction and to stop operation of the controller.

The proportionate type of controller corrects an error in the system, as shown by the indicator, by continuously adjusting the actuator to a speed that is in exact proportion to the displacement of the indicator. The diagram in Table III shows the proportionate controller in its simplest form as a direct link connection between the indicator and the actuating drive. However, the force amplification between the indicator and the drive is relatively low and hence limits this controller to applications where the indicator has sufficient operating force to adjust the speed of the vari-

Fluid Pressure

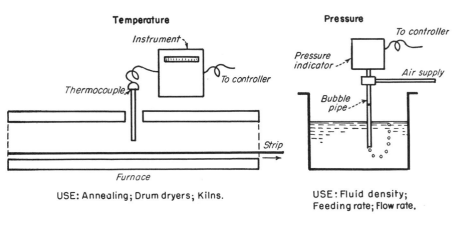

USE: Fluid level control; Constant pressure control; Filtering rate control.

Liquid Level

USE: Pumping rate control; System pressure control.

Temperature

USE: Annealing; Drum dryers; Kilns.

Pressure

USE: Fluid density; Feeding rate; Flow rate.

Proportioning-Throttling Controller

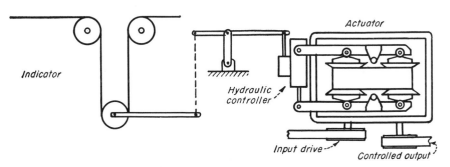

USE: Constant tension winding; Registry control; Exact section synchronizing.

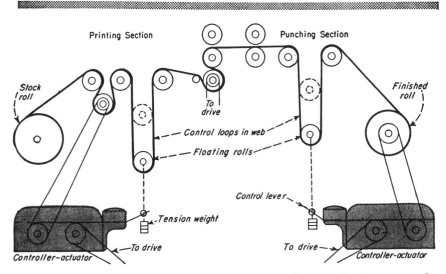

Fig. 1—Floating rolls are direct indicators of speed and tension in the paper web. Controller-actuators adjust feed and windup rolls to maintain registry during printing.

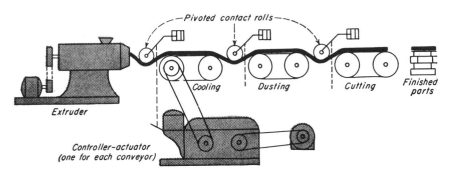

Fig. 2—Dimension control of extruded materials calls for primary indicators like the contact rolls shown. Their movements actuate conveyor control mechanisms.

able speed transmission directly.

The most accurate controller is the proportioning type with throttling action. Here, operation is in response to the rate of error indication. This type of controller, as shown in Table III, is connected to a throttling valve, which operates a hydraulic servo-mechanism for adjusting the variable speed transmission.

The throttling action of the valve provides a slow control action for small error correction or for continuous correction at a slow rate. For following large errors, as shown by the indicator, the valve opens to the full position and makes the correction as rapidly as the variable speed transmission will allow.

Many continuous processing systems can be automatically controlled by using a package unit consisting of a simple, mechanical, variable speed transmission and an accurate hydraulic controller.

This controller-transmission package can be used to change the speed relationship at the driving points in the

continuous system from any indicator that signals for correction by a displacement. It has anti-hunting characteristics because of the throttling action on the control valve, and is self-neutralizing because the control valve is part of the transmission adjustment system.

An example of a continuous processing system that requires automatic control is the rotary type printing press. When making billing forms on such a press, the printing plates are rubber and the forms are printed on a continuous web of paper. The paper varies in texture, moisture content, flatness, elasticity and finish. In addition, the length of the paper changes as the ink is applied.

In a typical application of this type, the accuracy required for proper registry of the printing and hole punching must be held to a differential of 1/32 in. in 15 ft of web. For this degree of accuracy, a floating or compensating roll, as shown in Fig. 1, is used as the indicator, because it is the most accurate means of indicating changes

in length of web by displacement. In this case, two floating rolls are used with two separate controllers and actuators, one to control the in-feed speed and tension of the paper stock, and the second to control the wind-up.

The in-feed is controlled by maintaining the turning speed of a set of feeding rolls that pull the paper off the stock roll. The second floating roll controls the speed of the wind-up mandrel. The web of paper is held to an exact value of tension between the feed rolls and the punching cylinder of the press by the in-feed control, and also between the punching cylinder and the wind-up roll. Hence, it is possible to not only control the tension in the web of different grades of paper, but also to adjust the relative length at these two points, thereby maintaining proper registry.

The secondary function of maintaining exact control of the tension in the paper as it is rewound after printing is to condition the paper and to obtain a uniformly wound roll so that the web is ready for subsequent operations.

Control of dimension or weight by tension and velocity regulation can be illustrated by applying the same general type of controller actuator to the take-off conveyors in an extruder line such as those used in rubber and plastics processing. The problem is twofold: First, to set the speed of the take-away conveyor at the extruder to match the variation in extrusion rate; and, second, to set the speeds of the subsequent conveyor sections to match the movement of the stock as it cools and tends to change dimension.

One way to handle this problem is to use as the indicators the pivoted idlers or contact rolls shown in Fig. 2. The rolls contact the extruded material between each of the conveyor sections and control the speed of the driving mechanism of the following section. The material forms a slight catenary between the stations and the change in the catenary length is used for indicating errors in driving speeds.

The plasticity of the material prevents the use of a complete control loop and hence the contact roll must operate with very little resistance or force through a small operating angle.

The problem of winding or coiling a strip of thin steel that has been plated or pre-coated for painting on a continuous basis is typical of processing systems in which primary indicators cannot be used. While it is important that no contact be made with the prepared surface of the steel, it is also desirable to rewind the strip after preparation in a coil that is sound and slip-free. An automatic, constant-tension winding control and a secondary type indicator are therefore used to initiate the control action.

The control system shown in Fig. 3 is used to wind coils from 16 in. core diameter to 48 in. maximum diameter. The power to wind the coil is used as the controlling medium because, by maintaining constant winding power as the coil builds up, a constant value of strip tension can be held within the limits required. Actually, this method is inaccurate to the extent that the losses in the driving equipment, which are a factor in the power being measured, are not constant and hence the strip tension changes slightly. This same factor enters into any control system that uses winding power as an index of control.

A torque measuring belt that operates a differential controller is used to measure the power of the winder. Then, in turn, the controller adjusts the variable speed transmission. The change in speed between the source of power and the transmission is measured by the three-shaft gear differential which is driven in tandem with the control belt. Any change in load across the control belt produces a change in speed between the driving and driven ends of the belt. The differential acts as the controller, since any change in speed between the two outside shafts of the differential results in a rotation or displacement of the center or control shaft. By connecting the control shaft of the differential directly to a screw-controlled variable speed transmission, a means is provided for adjusting the transmission to correct any change in speed and power as delivered by the belt.

This system is made completely automatic by establishing a neutralizing speed between the two input shafts of the differential (within the creep value of the belt). When there is no tension in the strip, for example when it is cut, the input speed to the actuator side of the differential is higher on the driven side than it is on the driving side of the differential. This unbalance reverses the rotation of the control shaft of the differential, which in turn resets the transmission to the high speed required for starting the next coil on the rewinding mandrel.

In operation, any element in the system that tends to change strip tension causes a change in winding power, which, in turn, is immediately compensated for by the rotation or tendency to rotate of the controlling shaft in the differential. Hence, the winding mandrel speed is continuously and automatically corrected to maintain

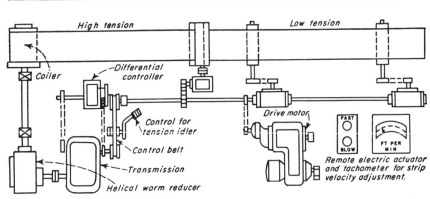

Fig. 3—Differential controller has third shaft that signals remote actuator when tension in sheet material changes. Coiler power is used as secondary control index.

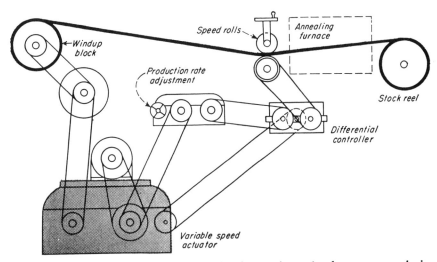

Fig. 4—Movement of wire through annealing furnace is regulated at constant velocity by continuously retarding the speed of the windup reels to allow for wire build-up.

constant tension in the strip.

When the correct speed relationships are established in the controller, the system operates automatically for all conditions of operation. In addition, tension in the strip can be adjusted to any value by moving the tension idler on the control belt to increase or decrease the load capacity of the belt to match a desired strip tension.

There are many continuous processing systems that require constant velocity of the material during processing, yet do not require accurate control of the tension in the material. An example of this type is the annealing of wire that is pulled off stock reels through an annealing furnace and then rewound on a wind-up block.

The problem is to pass the wire through the furnace at a constant rate, so that the annealing time is maintained at a fixed value. Since the wire is pulled through the furnace by the wind-up blocks shown in Fig. 4, its rate of movement through the furnace would increase as the wire builds up

on the reels unless a control is used to slow down the reels.

A solution to this problem is to use a constant velocity type control with the wire as a direct indication to an indicator that takes the speed of initiate a control action for adjusting the speed of the wind-up reel. In this case, the wire can be contacted directly and a primary type indicator in the form of a contact roll can be used to register any change in speed. The contact roll drives one input shaft of the differential controller. The second input shaft is connected to the driving shaft of the variable speed transmission to provide a reference speed. The third or control shaft will then rotate when any difference in speed exists between the two input shafts. Thus, if the control shaft is connected to a screw-regulated actuator, an adjustment is obtained for slowing down the wind-up blocks as the coils build up and the wire progresses through the furnace at a constant speed.

DRIVES FOR CONTROLLING TENSION

Mechanical, electrical and hydraulic methods for obtaining controlled tension on winding reel and similar drives, or for driving independent units of a machine in synchronism

Mechanical Drives

Band Brake—Used on coil winders, insulation winders and similar applications wherein maintaining the tension within close limits is not required.

Simple and economical but tension will vary considerably. Friction drag at start may be several times that during running by virtue of the difference between coefficient of friction at starting and the coefficient of sliding friction, which latter will also be affected by moisture, foreign matter, and wear of surfaces.

Capacity—limited by the heat radiating capacity of the brake at the maximum permissible running temperature.

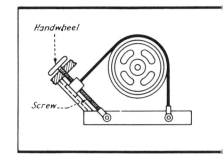

Differential Drives may be of various forms—epicyclic spur gears, bevel gear differentials or worm gear differentials.

The braking device on the ring gear or spider may be a band brake, a fan, an impeller, an electric generator or an electric drag such as a copper disk rotating in a powerful magnetic field. A brake will give a drag or tension reasonably constant over a wide speed range.

The other braking devices mentioned will exert a torque that will vary widely with speed but will be definite for any given speed of the ring gear or spider.

A definite advantage of any differential drive is that maximum driving torque can never exceed the torque developed by the braking device.

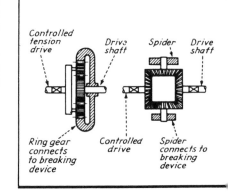

Differential gearing can be used to control a variable-speed transmission. With the ring gear and sun gear driven in opposite directions from the respective shafts to be held in synchronism, the gear train can be designed so that the spider on which the planetary gears are mounted will not rotate when the shafts are running at the desired relative speeds. If one or the other of the shafts speeds ahead, the spider rotates correspondingly. The spider rotation changes the ratio of the variable-speed transmission unit.

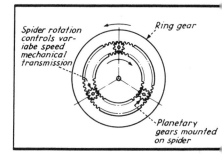

Electrical Drives

Shunt field rheostat in a d.c. motor drive can be used for synchronizing drives. Applied to a machine handling paper, cloth or similar material passing around a take-up roll, movement of the take-up roll moves a control arm that is connected to the rheostat. This type of drive is not suitable to wide changes of speed, above approximately 2½ to 1 ratio.

For wide ranges of speed, the rheostat is put in the shunt field of a d.c. generator which is driven by another motor. The voltage developed by the generator is controlled from zero to full voltage. The generator furnishes the current to the driving motor armature and the fields of the driving motor are separately excited. Thus the motor speed is controlled from zero to maximum.

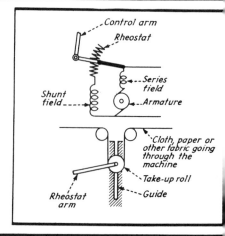

Selsyn motors can be used for direct drive to independent units in exact synchronism, provided inertias are not too great. But regardless of loads and speeds Selsyn motors can be used as the controlling units. As an example, variable-speed mechanical transmission units with built-in Selsyn motors are obtainable for furnishing constant-tension drives or synchronous driving of independent units.

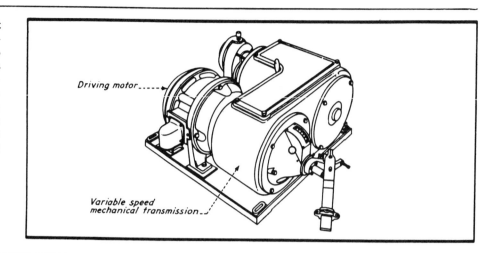

Hydraulic Drives

Hydraulic Control — Tension between successive pairs of rolls, or synchronism between successive units of a machine can be controlled automatically by hydraulic drives. Driving the variable delivery pump off of one of the pairs of rolls automatically maintains an approximately constant relative speed between the two units, at all speeds and loads. The variations caused by oil leakage and similar factors are compensated automatically by the idler roll and linkage which adjusts the pilot valve that controls the displacement of the variable delivery pump.

The counterweight on the idler roll is set for the desired tension in the felt, paper or other material. Increased tension resulting from the second pair of rolls going too fast, causes idler roll to be depressed, the control linkage thereby

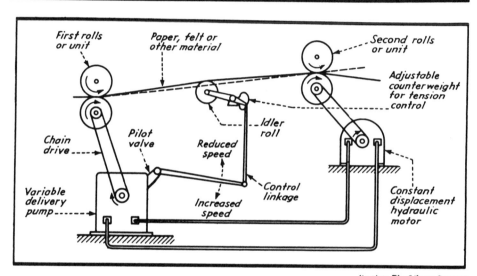

moving pilot valve to cause a decreased pump delivery which slows the speed of the second pair of rolls. The reverse

operations take place when the tension in the paper decreases, allowing the idler roll to move upwards.

If the material passing through the machine is too weak to operate a mechanical linkage, the desired control can be obtained by photo-electric cells. The hydraulic operation is exactly the same as that described above.

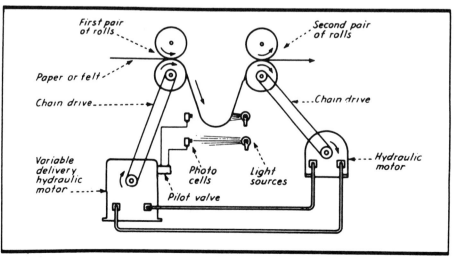

CONTROLLING TENSION (continued)

As mentioned, a band brake used to obtain a friction drag will give variable tension. In this hydraulic drive the winding tension is determined by the difference in torque exerted on the rewinder feed roll and the winding roll. The brake plays no part in establishing the tension.

The constant displacement hydraulic motor and variable displacement hydraulic motor are connected in series with the variable delivery pump. Thus the relative speeds of the two hydraulic motors will always remain substantially the same, the displacement of the variable speed motor being adjusted to an amount slightly greater than the displacement of the constant speed motor, thus tending to give the winding roll a speed slightly greater than the feed roll speed. This determines the tension, because the winding roll cannot go faster than the feed roll, both being in contact with the paper roll being wound. The pressure in the hydraulic line between the constant and variable displacement pump will increase correspondingly to the winding tension. For any setting of the winding

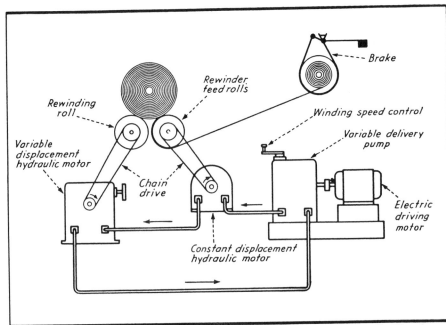

Courtesy The Oilgear Company

speed controller on the variable delivery hydraulic pump the motor speeds are practically constant, hence the surface speed of winding will remain substantially constant, regardless of the diameter of the roll being wound.

Hydraulic drive for fairly constant tension. The variable delivery constant speed pumping unit supplies the oil to two constant displacement motors, one driving the apparatus which carries the fabric through the bath at a constant speed and the other driving the winder. The two motors are in series. Motor *A* driving the winding reel, diameter of which ranges from about 5 in. when the reel is empty to about 33 in. when the reel is full, is geared to the reel so that even when the reel is empty the surface speed of the paper travel will tend to be somewhat faster than the mean rate of paper travel established by motor *B* driving the apparatus. When the reel is empty its speed nearly corresponds to that established by motor *B* driving the apparatus and only a small amount of oil will be by-passed through the choke interposed between the pressure and return line.

When roll is full the r.p.m. of the reel and its driving motor is only about one-seventh of the r.p.m. when the reel is empty. A much greater quantity of oil is forced through the choke when the reel is full because of the increased pressure in the line between the two motors. The pressure in this line increases as the reel diameter increases because the torque resistance encountered by the reel motor will be directly in proportion to the reel diameter, tension

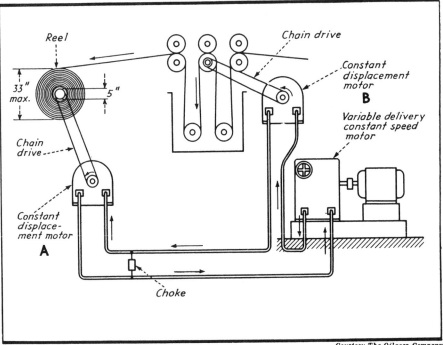

Courtesy The Oilgear Company

being constant. The greater the diameter of the fabric on the reel, the greater will be the torque exerted by the tension in the fabric. The installation is designed so that the torque developed by the motor driving the reel will be inversely proportional to the r.p.m. of the reel. Hence the tension on the fabric will remain at a fairly constant value regardless of the diameter of the reel. This drive is limited to about 3 hp. and is relatively inefficient.

Cutoff Prevents Overloading of Hoist

Fail-safe switch deactivates lifting circuit if load exceeds preset value. Split coupling permits quick attachment of cable.

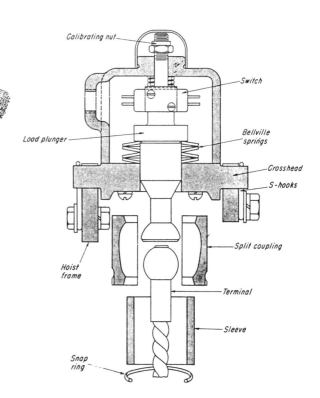

LOAD PLUNGER is inserted through the Belleville springs, which are supported on a swiveling crosshead. The crosshead is mounted on the hoist frame and retained by two S-hooks and bolts. Under load, the Belleville springs deflect and permit the load plunger to move axially. The end of the load plunger is connected to a normally closed switch. When the springs deflect beyond a preset value, the load plunger trips the switch opening the raising-coil circuit of the magnetic hoist-controller. The raising circuit becomes inoperative, but the lowering circuit is not affected. A second contact, normally open, is included in the switch to permit the inclusion of visual or audible overload signal devices.

The load plunger and the swaged-on cable termination have ball-and-socket seat sections to permit maximum free cable movement, reducing the possibility of fatigue failure. A split-coupling and sleeve permits quick attachment of the cable-ball terminal to the load plunger.

Ball-Type Transmission Is Self-Governing

The Gerritsen transmission, developed in England at the Tiltman Langley Laboratories Ltd., Redhill Aerodrome, Surrey, governs its own output speed within limits of plus/minus one percent. The usual difficulties of speed governing—lack of sensitivity, lag and hunting—associated with separate governor units are completely obviated because regulation is effected directly by the driving members through their own centrifugal force. The latter are precision bearing-steel balls that roll on four hardened-steel, cone-shaped rings, and these members can be used for different ratio arrangements.

The transmission can be used in three different ways: as a fixed "gear", as an externally controlled variable speed unit or as a self-governing drive that produces a constant output speed from varying input speeds.

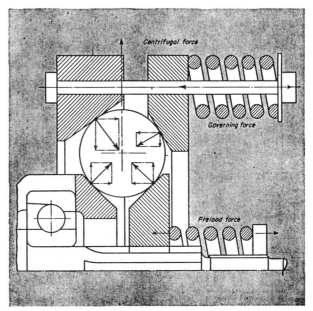

SELF-GOVERNING ACTION of the transmission is obtained by virtue of the centrifugal forces of the balls as they rotate. When the balls move outward radially, the input-output ratio changes. By properly arranging the rings and springs, the gear ratio can be controlled by the movement of the balls to maintain a constant value of output speed.

Mechanical, geared and cammed limit switches

E. L. RUDISILL AND F. D. YEAPLE

LIMIT SWITCHES can be defined as electric current switching devices that are operated by some type of mechanical motion. Limit switches find their greatest field of application on automatic machinery by controlling a complete operating cycle automatically by closing and opening electrical circuits in the proper sequence.

In addition to interlocking control circuits, limit switches have many other uses. For example, one of the most important is as safety devices to stop a machine, sound a warning signal, or illuminate a warning light when a dangerous operating condition develops. Thus, properly applied switches can both control highly efficient automatic electric machinery and protect it and the operator.

Actuator styles for linear mechanical switches

BASIC SPRING-RETURN

ADJUSTABLE LENGTH

TILT-TO-ACTUATE — Mercury switch

OVERSIZE WHEEL

FOUR-POSITION HEAD

EXTENDED HOUSING

ONE-WAY OVERRIDE

LINEAR CAM — Button

90-POSITION VERNIER

THREADED BUSHING

DUPLEX

GRAVITY (crane switches) — Slack in cable allows counterweight to fall — Unloading force

INFINITE-POSITION WORM

FLEXIBLE ROD — Spring

BELLCRANK

DIRECT-ACTING (hatchway) — Mechanical interlock — Contact

Latching switch with contact chamber

Actuator

Latch

Contact link

MECHANICAL HALF

Barrier

ELECTRICAL HALF

Contact arm

Geared rotary limit switches

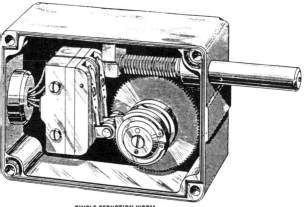

SINGLE-REDUCTION WORM

adjustment
tion

ing

Planet gears

t shaft

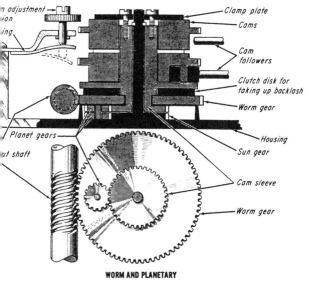

Clamp plate

Cams

Cam followers

Clutch disk for taking up backlash

Worm gear

Housing

Sun gear

Cam sleeve

Worm gear

WORM AND PLANETARY

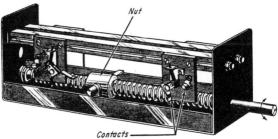

Nut

Contacts

TRAVELING NUT

Rotary-cam limit switches

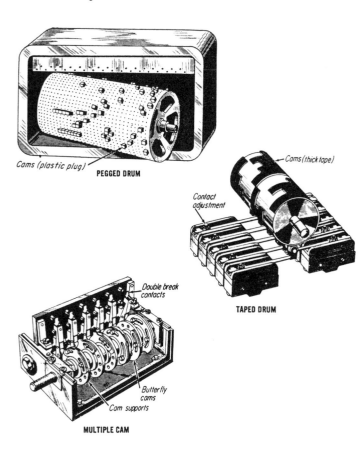

Cams (plastic plug)

PEGGED DRUM

Cams (thick tape)

Contact adjustment

TAPED DRUM

Double break contacts

Butterfly cams

Cam supports

MULTIPLE CAM

Adjust

Adjust

Switch

Switch

Cam

Cam

MICROMETER ADJUSTMENTS

Limit switches, used to confine or restrain the travel or rotation of moving parts within certain predetermined points, are actuated by varying methods. Some of these, such as cams, rollers, push-rods, and traveling nuts, are described and illustrated below.

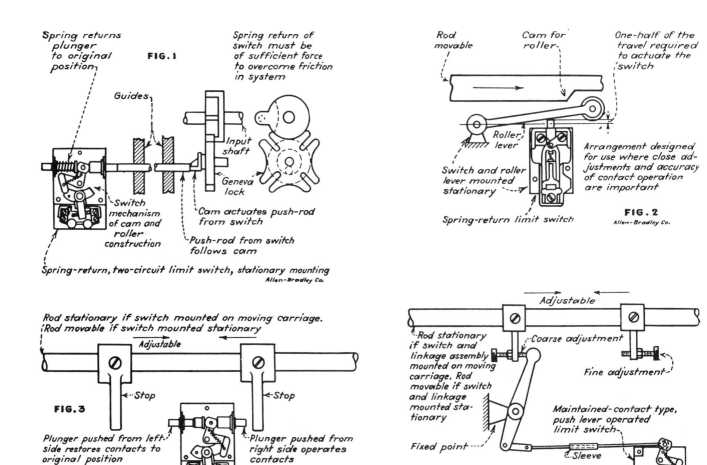

FIG. 1

Spring returns plunger to original position

Guides

Switch mechanism of cam and roller construction

Spring-return, two-circuit limit switch, stationary mounting
Allen-Bradley Co.

Input shaft

Geneva lock

Cam actuates push-rod from switch

Push-rod from switch follows cam

Spring return of switch must be of sufficient force to overcome friction in system

Rod movable

Cam for roller

One-half of the travel required to actuate the switch

Roller lever

Switch and roller lever mounted stationary

Spring-return limit switch

Arrangement designed for use where close adjustments and accuracy of contact operation are important

FIG. 2
Allen-Bradley Co.

Rod stationary if switch mounted on moving carriage. Rod movable if switch mounted stationary

Adjustable

FIG. 3

Stop

Stop

Plunger pushed from left side restores contacts to original position

Plunger pushed from right side operates contacts

Two-circuit, maintained-contact type limit switch mounted stationary, or mounted on moving carriage
Allen-Bradley Co.

Adjustable

Rod stationary if switch and linkage assembly mounted on moving carriage. Rod movable if switch and linkage mounted stationary

Coarse adjustment

Fine adjustment

Maintained-contact type, push lever operated limit switch

Fixed point

Sleeve

Switch and accompanying linkage mounted stationary, or mounted on moving carriage

FIG. 4
Allen-Bradley Co.

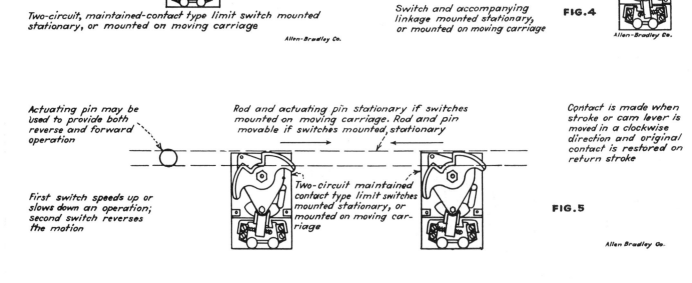

Actuating pin may be used to provide both reverse and forward operation

First switch speeds up or slows down an operation; second switch reverses the motion

Rod and actuating pin stationary if switches mounted on moving carriage. Rod and pin movable if switches mounted stationary

Two-circuit maintained contact type limit switches mounted stationary, or mounted on moving carriage

Contact is made when stroke or cam lever is moved in a clockwise direction and original contact is restored on return stroke

FIG. 5

Allen Bradley Co.

MACHINERY MECHANISMS

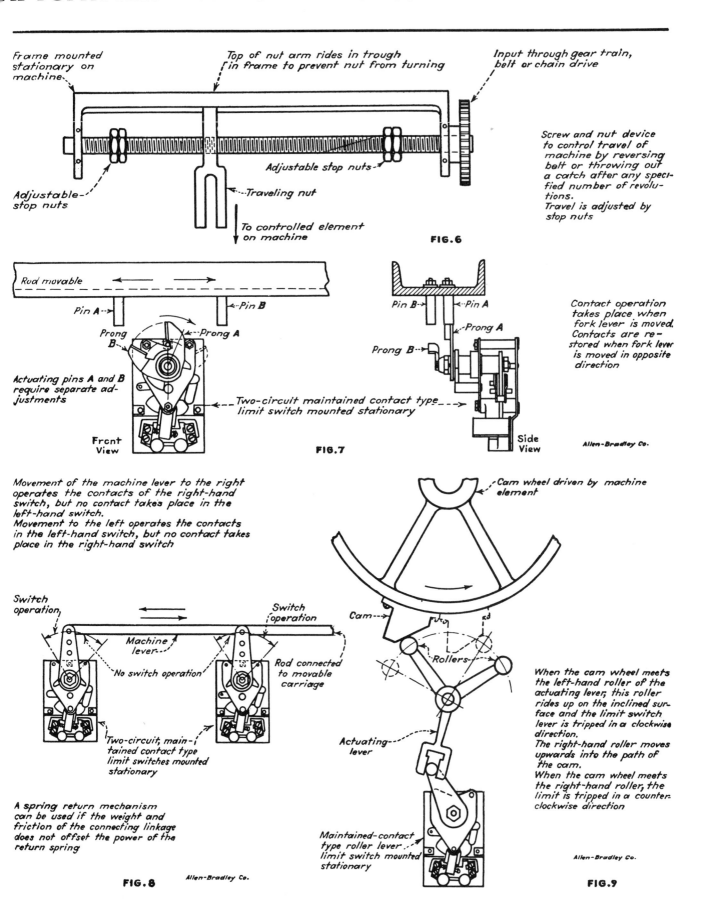

Frame mounted stationary on machine

Top of nut arm rides in trough in frame to prevent nut from turning

Input through gear train, belt or chain drive

Adjustable stop nuts

Adjustable stop nuts

Traveling nut

To controlled element on machine

Screw and nut device to control travel of machine by reversing belt or throwing out a catch after any specified number of revolutions.
Travel is adjusted by stop nuts

FIG.6

Rod movable

Pin A

Pin B

Prong B

Prong A

Actuating pins A and B require separate adjustments

Two-circuit maintained contact type limit switch mounted stationary

Front View

Pin B

Pin A

Prong A

Prong B

Contact operation takes place when fork lever is moved. Contacts are restored when fork lever is moved in opposite direction

Side View

Allen-Bradley Co.

FIG.7

Movement of the machine lever to the right operates the contacts of the right-hand switch, but no contact takes place in the left-hand switch.
Movement to the left operates the contacts in the left-hand switch, but no contact takes place in the right-hand switch

Cam wheel driven by machine element

Switch operation

Switch operation

Machine lever

No switch operation

Rod connected to movable carriage

Two-circuit, maintained contact type limit switches mounted stationary

A spring return mechanism can be used if the weight and friction of the connecting linkage does not offset the power of the return spring

Cam

Rollers

Actuating lever

Maintained-contact type roller lever limit switch mounted stationary

When the cam wheel meets the left-hand roller of the actuating lever, this roller rides up on the inclined surface and the limit switch lever is tripped in a clockwise direction.
The right-hand roller moves upwards into the path of the cam.
When the cam wheel meets the right-hand roller, the limit is tripped in a counterclockwise direction

Allen-Bradley Co.

FIG.8

Allen-Bradley Co.

FIG.9

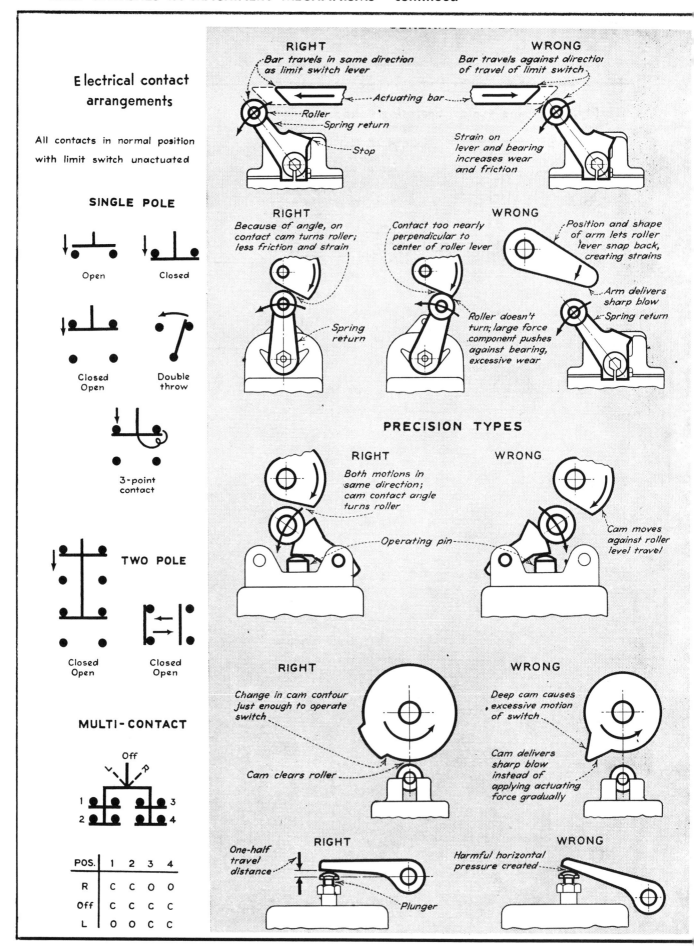

Electrical contact arrangements

All contacts in normal position with limit switch unactuated

SINGLE POLE

Open
Closed
Closed Open
Double throw
3-point contact

TWO POLE

Closed Open
Closed Open

MULTI-CONTACT

Off

1 2 3 4

POS.	1	2	3	4
R	C	C	O	O
Off	C	C	C	C
L	O	O	C	C

RIGHT
Bar travels in same direction as limit switch lever
Actuating bar
Roller
Spring return
Stop

WRONG
Bar travels against direction of travel of limit switch
Strain on lever and bearing increases wear and friction

RIGHT
Because of angle, on contact cam turns roller; less friction and strain
Spring return

WRONG
Contact too nearly perpendicular to center of roller lever
Roller doesn't turn; large force component pushes against bearing, excessive wear
Position and shape of arm lets roller lever snap back, creating strains
Arm delivers sharp blow
Spring return

PRECISION TYPES

RIGHT
Both motions in same direction; cam contact angle turns roller
Operating pin

WRONG
Cam moves against roller level travel

RIGHT
Change in cam contour just enough to operate switch
Cam clears roller

WRONG
Deep cam causes excessive motion of switch
Cam delivers sharp blow instead of applying actuating force gradually

RIGHT
One-half travel distance
Plunger

WRONG
Harmful horizontal pressure created

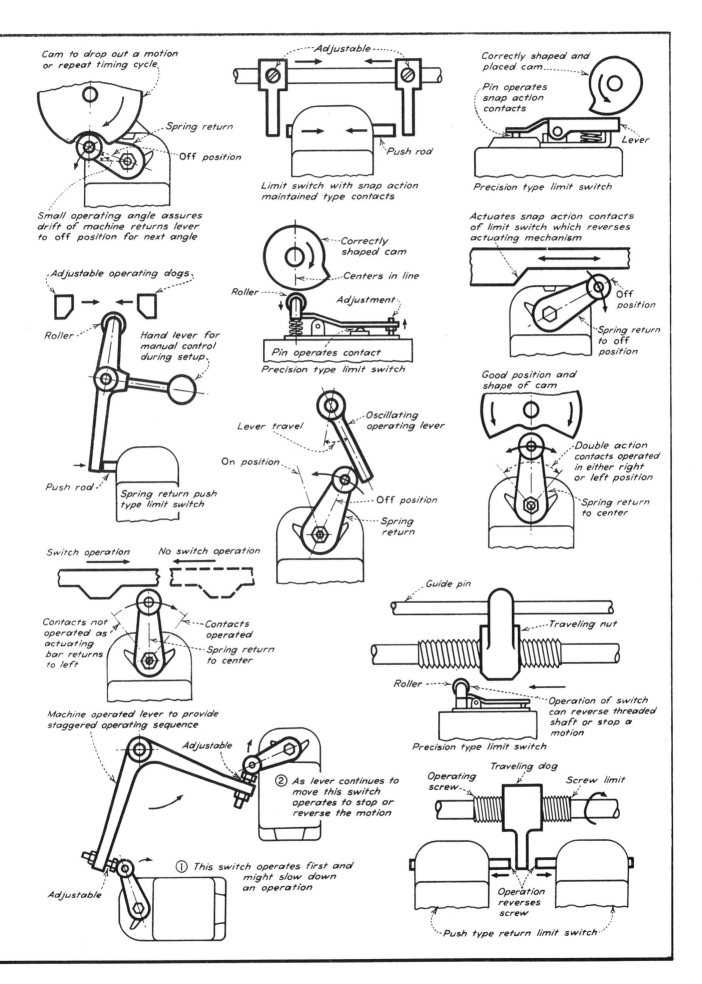

Cam to drop out a motion or repeat timing cycle

Spring return

Off position

Small operating angle assures drift of machine returns lever to off position for next angle

Adjustable

Push rod

Limit switch with snap action maintained type contacts

Correctly shaped and placed cam

Pin operates snap action contacts

Lever

Precision type limit switch

Adjustable operating dogs

Roller

Hand lever for manual control during setup

Push rod

Spring return push type limit switch

Correctly shaped cam

Centers in line

Roller

Adjustment

Pin operates contact

Precision type limit switch

Actuates snap action contacts of limit switch which reverses actuating mechanism

Off position

Spring return to off position

Good position and shape of cam

Double action contacts operated in either right or left position

Spring return to center

Lever travel

Oscillating operating lever

On position

Off position

Spring return

Switch operation

No switch operation

Contacts not operated as actuating bar returns to left

Contacts operated

Spring return to center

Guide pin

Traveling nut

Roller

Operation of switch can reverse threaded shaft or stop a motion

Precision type limit switch

Machine operated lever to provide staggered operating sequence

Adjustable

② As lever continues to move this switch operates to stop or reverse the motion

① This switch operates first and might slow down an operation

Adjustable

Operating screw

Traveling dog

Screw limit

Operation reverses screw

Push type return limit switch

357

Speed governors, designed to maintain speeds of machines within reasonably constant limits irrespective of loads, may depend for their action upon centrifugal force or cam linkages. Other types may utilize pressure differentials and fluid velocities as their actuating media. Examples of these governing devices are illustrated.

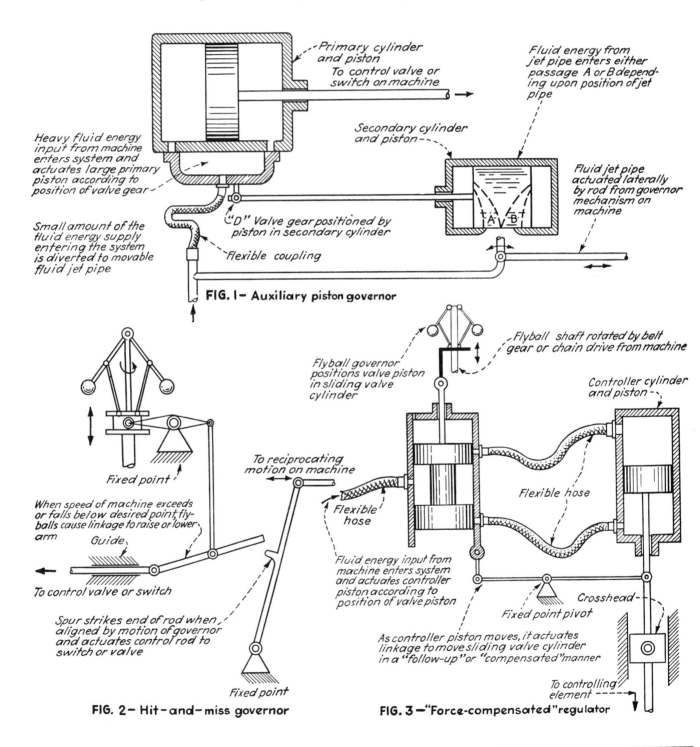

Primary cylinder and piston
To control valve or switch on machine

Fluid energy from jet pipe enters either passage A or B depending upon position of jet pipe

Heavy fluid energy input from machine enters system and actuates large primary piston according to position of valve gear

Secondary cylinder and piston

Fluid jet pipe actuated laterally by rod from governor mechanism on machine

Small amount of the fluid energy supply entering the system is diverted to movable fluid jet pipe

"D" Valve gear positioned by piston in secondary cylinder

Flexible coupling

FIG. I– Auxiliary piston governor

Flyball governor positions valve piston in sliding valve cylinder

Flyball shaft rotated by belt gear or chain drive from machine

Controller cylinder and piston

Fixed point

To reciprocating motion on machine

Flexible hose

When speed of machine exceeds or falls below desired point, flyballs cause linkage to raise or lower arm

Guide

Flexible hose

To control valve or switch

Fluid energy input from machine enters system and actuates controller piston according to position of valve piston

Spur strikes end of rod when aligned by motion of governor and actuates control rod to switch or valve

As controller piston moves, it actuates linkage to move sliding valve cylinder in a "follow-up" or "compensated" manner

Crosshead

Fixed point pivot

Fixed point

To controlling element

FIG. 2– Hit-and-miss governor

FIG. 3–"Force-compensated" regulator

AUTOMATICALLY GOVERNING SPEED

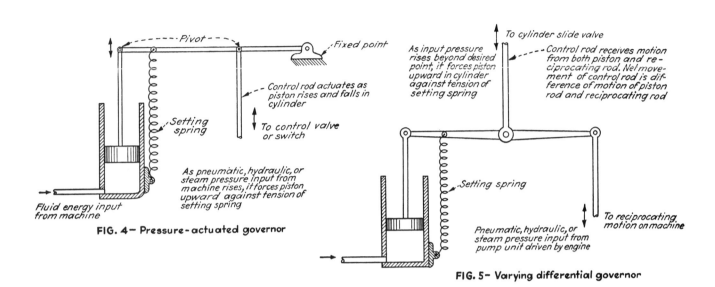

As pneumatic, hydraulic, or steam pressure input from machine rises, it forces piston upward against tension of setting spring

Pivot

Fixed point

Control rod actuates as piston rises and falls in cylinder

Setting spring

To control valve or switch

Fluid energy input from machine

FIG. 4— Pressure-actuated governor

To cylinder slide valve

As input pressure rises beyond desired point, it forces piston upward in cylinder against tension of setting spring

Control rod receives motion from both piston and reciprocating rod. Net movement of control rod is difference of motion of piston rod and reciprocating rod

Setting spring

Pneumatic, hydraulic, or steam pressure input from pump unit driven by engine

To reciprocating motion on machine

FIG. 5— Varying differential governor

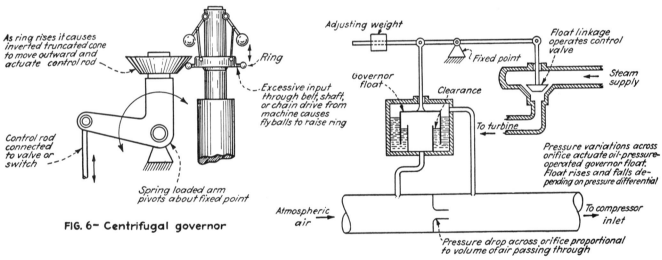

As ring rises it causes inverted truncated cone to move outward and actuate control rod

Ring

Excessive input through belt, shaft, or chain drive from machine causes flyballs to raise ring

Control rod connected to valve or switch

Spring loaded arm pivots about fixed point

FIG. 6— Centrifugal governor

Adjusting weight

Fixed point

Float linkage operates control valve

Governor float

Clearance

Steam supply

To turbine

Pressure variations across orifice actuate oil-pressure-operated governor float. Float rises and falls depending on pressure differential

Atmospheric air

To compressor inlet

Pressure drop across orifice proportional to volume of air passing through

FIG. 7— Constant volume governor

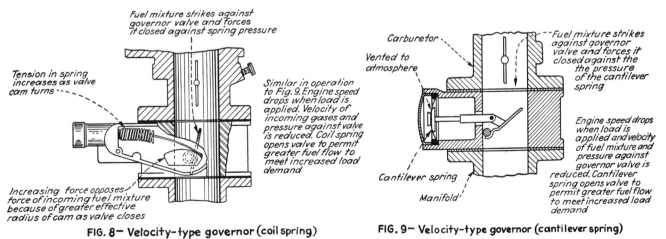

Fuel mixture strikes against governor valve and forces it closed against spring pressure

Tension in spring increases as valve cam turns

Similar in operation to Fig. 9. Engine speed drops when load is applied. Velocity of incoming gases and pressure against valve is reduced. Coil spring opens valve to permit greater fuel flow to meet increased load demand

Increasing force opposes force of incoming fuel mixture because of greater effective radius of cam as valve closes

FIG. 8— Velocity-type governor (coil spring)

Carburetor

Vented to atmosphere

Fuel mixture strikes against governor valve, it forces it closed against the the pressure of the cantilever spring

Cantilever spring

Manifold

Engine speed drops when load is applied and velocity of fuel mixture and pressure against governor valve is reduced. Cantilever spring opens valve to permit greater fuel flow to meet increased load demand

FIG. 9— Velocity-type governor (cantilever spring)

Centrifugal, pneumatic, hydraulic and electric governors

BERYL A. BOGGS

Centrifugal sensors are the most common—they are simple and sensitive and have high output force. There is more published information on centrifugal governors than on all other types combined.

In operation, centrifugal flyweights develop a force proportional to the square of the speed, modified by linkages as required. In small engines the flyweight movement can actuate the fuel throttle directly. Larger engines require amplifiers or relays, which gives rise to innumerable combinations of pilot pistons, linear actuators, dashpots, compensators, and gear boxes.

Centrifugal governors

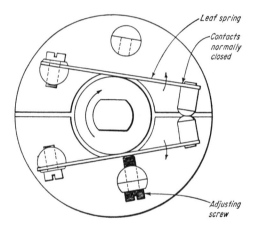

ACCELERATION GOVERNOR
(steam engine)

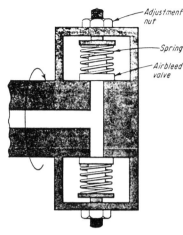

CENTRIFUGAL VALVE

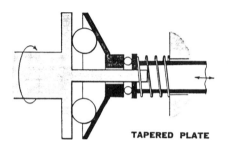

TAPERED PLATE

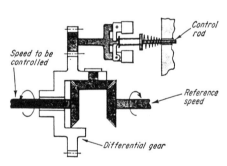

DIFFERENTIAL CENTRIFUGAL

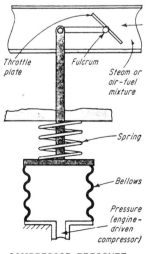

CENTRIFUGAL CONTACTS

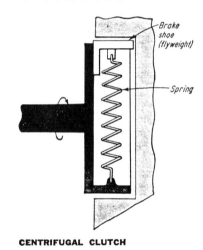

CENTRIFUGAL CLUTCH

Pneumatic governors

Pneumatic sensing devices are the most inexpensive, and also the most inaccurate, of all speed-measuring and governing methods, yet they are entirely adequate for many applications. The pressure or velocity of cooling or combustion air is used to measure and govern the speed of the engine.

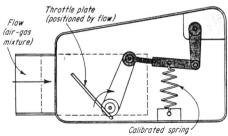

CARBURETOR-FLOW VELOCITY
(linkage)

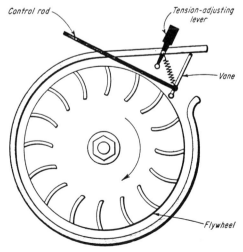

COMPRESSOR PRESSURE
(direct)

FAN-FLOW VELOCITY

More pneumatic governors

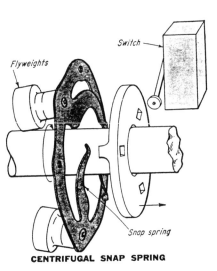

CENTRIFUGAL SNAP SPRING

Flyweights
Switch
Snap spring

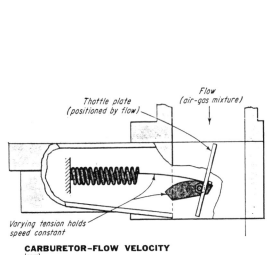

CARBURETOR–FLOW VELOCITY
(cam)

Thottle plate
(positioned by flow)
Flow
(air-gas mixture)
Varying tension holds
speed constant

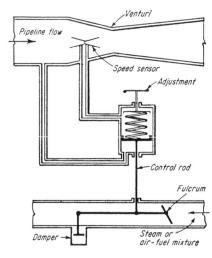

COMPRESSOR PRESSURE
(differential)

Venturi
Pipeline flow
Speed sensor
Adjustment
Control rod
Fulcrum
Damper
Steam or
air-fuel mixture

Hydraulic governors

Hydraulic sensors

These measure discharge pressure of an engine-driven pump. Pressure is proportional to the square of the speed of the pump in most designs, although there are special impellers with linear pressure-speed characteristics.

Straight vanes are better than curved vanes because the pressure is less affected by the volume flow. Low pressures are preferred over high because fluid friction is less.

Typical applications include farm tractors using diesel or gas engines, larger diesel engines, and small steam turbines. Hydraulic-governor sensing using other than pressure has had limited application and success.

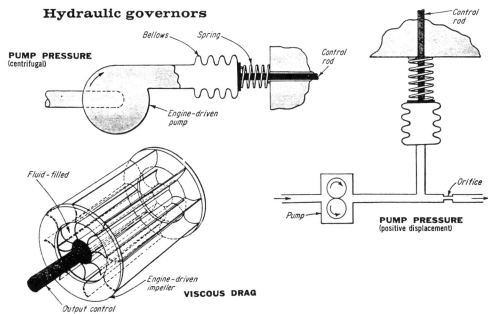

PUMP PRESSURE
(centrifugal)

Bellows
Spring
Control rod
Engine-driven pump

VISCOUS DRAG

Fluid-filled
Output control
Engine-driven impeller

PUMP PRESSURE
(positive displacement)

Control rod
Pump
Orifice

Electric governors

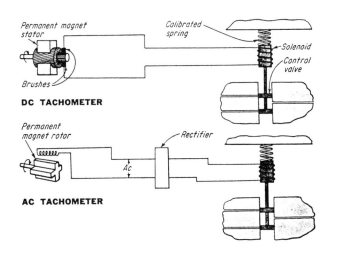

DC TACHOMETER

Permanent magnet stator
Brushes
Calibrated spring
Solenoid
Control valve

AC TACHOMETER

Permanent magnet rotor
Rectifier
Ac

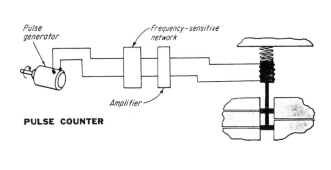

PULSE COUNTER

Pulse generator
Frequency-sensitive network
Amplifier

Speed control for small mechanisms

Friction devices, actuated by centrifugal force, automatically keep speed constant regardless of variation of load or driving force.

FEDERICO STRASSE

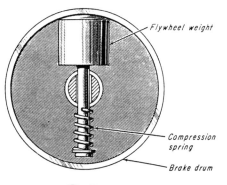

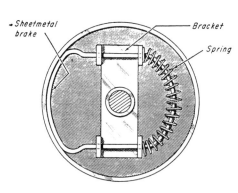

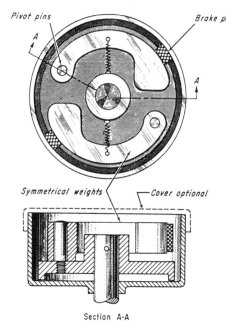

Section A-A

WEIGHT counterbalanced by a spring brakes the shaft when rotation speed becomes too high. Braking area is small.

SHEETMETAL BRAKE provides larger braking area than previous design. Operation is thus more even and cooler.

SYMMETRICAL WEIGHTS give even braking action when they pivot outward. Entire action can be enclosed.

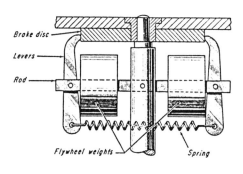

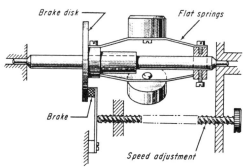

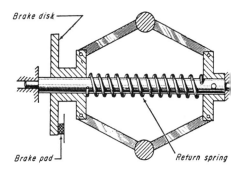

WEIGHT-ACTUATED LEVERS make this arrangement suitable where high braking moments are required.

THREE FLAT SPRINGS carry weights that provide brake force upon rotation. Device can be provided with adjustment.

TYPICAL GOVERNOR action of swinging weights is utilized here. As in the previous device, adjustment is optional.

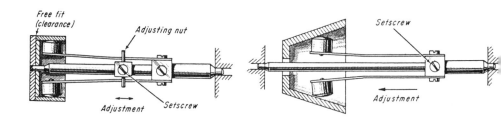

ADJUSTMENT of speed at which this device starts to brake is quick and easy. Adjusting nut is locked in place with setscrew.

TAPERED BRAKE DRUM is another way of providing for varying speed control. The adjustment is again locked

Four designs for mechanical timing
Mechanical timer for short-cycle operation

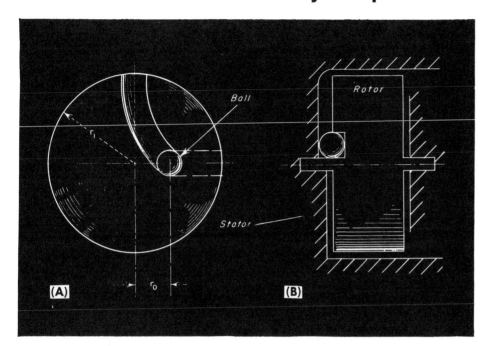

Fig. 1. Three basic components of this timing device are a stator with a radial groove, a rotor with a spiral groove, and a ball that rides in these grooves. When the mechanism rotates, the ball tends to move radially in the stator groove because of centrifugal force. As it does so, it is constrained by the spiral groove and thereby exerts a driving force on the rotor. With a constant rate spiral, curves in Fig. 2 give the time characteristics plotted against angular velocity.

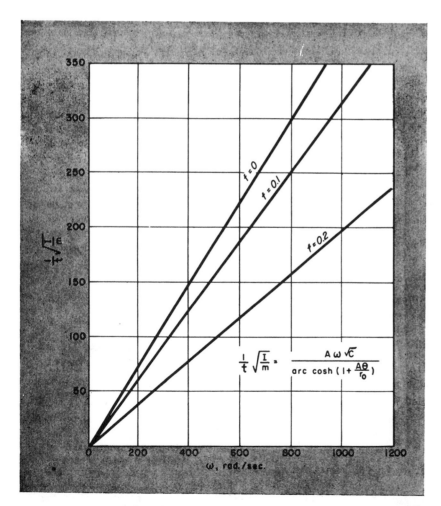

Fig. 2— Speed-time relationship plotted for three values of the coefficient of friction. Knowing ω, $\frac{1}{t}\sqrt{I/m}$ can be read from the appropriate f curve. Then, numerical values of I and m are chosen to get the desired value of time delay.

THIS MECHANISM, shown in Fig. 1(A) and (B), is a simple means of obtaining a timing cycle inversely proportional to the rotational velocity of a member. The principal advantage of this unit is that the cycle is completed with the same angular displacement of the stator housing regardless of the value of the angular velocity of the housing. Thus, such a mechanism can be used to meter the distance travelled by a rotating body. At speeds of 1,000 to 100,000 rpm, it has a practical timing range of approximately 1/1,000 to 1.0 sec.

$$\frac{1}{t}\sqrt{\frac{I}{M}} = \frac{A\,\omega\,\sqrt{C}}{\operatorname{arc\,cosh}\left(1 + \frac{A\theta}{r_0}\right)}$$

where

$$C = \frac{\left(1 - \dfrac{f}{\tan\phi}\right)}{1 + 2f\tan\phi}$$

$$A = \frac{r_1 - r_0}{\theta}$$

f—Coefficient of friction for ball in rotor-stator slot
F—Driving ball centrifugal force
I—Moment of inertia of rotor
m—Mass of driving ball
r_0—Initial spiral radius
r_1—Final spiral radius
t—Time
ϕ—Spiral angle
θ—Rotor face angle
ω—Angular velocity of mechanism

continued, next page

Cam-operated mechanical timer

2-PART CAM is heart of this timer. Lower disk can move through a short arc relative to the upper disk which is attached to the shaft, mainspring and friction clutch. When turned, the upper disk lifts the follower out of the slot, then the cam pin picks up the lower disk and carries it around with it. The upper disk is rotated for the desired time interval. When released the dual cams return to their original relative position and present a sharp-edged slot for the follower to snap into.

The **time interval** possible with mechanical timers is very flexible; it can be a few seconds or up to 12 hours. Most commercial applications have a range of either a few minutes or a few hours with any desired adjustability. Time intervals of a few seconds are achieved by attaching the cam mechanism to the "seconds" (high-speed) gear in the clock mechanism, which is omitted in timers that mark only minutes and hours. Similarly, intervals of more than 12 hours are clocked by attaching the cam mechanism to a still slower gear in the clock mechanism.

Accuracy is a function of workmanship and comparable to that of any watch mechanism. For most commercial applications, accuracy ranges from 0.5 to 5% of the time interval.

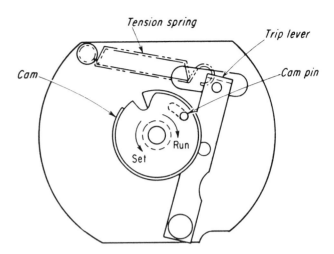

Automatic recycling of a mechanical timer is possible, but generally other timing devices are more suitable where manual resetting is objectionable. This is particularly true where programming of multiple automatic operations is needed. Mechanical timers may be reset by a solenoid. This can be just as effective as manual resetting, but may not be as economical as using timers more readily adapted to recycling.

Double-operation timers

Mechanical timers of commercial quality can also be adapted to timing of two sequential operations in one cycle. Usually this is done by arranging two cams in series with simple friction clutches, or a special gear segment and cam arrangement. Timing of the first operation is determined by the angular displacement of the gear segment. The second operation is performed as fast as the inertia of the load and the remaining energy in the spring will permit.

Sometimes, this second operation is used to trigger an auxiliary spring or mechanism, as in photo below. The multiple-operation timer that releases spring energy in more than two steps is generally considered impractical for conventional timing jobs.

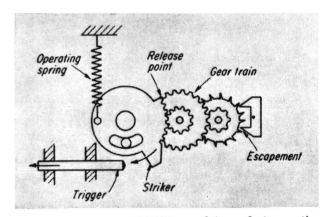

DOUBLE-OPERATION TIMER regulates a first operation for a predetermined time interval until striker contacts trigger. Then trigger is pushed with remaining spring energy.

Varying mechanical timer

mounted on the fuel-injector drive has jaws separated by a pair of centrifugal weights. These move outward as engine speed increases, forcing the jaws apart against the clutch springs. Result is the driven member of the timer is advanced in relation to driving member (A).

The drive jaws are bolted to a chain-drive sprocket that operates the timer. The driven jaws (C) are bolted to the end of a camshaft that operates individual injection pumps. Torque is transmitted through the jaws by the flyweights inserted between them. Since the jaws are bolted to their backing plates they cannot move outward but have to rotate with respect to each other. This produces the required fuel-injection advance as engine speed increases.

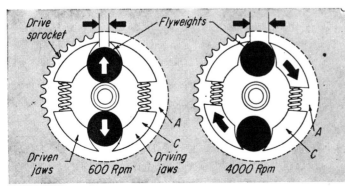

CHAPTER 9.

NONMECHANICAL METHODS OF MACHINE AND MECHANISM CONTROL

Providing freedom to move six ways

Expandable jacks give complete control of a platform's velocity, acceleration, location. Linkage is useful for simulators, maybe tools

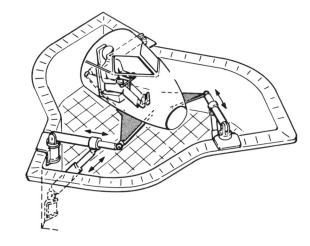

With modern computer control, it is no more complex to adjust the position and extension of three telescoping legs of a platform than to actuate the crank of the old work horse among mechanisms: the four-bar linkage. Now a British engineer is taking advantage of this flexibility in his design of a fundamentally new kind of linkage.

D. E. Stewart, at Elliott-Automation's Space Development & Weapons Research Laboratory in London, was designing a helicopter training simulator when he hit on his new concept. But he feels the linkage has immediate application in simulating the accelerations and forces that affect an airplane, ship, or spacecraft in motion. He sees it also as the basis of a whole new class of machine tools and automatic assembly equipment, medical instrumentation, and tunneling and earth-moving machines.

Stewart claims his eight-bar spatial linkage (six legs, base, and platform—see pictures) will enable a platform to move simultaneously in all six degrees of freedom, with greater amplitude than any other known system permits. (The six degrees comprise three linear movements—vertical, lateral, and longitudinal—and three angular movements—pitch, roll, and yaw.)

Jacks and joints. Stewart's basic design employs a triangular platform connected at each corner to a leg through a three-axis joint (drawing). Each leg is connected to the ground mounting by a two-axis joint, one axis normal to the leg and controlled and the other normal to the first but not controlled. Each leg is also controllable for length.

In another form (pictures), each leg consists of two hydraulic jacks or ball-screws. The outer end of the piston rod of one jack is connected to the platform by a three-axis joint, and the cylinder end is connected to the foundation by a two-axis joint. The second jack in each leg is connected by a single-axis joint to the end of the cylinder of the first jack, and the other end by a two-axis joint to the founda-

Coordinated six-leg system . . .

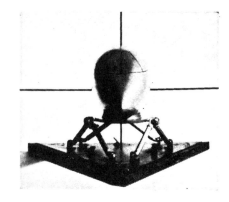

provides perfect vertical motion . . .

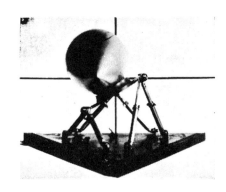

. . . or more complex pitch and roll

Control equations. Displacements fx, fy, fz of the corners of the equilateral-triangle platform in terms of three angular motions, θ, ϕ and ψ:

L = length of leg
l = initial length of leg (constant)
S = altitude of equilateral triangle
$X = x + fx$
$Y = y + fy$
$Z = z + fz$

For leg 1

$$L_1 = [l^2 + 2lX + X^2 + Y^2 + Z^2]^{1/2}$$
$$fx_1 = +S \sin \psi \cos \theta$$
$$fy_1 = +S(1 - \cos \psi \cos \theta)$$
$$fz_1 = -S \sin \theta$$

For leg 2

$$L_2 = [l^2 - lX - 1.73lY + X^2 + Y^2 + Z^2]^{1/2}$$
$$fx_2 = +0.577S(1 - \cos \psi \cos \phi + \sin \phi \sin \theta \sin \psi)$$
$$fy_2 = -0.577S(\sin \psi \cos \phi + \sin \phi \sin \theta \cos \psi)$$
$$fz_2 = +0.577S \sin \phi \cos \theta$$

For leg 3

$$L_3 = [l^2 - lX + 1.73lY + X^2 + Y^2 + Z^2]^{1/2}$$
$$fx_3 = -0.577S(1 - \cos \psi \cos \phi + \sin \phi \sin \theta \sin \psi)$$
$$fy_3 = +0.577S(\sin \psi \cos \phi + \sin \phi \sin \theta \cos \psi)$$
$$fz_3 = -0.577S \sin \phi \cos \theta$$

How joints, legs, and platform are arranged in typical application

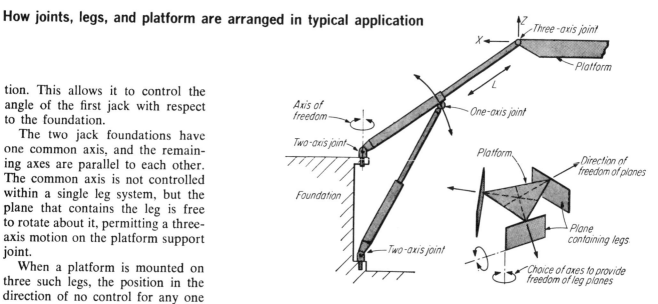

tion. This allows it to control the angle of the first jack with respect to the foundation.

The two jack foundations have one common axis, and the remaining axes are parallel to each other. The common axis is not controlled within a single leg system, but the plane that contains the leg is free to rotate about it, permitting a three-axis motion on the platform support joint.

When a platform is mounted on three such legs, the position in the direction of no control for any one leg is determined by the controlled position of the other two. Each leg defines the position of its platform point in two dimensions; the three legs define three platform points in three dimensions. If the three points are moved similarly in the x-y-z coordinates, the three linear motions are obtained; if they are moved differentially in the same coordinates, the three angular motions are produced. The smaller the platform relative to the leg stroke, the larger the angular motions will be.

Control system. The new linkage offers structural rigidity and flexibility of response that makes up for failure to reach the ideal of using

one motor per motion. In fact, control of the new system is comparatively simple. By giving signals to the various jacks, an engineer can operate the platform from a predetermined program involving both linear and angular accelerations.

Equations for the related movements have been worked out (at left) and stored in a computer, which translates any instruction into the component movements of the various jacks. Each jack feeds data back to the computer on its position and velocity, completing the servo loop.

Stewart claims that this control is unique in permitting the mecha-

nism to truly simulate any accelerating body in any direction and attitude. A machine is not tied to earth-fixed axes but can be controlled in any coordinate system—earth-based, body-based, or any combination of the two.

Stewart's prototype is driven by hydraulic jacks, but he says these could be replaced by screw jacks to give a longer stroke for a given size. Also, rotary actuators or electric motors could be substituted for the jacks that control the leg angles, reducing the number of foundation fixings but subjecting the remaining jack in each leg to a greater bending moment. □

Hydrostatic drive teams with swashplate in all-terrain vehicle

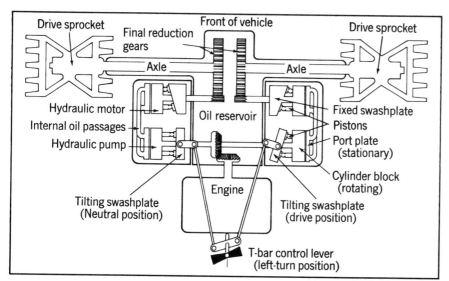

Trackster's remarkable hill-climbing ability and controlled descent of slopes is credited to its dual hydrostatic drive, blending power with fingertip control. The always-engaged drive keeps power on both tracks at all times on a straight course.

The drive comprises two independent hydrostatic transmissions, each consisting of a hydraulic motor and a pump. These units are plugged into the center axle, meshing with the driver gear and with the final drive gear, which gives a 6:1 reduction through spur gears to the drive axles. Maximum input speed of the hydrostatics is 4200 rpm.

To control these transmissions, movement of the T-bar tiller tilts the swash plate of each hydraulic pump. If the T-bar is pushed straight forward, both swash plates will rotate at the same angle, the pumps will produce the same output, and both transmissions will turn at the same speed. When the T-bar is turned, one swash plate will tilt more than the other, increasing the speed of that transmission and track.

Mechanisms Actuated by Air or

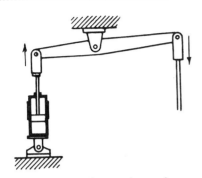

Fig. 1—Cylinder can be used with a first class lever.

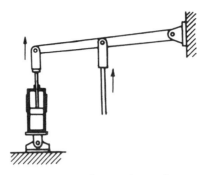

Fig. 2—Cylinder can be used with a second class lever.

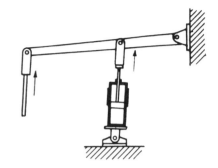

Fig. 3—Cylinder can be used with a third class lever.

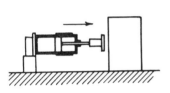

Fig. 4—Cylinder can be linked up directly to the load.

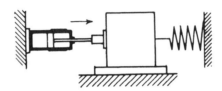

Fig. 5—Spring reduces the thrust at the end of the stroke.

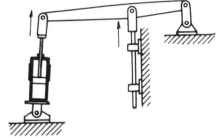

Fig. 6—Point of application of force follows the direction of thrust.

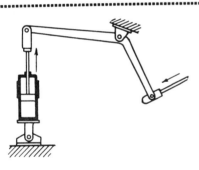

Fig. 7—Cylinder can be used with a bent lever.

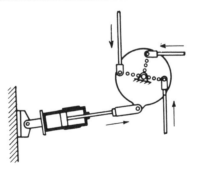

Fig. 8—Cylinder can be used with a trammel plate.

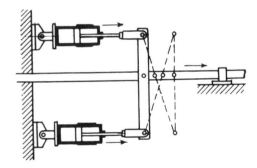

Fig. 9—Two pistons with fixed strokes position load in any of four stations.

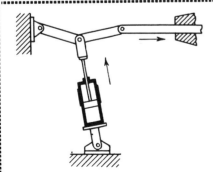

Fig. 10—A toggle can be actuated by the cylinder.

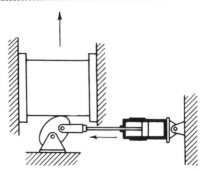

Fig. 11—The cam supports the load after completion of the stroke.

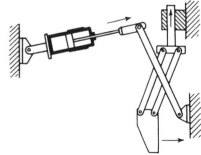

Fig. 12—Simultaneous thrusts in two different directions are obtained.

Hydraulic Cylinders

Acknowledgment is made to Adel Precision Product Corporation, Blackhawk Manufacturing Company, Hydraulic Equipment Company, Mead Specialties Company, Westinghouse Air Brake Co., and especially to Hanna Engineering Works.

(Note: In place of cylinders, electrically powered thrust units or solenoids can be used.)

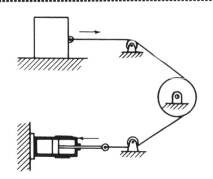

Fig. 13—Force is transmitted by a cable.

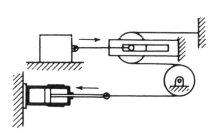

Fig. 14—Force can be modified by a system of pulleys.

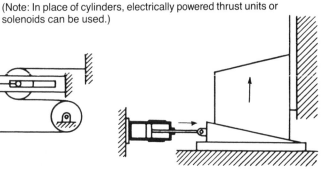

Fig. 15—Force can be modified by wedges.

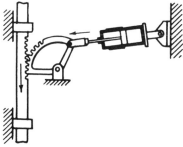

Fig. 16—Gear sector moves rack perpendicular to stroke of piston.

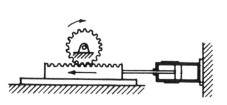

Fig. 17—Rack turns gear sector.

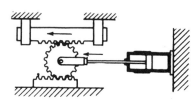

Fig. 18—Motion of movable rack is twice that of piston.

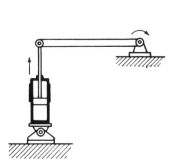

Fig. 19—Torque applied to the shaft can be transmitted to a distant point.

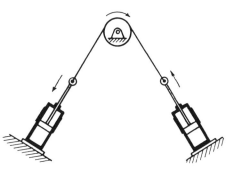

Fig. 20—Torque can also be applied to a shaft by a belt and pulley.

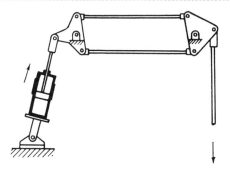

Fig. 21—Motion is transmitted to a distant point in the plane of motion.

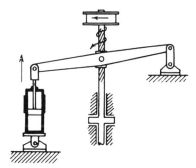

Fig. 22—A steep screw nut produces a rotation of shaft.

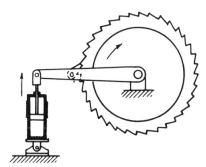

Fig. 23—Single sprocket wheel produces rotation in the plane of motion.

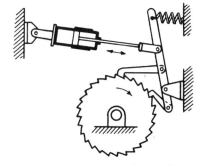

Fig. 24—Double sprocket wheel makes the rotation more nearly continuous.

MORE APPLICATIONS OF FLUID POWER

LIFT TRUCK STEERS HYDRAULICALLY

Hydraulically steered rear wheels are featured on a 35,000-lb-capacity fork truck designed for heavy outdoor lifting and tiering work by the Industrial Truck Division of Clark Equipment Co., Battle Creek, Mich. The power steering made it feasible to place driver and controls over fender of left front wheel. This position gives the driver increased visibility when raising and placing loads.

Extended pitman arm shaft . . .

offsets the steering column from the centrally mounted pitman arm and drag links which carry steering action to rear of truck. Control valve is mounted in the final drag link and is spring centered. Dual hydraulic cylinders equalize turning effort in each direction, make possible a 70° turning angle and prevent mechanism from locking in either extreme position. Truck is called the CY-350.

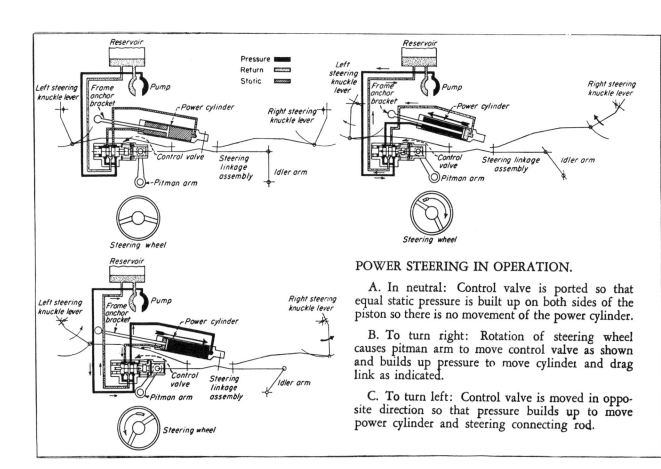

POWER STEERING IN OPERATION.

A. In neutral: Control valve is ported so that equal static pressure is built up on both sides of the piston so there is no movement of the power cylinder.

B. To turn right: Rotation of steering wheel causes pitman arm to move control valve as shown and builds up pressure to move cylinder and drag link as indicated.

C. To turn left: Control valve is moved in opposite direction so that pressure builds up to move power cylinder and steering connecting rod.

SELF-POWERED HATCH COVER

Lewis Welding & Engineering Corp,

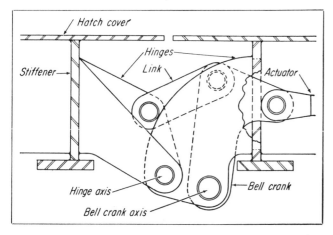

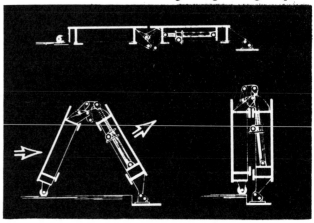

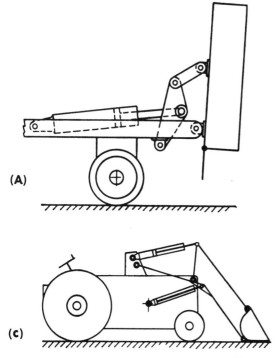

A unique hinge linkage actuated by a screw jack makes it possible for a single operator to open or close holds—in minutes

When the drive motor is energized the actuators extend, pivoting the bellcrank counterclockwise. Connected to the outer panel by the link, the bellcrank forces the panels to pivot on the hinge axis. The panels, supported by rollers on their outer ends, break upward as the screw jacks continue to extend. Function of the bellcrank and link is to swing the outer panels 180 deg without imposing bending moments on the screw jack.

Drive. A single electric motor in each pair of panels is connected to the jacks by line shafts. V-belt, timing belt, or a right-angle gear reducer drive can be specified to connect the motor to the line shaft; with the right-angle reducer drive an opening permits the panels to be articulated with a portable air wrench in case of electric power failure. Nyon couplings permit the shafts to flex, during opening and closing, when the jack screws pivot. Maintenance-free, the nylon flex couplings were selected because of their resistance to the corrosive conditions met by marine equipment.

HYDRAULIC DUMPERS AND LOADERS

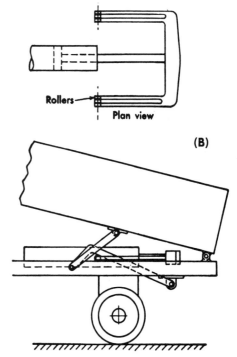

Dump motion in (A) is obtained by means of two four-bar linkages in series. Cam rollers in (B) help start heavy loads. Top cylinder in (C) lifts bucket while second dumps load. From J. S. Beggs, *Mechanisms*, McGraw-Hill Book Co.

Rotary-Pump Mechanisms SIGMUND RAPPAPORT

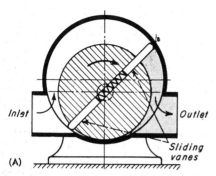

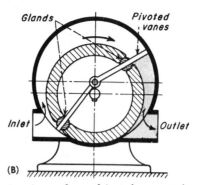

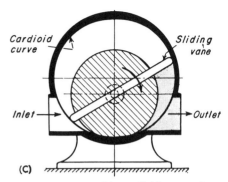

Fig. 1 — (A) Ramelli pump with spring-loaded vanes to insure contact with wall; vane ends rounded for line contact. (B) Two vanes pivot in

housing and are driven by eccentrically mounted disk; vanes slide in glands and are always radial to housing, thus providing surface contact.

(C) Housing with cardioid curve allows single vane to be used, because opposing points on housing in line with disk center are equidistant.

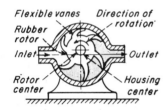

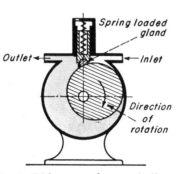

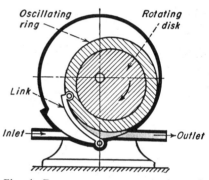

Fig. 2—Flexible vanes on eccentric rubber rotor displace liquid as in sliding-vane pumps. Instead of vanes sliding in and out, they bend against casing to pump.

Fig. 3—Disk mounted eccentrically on drive shaft displaces liquid in continuous flow. Spring-loaded gland separates inlet from outlet except when disk is at top of stroke.

Fig. 4—Rotary compressor pump has link separating suction and compression sides. Link is hinged to ring which oscillates while driven by disk. Oscillating action pumps liquid in continuous flow.

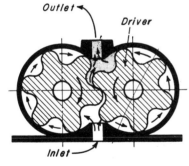

Fig. 5—Gear pump transports liquid between tooth spaces and housing wall. Circular tooth shape has only one tooth making contact and is more efficient than an involute shape which may enclose a pocket between two adjoining teeth, recirculating part of liquid. Helical teeth are also used.

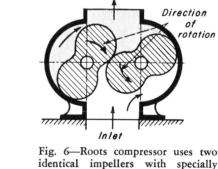

Fig. 6—Roots compressor uses two identical impellers with specially shaped teeth. Shafts connected by external gearing to insure constant contact between impellers.

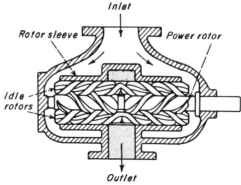

Fig. 7—Three-screw pump drives liquid between screw threads along axis of screws. Two idle rotors are driven by fluid pressure, not by metallic contact with power rotor.

Data based on material and sketches in AWF und VDMA Getriebleatter, published by Ausschuss fúer Getriebe beim Ausschuss fuer Wirtschaftliche Fertigung, Leipzig, Germany.

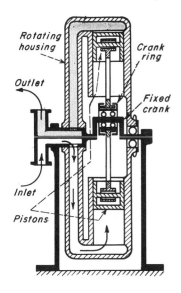

Fig. 8—Housing of Hele-Shaw-Beacham pump rotates round cranked shaft. Connecting rods attached to crank ring cause pistons to oscillate as housing rotates. No valves necessary since fixed hollow shaft, divided by wall, has suction and compression sides always in correct register with inlet and outlet ports.

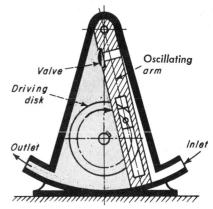

Fig. 9—Disk drives oscillating arm which acts as piston. Velocity of arm varies because of quick-return type mechanism. Liquid slowly sucked in and expelled during clockwise rotation of arm; return stroke transfers liquid rapidly.

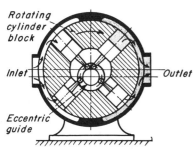

Fig. 10—Rotating cylinder block mounted concentrically in housing. Connecting-rod ends slide around eccentric guide as cylinders rotate and cause pistons to reciprocate. Housing divided into suction and compression compartments.

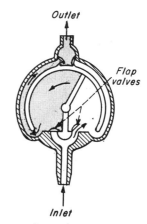

Fig. 11—Rotary-reciprocating pump usually operated manually to pump high-viscosity liquids such as oil,

Offset planetary gears induce rotary-pump action

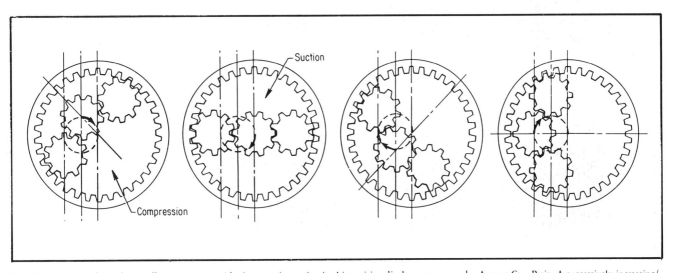

Two planetary gears driven by an offset sun gear provide the pumping action in this positive-displacement pump by Acrome Co., Paris. A successively increasing/ decreasing (suction/compression) is formed on either side of the sun and planet gears (from *Machine Design*).

NEW CONFIGURATIONS OF GEROTOR PUMPS AND HYDRAULIC MOTORS

The gerotor principle, which uses an inner rotor with one less tooth than the outer stator, is as old as the hydraulic industry. In the past, one obstacle to its widespread use has been the complexity of making the precision rotors and universal joints. Now, there is an exceptional amount of activity to modify gerotor designs to reduce cost, to tighten up clearances between meshing lobes, and to minimize wear (drawings right).

The system's attraction is that the eccentric orbit sucks in fluid and pumps it to discharge it. Furthermore, if the fluid is pumped into it, the rotor is made to orbit. A full orbit of the inner rotor results in relative rotary displacement of only one tooth between the rotor and the stator. Thus, there is a mechanical advantage of 6:1 for a 6-lobe/7-lobe set. There are thousands of applications in pumps, motors, and drives where this advantage is exploited.

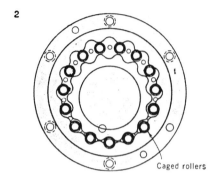

2

Caged rollers

Caged rollers seal between mating lobes of gerotor pump.

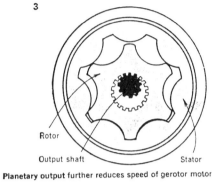

3

Rotor

Output shaft

Stator

Planetary output further reduces speed of gerotor motor

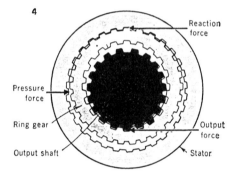

4

Reaction force

Pressure force

Ring gear

Output shaft

Output force

Stator

Multiple mesh gearing is both fluid motor and reducer.

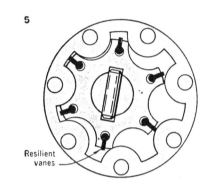

5

Resilient vanes

Resilient vanes press hardest at closest point of mesh.

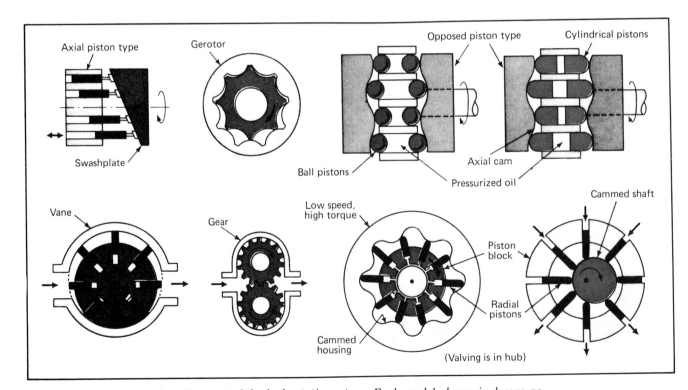

Eight basic concepts describe most of the hydrostatic motors. Each model shown is drawn as fixed displacement, but many of them are available with adustable displacement. (from *Design Engineering*)

374

Automatic positioning system . .

uses a direct-dialing analog programmer for point-to-point machine setup on short production runs in operations such as punching, drilling, inserting. System is said to be time-saving in programming for repetitive and symmetrical motion patterns. Variations can be made by resetting dials. Positioning is by electrohydraulic servo systems, capable of positioning speed of 0.2 sec per basic motion with positioning accuracy of 1 part in 1000. Modulated signal permits adjustment for longitudinal and transverse axis movements.

CDC Control Services, Inc.

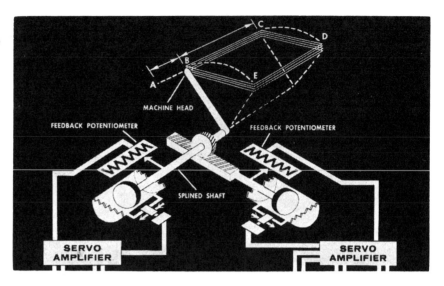

Foot-controlled braking system

The crane braking system (right) operates when the main line switch closes. Full application of the master-cylinder foot-pedal compresses the brake-setting spring mounted on the hydraulic releasing cylinder. After the setting spring is fully compressed, the hydraulic pressure switch closes, completing the electric circuit and energizing the magnetic check valve. The setting spring remains compressed as long as the magnetic check valve is energized, since the check valve traps the fluid in the hydraulic releasing cylinder. Upon release of the foot pedal, brake lever arm is pulled down by the brake releasing spring, thus releasing the brake shoes.

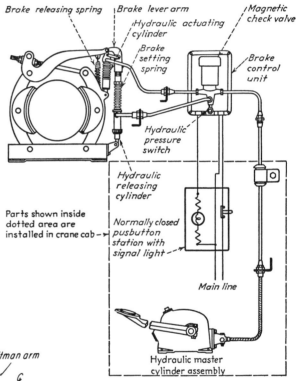

Team of linkages actuates steering in 300 hp diesel tractor

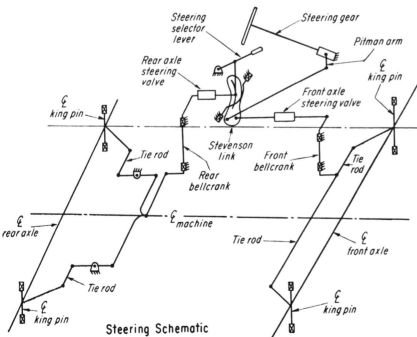

Steering Schematic

Hydraulic power for operating brakes, clutch and steering is supplied by an engine-driven pump delivering 55 gpm at 1200 psi. System is designed to give 15-gpm preference to the steering system. Steering drive to each wheel is mechanical for synchronization, with mechanical selection of front-wheel, four-wheel or crab steering hookup; hydraulic power amplifies manual steering effort.

15 JOBS for PNEUMATIC POWER

Here's how suction can be employed to feed, hold, position and lift parts, form plastic sheet, sample gases, test for leaks, convey solids, and de-aerate liquids; and compressed air can convey materials, atomize and agitate liquids, speed heat transfer, support combustion and protect cable.

REFERENCE:
Air Motor Applications. Published by Gast Manufacturing Corporation, Benton Harbor, Michigan.

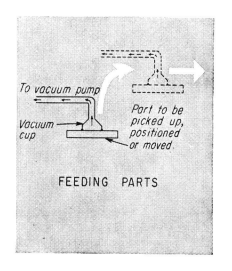

FEEDING PARTS

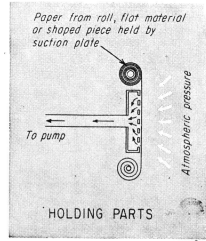

HOLDING PARTS

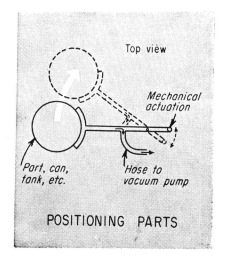

POSITIONING PARTS

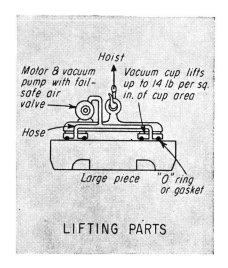

LIFTING PARTS

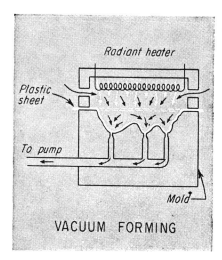

VACUUM FORMING

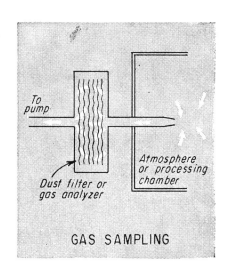

GAS SAMPLING

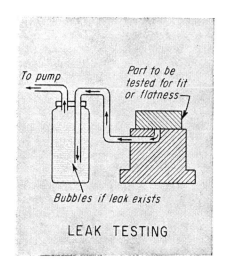

To pump

Part to be tested for fit or flatness

Bubbles if leak exists

LEAK TESTING

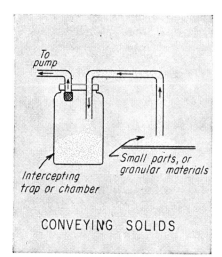

To pump

Intercepting trap or chamber

Small parts, or granular materials

CONVEYING SOLIDS

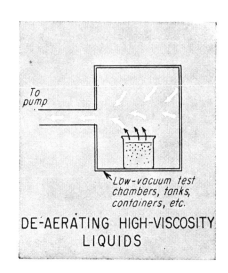

To pump

Low-vacuum test chambers, tanks, containers, etc.

DE-AERATING HIGH-VISCOSITY LIQUIDS

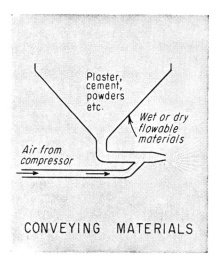

Plaster, cement, powders etc.

Wet or dry flowable materials

Air from compressor

CONVEYING MATERIALS

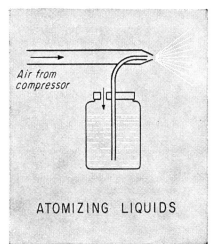

Air from compressor

ATOMIZING LIQUIDS

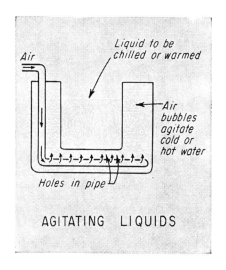

Air

Liquid to be chilled or warmed

Air bubbles agitate cold or hot water

Holes in pipe

AGITATING LIQUIDS

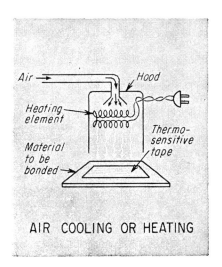

Air

Hood

Heating element

Thermo-sensitive tape

Material to be bonded

AIR COOLING OR HEATING

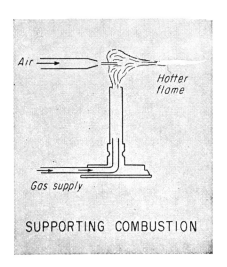

Air

Hotter flame

Gas supply

SUPPORTING COMBUSTION

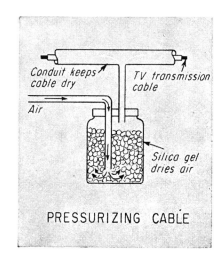

Conduit keeps cable dry

TV transmission cable

Air

Silica gel dries air

PRESSURIZING CABLE

10 ways to use METAL DIAPHRAGMS and CAPSULES

D. C. WHITTEN

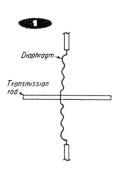

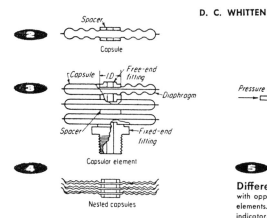

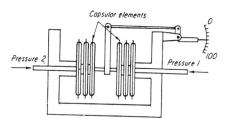

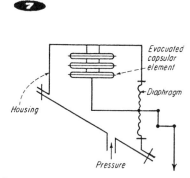

Metal diaphragm . . .

is usually corrugated (1) or formed to some irregular profile. It can be used as a flexible seal for actuating rod. Capsule (2) is an assembly of two diaphragms sealed together at their outer edges, usually by soldering, bronzing or welding. Two or more capsules assembled together

are known as a capsular element (3). End fittings for capsules vary according to their function; the "fixed end" is fixed to the equipment. The "free end" moves the related components and linkages. Nested capsule (4) requires less space and can be designed to withstand large external overpressures without damage.

Differential pressure gage . . .

with opposing capsules can have either single or multicapsular elements. The multicapsular type gives greater movement to indicator. Capsules give improved linearity over bellows for such applications of pressure measuring devices. Force exerted by any capsule is equal to the total effective area of the diaphragms (about 40% actual area) multiplied by the pressure exerted on it. Safe pressure is the max pressure that can be applied to a diaphragm before hysteresis or set become objectionable.

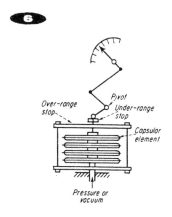

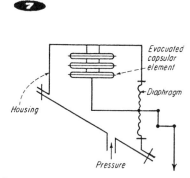

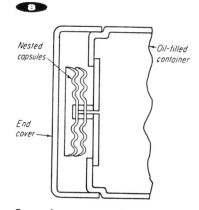

Pressure gage . . .

has capsular element linked to dial indicator by three-bar linkage. Such a gage measures pressure or vacuum relative to prevailing atmospheric pressure. If greater angular motion of indicator is required than can be obtained from three-bar linkage, a quadrant and gear can be used.

Absolute pressure gage . . .

has an evacuated capsular element inside an enclosure that is connected to pressure source only. Diaphragm allows linkage movement from capsule to pass through sealed chamber. This arrangement can also be used as a differential pressure gage by making a second pressure connection to the interior of the element.

Expansion compensator . . .

for oil-filled systems takes up less space when capsules are nested. In this application, one end of capsule is open and connected to oil in system; other end is sealed. Capsule expansion prevents internal oil pressure from increasing dangerously from thermal expansion. Capsule is protected by end cover.

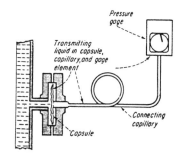

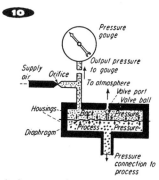

Capsule pressure-seal . . .

works like a thermometer system except that the bulb is replaced by a pressure-sensitive capsule. The capsule system is filled with liquid such as silicone-oil and is self-compensating for ambient and operating temperatures. When subjected to external pressure changes, the capsule expands or contracts until the internal system pressure just balances the external pressure surrounding the capsule.

Force-balance seal . . .

solves the problem, as in seal 9, of keeping corrosive, viscous or solids-bearing fluids out of the pressure gage. The air pressure on one side of a diaphragm is controlled so as to exactly balance the other side of the diaphragm. The pressure gage is connected to measure this balancing air pressure. Gage, therefore, reads an air pressure that is always exactly equal to the process pressure.

Differential Transformer Sensing Devices

W. D. MACGEORGE

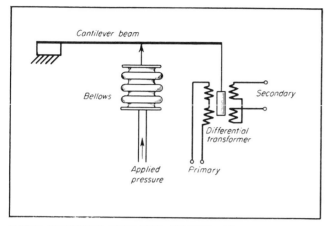

GAGE PRESSURE BELLOWS TRANSMITTER. Bellows is connected to cantilever beam with a needle bearing. Beam adopts a different position for every pressure; transformer output varies with beam position. Bellows are available for ranges from 0-10 in. to 0-200 in. of water for pressure indication or control.

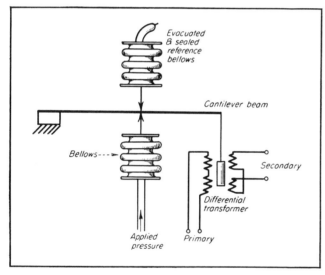

ABSOLUTE PRESSURE BELLOWS TRANSMITTER. Similar to above except for addition of reference bellows which is evacuated and sealed. Used for measuring negative gage pressures with ranges from 0-50 mm to 0-30 in. of mercury. Reference bellows compensates for variations in atmospheric pressure.

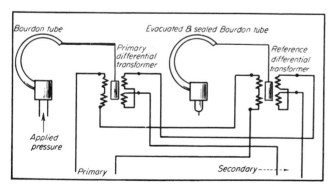

ABSOLUTE PRESSURE BOURDON TUBE TRANSMITTER. Device is used to indicate or control absolute pressures from 15 to 10,000 psi, depending on tube rating. Reference tube is evacuated and sealed, and compensates for variations in atmospheric pressure by changing output of reference differential transformer. Signal output consists of algebraic sum of outputs of primary and reference differential transformers.

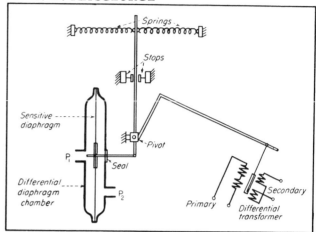

DIFFERENTIAL DIAPHRAGM PRESSURE TRANSMITTER. Differential pressures P_1 and P_2 act on opposite sides of sensitive diaphragm and move the diaphragm against the spring load. Diaphragm displacement, spring extension, and transformer core movement are proportional to difference in pressure. Device can be used to measure differentials as low as 0.005 in. of water. It can be used as the primary element in a differential pressure flow meter, or in a boiler windbox to furnace draft regulator.

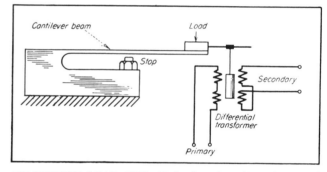

CANTILEVER LOAD CELL. Deflection of cantilever beam and displacement of differential transformer core are proportional to applied load. Stop prevents damage to beam in the event of overload. Beams are available for ranges from 0-5 to 0-500 pounds and can be used for the precise measurement of either tension or compression forces.

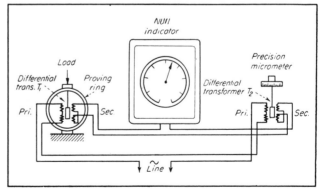

PROVING RING. The core of the transmitting transformer, T_1, is fastened to the top of the proving ring, while the windings are stationary. The proving ring and transformer core deflect in proportion to applied load. The signal output of the balancing transformer, T_2, opposes the output of T_1, so that at the balance point the null point indicator reads zero. The core of the balancing transformer is actuated by a calibrated micrometer screw which indicates the proving ring deflection when the differential transformer outputs are equal and balanced.

Differential Transformer Sensing Devices—Continued

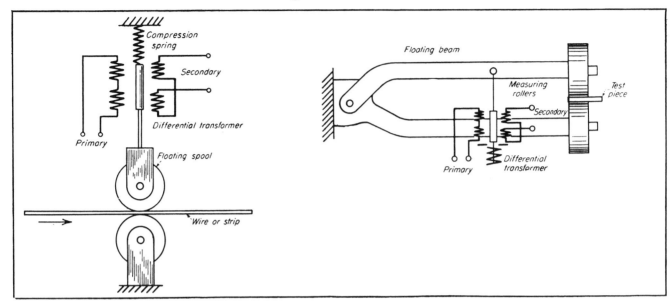

GAGING AND CALIPERING. Above, left, the thickness of a moving wire or strip is gaged by the position of the floating spool and transformer core. If the core is at the null point for proper material thickness, the transformer output phase and magnitude indicate whether the material is too thick or thin and the amount of the error. The signal may be amplified to operate a controller, recorder, or indicator. The device at the right can be used as a production caliper or as an accurate micrometer. If the transformer output is fed into a meter indicator with *go* and *no-go* bands, it is a convenient device for gaging items.

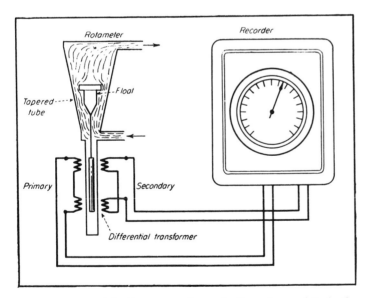

FLOW METER. The flow area varies as the float rises or falls in the tapered tube. High flows cause the float to rise, and low flows cause it to drop. The differential transformer core follows the float travel and generates an a-c signal which is fed into a square root recorder. A servo can be equipped with a square root cam to read on a linear chart. The transformer output can also be amplified and used to actuate a flow regulating valve so that the flow meter becomes the primary element in a flow controller. Normally meter has an accuracy of better than 2 percent, but the flow range is limited.

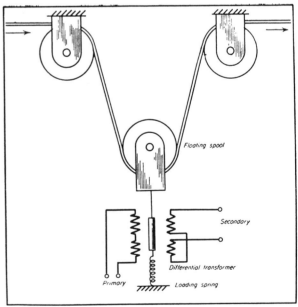

TENSION CONTROL. The loading spring is adjusted so that when the transformer core is at the null point, the proper tension is maintained in the wire. The amplified output of the transformer is transmitted to some type of tension controlling device which increases or reduces the tension in the wire depending on the phase and magnitude of the differential transformer signal.

High-speed counters

WILLIAM FOLEY

The electronic counter simply counts electrical pulses and gives a running display of accumulated pulses at any instant. Since the input is an electrical signal, generally a transducer is required to transform the nonelectrical signal into a usable input for the counter.

With a preset function on the counter, you can select any number within the count capacity of the device. Once the counter reaches the preset number, it can open or close the relay to control some operation. The counter will either reset automatically or stop. A dual unit permits continuous control over two different count sequence operations. Two sets of predetermining switches are usually mounted on the front panel of the counter, but they can be remote. If two different numbers are programmed into the counter, it will alternately count the two selected numbers. Multiple presets are also available, but at higher cost.

Besides performing two separate operations, a dual preset can control speeds as shown in Fig. 1. In the metal shearing operation run at high speed, one preset switch can be used to slow the material down at a given distance before the second preset actuates the shearing, then both automatically reset and start to measure again. The same presets could be used to alternately shear the material into two different lengths.

One area of measurement well adapted to high-speed counters is measuring continuous materials such as wire, rope, paper, textiles, or steel. Fig. 2 shows a coil-winding operation in which a counter stops the machine at a predetermined number of turns of wire.

Another application is shown in Fig 3 where magazines are counted as they run off a press. A photoelectric pickup senses the alternate light and dark lines formed by the shadow of the folded edge of each magazine. At the predetermined number, a knife edge, actuated by the counter, separates the magazines into exact batches.

A third application is machine-tool control. A preset counter can be used with a transducer or pulse generator mounted on the feed mechanism. It could, for example, convert revolutions of screw feed, hence displacement, into pulses to be fed into the counter. A feed of 0.129 in. might represent a count of 129 to the counter, which when preset at that number, could stop, advance, or reverse the feed mechanism.

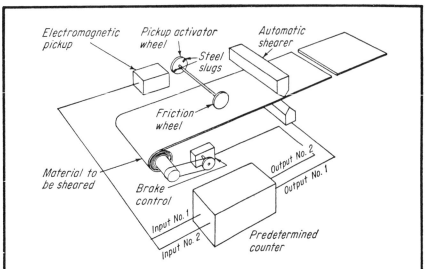

1. Dual preset function on high-speed counter controls high-speed shearing operation. If material is to be cut in 10-ft lengths and each pulse of electromagnetic pickup represents 0.1 ft, operator presets 100 into first input channel. Second input is set to 90. When 90 pulses are counted, second channel slows the material, then when counter reaches 100, first channel actuates shear. Both channels reset instantaneously and start next cycle.

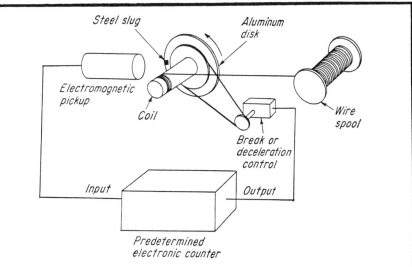

2. Coil-winding machine with electronic counting for measuring length.

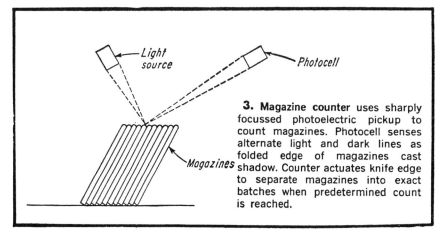

3. Magazine counter uses sharply focussed photoelectric pickup to count magazines. Photocell senses alternate light and dark lines as folded edge of magazines cast shadow. Counter actuates knife edge to separate magazines into exact batches when predetermined count is reached.

Permanent Magnet Mechanisms

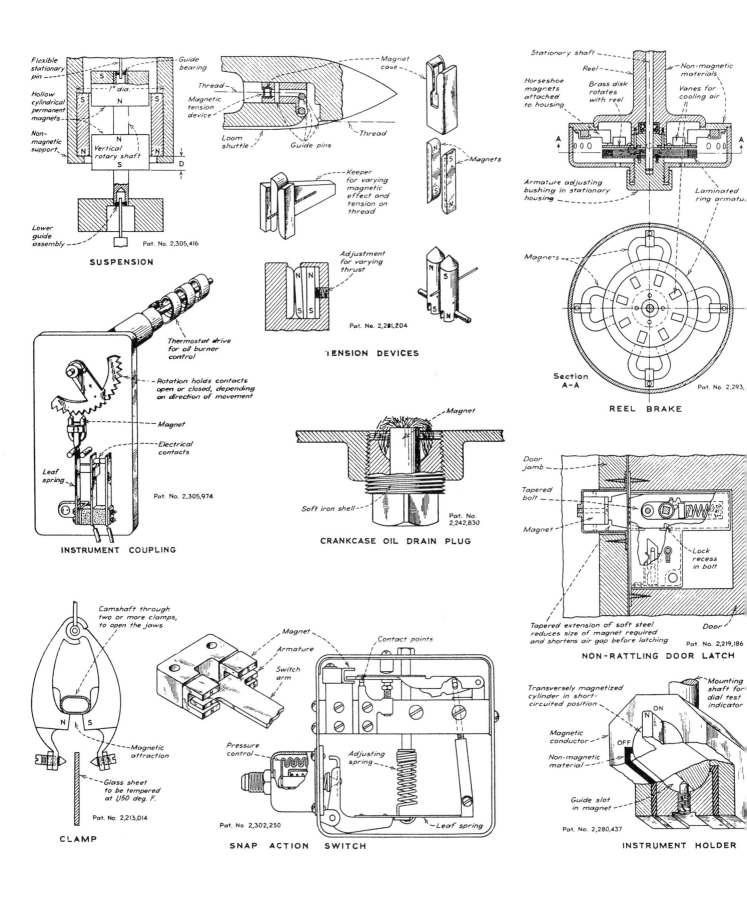

SUSPENSION

Flexible stationary pin
Guide bearing
1" dia.
Hollow cylindrical permanent magnets
Non-magnetic support
Vertical rotary shaft
D
Lower guide assembly
Pat. No. 2,305,416

Magnet case
Thread
Magnetic tension device
Loom shuttle
Guide pins
Thread

Keeper for varying magnetic effect and tension on thread
Magnets
Adjustment for varying thrust

TENSION DEVICES
Pat. No. 2,281,204

Stationary shaft
Reel
Non-magnetic materials
Horseshoe magnets attached to housing
Brass disk rotates with reel
Vanes for cooling air
A
A
Armature adjusting bushing in stationary housing
Laminated ring armature
Magnets
Section A-A
Pat. No. 2,293,
REEL BRAKE

Thermostat drive for oil burner control
Rotation holds contacts open or closed, depending on direction of movement
Magnet
Electrical contacts
Leaf spring
Pat. No. 2,305,974
INSTRUMENT COUPLING

Magnet
Soft iron shell
Pat. No. 2,242,830
CRANKCASE OIL DRAIN PLUG

Door jamb
Tapered bolt
Magnet
Lock recess in bolt
Tapered extension of soft steel reduces size of magnet required and shortens air gap before latching
Door
Pat. No. 2,219,186
NON-RATTLING DOOR LATCH

Camshaft through two or more clamps, to open the jaws
N   S
Magnetic attraction
Glass sheet to be tempered at 1,150 deg. F.
Pat. No. 2,213,014
CLAMP

Magnet
Armature
Switch arm
Contact points
Pressure control
Adjusting spring
Leaf spring
Pat. No. 2,302,250
SNAP ACTION SWITCH

Transversely magnetized cylinder in short-circuited position
Mounting shaft for dial test indicator
ON
OFF
Magnetic conductor
Non-magnetic material
Guide slot in magnet
Pat. No. 2,280,437
INSTRUMENT HOLDER

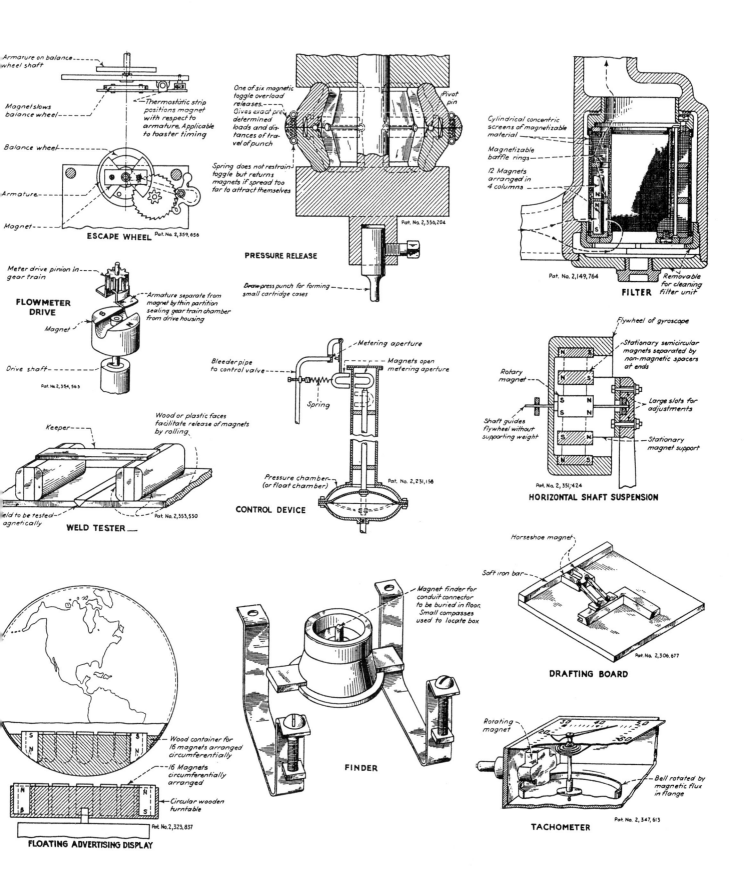

Armature on balance wheel shaft

Magnet slows balance wheel

Balance wheel

Armature

Magnet

Thermostatic strip positions magnet with respect to armature. Applicable to toaster timing

ESCAPE WHEEL Pat. No. 2,359,656

One of six magnetic toggle overload releases. Gives exact predetermined loads and distances of travel of punch

Pivot pin

Spring does not restrain toggle but returns magnets if spread too far to attract themselves

Pat. No. 2,356,204

PRESSURE RELEASE

Draw-press punch for forming small cartridge cases

Cylindrical concentric screens of magnetizable material

Magnetizable baffle rings

12 Magnets arranged in 4 columns

Pat. No. 2,149,764

FILTER Removable for cleaning filter unit

Meter drive pinion in gear train

FLOWMETER DRIVE

Magnet

Armature separate from magnet by thin partition sealing gear train chamber from drive housing

Drive shaft

Pat. No. 2,354,563

Metering aperture

Magnets open metering aperture

Bleeder pipe to control valve

Spring

Pressure chamber (or float chamber)

Pat. No. 2,231,158

CONTROL DEVICE

Flywheel of gyroscope

Stationary semicircular magnets separated by non-magnetic spacers at ends

Rotary magnet

Large slots for adjustments

Shaft guides flywheel without supporting weight

Stationary magnet support

Pat. No. 2,351,424

HORIZONTAL SHAFT SUSPENSION

Wood or plastic faces facilitate release of magnets by rolling

Keeper

Weld to be tested magnetically

Pat. No. 2,353,550

WELD TESTER

Magnet finder for conduit connector to be buried in floor. Small compasses used to locate box

FINDER

Horseshoe magnet

Soft iron bar

Pat. No. 2,306,677

DRAFTING BOARD

Wood container for 16 magnets arranged circumferentially

16 Magnets circumferentially arranged

Circular wooden turntable

Pat. No. 2,323,837

FLOATING ADVERTISING DISPLAY

Rotating magnet

Bell rotated by magnetic flux in flange

Pat. No. 2,347,613

TACHOMETER

383

THE APPLICATION of controlled impact forces can be as practical in specialized stationary machinery as in the portable electric hammers shown here. Such mechanisms have been employed in vibrating concrete forms, in nailing machines and other special machinery. In portable hammers they are efficient

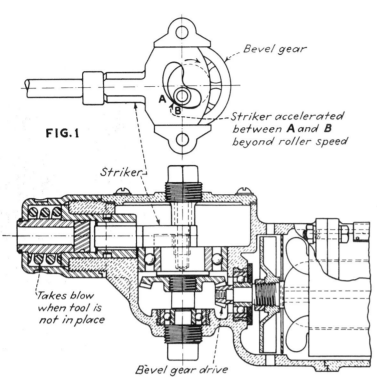

FIG.1

Bevel gear

Striker accelerated between **A** and **B** beyond roller speed

Striker

Takes blow when tool is not in place

Bevel gear drive

Fig. 1—Free driving *throw* of cam slotted striker is produced by eccentric stud roller during contact between points A and B of slot. This accelerates striker beyond tangential speed of roller for an instant before being picked up for return stroke

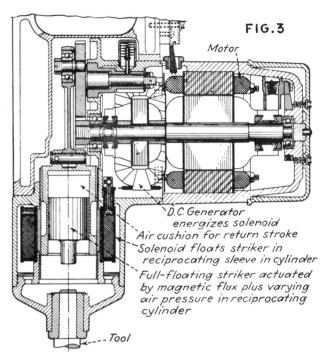

FIG.3

Motor

D.C. Generator energizes solenoid
Air cushion for return stroke
Solenoid floats striker in reciprocating sleeve in cylinder
Full-floating striker actuated by magnetic flux plus varying air pressure in reciprocating cylinder

Tool

Fig. 3—Striker has no mechanical connection with reciprocating drive in this Speedway hammer

Striker

FIG.2

Sliding spline shaft

Fan

Bevel gears with unbalanced weights rotate in opposite directions

Striker

Spline plate

Striker assembly and spline shaft reciprocate in housing

Fig. 2—Centrifugal force of two oppositely rotating weights throw striker assembly of this Master hammer. Power connection is maintained by sliding splined shaft. Guide, not shown, prevents rotation of the striker assembly

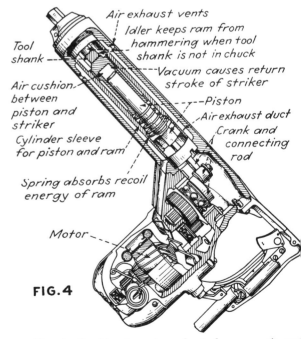

Air exhaust vents
Idler keeps ram from hammering when tool shank is not in chuck
Tool shank
Vacuum causes return stroke of striker
Air cushion between piston and striker
Piston
Air exhaust duct
Crank and connecting rod
Cylinder sleeve for piston and ram
Spring absorbs recoil energy of ram
Motor

FIG.4

Fig. 4—Combination of mechanical, pneumatic and spring action is used in this Van Dorn hammer

HAMMER MECHANISMS

in drilling, chiseling, digging, chipping. tamping, riveting and similar operations where quick concentrated blows are required. Striker mechanisms illustrated are operated by means of springs, cams, magnetic force, air and vacuum chambers, and centrifugal force. Drawings show only striking mechanism.

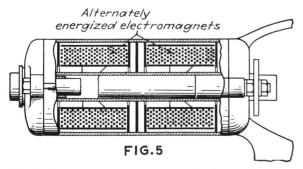

FIG.5

Fig. 5—Two electromagnets operate the Syntron hammer. Weight of blows may be controlled by varying electric current in coils or timing current reversals by air gap adjustment of the contacts

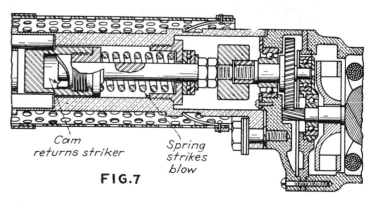

FIG.7

Fig. 7—Spring operated Milwaukee hammer employs shaft rotating in female cam to return striker

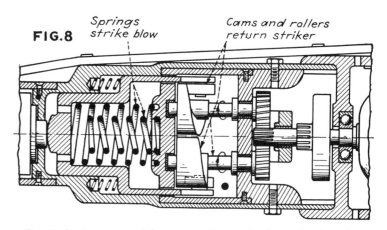

FIG.8

Fig. 8—Spring operated hammer with two fixed rotating barrel cams which return striker by means of two rollers on opposite sides of the striker sleeve. Auxiliary springs prevent striker from hitting the retaining cylinder. Means of rotating the tool, not shown, are also incorporated in this hammer

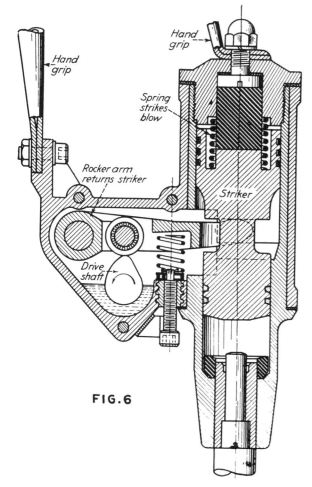

FIG.6

Fig. 6—Spring operated hammer with cam and rocker for return stroke has screw adjustment of blow

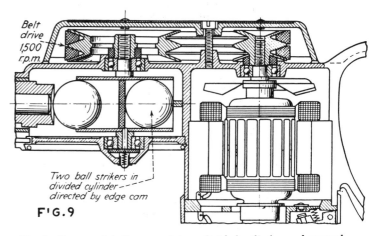

FIG.9

Fig. 9 Two steel balls rotated in a divided cylinder and steered by an edge cam develop centrifugal force to strike blows against tool holder. Collar is held clear of hammer by compression spring when no tool is in holder. A second spring cushions blows when motor is running but tool is not held against work

385

THERMOSTATIC MECHANISMS

Sensitivity or change in deflection for a given temperature change depends upon the combination of metals selected as well as the dimensions of the bimetal element. Sensitivity increases with the square of the length and inversely with the thickness. The force developed for a given temperature change also depends on the type of bimetal, whereas the allowable working load for the thermostatic strip increases with the width and the square of the thickness. Thus, the design of bimetal elements depends upon the relative importance of sensitivity and working load.

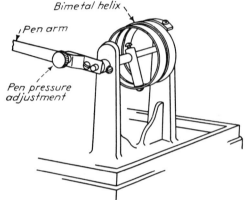

Fig. 1 —In the Taylor recording thermometer, a pen is moved vertically across a revolving chart by a brass-invar bimetal element. To obtain sensitivity, the long movement of the pen requires a long strip of bimetal, which is coiled into a helix to save space. For accuracy, a relatively large cross section gives stiffness, although the large thickness requires increased length to obtain the desired sensitivity.

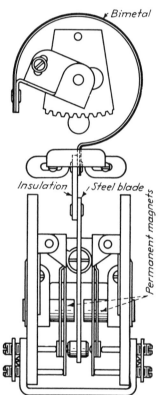

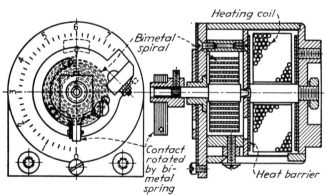

Fig. 3 —In this Westinghouse overload relay for large motors, a portion of the motor current is passed through a heating coil within the relay. Heat from the coil raises the temperature of a bimetal spiral which rotates a shaft carrying an electrical contact. To withstand the operating temperature, a heat-resistant bimetal is used, coiled into the spiral form for compactness. Because of the large deflection needed, the spiral is long and thin, whereas the width is made large to provide the required contact pressure.

By the use of heat barriers between the bimetal spiral and the heating coil, temperature rise of the bimetal can be made to follow closely the increase in temperature within the motor. Thus, momentary overloads do not cause sufficient heating to close the contacts, whereas a continued overload will in time cause the bimetal to rotate the contact arm around to the adjustable stationary contact, causing a relay to shut down the motor.

Fig. 2 —Room temperatures in summer as well as winter are controlled over a wide range by a single large-diameter coil of brass-invar in the Friez thermometer. To prevent chattering, a small permanent magnet is mounted on each side of the steel contact blade. The magnetic attraction on the blade, increasing inversely with the square of the distance from the magnet, gives a snap action to the contacts.

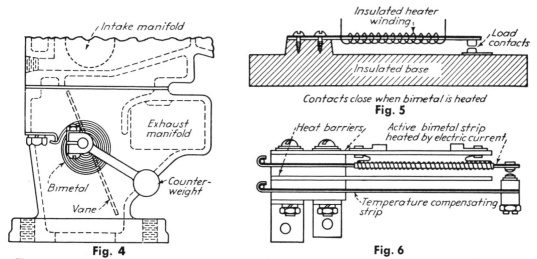

Fig. 4

Fig. 5
Contacts close when bimetal is heated

Fig. 6

Fig. 4 —On the Dodge carburetor, when the engine is cold, a vane in the exhaust passage to the "hot spot" is held open by a bimetal spring against the force of a small counterweight. When the thermostatic spiral is heated by the outside air or by the warm air stream from the radiator, the spring coils up and allows the weight to close the vane. Since high accuracy is not needed, a thin, flexible cross section is used with a long length to give the desired sensitivity.

Fig. 5 —In the Friez relay, a constant current through an electrical heating coil around a straight bimetal strip gives a time-delay action. Since the temperature range is relatively large, high sensitivity is not necessary, hence a short straight strip of bimetal is suitable. Because of the relatively heavy thickness used, the strip is sufficiently stiff to close the contact firmly without chattering.

Fig. 6 —A similar type of bimetal element is used in the Ward Leonard time-delay relay for mercury-vapor rectifiers. This relay closes the potential circuit to the mercury tube only after the filament has had time to reach its normal operating temperature. To eliminate the effect of changes in room temperature on the length of the contact gap, and therefore the time interval, the stationary contact is carried by a second bimetal strip similar to the heated element. Barriers of laminated plastic on both sides of the active bimetal strip shield the compensating strip and prevent air currents from affecting the heating rate. The relatively high temperature range allows the use of a straight thick strip, whereas the addition of the compensating strip makes accurate timing possible with a short travel.

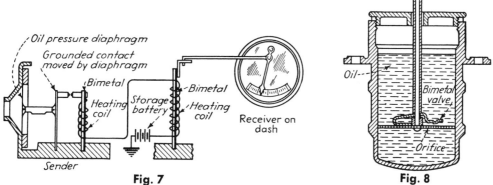

Fig. 7

Fig. 8

Fig. 7 —Oil pressure, engine temperature, or gasoline level are indicated electrically on automobile dashboard instruments built by King-Seeley in which a bimetal element is used in both the sender and receiver. A grounded contact at the sender completes an electric circuit through heaters around two similar bimetal strips. Since the same current flows around the two bimetal elements, their deflections are the same. But the sender element when heated will bend away from the grounded contact until the circuit is broken. Upon cooling, the bimetal again makes contact and the cycle continues, allowing the bimetal to follow the movement of the grounded contact. For the oil-pressure gage, the grounded contact is attached to a diaphragm; for the temperature indicator, the contact is carried by another thermostatic bimetal strip; in the gasoline-level device, the contact is shifted by a cam on a shaft rotated by a float. Deflections of the receiving bimetal are amplified through a linkage that operates a pointer over the scale of the receiving instrument. Since only small deflections are needed, the bimetal element is in the form of a short stiff strip.

Fig. 8 —Oil dashpots used in heavy-capacity Toledo scales have a thermostatic control to compensate for changes in oil viscosity with temperature. A rectangular orifice in the plunger is covered by a swaged projection on the bimetal element. With a decrease in oil temperature, the oil viscosity increases, tending to increase the damping effect; but the bimetal deflects upward, enlarging the orifice enough to keep the damping force constant. A wide bimetal strip is used for stiffness so that the orifice will not be altered by the force of the flowing oil.

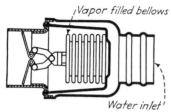

Fig. 9 —Automobile cooling water temperature is controlled by a self-contained bellows in the thermostat made by the Bridgeport Brass Company. As in the radiator air valve, the bellows itself is subjected to the temperature to be controlled. As the temperature of the water increases to about 140°F., the valve starts to open; at approximately 180°F., free flow is permitted. At intermediate temperatures, the valve opening is in proportion to the temperature.

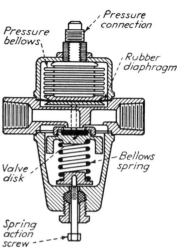

Fig. 10—In a throttling type of circulating water control valve made by C. J. Tagliabue Manufacturing Company for use in refrigeration plants, the valve opening varies with the pressure on the bellows. This valve controls the rate of flow of the cooling water through the condenser, a greater amount of water being required when the temperature, and therefore the pressure, increases. The pressure in the condenser is transmitted through a pipe to the valve bellows thereby adjusting the flow of cooling water. The bronze bellows is protected from contact with the water by a rubber diaphragm.

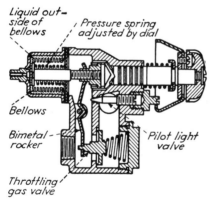

Fig. 11 —An automatic gas-range control made by the Wilcolator Company has a sealed thermostatic element consisting of a bulb, capillary tube, and bellows. As food is often placed near the bulb, a nontoxic liquid, chlorinated diphenyl, is used in the liquid expansion system. The liquid is also non-inflammable and has no corrosive effect upon the phosphor bronze bellows. By placing the liquid outside instead of inside the bellows, the working stresses are maximum at normal temperatures when the bellows bottoms on the cup. At elevated working temperatures, the expansion of the liquid compresses the bellows against the action of the extended spring which, in turn, is adjusted by the knob. Changes in calibration caused by variations in ambient temperature are compensated by making the rocker arm of bimetal suitable for high-temperature service.

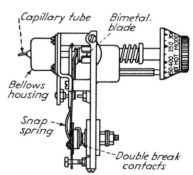

Fig. 12 —For electric ranges, the Wilcolator thermostat has the same bellows unit as is used on the gas-type control. But, instead of a throttling action, the thermostat opens and closes the electrical contacts with a snap action. To obtain sufficient force for the snap action, the control requires a temperature difference between "on" and "off" positions. For a control range from room temperature to 550°F., the differential in this device is plus or minus 10°F.; with a smaller control range, the differential is proportionately less. The snap-action switch is made of beryllium copper, giving high strength, better snap action, and longer life than obtainable with phosphor bronze, and because of its corrosion resistance the beryllium-copper blade requires no protective finish.

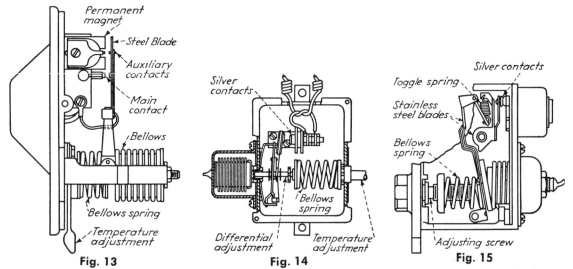

Fig. 13

Fig. 14

Fig. 15

Fig. 13 —For heavy-duty room-temperature controls, the Penn thermostat uses a bellows mechanism that develops a high force with small changes in temperature. The bellows is partly filled with liquid butane, which at room temperatures is a gas having a large change in vapor pressure for small temperature differentials. Snap action of the electrical contact is obtained from a small permanent magnet that pulls the steel contact blade into firm contact when the bellows cools. Because of the firm contact, the device is rated at 20 amp. for noninductive loads. To avoid chattering or bounce under the impact delivered by the rapid magnetic closing action, small auxiliary contacts are carried on light spring blades. With the large force developed by the bellows, a temperature differential of only 2°F. is obtained.

Fig. 14 —Snap action in the Tagliabue refrigerator control is obtained from a bowed flat spring. The silver contacts carried on an extended end of the spring open or close rapidly when movement of the bellows actuates the spring. With this snap action, the contacts can control an alternating-current motor as large as $1\frac{1}{2}$ hp. without the use of auxiliary relays. Temperature differential is adjusted by changing the spacing between two collars on the bellows shaft passing through the contact spring. For temperatures used in freezing ice, the bellows system is partly filled with butane.

Fig. 15 —In the General Electric refrigerator control, the necessary snap action is obtained from a toggle spring supported from a long arm moved by the bellows. With this type of toggle action, the contact pressure is a maximum at the instant the contacts start to open. Thermostatic action is obtained from a vapor-filled system using sulphur dioxide for usual refrigerating service or methyl chloride where lower temperatures are required. To reduce friction, the bellows makes point contact with the bellows cup. Operating temperature is adjusted by changing the initial compression in the bellows spring. For resistance to corrosion, levers and blades are stainless steel with bronze pin bearings.

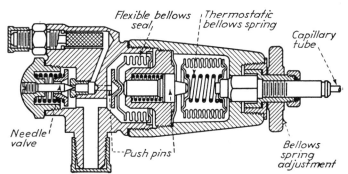

Fig. 16 —Two bellows units are used in the Fedders thermostatic expansion valve for controlling large refrigeration systems. A removable power bellows unit is operated by vapor pressure in a bulb attached to the evaporator output line. The second bellows serves as a flexible, gastight seal for the gas valve. A stainless steel spring holds the valve closed until opened by pressure transmitted from the thermostatic bellows through a molded push pin.

Temperature-Regulating Mechanisms

L. C. BLAUVELT

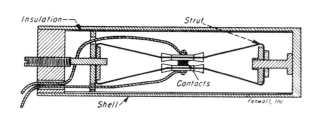

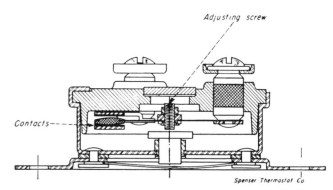

1. Bimetallic device is simple, compact and precise. Contacts mounted on low-expansion struts determine slow make-and-break action. Shell contracts or expands with temperature changes, opening or closing the electrical circuit that controls a heating or cooling unit. Adjustable and resistant to shock and vibration. Range: —100 to 1,500 F. Accuracy: Operates on less than 0.5 deg. temperature change.

2. Typical inclosed disk-type, snap-action control has fixed operating temperature. Suitable for unit and space heaters, small hot water heaters, clothes dryers and other applications requiring non-adjustable temperature control. Useful where dirt, dust, oil or corrosive atmosphere is involved. Available with various temperature differentials and with manual reset. Depending on model, temperature setting range is from —10 to 550F and minimum differential may be 10, 20, 30, 40 or 50F.

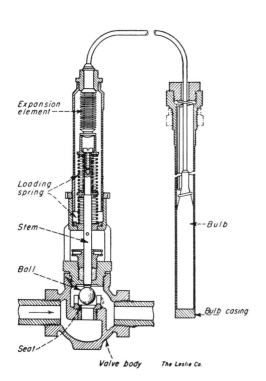

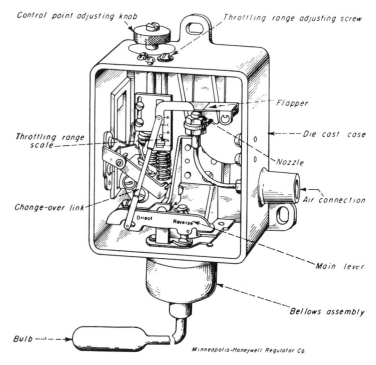

5. Self-contained regulator is actuated by expansion or contraction of liquid or gas in temperature sensitive bulb which is immersed in medium being controlled. Signal is transmitted from bulb to sealed expansion element which opens or closes the ball-valve. Range: 20 to 270 F. Accuracy: ±1 deg. Max. press. 100 psi. for dead end service, 200 psi continuous flow.

6. Remote bulb, non-indicating regulator uses a bellows assembly to operate a flapper. This allows air pressure in the control system to build up or bleed depending upon the position of the change-over link. Unit can be direct or reverse acting. Control knob adjusts the setting and the throttling range adjustment determines the percentage of the control range in which full output pressure (3-15 psi) is obtained. Range: 0-700 F. Accuracy: About ±0.5 percent of full scale range depending upon installation factors.

Temperature regulators are either of the on-off or throttling type. The characteristics of the process determine which should be used. Within each group, selection of a device is governed by the accuracy required, space limitations, simplicity and cost.

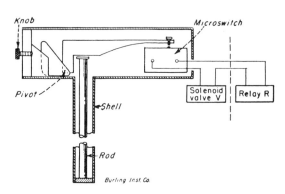

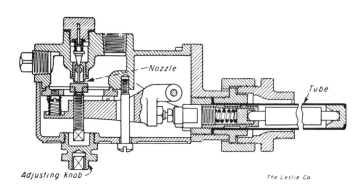

3. Bimetallic unit has rod with a low coefficient of expansion and a shell with a high coefficient. Microswitch gives snap action to the electrical control circuit. Current can be large enough to operate a solenoid valve or relay directly. Set point is adjusted by knob which moves the pivot point of the lever. Range: —20 to 1,750 F. Accuracy: 0.25 to 0.50 degrees.

4. Bimetallic actuated, air piloted control. Expansion of rod causes air signal (3-15 psi) to be transmitted to a heating or cooling pneumatic valve. Position of pneumatic valve depends upon the amount of air bled through the pilot valve of the control. This produces a throttling type of temperature control as contrasted to the on-off characteristic that is obtained with the three units described previously. Range: 32 to 600 F. Accuracy: ± 1 to ±3 F depending upon the range.

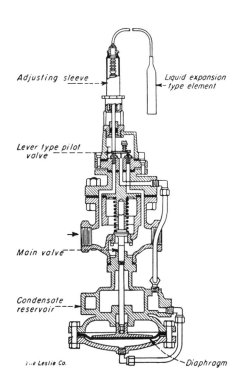

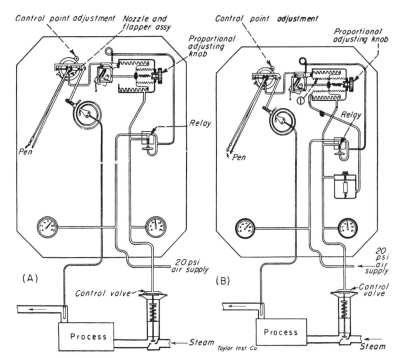

7. Lever-type pilot valve is actuated by temperature sensitive bulb. Motion of lever causes water or steam being controlled to exert pressure on a diaphragm which opens or closes the main valve. Range: 20 to 270 F. Accuracy: ±1 to 4 degrees. Pressure: 5-125 psi, steam; 5-175 psi, water.

8. Two recording and controlling instruments with adjustable proportional ranges. In both, air supply is divided by a relay valve. A small part goes through nozzle and flapper assembly. The main part goes to the control valve. Unit B has an extra bellows for automatic resetting. It is designed for systems with continuously changing control points and can be used where both heating and cooling are required for one process. Both A and B are easily changed from direct to reverse acting. Accuracy: One per cent of range of —40 to 800 F.

Photoelectric Controls

PHOTOSWITCH DIVISION, ELECTRONICS CORPORATION OF AMERICA

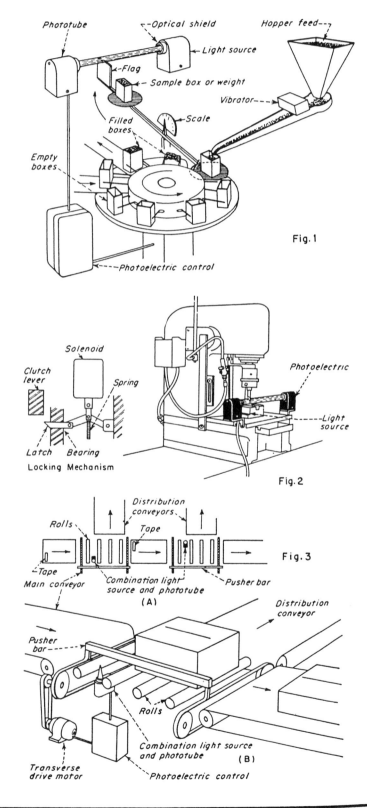

FIG. 1—Automatic weighing and filling. Problem is to fill each box with exact quantity of products such as screws. Electric feeder vibrates parts through chute and into box on small balance. Photoelectric control is mounted at rear of scale. Light beam is restricted to very small dimensions by optical slit. Control is positioned so that light is interrupted by balanced cantilever arm attached to scale when proper box weight is reached. Photoelectric control then stops flow of parts by de-energizing feeder. Simultaneously, indexing mechanism is activated to remove filled box and replace with empty one. Completion of indexing re-energizes feeder which starts flow of screws.

FIG. 2—Operator safeguard. Most presses operate by foot pedal leaving hands free for loading and unloading operations. This creates a safety hazard. Use of mechanical gate systems reduce production speeds. With photoelectric controls, curtain of light is set up with multiple series of photoelectric scanners and light sources. When light is broken at any point by operator's hand, control energizes a locking mechanism which prevents punch press drive from being energized. Wiring is such that power or tube failure causes control to function as though light beam was broken. In addition, the light is frequently used as the actuating control since clutch is thrown as soon as operator removes his hand from the die on the press table.

FIG. 3—Sorting cartons of three types of electronic tubes. Since cartons containing one type differ widely as to size, it is not feasible to sort by carton size and shape. Solution: small strip of reflecting tape is placed on cartons by a packer during assembly. For one type of tube, strip is placed along one edge of bottom and extends almost to the middle. For second type, strip is located along same edge but from middle to opposite side. No tape is used for third type. Cartons are placed on conveyor so that tape is at right angle to direction of travel. Photoelectric controls, shown in A, "see" the reflecting tape pass and operate pusher bar mechanism, shown in B, which pushes carton on to proper distribution conveyor. Cartons without tape pass through.

FIG. 4—Cut-off machine uses photoelectric control for strip material which does not have sufficient mass to operate a mechanical limit switch satisfac-

Some typical applications for reducing production costs and increasing operator safeguards by precisely and automatically controlling the feed, transfer or inspection of products from one process stage to another.

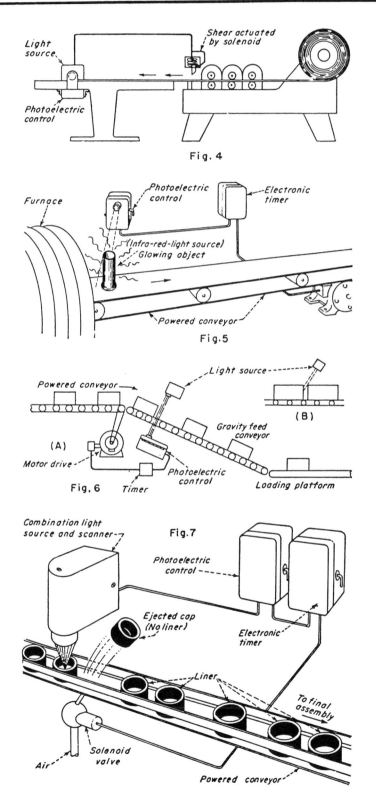

Fig. 4

Fig. 5

Fig. 6

Fig. 7

torily. Forward end of strip breaks light beam thus actuating the cut-off operation. Light source and control is mounted on adjustable stand at end of machine to vary length of finished stock.

FIG. 5—Heat-treating conveyor uses electronic timer in conjunction with photoelectric control to carry parts emerging from furnace at 2300 F. Problem is to operate conveyor only when a part is placed on it and only for distance required to reach next process stage. Parts are ejected on to conveyor at varying rates. High temperatures caused failures when mechanical switches were used. Glowing white-hot part radiates infra-red rays which actuates photoelectric control as soon as part comes in view. Control operates conveyor which carries part away from furnace and simultaneously starts timer. Conveyor is kept running by timer for pre-determined length of time required to position part for next operation.

FIG. 6—Jam detector. Cartons jamming on conveyor cause loss of production time and damage to cartons, products and conveyors. Detection is accomplished with a photoelectric control using a timer as shown in (A). Each time a carton passes light source, control beam is broken which starts timing interval in the timer. Timing circuit is reset without relay action each time beam is restored before preset timing interval has elapsed. If jam occurs causing cartons to butt one against the other, light beam cannot reach control. Timing circuit will then time out, opening load circuit which stops conveyor motor. By locating light source at an angle to conveyor, as shown in (B), power conveyor can be delayed if cartons are not butting but are too close to each other.

FIG. 7—Automatic inspection. As steel caps are conveyed to final assembly, they pass intermediate stage where an assembler inserts insulation liner into cap. Inspection point for missing liners has reflection-type photoelectric scanner which incorporates both light source and phototube with common lens system to instantly recognize difference in reflection between dark liner and light steel cap. When it detects a cap without a liner, a relay operates ejector device composed of air blast controlled by solenoid valve. Start and duration of air blast is accurately controlled by timer so that no other caps are displaced.

Liquid Level Indicators and

Thirteen different systems of operation are shown.   Each one represents at least

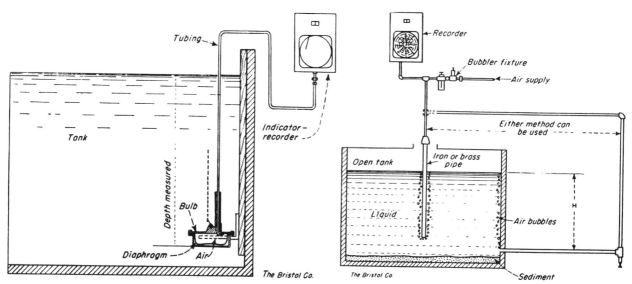

DIAPHRAGM ACTUATED INDICATOR. Can be used with any kind of liquid, whether it be flowing, turbulent, or carrying solid matter. Recorder can be mounted above or below the level of the tank or reservoir.

BUBBLER TYPE RECORDER measures height *H*. Can be used with all kinds of liquids, including those carrying solids. Small amount of air is bled into submerged pipe. Gage measures pressure of air that displaces fluid.

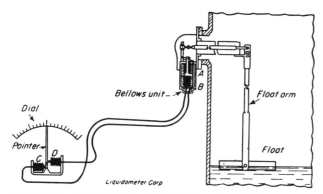

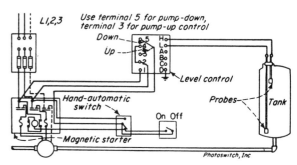

BELLOWS ACTUATED INDICATOR. Two bellows and connecting tubing are filled with incompressible fluid. Change in liquid level displaces transmitting bellows and pointer.

ELECTRICAL TYPE LEVEL CONTROLLER. Positions of probes determine duration of pump operation. When liquid touches upper probe, relay operates and pump stops. Through auxiliary contacts, lower probe provides relay holding current until liquid drops below it.

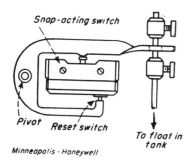

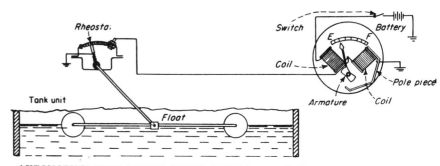

FLOAT-SWITCH TYPE CONTROL-LER. When liquid reaches predetermined level, float actuates switch through horseshoe-shape arm. Switch can operate valve or pump, as required.

AUTOMOTIVE TYPE LIQUID LEVEL INDICATOR. Indicator and tank unit are connected by a single wire. As liquid level in tank increases, brush contact on tank rheostat moves to the right, introducing an increasing amount of resistance into circuit that grounds the "F" coil. Displacement of needle from empty mark is proportional to the amount of resistance introduced into this circuit.

Controllers

H. W. HAMM

one commercial instrument.   Some of them are available in several modified forms.

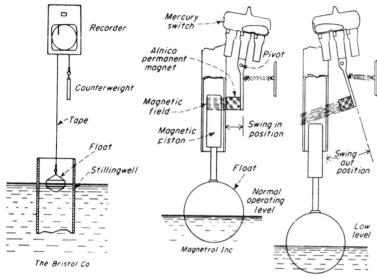

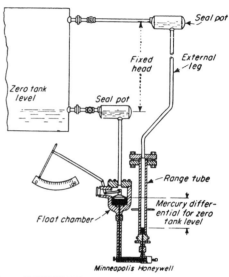

FLOAT TYPE RECORDER. Pointer can be attached to a calibrated float tape to give an approximate instantaneous indication of fluid level.

MAGNETIC LIQUID LEVEL CONTROLLER. When liquid level is normal, common-to-right leg circuit of mercury switch is closed. When level drops to predetermined level, magnetic piston is drawn below the magnetic field.

DIFFERENTIAL PRESSURE SYSTEM. Applicable to liquids under pressure. Measuring element is mercury manometer. Mechanical or electric meter body can be used. Seal pots protect meter body.

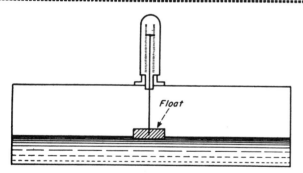

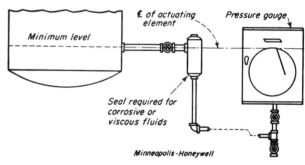

DIRECT READING FLOAT TYPE GAGE. Inexpensive, direct-reading gage has dial calibrated to tank volume. Comparable type as far as simplicity is concerned has needle connected through a right-angle arm to float. As liquid level drops, float rotates the arm and the needle.

PRESSURE GAGE INDICATOR for open vessels. Pressure of liquid head is imposed directly upon actuating element of pressure gage. Center line of the actuating element must coincide with the minimum level line, if the gage is to read zero when the liquid reaches the minimum level.

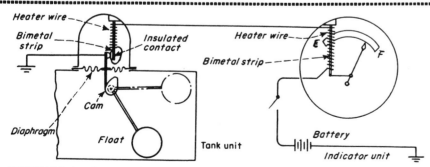

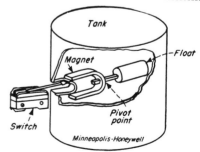

BIMETALLIC TYPE INDICATOR. When tank is empty, contacts in tank unit just touch. With switch closed, heaters cause both bimetallic strips to bend. This opens contacts in tank and bimetals cool, closing circuit again. Cycle repeats about once per sec. As liquid level increases, float forces cam to bend tank bimetal. Action is similar to previous case, but current and needle displacement are increased.

SWITCH ACTUATED LEVEL CONTROLLER. Pump is actuated by switch. Float pivots magnet so that upper pole attracts switch contact. Tank wall serves as other contact.

Instant muscle with pyrotechnic power

Cartridge-actuated devices generate a punch
that cuts cable and pipe, shears bolts for
fast release, and provides emergency thrust

ARMOND W. SCHELLMAN

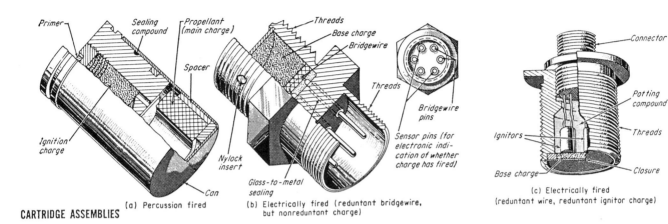

CARTRIDGE ASSEMBLIES

(a) Percussion fired

(b) Electrically fired (redundant bridgewire,
but nonredundant charge)

(c) Electrically fired
(redundant wire, redundant ignitor charge)

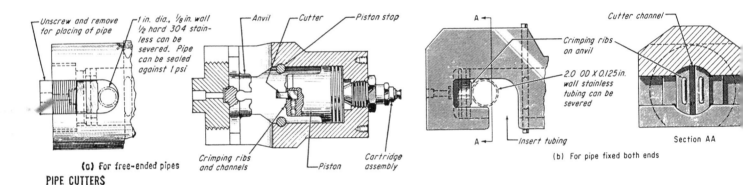

PIPE CUTTERS

(a) For free-ended pipes

(b) For pipe fixed both ends

Section AA

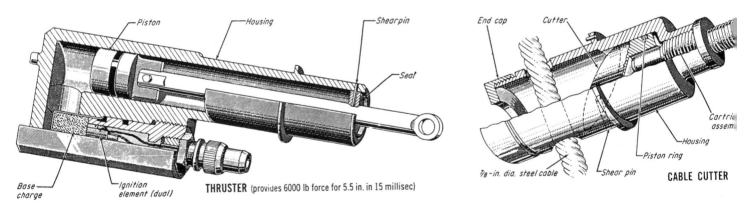

THRUSTER (provides 6000 lb force for 5.5 in. in 15 millisec)

CABLE CUTTER

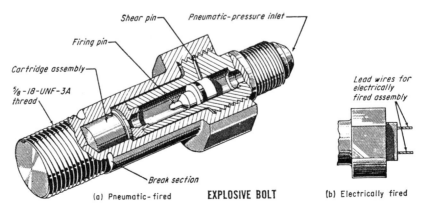

(a) Pneumatic-fired   **EXPLOSIVE BOLT**   (b) Electrically fired

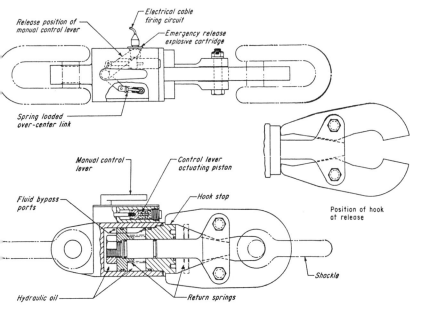

EMERGENCY HOOK RELEASE lets loads be jettisoned at any time. Hook is designed to release automatically if overloaded.

PIN RETRACTS to release load or clear a channel for free movement.

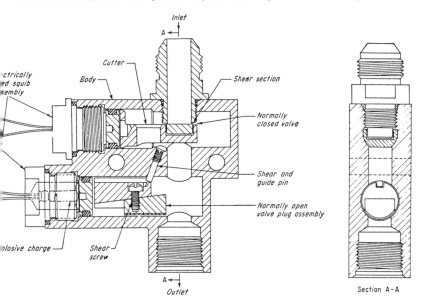

DUAL VALVE is so designed to let flow be started and stopped by same unit. Firing one squib starts flow; firing the other squib stops flow.

Section A-A

Quick disconnector

A tube joint can be separated almost instantaneously by remote control by use of an explosive bolt and a split threaded ring, in a design developed by James Mayo of Langley Research Center, Hampton, Va.

External threads of the ring mesh with the internal threads of the members that are joined—and must be separated quickly. The ring has a built-in spring tendency to assume a helically wound shape and to reduce to a smaller diameter when not laterally constrained. During assembly, it is held to its expanded size by two spring plates, whose rims fit into internal grooves machined in the split ring. The plates are tightened together by an explosive bolt and nut.

Upon ignition of the explosive bolt, the plates fly apart from the axial spring tension of the ring. The ring then contracts to its normally smaller diameter, releasing the two structural members.

The concept lends itself to any size and configuration. The retaining media need not be limited to V-type screw threads.

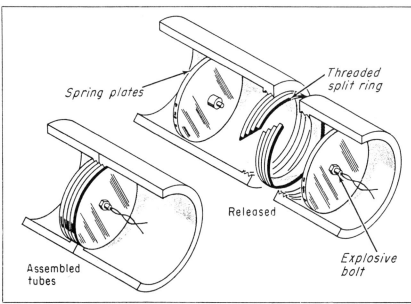

Threaded split ring with helical-spring effect holds ends of tubes at joint until explosive bolt is fired; then it releases instantly

Bellows Expansion Changes Leverage

Air flowing through holes in a bellows dashpot sets exposure time in an automatic camera. The pointer of a meter movement energized by a photocell controls flow rate by covering more holes to increase exposure time.

Exposure time . . .

is automatically controlled, after aperture is set for desired depth of focus, and louvers on photocell are set for film speed. Shunt resistance across meter element, adjusted by aperture lever, compensates for lens stop. When shutter is released, the spring-loaded actuating lever moves down against restraint of dashpot. Increasing the number of holes covered increases the exposure time; with all holes open, shutter speed is 1/250 sec; all holes covered, 1/15 sec. Design is by Agfa Camera Works, Munich, Germany.

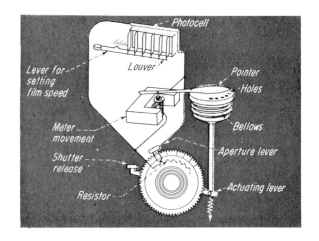

Harvester Stays Level

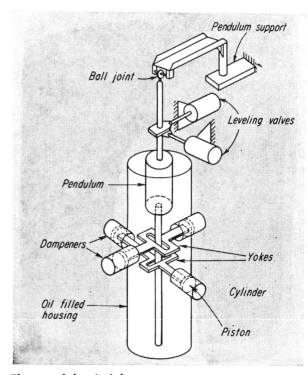

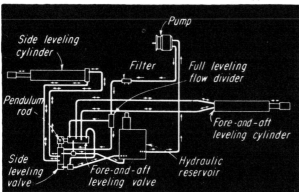

Hydraulic power . . .

for leveling is provided by engine-driven pump supplied from a central reservoir. Displaced pendulum rod moves the leveling valves to actuate the leveling cylinders. When thresher platform is leveled, pendulum is centered and valves closed.

The pendulum's job . . .

is to serve as vertical reference line and actuator that levels the combine platform no matter whether a hill goes up and down, or sideways. Two leveling valves connected to the pendulum-suspension rod are activated when the combine tilts. Oil pressure through these valves is directed to the appropriate leveling cylinder. This cylinder then levels the work surfaces, centers the pendulum, and closes the leveling valves.

Four pistons, in an oil-filled dashpot below the pendulum bob, provide damping in two planes. They are worked by an extension of the pendulum rod, which swings with the bob. Simple yokes allow the pendulum assembly to move in any direction and provide proportional displacement of both leveling valves for various tilt conditions.

CHAPTER 10.

FASTENING, LATCHING, CLAMPING AND CHUCKING DEVICES

Remotely Controlled Latch

A simple mechanism engages and disengages parallel plates carrying couplings and connectors.

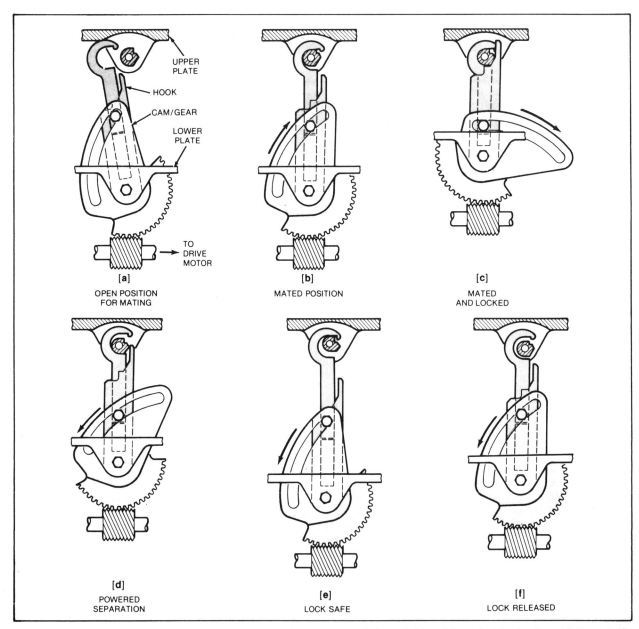

Figure 1. The **Latch Operation Sequence** is shown for locking in steps (a) through (c) and for unlocking in steps (d) through (f).

A new latch mates two parallel plates in one continuous motion (see Figure 1). On the Space Shuttle, the latch connects (and disconnects) plates carrying 20 fluid couplings and electrical connectors. (The coupling/connector receptacles are on one plate, and mating plugs are on the other plate.) Designed to lock items in place for handling, storage, or processing under remote control, the mechanism also has a fail-safe feature: It does not allow the plates to separate completely unless both are supported. Thus, plates cannot fall apart and injure people or damage equipment.

The mechanism employs four cam/gear assemblies, one at each corner of the lower plate (the photo in Figure 2 shows two cam/gears). The gears on each side of the plate face inward to balance the loading and help aline the plates. Worm gears on the cam/gear assemblies are connected to a common drive motor.

Figure 1 illustrates the sequence of movements as a pair of plates is latched and unlatched. Initially, the hook is extended and tilted out. The two plates are brought together, and when they are 4.7 in. (11.9 cm) apart, the drive motor is started (a). The worm gear rotates the hook until it closes on a pin on the opposite plate (b). Further rotation of the worm gear shortens the hook extension and raises the lower plate (c). At that point, the couplings and connectors on the two plates are fully engaged and locked.

To disconnect the plates, the worm gear is turned in the opposite direction. This motion lowers the bottom plate and pulls the couplings apart (d). However, if the bottom plate is unsupported, the latch safety feature operates. The hook cannot clear the pin if the lower plate hangs freely (e). If the bottom plate is supported, the hook extension lifts the hook clear of the pin (f) so that the plates are completely separated.

This work was done by Clifford J. Barnett, Paul Castiglione, and Leo R. Coda of Rockwell International Corp. for **Johnson Space Center**. *MSC-18365*

Latch clamps, toggles, then locks

A simple, compact arrangement of links combines high force-multiplication with locking reliability in latches that must operate in limited-access areas. One sweep of the handle (counterclockwise in the drawing below) hooks the latch into one of the members to be coupled, then tightly draws it to the other with a force that is highly magnified by means of the design's double toggle action.

Developed by Earl V. Holman of North American Aviation, Inc., Downey, Calif., under contract to the Manned Spacecraft Center in Houston, the latch will be one of 12 that will clamp the Lunar Excursion Module to the Command & Service Module in the Apollo program.

Four of the latches will hook automatically during docking, a connecting tunnel will pressurize, and an astronaut will crawl through to lock all 12 latches manually. A 20-lb. push provides a 4000-lb. clamping force. The tightening force can be applied almost directly downward—a desired characteristic in manually operated latches.

How a latch operates. As the handle is rotated, the hook, GB in the drawing on page 104, pivots downward into the notch of the coupling member. When GB cannot progress further, the continual rotation of EC causes F and C to spread apart in toggle fashion. Point F, however, is prevented from shifting to the left by Catch H, so Point C moves to the right to force a double toggle action (toggles EDCF and ABCF).

Point C also draws Point G to the right, with the added mechanical advantage of proportion AC to AB. Toward the end of the stroke, the toggles overcenter (that is, Point D falls below Line FC), and the latch

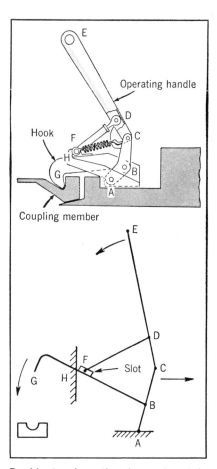

Double toggle action is produced by design at top. Schematic simplifies the interaction of locking forces.

is securely locked. Snapping the operating handle upward causes the hook to quickly pivot up and out of the notch. This release motion is aided by the spring.

The latch drawing shows Point F mounted on a small bracket that can slide on the top of the hook. A later modification, however, employs a slot in place of the bracket, in a manner similar to that shown in the schematic diagram (lower half of illustration above).

Inquiries concerning royalty-free licenses for commercial use of the latch should be addressed to NASA, Code GP, Washington, D. C. 20546. □

Toggle fastener inserts, locks, releases easily

For easy insertion and removal from one side of the work when access to the other side is difficult or impossible, a pin-type toggle fastener has been devised by C. C. Kubokawa at NASA's Ames Research Center. It can be used to fasten plates together, fasten things to walls or decks, or fasten units with surfaces of different curvature, such as a concave shape to a convex surface.

With actuator pin. The cylindrical body of the fastener has a tapered end for easy entry into the hole; the head is threaded to receive a winged locknut and, if desired, a ring for pulling the fastener out again after release. Slots in the body hold two or more toggle wings that respond to an actuator pin. These wings are extended except when the spring-loaded pin is depressed.

For installation, the actuator pin is depressed, retracting the toggle wings. When the fastener is in place, the pin is released, and the unit is then tightened by screwing the locknut down firmly, exerting a compressive force on the now-expanded toggle wings. For removal, the locknut is loosened and the pin is again depressed to retract the toggle wings. Meanwhile, the threaded outer end of the

cylindrical body functions as a stud to which a suitable pull ring can be screwed to facilitate removal of the fastener.

This invention has been patented by NASA (U.S. Patent No. 3,534,650). Royalty-free rights for commercial development may be negotiated through inquiry to the Patent Counsel, Ames Research Center, Moffett Field, Calif. 94035. ■

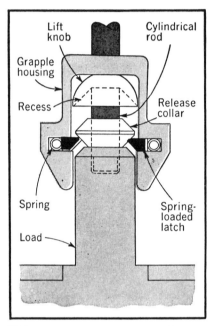

Sliding release collar is key feature.

New grapple frees loads automatically

A simple grapple mechanism designed at Argonne National Laboratory in Illinois engages and releases loads from overhead cranes automatically. This self-releasing device, which is used for highly radioactive loads in a reactor, lends itself to jobs where human intervention is impossible or inefficient, such as lowering and releasing loads from helicopters.

The mechanism (drawing, above) consists of two pieces: a lift knob secured to the load and a grapple member attached to the crane. The sliding latch-release collar under the lift knob is the design's key feature.

Spring magic. The grapple housing, which has a cylindrical inner

surface, contains a machined groove fitted with a garter spring and three metal latches. When the grapple is lowered over the lift knob, these latches recede into the groove as their edges come into contact with the knob. After passing the knob, they spring forward again, locking the grapple to the knob. Now the load can be lifted.

When the load is lowered to the ground again, gravity pull or pressure from above forces the grapple housing down until the latches come into contact with a double-cone-shaped release collar. The latches move back into the groove as they pass over the upper cone's surface and move forward again when they slide over the lower cone.

The grapple is then lifted so that the release collar moves up the cylindrical rod until it is housed in a recess in the lift knob. Since the collar can move no farther, the latches are forced by the upward pull to recede again into the groove —allowing the grapple to be lifted.

Developed to remove radioactive fuel rods, this mechanism was designed by John A. Froehlich and George A. Karastas, engineers at Argonne's Particle Accelerator Div.

Quick-release lock pin has foolproof ball device

A novel quick-release locking pin has been developed that can be withdrawn to separate the linked members only when stresses on the joint are negligible.

The pin may be the answer to the increasing demand for locking pins and fasteners that will pull out quickly and easily when desired, yet will stay securely in place without chance of unintentional release.

Key factor in the foolproof pin is a group of detent balls and a matching groove. The balls must be in the groove whenever the pin is either installed or pulled out of the assembly. This is simple to do during installation, but during removal the load must be off the pin to get the balls to drop into the groove.

How it works. The locking pin, developed by T.E. Othman, E.P. Nelson, and L.J. Zmuda of North American Rockwell Corp. under contract to Marshall Space Flight Center, consists of a forward-

Fastener with controllable toggles can be inserted and locked from only one side

pointing sleeve, with a spring-loaded sliding handle at its rear end, housing a sliding plunger that is urged backward (to its locking position) by a spring within the handle.

To some extent the plunger can slide forward against the plunger spring, and the handle backward against the handle spring. A groove near the front end of the plunger accommodates the detent balls when the plunger is pushed forward by compression of its spring. When the plunger is released backward, the balls are forced outward into holes in the sleeve, preventing withdrawal of the pin.

To install the pin, the plunger is pressed forward so the balls fall into their groove and the pin is pushed into the hole. When the plunger is released, the balls lock the sleeve against withdrawal.

To withdraw the pin, the plunger is pressed forward to accommodate the locking balls, and at the same time the handle is pulled backward. If the loading on the pin is negligible, the pin is withdrawn from the joint; if it is considerable, the handle spring is compressed and the plunger forced backward by the handle so the balls will return to their locking position.

The allowable amount of stress on the joint to permit removal can be varied by adjusting the pressure required for compression of the handle spring. If the stresses on the joint are too great for the pin to be withdrawn in the normal manner, hammering the forward end of the plunger simply ensures that the plunger is in its rearward position, with the locking balls preventing withdrawal of the pin. A stop on the forward end of the plunger prevents the plunger from being driven backward.

Although a patent application has been filed, royalty-free licenses may be granted. Inquiries should be made to NASA, Code GP, Washington, D.C. ∎

Cables operate a quick-release latch

Landing a craft on the moon is no easy matter, but once there, neither is getting that craft back to earth. This latter problem is what confronted engineers R. T. Barbour and D. E. Necker of the Space Div., North

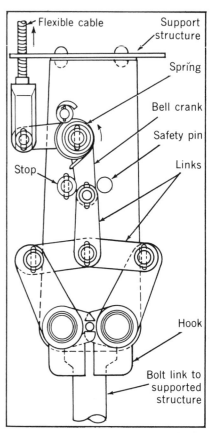

Instant-release latch is opened by cutting a preloaded flexible cable.

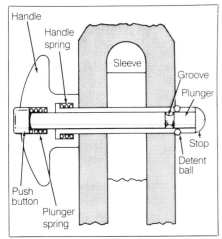

Foolproof locking pin releases quickly only when stress on joint is negligible

Designs of clamp mechanisms used as robot hands

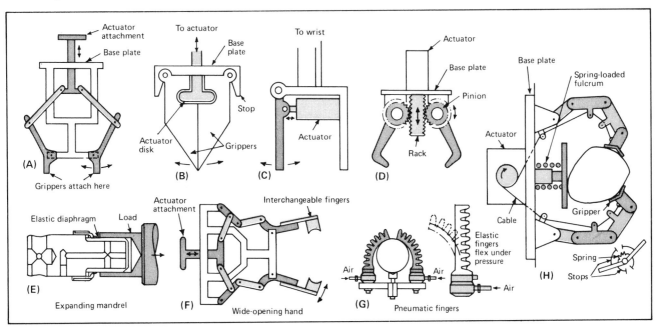

American Rockwell, Downey, Calif., who had to develop means to test the earth landing system of NASA's Apollo command module to make sure it would work on its return.

To test the landing system, a vehicle complete with a parachute deployment module was carried aloft in a B-52 aircraft to an altitude of about 60,000 ft., where it was released for a trial descent. The vehicle was supported at two points, which had to release it simultaneously.

Means to the end. Barbour and Necker solved the problem of the release by using a pair of double-hook latches held closed by a toggle linkage, which was connected to a flexible cable through a spring-loaded bell crank. Both cables were routed under tension to an explosive-actuated cutter, where they were interconnected.

Bolts with mushroom-like heads were attached to the test vehicle and gripped by the double hooks at each of the two ends. By torquing the nut-and-bolt assembly, a preload was produced, assuring the opening of the hooks under even adverse conditions. A coil spring also added an opening force on the toggle. The hooks were kept closed at first by a safety pin and then by preloading the interconnecting cable.

Upon command to release the test vehicle, the guillotine's cutter action was electrically initiated to cut the cable, which then released the hooks.

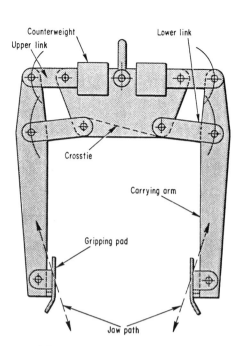

Lift-tong mechanism firmly grips objects

Twin four-bar linkages are the key components in this tong mechanism that can grip with a constant weight-to-grip force ratio for any size object within its grip range. Developed by Gripall Corp., the tong mechanism relies on a cross-tie between the two sets of linkages to produce equal and opposite linkage movement. The vertical links have extensions with grip pads mounted at their ends, while the horizontal links are so proportioned that their pads move in an inclined straight-line path. The weight of the load being lifted, therefore, wedges the pads against the load with a force that is proportional to the object's weight and independent of its size.

from MACHINE DESIGN

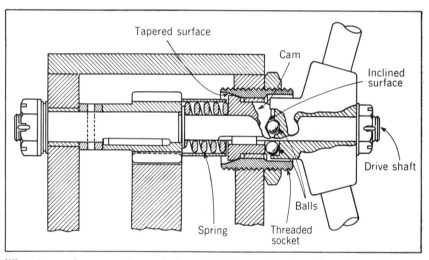

When torque is removed, cam is forced into tapered surface for brake action.

Automatic brake locks hoist when driving torque ceases

A brake mechanism attached to a chain hoist is helping engineers at North American Aviation to lift and align equipment accurately by automatically locking in position when the driving torque is removed from the hoist.

According to the designer, Joseph Pizzo, the brake could also be used on wheeled equipment operating on slopes, to act as an auxiliary brake system.

How it works. When torque is applied to the driveshaft (as shown in the sketch), four steel balls try to move up the inclined surfaces of the cam. Although called a cam by the designers, it is really a concentric collar with a cam-like surface on one of the end faces. Since the balls are contained by four cups in the hub, the cam is forced to move forward axially to the left. Because the cam moves away from the tapered surface, the cam and the driveshaft **that is keyed to it** are now free to rotate.

If the torque is removed, a spring resting against the cam and the driveshaft gear forces the cam back into the tapered surface of the **threaded socket** for instant braking.

Although this particular mechanism (which can rotate in either direction) was designed for manual operation, the principle can be used on powered systems.

Perpendicular-Force Latch

Installation and removal of equipment modules are simplified.

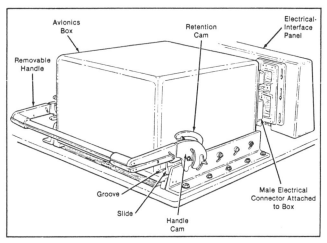

Figure 1. An **Avionics Box** mates with electrical connectors in the rear and is locked in position on the cold plate when installed with the latching mechanism.

A latching mechanism simultaneously applies force in two perpendicular directions to install or remove electronic-equipment modules. The mechanism (see Figure 1) requires only the simple motion of a handle to push or pull an avionic module to insert or withdraw connectors on its rear face into or from spring-loaded mating connectors on a panel and to force the box downward onto or release the box from a mating cold plate that is part of the panel assembly. The concept is also adaptable to hydraulic, pneumatic, and mechanical systems. Mechanisms of this type can be used to simplify the installation and removal of modular equipment where movement is restricted by protective clothing as in hazardous environments or where the installation and removal are to be performed by robots or remote manipulators.

Figure 2 shows an installation sequence. In step 1, the handle has been installed on the handle cam and turned downward. In step 2, the technician or robot pushes the box rearward as slides attached to the rails enter grooves near the bottom of the box. In step 3, as the box continues to move to the rear, the handle cam automatically aligns with the slot in the rail and engages the rail roller.

In step 4, the handle is rotated upward 75°, forcing the box rearward to mate with the electrical connectors. In step 5, the handle is pushed upward an additional 15°, locking the handle cam and the slide. In step 6, the handle is rotated an additional 30°, forcing the box and the mating spring-loaded electrical connectors downward so that the box engages the locking pin and becomes clamped to the cold plate. The sequence for removal is identical except that the motions are reversed.

This work was done by John P. Mattei, Peter A. Buck, and Michael D. Williams of Rockwell International Corp. for **Johnson Space Center**. *For further information, Circle 90 on the TSP Request Card.* MSC-21406

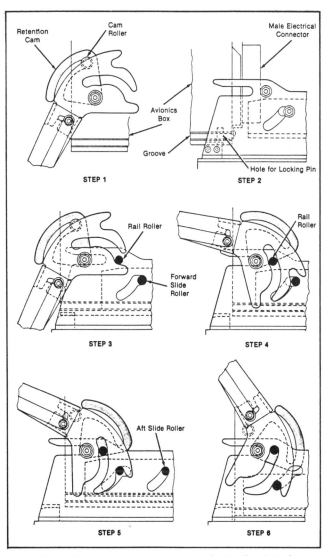

Figure 2. This **Installation Sequence** shows the positions of the handle and retention cams as the box is moved rearward and downward.

Quick Release Mechanisms

Quick Release Mechanism

George A. Fries

Quick release mechanisms have numerous applications. Although the design shown here operates as a tripping device for a quick release hook, mechanical principles involved undoubtedly have many other applications. Fundamentally, it is a toggle-type mechanism featured by the fact that the greater the load the more effective the toggle.

The hook is suspended from the shackle, and the load or work is supported by the latch which is machined to fit the fingers C. The fingers C are pivoted about a pin. Assembled to the fingers are the arms E, pinned at one end and joined at the other by the sliding pin G. Inclosing the entire unit are the side plates H containing the slot J for guiding the pin G in a vertical movement when the hook release. The helical spring returns the arms to the bottom position after they have been released.

To trip the hook, the tripping lever is pulled by the cable M until the arms E pass their horizontal center line. The toggle effect is then broken, thereby releasing the load.

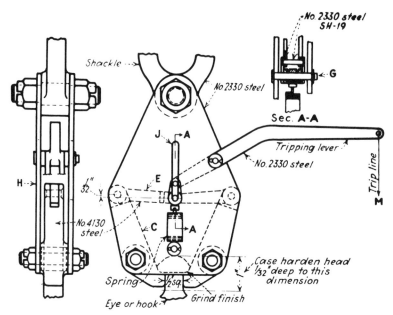

Simple quick-release toggle mechanism as designed for tripping a lifting hook

Positive Locking and Quick Release Mechanism (Brief 63-10420)

The objective was to design a simple device which would hold two objects together securely and quickly release them on demand.

One object, such as a plate, is held to another object, such as a vehicle, by a spring-loaded slotted bolt, which is locked in position by two retainer arms. The retainer arms are constrained from movement by a locking cylinder. To release the plate, a detent is actuated to lift the locking cylinder and rotate the retainer arms free from contact with the slotted bolt head. As a result of this action, the spring-loaded bolt is ejected and the plate is released from the vehicle.

Actuation of the slidable detent can be initiated by a squib, a fluid-pressure device, or a solenoid and the principle of this device can be employed wherever a positive engagement that can be quickly released on demand is required. Some suggested applications of this principle are in coupling devices for load-carrying carts or trucks, hooks or pick-up attachments for cranes, and quick-release mechanisms for remotely controlled manipulators. No patent application is involved in this idea.

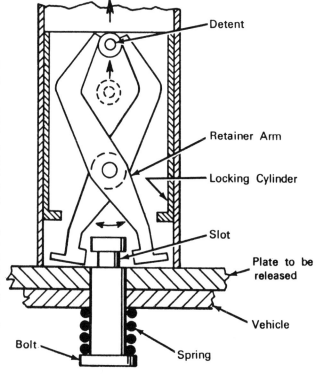

Diagram of the mechanism locking a vehicle and plate.

Ring springs clamp platform elevator into position

A simple yet effective technique keeps a platform elevator locked safely in position without an external clamping force. The platform (drawing, right), being built by Japan's Mitsui Shipbuilding Co., contains special ring assemblies that grip the four column-shafts with a strong force by the simple physical interaction of two tapered rings.

Thus, unlike conventional platform elevators, no outside power supply is required to hold the platform in position. Conventional jacking power is employed, however, in raising the platform from one position to another.

How the rings work. The ring assemblies are larger versions of the ring springs sometimes employed for shock absorption. In the Japanese version, the assembly is made up of an inner nonmetallic ring tapering upward and an outer steel ring tapering downward (drawing, right).

The outside ring is linked to the platform, and the inside ring is positioned against the circumference of the column shaft. When the platform is raised to the designed height, the jack force is removed, and the full weight of the platform bears downward on the outside ring with a force that, through a wedging action, is transferred into a horizontal inward force of the inside ring.

Thus, the column shaft is gripped tightly by the inside ring; the heavier the platform the larger the gripping force produced.

Advantage of the technique is that the shafts do not need notches or threads, and cost is reduced. Moreover, the shafts can be made of reinforced concrete.

Mitsui has successfully tested a full-size ring assembly and column shaft by subjecting the unit to a 400-ton load.

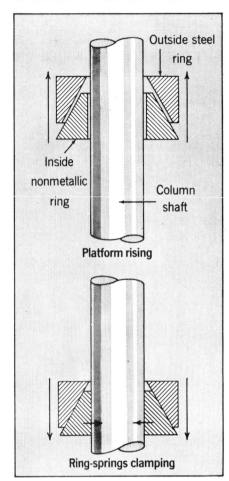

Platform rising

Ring-springs clamping

Ring springs unclamp the column as platform is being raised (upper). As soon as the jack power is removed (lower), the column is gripped by the inner ring.

Cammed jaws in hydraulic cylinder grip sheets

A single, double-acting hydraulic cylinder in each work holder clamps and unclamps the work and retracts or advances the jaws as required. With the piston rod fully withdrawn into the hydraulic cylinder (A), the jaws of the holder are retracted and open. When the control valve atop the work holder is actuated, the piston rod moves forward a total of 12 in. The first 10 in. of movement (B) brings the sheet-locater bumper into contact with the work. The cammed surface on the rod extension starts to move the trip block upward and the locking pin starts to drop into position. The next ¾ in. of piston-rod travel (C) fully engages the work-holder locking pin and brings the lower jaw of the clamp up to the bottom of the work. The work holder slide is now locked between the forward stop and the locking pin. The last 1¼ in. of piston travel (D) clamps the workpiece between the jaws with a pressure of 2500 lbs. No adjustment for work thickness is necessary. A jaws-open limit switch is used to hold the work holder in position (C) for loading and unloading operations.

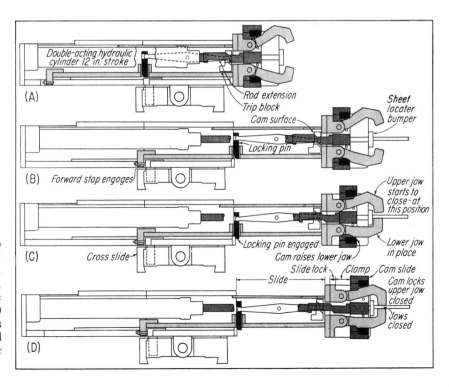

Quick-Acting Clamps for Machines and Fixtures

(A) An eccentric clamp. (B) Spindle clamping bolt. (C) Method of clamping hollow column to structure. Permits quick rotary adjustment of column. (D) (a) Cam catch for clamping rod or rope. (b) Method of fastening small cylindrical member to structure using thumb nut and clamp jaws. Permits quick longitudinal adjustment of shaft in structure. (E) Cam catch to lock a wheel or spindle. (F) Spring handle. Movement of handle in vertical or horizontal position provides movement at A. (G) Roller and inclined slot for locking a rod or rope. (H) Method of clamping light member to structure. Serrated edge on structure provides for quickly accommodating members of different thicknesses. (I) Spring taper holder with sliding ring. (J) Special clamp for holding member a. (K) Cone, nut, and levers grip member a. Grip may have two or more jaws. With only two jaws, device serves as a small vise. (L) Two different types of cam clamps. (M) Cam cover catch. Movement of handle downward locks cover tightly. (N) Clamping sliding member to slotted structure using wedge bolt. Provides means of quick adjustment of member on structure.

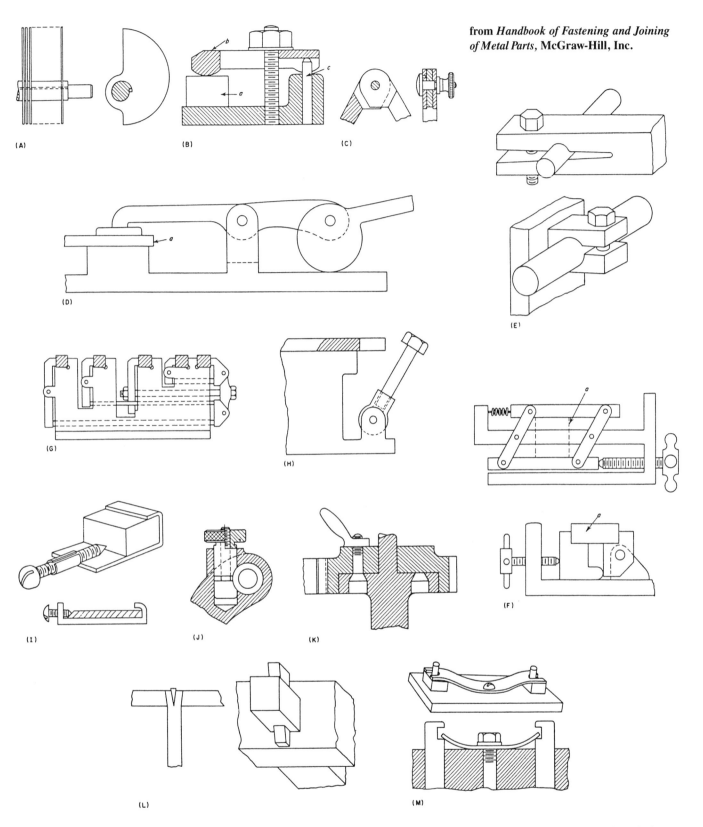

from *Handbook of Fastening and Joining of Metal Parts*, McGraw-Hill, Inc.

(A) Method of fastening condenser plates to structure using circular wedge. Rotation of plates in clockwise direction locks plates on structure. (B) Method of clamping member **a** using special clamp. Detail **b** pivots on pin **c**. (C) Method of clamping two movable parts so that they may be held in any angular position by use of clamping screw. (D) Cam clamp for clamping member **a**. (E) Methods of clamping cylindrical member. (F) Method of clamping member **a** using special clamps. (G) Special clamping device which permits parallel clamping of five parts by the tightening of one bolt. (H) Method of securing structure by use of bolt and movable detail which provides a quick method of fastening cover. (I) Method of quickly securing, adjusting, or releasing center member. (J) Method of securing bushing in structure by the use of clamp screw and thumb nut. (K) Method of securing attachment to structure by use of bolt and hand lever used as a nut. (L) Method of fastening member to structure using wedge. (M) Methods of fastening two members to structure using spring and one screw. Members may be removed without loosening screw.

FRICTION CLAMPING DEVICES

BERNARD J. WOLFE

ALL TYPES of mechanisms used for gaining mechanical advantage have probably been used in the design of friction clamps. This type of clamp can hold moderately large loads by friction grip on smooth surfaces even of comparatively small area and, in some designs, tightened or released with little effort and movement of the control. In the clamps illustrated here the mechanical advantage is gained by the use of the common devices: lever, toggle, screw, wedge, and combinations of these means.

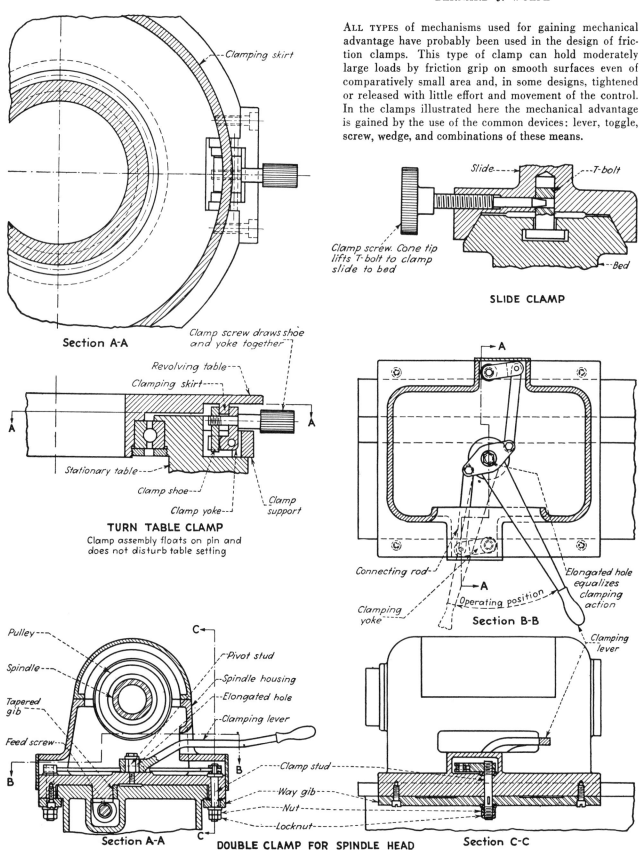

Section A-A

Clamping skirt

Clamp screw. Cone tip lifts T-bolt to clamp slide to bed

Slide — T-bolt — Bed

SLIDE CLAMP

Clamp screw draws shoe and yoke together
Revolving table
Clamping skirt
Stationary table
Clamp shoe
Clamp yoke
Clamp support

TURN TABLE CLAMP
Clamp assembly floats on pin and does not disturb table setting

Connecting rod
Clamping yoke
Operating position
Elongated hole equalizes clamping action
Clamping lever

Section B-B

Pulley
Spindle
Tapered gib
Feed screw
Pivot stud
Spindle housing
Elongated hole
Clamping lever
Clamp stud
Way gib
Nut
Locknut

Section A-A

DOUBLE CLAMP FOR SPINDLE HEAD

Section C-C

AND PRINCIPLES OF DESIGN

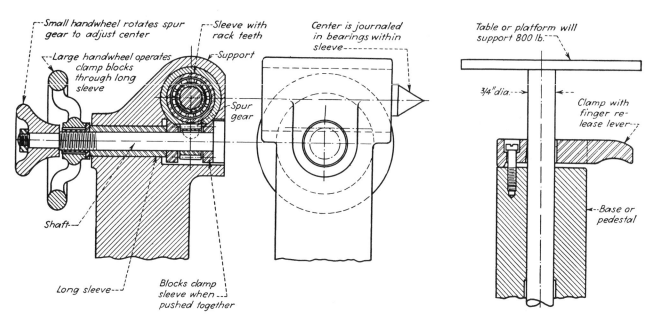

Small handwheel rotates spur gear to adjust center

Large handwheel operates clamp blocks through long sleeve

Sleeve with rack teeth

Support

Center is journaled in bearings within sleeve

Spur gear

Shaft

Long sleeve

Blocks clamp sleeve when pushed together

CENTER SUPPORT CLAMP

Table or platform will support 800 lb.

3/4" dia.

Clamp with finger release lever

Base or pedestal

PEDESTAL CLAMP

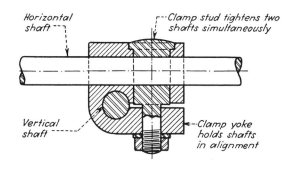

Horizontal shaft

Clamp stud tightens two shafts simultaneously

Vertical shaft

Clamp yoke holds shafts in alignment

RIGHT ANGLE CLAMP

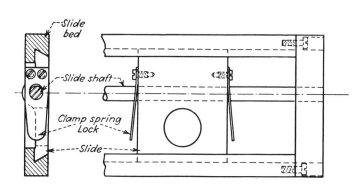

Slide bed

Slide shaft

Clamp spring Lock

Slide

SLIDE CLAMP

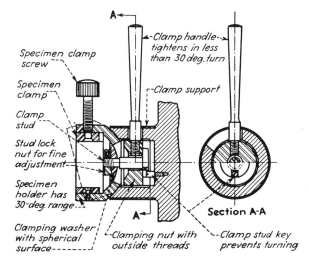

Clamp handle tightens in less than 30 deg. turn

Clamp support

Specimen clamp screw

Specimen clamp

Clamp stud

Stud lock nut for fine adjustment

Specimen holder has 30-deg. range

Clamping washer with spherical surface

Clamping nut with outside threads

Clamp stud key prevents turning

Section A-A

SPECIMEN HOLDER CLAMP

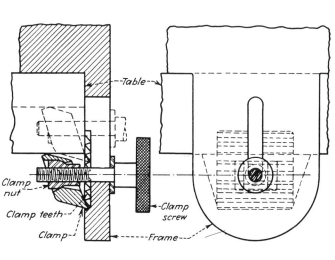

Table

Clamp nut

Clamp teeth

Clamp

Clamp screw

Frame

TABLE CLAMP

Detent designs for stopping mechanical movements

Some of the more robust and practical devices for locating or holding mechanical movements are surveyed by the author.

LOUIS DODGE

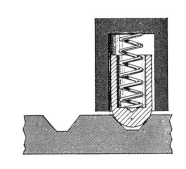

FIXED HOLDING POWER IS CONSTANT IN BOTH DIRECTIONS

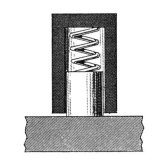

DOMED PLUNGER HAS LONG LIFE

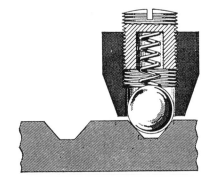

ADJUSTABLE HOLDING POWER

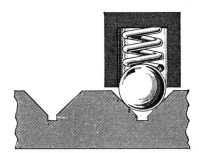

WEDGE ACTION LOCKS MOVEMENT IN DIRECTION OF ARROW

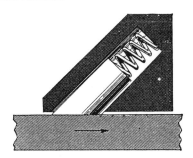

FRICTION RESULTS IN HOLDING FORCE

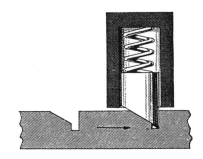

NOTCH SHAPE DICTATES DIRECTION OF ROD MOTION

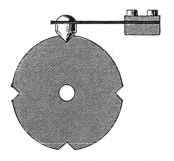

LEAF SPRING PROVIDES LIMITED HOLDING POWER

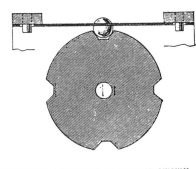

LEAF SPRING DETENT CAN BE REMOVED QUICKLY

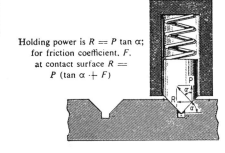

Holding power is $R = P \tan \alpha$; for friction coefficient, F, at contact surface $R = P (\tan \alpha + F)$

CONICAL OR WEDGE-ENDED DETENT

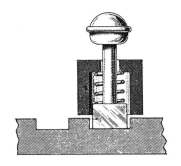

POSITIVE DETENT HAS MANUAL RELEASE

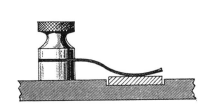

LEAF SPRING FOR HOLDING FLAT PIECES

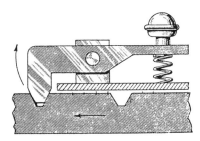

AUTOMATIC RELEASE OCCURS IN ONE DIRECTION, MANUAL RELEASE NEEDED IN OTHER DIRECTION

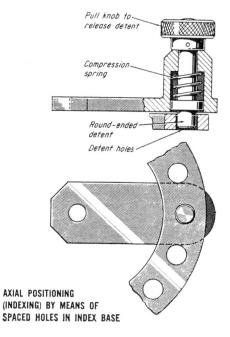

Pull knob to release detent

Compression spring

Round-ended detent

Detent holes

AXIAL POSITIONING
(INDEXING) BY MEANS OF
SPACED HOLES IN INDEX BASE

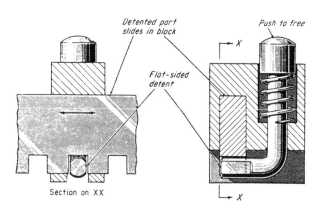

Detented part slides in block

Flat-sided detent

Push to free

Section on XX

X — X

POSITIVE DETENT HAS PUSH
BUTTON RELEASE FOR
STRAIGHT RODS

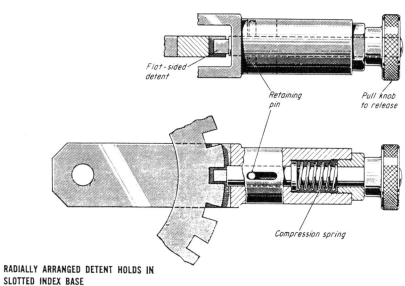

Flat-sided detent

Retaining pin

Pull knob to release

Compression spring

RADIALLY ARRANGED DETENT HOLDS IN
SLOTTED INDEX BASE

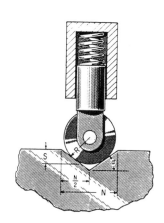

ROLLER DETENT POSITIONS IN A NOTCH:

$$\text{RISE, } S = \frac{N \tan a}{2} - R \times \frac{1 - \cos}{\cos a}$$

$$\text{ROLLER RADIUS, } R = \left(\frac{N \tan a}{2} - S \right) \left(\frac{\cos a}{1 - \cos a} \right)$$

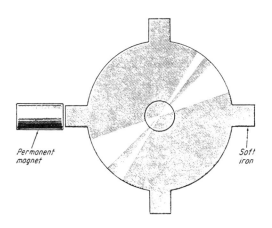

Permanent magnet

Soft iron

MAGNETIC DETENT

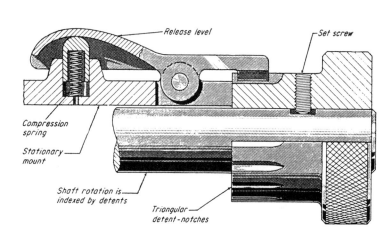

Release level

Set screw

Compression spring

Stationary mount

Shaft rotation is indexed by detents

Triangular detent-notches

AXIAL DETENT FOR POSITIONING OF
ADJUSTMENT KNOB WITH MANUAL RELEASE

SPRING-LOADED CHUCKS and HOLDING FIXTURES

SIGMUND RAPPAPORT

Spring-loaded fixtures for work-holding are sometimes preferable to other types. Their advantages are: Shorter setup time and quick workpiece change. And work distortion is much reduced because spring force can be easily and accurately adjusted.

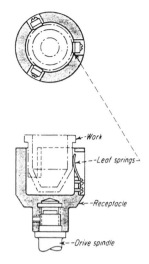

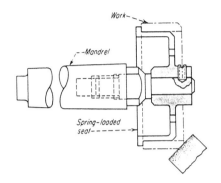

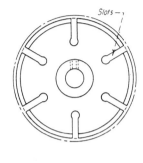

Spring-loaded Nest . . .

has radial slots extending into face. These ensure even grip on work, which is pushed over rim. Slight lead on rim makes work-mounting easier. Chief use of this fixture is for ball-bearing race grinding where only light cutting forces are necessary.

Cupped Fixture . . .

has three leaf-springs equally spaced in wall. Work, usually to be lacquered, is inserted into cup during rotation. Because work is placed in fixture by hand, spindle is usually friction driven for safety.

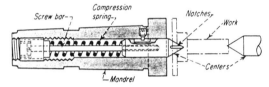

Lathe Center . . .

is spring loaded and holds work by spring pressure alone. Eight sharp-edged notches on the conical surface of driving center bite into the work and drive it. Spring tension is adjustable.

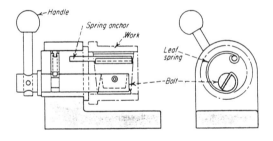

Leaf-spring Gripper . . .

is used mainly to hold work during assembly. One end of flat coil-spring is anchored in housing; other end is held in bolt. When bolt is turned, spring tightened and its OD decreased. After work is slid over spring, bolt handle is released. Spring then presses against work, holding it tight.

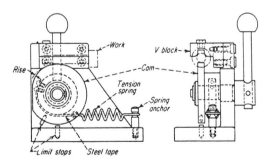

Spring Clamp . . .

has cam and tension spring to apply clamping force. Tension spring activates cam through steel band. When handle is released, cam clamps work against V-bar. Two stop-pins limit travel when there is no work in the fixture.

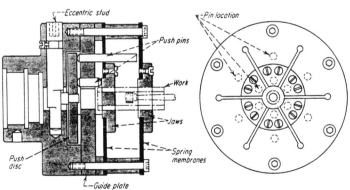

Slotted Membranes . .

have holding jaws attached. ID of holding jaws hold work. When eccentric stud is turned it forces two sets of push-pins against the spring-membrane. This action deflects membrane, opening jaws for receiving or releasing work. Turning the eccentric stud back relieves the push-pins, allowing spring to snap back and grip work.

CHAPTER 11

KEY EQUATIONS AND CHARTS FOR DESIGNING MECHANISMS

Designing geared five-bar mechanisms

Geared five-bar mechanisms have excellent force-transmission characteristics and can produce more complex output motions—including dwells—than conventional four-bar designs

from Design Engineering

In the design of machinery, it is often necessary to use a mechanism to convert uniform input rotational motion into non-uniform output rotation or reciprocation. Mechanisms designed for such purposes are almost invariably based on four-bar linkages. Such linkages produce a sinusoidal output that can be modified to yield a variety of motions.

Four-bar linkages have their limitations, however. Since they cannot produce dwells of useful duration, the designer may have to use a cam when a dwell is desired and accept the inherent speed restrictions and vibration associated with cams. A further limitation of four-bar linkages is that only a few types have efficient force-transmission capabilities.

One way to increase the variety of output motions of a four-bar linkage, and obtain longer dwells and better force transmissions is to add a link. The resulting five-bar linkage would be operationally impractical, however, because it would be endowed with two degrees of freedom and would thus require two inputs to control the output.

Simply constraining two adjacent links would not solve the problem. The five-bar chain would then merely degenerate back to a four-bar linkage. If, on the other hand, any two nonadjacent links are constrained in a manner that removes only one degree of freedom, the five-bar chain becomes a functionally useful mechanism.

Gearing provides solution. There are several ways to constrain two nonadjacent links in a five-bar chain. Some possibilities are using gears, slot-and-pin joints or nonlinear band mechanisms. Of these three possibilities, gearing is the most attractive. Some practical gearing systems (Fig 1) include paired external gears, planet gears revolving within an external ring gear and planet gears driving slotted cranks.

In one successful system (Fig 1A) each of the two external gears has a fixed crank that is connected to a crossbar by a rod. The system has been successful in high-speed machines, where it transforms rotary motion into high-impact linear motion. A similar system (Fig 1B) is used in a Stirling engine.

In a different type of system (Fig 1C) a pin on a planet gear traces an epicyclic

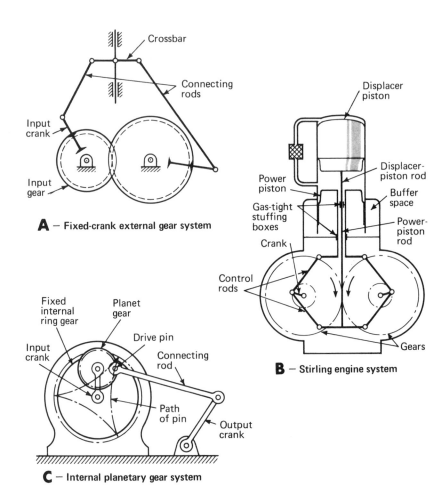

A — Fixed-crank external gear system

B — Stirling engine system

C — Internal planetary gear system

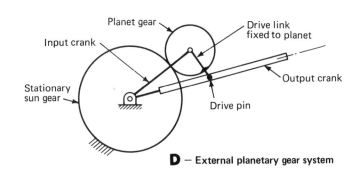

D — External planetary gear system

1. Five-bar mechanism designs can be based on paired external gears or planetary gears. They convert simple input motions into complex outputs

three-lobe curve to drive an output crank back and forth with a long dwell at the extreme right-hand position. A slotted output crank (Fig 1D) can be used to obtain a similar type of output.

Which type is best? Intrigued by the potentialities of geared five-bar mechanisms, two associate professors of mechanical engineering—Daniel H. Suchora of Youngstown State University (Youngstown, OH) and Michael Savage of the University of Akron (Akron, OH)—decided to study one form of this type of mechanism in detail. The work was initiated as part of Suchora's dissertation while he was at Case Western Reserve University.

Five kinematic inversions of this form (Fig 2) were established by the two researchers. As an aid in distinguishing between the five, each type is named according to which link acts as the fixed link. The study showed that the Type 5 mechanism could be of the greatest practical value.

In the Type 5 mechanism (Fig 3A) one gear is stationary and acts as a sun gear. The input shaft at Point E drives the input crank which, in turn, causes the planet gear to revolve around the sun gear. Link a_2, fixed to the planet, then drives the output crank, Link a_4, by means of the connecting link, Link a_3. At any input position, the third and fourth links may be assembled in either of two distinct positions or "phases" (Fig 3B).

Variety of outputs. The different types of output motions that may be obtained from a Type 5 mechanism are based on the different epicyclic curves traced by Link Joint B. The variables that control the shape of a "B-curve" are the gear ratio GR $(GR = N_2/N_5)$, the link ratio a_2/a_1 and the initial position of the gear set, defined by the initial positions of θ_1 and θ_2, designated as θ_{10} and θ_{20}, respectively.

Typical B-curve shapes (Fig 4) include ovals, cusps and loops. When the B-curve is oval (Fig 4B) or semioval (Fig 4C) the resulting B-curve is similar to the true-circle B-curve produced by a four-bar linkage. The resulting output motion of Link a_4 will be a sinusoidal type of oscillation, similar to that produced by a four-bar linkage.

When the B-curve is cusped (Fig 4A), dwells are obtained. When the B-curve is looped (Figs 4D and 4E), a double oscillation is obtained.

In the case of the cusped B-curve (Fig 4A) by selecting a_2 to be equal to the pitch radius of the planet gear r_2, Link Joint B becomes located at the pitch circle of the planet gear. The gear ratio in all the cases illustrated is unity $(GR = 1)$.

Researchers Suchora and Savage analyzed the type of output motion pro-

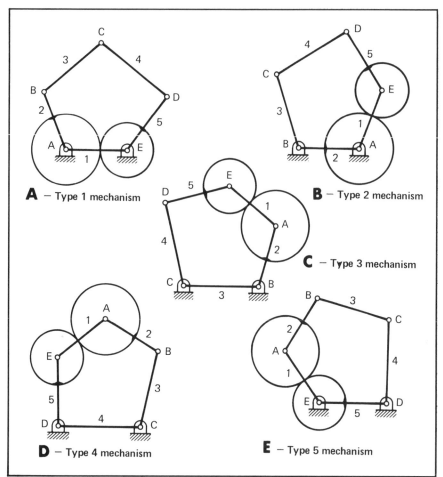

2. Five types of geared five-bar mechanisms. A different link acts as the fixed link in each type. Type 5 may be the most useful in machinery design

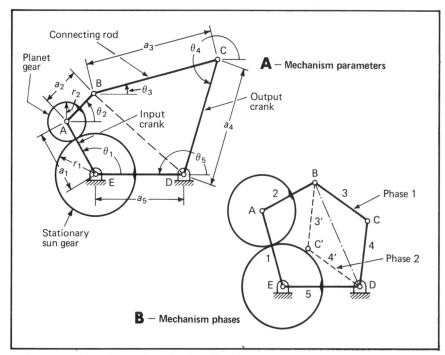

3. Detailed design of Type 5 mechanism. The input crank causes the planet gear to revolve around the sun gear, which is always stationary

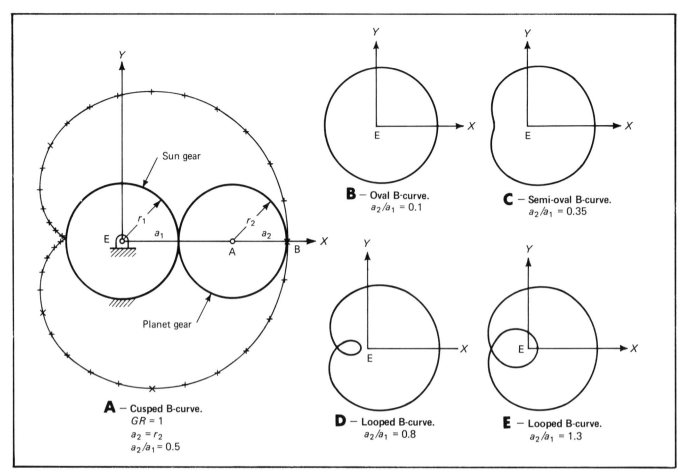

4. Typical B-curve shapes obtained from various Type 5 geared five-bar mechanisms. The shape of the epicyclic curve is changed by the link ratio a_2/a_1 and other parameters, as described in the text of this article

Calculating displacement, velocity and accleration

Displacement θ_4 can be found from the following equation:

$$\theta_4 = 2 \tan^{-1} \left(\frac{I \pm \sqrt{I^2 + H^2 - J^2}}{H + J} \right)$$

where $H = a_1 \cos \theta_1 + a_2 \cos \theta_2 - a_5$; $I = a_1 \sin \theta_1 + a_2 \sin \theta_2$; and $J = 1/2a_4 (a^2_3 - a^2_4 - H^2 - I^2)$; and where $\theta_2 = \theta_{20} + (1 + 1/GR) (\theta_1 - \theta_{10})$, where θ_{10} and θ_{20} are the initial values of the angles θ_1 and θ_2, respectively.

For layout purposes, once θ_4 is determined, θ_3 can be found from:

$$\theta_3 = \tan^{-1} \left(\frac{a_4 \sin \theta_4 + I}{a_4 \cos \theta_4 + H} \right)$$

To find velocities θ'_4 and θ'_3, use these equations:

$$\theta'_4 = \frac{a_1 \sin (\theta_3 - \theta_1) + a_2 \sin (\theta_3 - \theta_2) \theta'_2}{a_4 \sin (\theta_4 - \theta_3)}$$

$$\theta'_3 = \frac{a_1 \sin (\theta_1 - \theta_4) + a_2 \sin (\theta_2 - \theta_4) \theta'_2}{a_4 \sin (\theta_4 - \theta_3)}$$

where $\theta'_2 = (1 + 1/GR)$.

Use these equations to determine accelerations θ''_4 and θ''_3:

$$\theta''_4 = \frac{L}{a_3 a_4 \sin (\theta_4 - \theta_3)}$$

$$\theta''_3 = \frac{K}{a_3 a_4 \sin (\theta_4 - \theta_3)}$$

where $K = a_3a_4\cos(\theta_3 - \theta_4) \theta'_3{}^2 + a_4{}^2\theta'_4{}^2 + a_1a_2\cos(\theta_1 - \theta_4) + a_2a_4\cos(\theta_2 - \theta_4)\theta'_2{}^2$ and $L = a_3{}^2\theta'_3{}^2 - a_3a_1\cos(\theta_3 - \theta_4)\theta'_4{}^2 - a_1a_3\cos(\theta_3 - \theta_1) + a_2a_3\cos(\theta_3 - \theta_2)\theta'_2{}^2$.

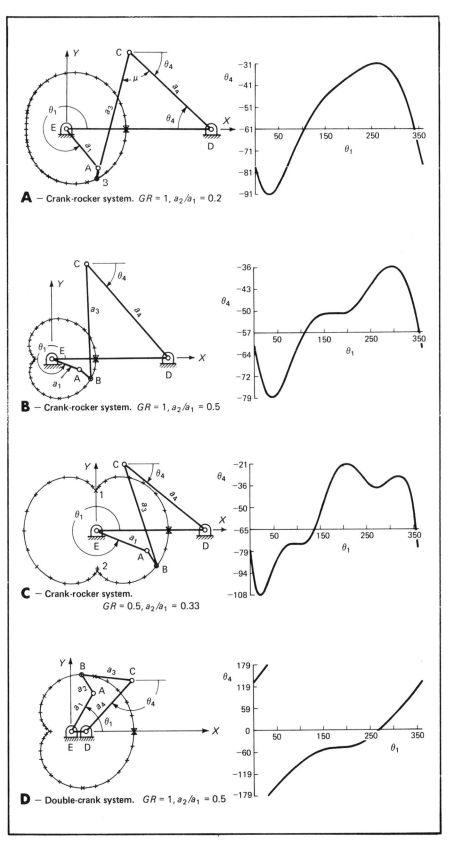

A — Crank-rocker system. $GR = 1$, $a_2/a_1 = 0.2$

B — Crank-rocker system. $GR = 1$, $a_2/a_1 = 0.5$

C — Crank-rocker system.
$GR = 0.5$, $a_2/a_1 = 0.33$

D — Double-crank system. $GR = 1$, $a_2/a_1 = 0.5$

5. Variety of output motions can be produced by varying design of five-bar geared mechanisms. Dwells are obtainable with proper design. Force transmission is excellent. In these diagrams, the angular position of the output link is plotted against the angular position of the input link for various five-bar mechanism designs

duced by the geared five-bar mechanisms by plotting the angular position θ_4 of the output link a_4 against the angular position of the input link θ_1 for a variety of mechanism configurations (Fig 5).

In three of the four cases illustrated, $GR = 1$, although the gear pairs are not shown. Thus one input rotation generates the entire path of the B-curve. Each mechanism configuration produces a different output.

One configuration (Fig 5A) produces an approximately sinusoidal reciprocating output motion that usually has better force-transmission capabilities than equivalent four-bar outputs. The transmission angle μ should be within 45 to 135 deg during the entire rotation for best results.

Another configuration (Fig 5B) produces a horizontal or almost-horizontal portion of the output curve. The output link, Link a_4, is virtually stationary during this period of input rotation—from about 150 to 200 deg of input rotation θ_1 in the case illustrated. Dwells of longer duration are possible with proper design.

Changing the gear ratio to 0.5 (Fig 5C) a complex motion is obtained, with two intermediate dwells that occur at Cusps 1 and 2 in the path of the B-curve. One dwell, from $\theta_1 = 80$ deg to 110 deg, is of good quality. The dwell from 240 deg to 330 deg is actually a small oscillation.

Dwell quality is affected by the location of Point D with respect to the cusp, and by the lengths of Links a_3 and a_4. It is possible to design this form of mechanism to produce two usable dwells per rotation of input.

In a double-crank version of the geared five-bar mechanism (Fig 5D), the output link makes full rotations. The output motion is approximately linear, with a usable intermediate dwell caused by the cusp in the path of the B-curve.

From the foregoing, it's apparent that the Type 5 geared mechanism with $GR = 1$ offers many useful motions to machinery designers. Researchers Suchora and Savage have derived the necessary displacement, velocity and acceleration equations (see "Calculating displacement, velocity and acceleration" box). Although the equations may seem complex, they are quickly and easily solved using a small electronic calculator.

Ted Black

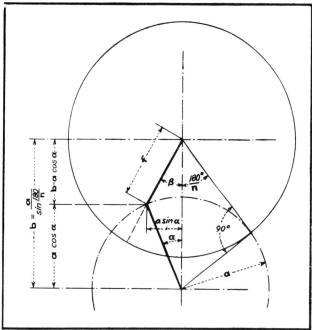

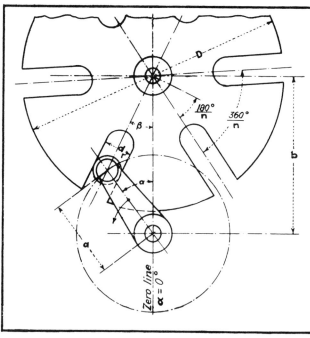

Fig. 1—Basic outline sketch for the external Geneva wheel. The symbols are identified for application in the basic equations.

Fig. 2—Schematic sketch of a six slot Geneva wheel. Roller diameter, d_r, must be considered when determining D.

Kinematics of Intermittent Mechanisms
I–The External Geneva Wheel

Table I—Notation and Formulas for the External Geneva Wheel

Assumed or given: a, n, d and p

a = crank radius of driving member

n = number of slots

d_r = roller diameter

p = constant velocity of driving crank in rpm

$$m = \frac{1}{\sin \frac{180}{n}}$$

b = center distance = am

D = diameter of driven member = $2\sqrt{\dfrac{d^2_r}{4} + a^2 \cot^2 \dfrac{180}{n}}$

ω = constant angular velocity of driving crank = $\dfrac{p\pi}{30}$ radians per sec

α = angular position of driving crank at any time

β = angular displacement of driven member corresponding to crank angle α

$$\cos\beta = \frac{m - \cos \alpha}{\sqrt{1 + m^2 - 2m \cos \alpha}}$$

Angular Velocity of driven member = $\dfrac{d\beta}{dt}$ = $\omega \left(\dfrac{m \cos \alpha - 1}{1 + m^2 - 2m \cos \alpha} \right)$

Angular Acceleration of driven member = $\dfrac{d^2\beta}{dt^2}$ = $\omega^2 \left(\dfrac{m \sin \alpha (1 - m^2)}{(1 + m^2 - 2m \cos \alpha)^2} \right)$

Maximum Angular Acceleration occurs when $\cos \alpha$ =

$$\sqrt{\left(\frac{1 + m^2}{4\,m} \right)^2 + 2} - \left(\frac{1 + m^2}{4\,m} \right)$$

Maximum Angular Velocity occurs at α = 0 deg, and equals

$\dfrac{\omega}{m - 1}$ radians per sec

ONE OF THE MOST commonly used mechanisms for producing intermittent rotary motion from a uniform input speed is the external Geneva wheel.

The driven member, or star wheel, contains a number of slots into which the roller of the driving crank fits. The number of slots determines the ratio between dwell and motion period of the driven shaft. Lowest possible number of slots is three, while the highest number is theoretically unlimited. In practice the 3 slot Geneva is seldom used because of the extremely high acceleration values encountered. Genevas with more than 18 slots also are infrequently used, since they necessitate wheels of comparatively large diameters.

In external Genevas of any number of slots, the dwell period always exceeds the motion period. The opposite is true of the internal Geneva, while for the spherical Geneva both dwell and motion periods are 180 degrees.

For proper operation of the external Geneva, the roller must enter the slot tangentially. In other words, the centerline of the slot and the line connecting roller center and crank rotation center must compose a right angle when the roller enters or leaves the slot.

Calculations that follow below are

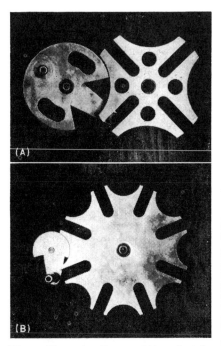

Fig. 3—Four slot Geneva (A) and eight slot (B). Both have locking devices.

S. RAPPAPORT

based upon these conditions stated.

Consider an external Geneva wheel, shown in Fig. 1, in which

n = number of slots
a = crank radius

From Fig 1, b = center distance = $\dfrac{a}{\sin \dfrac{180}{n}}$

Let $\dfrac{1}{\sin \dfrac{180}{n}} = m$

then $b = a\,m$

It will simplify the development of the equations of motion to designate the connecting line of wheel and crank centers as the zero line. This is contrary to the practice of assigning the zero value of α, representing the angular position of the driving crank, to that position of the crank where the roller enters the slot.

Thus, from Fig. 1, the driven crank radius f at any angle is

$$f = \sqrt{(am - a\cos\alpha)^2 + a^2\sin^2\alpha} = a\sqrt{1 + m^2 - 2m\cos\alpha} \qquad (1)$$

and the angular displacement β can be found from

$$\cos\beta = \frac{m - \cos\alpha}{\sqrt{1 + m^2 - 2m\cos\alpha}} \qquad (2)$$

A six slot Geneva is shown schematically in Fig. 2. The outside diameter

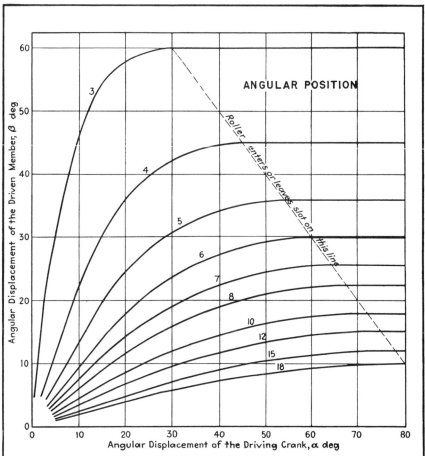

Fig. 4—Chart for determining the angular displacement of the driven member.

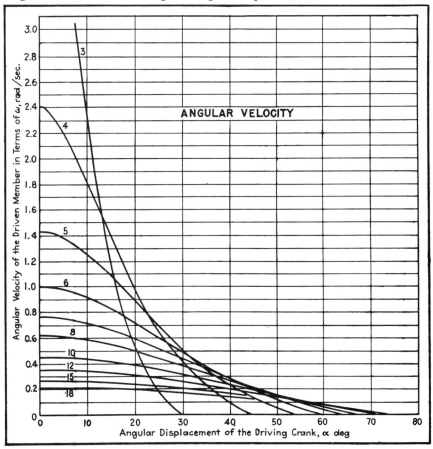

Fig. 5—Chart for determining the angular velocity of the driven member.

421

Table II—Principal Kinematic Data for External Geneva Wheel

No. of Slots	$\dfrac{360°}{n}$	Dwell period	Motion period	m and center-distance for $\alpha = 1$	Maximum angular velocity of driven member, radians per sec. equals ω multiplied by values tabulated. Crank at 0° position	Angular acceleration of driven member when roller enters slot, radians² per sec², equals ω^2 multiplied by values tabulated.			Maximum angular Acceleration of driven member, radians² per sec², equals ω^2 multiplied by values tabulated		
						α	β	Multiplier	α	β	Multiplier
3	120°	300°	60°	1.155	6.458	30°	60°	1.729	4°	27° 58′	29.10
4	90°	270°	90°	1.414	2.407	45°	45°	1.000	11° 28′	25° 11′	5.314
5	72°	252°	108°	1.701	1.425	54°	36°	0.727	17° 31′	21° 53′	2.310
6	60°	240°	120°	2.000	1.000	60°	30°	0.577	22° 55′	19° 51′	1.349
7	51° 25′ 43″	231° 30′	128° 30′	2.305	0.766	64° 17′ 8″	25° 42′ 52″	0.481	27° 41′	18° 11′	0.928
8	45°	225°	135°	2.613	0.620	67° 30′	22° 30′	0.414	31° 38′	16° 32′	0.700
9	40°	220°	140°	2.924	0.520	70°	20°	0.364	35° 16′	15° 15′	0.559
10	36°	216°	144°	3.236	0.447	72°	18°	0.325	38° 30′	14° 16′	0.465
11	32° 43′ 38″	212° 45′	147° 15′	3.549	0.392	73° 38′ 11″	16° 21′ 49″	0.294	41° 22′	13° 16′	0.398
12	30°	210°	150°	3.864	0.349	75°	15°	0.268	44°	12° 26′	0.348
13	27° 41′ 32″	207° 45′	152° 15′	4.179	0.315	76° 9′ 14″	13° 50′ 46″	0.246	46° 23′	11° 44′	0.309
14	25° 42′ 52″	205° 45′	154° 15′	4.494	0.286	77° 8′ 34″	21° 51′ 26″	0.228	48° 32′	11° 3′	0.278
15	24°	204°	156°	4.810	0.263	78°	12°	0.213	50° 30′	10° 27′	0.253
16	22° 30′	202° 30′	157° 30′	5.126	0.242	78° 45′	11° 15′	0.199	52° 24′	9° 57′	0.232
17	21° 10′ 35″	201°	159°	5.442	0.225	79° 24′ 43″	10° 35′ 17″	0.187	53° 58′	9° 26′	0.215
18	20°	200°	160°	5.759	0.210	80°	10°	0.176	55° 30′	8° 59′	0.200

D of the wheel (when taking the effect of the roller diameter d_r into account) is found to be

$$D = 2 \sqrt{\frac{d_r^2}{4} + a^2 \cot^2 \frac{180}{n}} \qquad (3)$$

Differentiating Eq (2) and dividing by the differential of time, dt, the angular velocity of the driven member is

$$\frac{d\beta}{dt} = \omega \left(\frac{m \cos \alpha - 1}{1 + m^2 - 2m \cos \alpha} \right) \qquad (4)$$

where ω represents the constant angular velocity of the crank.

By differentiation of Eq (4) the acceleration of the driven member is found to be

$$\frac{d^2\beta}{dt^2} = \omega^2 \left(\frac{m \sin \alpha (1 - m^2)}{(1 + m^2 - 2m \cos \alpha)^2} \right) \qquad (5)$$

All notations and principal formulas are given in Table I for easy reference. Table II contains all the data of principal interest for external Geneva wheels having from 3 to 18 slots. All other data can be read from the charts: Fig. 4 for angular position, Fig. 5 for angular velocity, and Fig. 6 for angular acceleration.

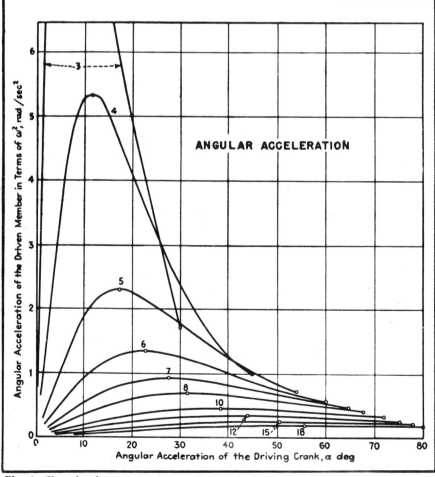

Fig. 6—Chart for determining the angular acceleration of the driven member.

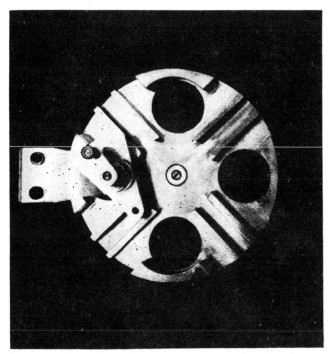

Fig. 1—A four slot internal Geneva wheel incorporating a locking mechanism. The basic sketch is shown in Fig. 3.

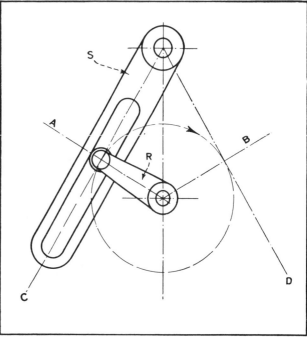

Fig. 2—Slot-crank motion from A to B represents external Geneva action; from B to A represents internal Geneva motion.

Kinematics of Intermittent Mechanisms
II—The Internal Geneva Wheel

S. RAPPAPORT

WHERE INTERMITTENT DRIVES must provide dwell periods of more than 180 deg, the external Geneva wheel design is quite satisfactory and is almost the standard device employed. But where the dwell period has to be less than 180 deg, other intermittent drive mechanisms must be used. The internal Geneva wheel is one way of obtaining this type of motion.

Dwell period of all internal Genevas is always smaller than 180 deg. Thus more time is left for the star to achieve maximum velocity, and acceleration is lower. The highest value of the angular acceleration occurs when the roller enters or leaves the slot. However, the acceleration curve does not reach a peak within the range of motion of the driven wheel. The geometrical maximum would occur in the continuation of the curve, but this continuation has no significance, since the driven member will have entered the dwell phase associated with the high angular displacement of the driving member.

The geometrical maximum lies in the continuation of the curve, falling into the region representing the mo-

Table I—Notation and Formulas for the Internal Geneva Wheel

Assumed or given: a, n, d and p

a = crank radius of driving member
n = number of slots
d = roller diameter
p = constant velocity of driving crank in rpm

$$m = \frac{1}{\sin \dfrac{180°}{n}}$$

b = center distance = $a\,m$

$$D = \text{inside diameter of driven member} = 2\sqrt{\frac{d^2}{4} + a^2 \cot^2 \frac{180°}{n}}$$

ω = constant angular velocity of driving crank in radians per sec =

$$\frac{p\pi}{30} \text{ radians per sec}$$

α = angular position of driving crank at any time
β = angular displacement of driven member corresponding to crank angle α

$$\cos\beta = \frac{m + \cos\alpha}{\sqrt{1 + m^2 + 2m\cos\alpha}}$$

Angular velocity of driven member = $\dfrac{d\beta}{dt} = \omega\left(\dfrac{1 + m\cos\alpha}{1 + m^2 + 2m\cos\alpha}\right)$

Angular acceleration of driven member = $\dfrac{d^2\beta}{dt^2} = \omega^2\left[\dfrac{m\sin\alpha\,(1 - m^2)}{(1 + m^2 + 2m\cos\alpha)^2}\right]$

Maximum angular velocity occurs at $\alpha = 0°$ and equals = $\dfrac{\omega}{1 + m}$ radians per sec

Maximum angular acceleration occurs when roller enters slot and equals =

$$\frac{\omega^2}{\sqrt{m^2 - 1}} \text{ radians}^2 \text{ per sec}^2$$

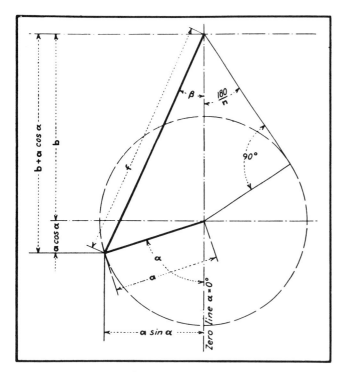

Fig. 3—Basic outline sketch for developing the equations of the internal Geneva wheel, using the notations as shown.

Fig. 4—Schematic sketch of a six slot internal Geneva wheel. Symbols are identified and motion equations given in Table I.

tion of the external Geneva wheel. This can be seen by the following considerations of a crank and slot drive, sketched in Fig. 2.

When the roller crank R rotates, slot link S will perform an oscillating movement, for which the displacement, angular velocity and acceleration can be given in continuous curves.

When the crank R rotates from A to B, then the slot link S will move from C to D, exactly reproducing all moving conditions of an external Geneva of equal slot angle. When crank R continues its movement from B back to A, then the slot link S will move from D back to C, this time reproducing exactly (though in a mirror picture with the direction of motion being reversed) the moving conditions of an internal Geneva.

Therefore, the characteristic curves of this motion contain both the external and internal Geneva wheel conditions; the region of the external Geneva lying between A and B, the region of the internal Geneva lying between B and A.

The geometrical maxima of the acceleration curves lie only in the region between A and B, representing that portion of the curves which belongs to the external Geneva.

Principal advantage of the internal Geneva, other than its smooth operation, is the sharply defined dwell period. A disadvantage is the relatively large size of the driven member, which increases the force resisting acceleration. Another feature, which is

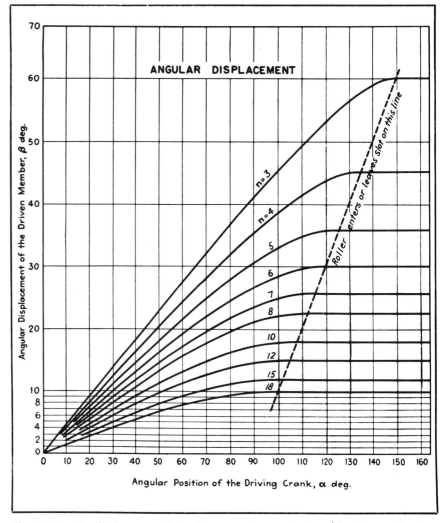

Fig. 5—Angular displacement of the driven member can be determined from this chart.

sometimes a disadvantage, is the cantilever arrangement of the roller crank shaft. This shaft cannot be a through shaft because the crank has to be fastened to the overhanging end of the input shaft.

To simplify the equations, the connecting line of wheel and crank centers is taken as the zero line. The angular position of the driving crank, α, is zero when on this line. Then the following relations are developed, based on Fig. 3:

n = number of slots
a = crank radius

$$b = \text{center distance} = \frac{a}{\sin \frac{180°}{n}}$$

Let

$$\frac{1}{\sin \frac{180°}{n}} = m,$$

then

$$b = am$$

To find the angular displacement, β, of the driven member, the driven crank radius, f, is first calculated from

$$f = \sqrt{a^2 \sin^2 \alpha + (am + a \cos \alpha)^2} = a \sqrt{1 + m^2 + 2m \cos \alpha} \quad (1)$$

and since

$$\cos \beta = \frac{m + \cos \alpha}{f}$$

it follows:

$$\cos \beta = \frac{m + \cos \alpha}{\sqrt{1 + m^2 + 2m \cos \alpha}} \quad (2)$$

From this formula, β, the angular displacement, can be calculated for any angle α, the angle of the driving member.

The first derivative of Eq (2) gives the angular velocity as

$$\frac{d\beta}{dt} = \omega \left(\frac{1 + m \cos \alpha}{1 + m^2 + 2m \cos \alpha} \right) \quad (3)$$

where ω designates the uniform speed of the driving crank shaft, namely

$$\omega = \frac{p\pi}{30}$$

if p equals its number of revolutions per minute.

Differentiating Eq (3) once more develops the equation for the angular acceleration:

$$\frac{d^2\beta}{dt^2} = \omega^2 \left[\frac{m \sin \alpha (1 - m^2)}{(1 + m^2 + 2m \cos \alpha)^2} \right] \quad (4)$$

The maximum angular velocity occurs, obviously, at $\alpha = 0$ deg. Its value is found by substituting 0 deg for α in Eq (3). It is

$$\frac{d\beta}{dt}_{\max} = \frac{\omega}{1 + m} \quad (5)$$

The highest value of the accelera-

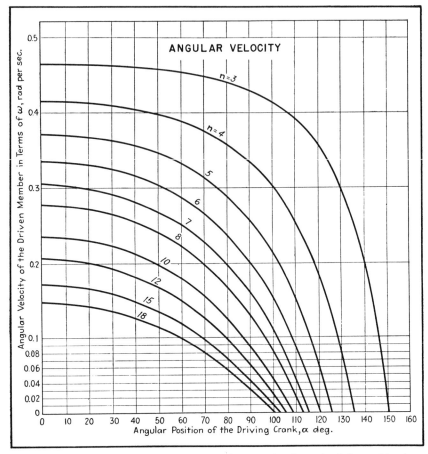

Fig. 6—Angular velocity of the driven member can be determined from this chart.

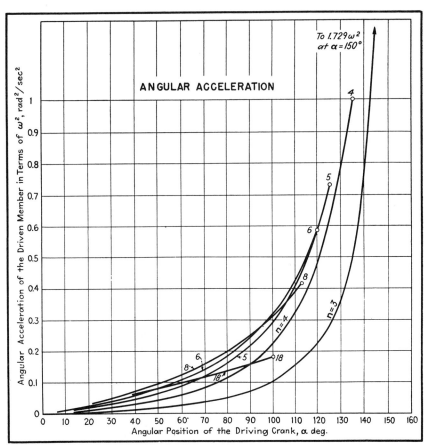

Fig. 7—Angular acceleration of the driven member can be determined from this chart.

425

Table II—Kinematic Data For the Internal Geneva Wheel

Number of slots, n	$\dfrac{360°}{n}$	Dwell period	Motion period	m and center-distance for $a=1$	Maximum angular velocity of driven member equals ω radians per sec, multiplied by values tabulated. Both α and β in 0° position	Angular acceleration of driven member when roller enters slot equals ω^2 radians^2 per sec^2 multiplied by values tabulated		
						α	β	Multiplier
3	120°	60°	300°	1.155	0.464	150°	60°	1.729
4	90°	90°	270°	1.414	0.414	135°	45°	1.000
5	72°	108°	252°	1.701	0.370	126°	36°	0.727
6	60°	120°	240°	2.000	0.333	120°	30°	0.577
7	51° 25′ 43″	128° 30′	231° 30′	2.305	0.303	115° 42′ 52″	25° 42′ 52″	0.481
8	45°	135°	225°	2.613	0.277	112° 30′	22° 30′	0.414
9	40°	140°	220°	2.924	0.255	110°	20°	0.364
10	36°	144°	216°	3.236	0.236	108°	18°	0.325
11	32° 43′ 38″	147° 15′	212° 45′	3.549	0.220	106° 21′ 49″	16° 21′ 49″	0.294
12	30°	150°	210°	3.864	0.206	105°	15°	0.268

tion is found by substituting $180/n + 90$ for α in Eq (4):

$$\frac{d^2\beta}{dt^2}_{\max} = \frac{\omega^2}{\sqrt{m^2-1}} \qquad (6)$$

A schematic sketch for a six slot internal Geneva wheel is shown in Fig. 4. All the symbols used in this sketch, and throughout the text, are compiled in Table I for easy reference.

Table II contains all the data of principal interest on the performance of internal Geneva wheels having from 3 to 18 slots. Other data can be read from the charts: Fig. 5 for angular position, Fig. 6 for angular velocity and Fig. 7 for angular acceleration.

Equations for designing cycloid mechanisms

E. H. SCHMIDT

1. Equations for epicycloid drives

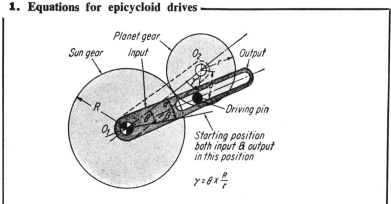

Planet gear
Sun gear
Input
O_2
Output
Driving pin
Starting position both input & output in this position
$\gamma = \theta \times \dfrac{R}{r}$
R
O_1

The equations for angular displacement, velocity and acceleration for basic epicyclic drive are given below.

Angular displacement

$$\tan \beta = \frac{(R+r)\sin\theta - b\sin(\theta+\gamma)}{(R+r)\cos\theta - b\cos(\theta+\gamma)} \qquad (1)$$

Angular velocity

$$V = \omega \frac{1 + \dfrac{b^2}{r(R+r)} - \left(\dfrac{2r+R}{r}\right)\left(\dfrac{b}{R+r}\right)\left(\cos\dfrac{R}{r}\theta\right)}{1 + \left(\dfrac{b}{R+r}\right)^2 - \left(\dfrac{2b}{R+r}\right)\left(\cos\dfrac{R}{r}\theta\right)} \qquad (2)$$

Angular acceleration

$$A = \omega^2 \frac{\left(1 - \dfrac{b^2}{(R+r)^2}\right)\left(\dfrac{R^2}{r^2}\right)\left(\dfrac{b}{R+r}\right)\left(\sin\dfrac{R}{r}\theta\right)}{\left[1 + \dfrac{b^2}{(R+r)^2} - \left(\dfrac{2b}{R+r}\right)\left(\cos\dfrac{R}{r}\theta\right)\right]^2} \qquad (3)$$

Symbols

A = angular acceleration of output, deg/sec^2

b = radius of driving pin from center of planet gear

r = pitch radius of planet gear

R = pitch radius of fixed sun gear

V = angular velocity of output, deg/sec

β = angular displacement of output, deg

$\gamma = \theta R/r$

θ = input displacement, deg

ω = angular velocity of input, deg/sec

426

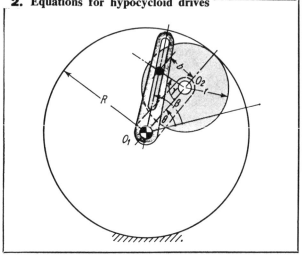

$$\tan \beta = \frac{\sin \theta - \left(\dfrac{b}{R-r}\right)\left(\sin \dfrac{R-r}{r}\theta\right)}{\cos \theta + \left(\dfrac{b}{R-r}\right)\left(\cos \dfrac{R-r}{r}\theta\right)} \qquad (4)$$

$$V = \omega \frac{1 - \left(\dfrac{R-r}{r}\right)\left(\dfrac{b^2}{(R-r)^2}\right) + \left(\dfrac{2r-R}{r}\right)\left(\dfrac{b}{R-r}\right)\left(\cos \dfrac{R}{r}\theta\right)}{1 + \dfrac{b^2}{(R-r)^2} + \left(\dfrac{2b}{R-r}\right)\left(\cos \dfrac{R}{r}\theta\right)} \qquad (5)$$

$$A = \omega^2 \frac{\left(1 - \dfrac{b^2}{(R-r)^2}\right)\left(\dfrac{b}{R-r}\right)\left(\dfrac{R^2}{r^2}\right)\left(\sin \dfrac{R}{r}\theta\right)}{\left[1 + \dfrac{b^2}{(R-r)^2} + \left(\dfrac{2b}{R-r}\right)\left(\cos \dfrac{R}{r}\theta\right)\right]^2} \qquad (6)$$

DESCRIBING APPROXIMATE STRAIGHT LINES

3. Gear rolling on a gear—flatten curves

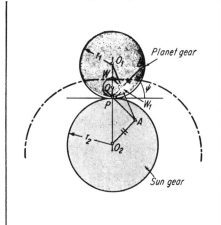

It is frequently desirable to find points on the planet gear that will describe approximately straight lines for portions of the output curve. Such points will yield dwell mechanisms, as shown on pp. 10 to 12. Construction is as follows (shown at left):

1. Draw an arbitrary line PB.
2. Draw its parallel O_2A.
3. Draw its perpendicular PA at P. Locate point A.
4. Draw O_1A. Locate W_1.
5. Draw perpendicular to PW_1 at W_1 to locate W.
6. Draw a circle with PW as the diameter.

All points on this circle describe curves with portions that are approximately straight. This circle is also called the inflection circle because all points describe curves which have a point of inflection at the position illustrated. (Shown is the curve passing through point W.)

4. Gear rolling on a rack—vee curves

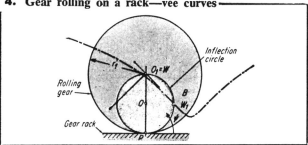

This is a special case. Draw a circle with a diameter half that of the gear (diameter O_1P). This is the inflection circle. Any point, such as point W_1, will describe a curve that is almost straight in the vicinity selected. Tangents to the curves will always pass through the center of the gear, O_1 (as shown).

5. Gear rolling inside a gear—zig-zag

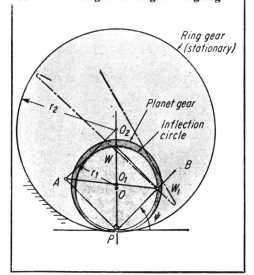

To find the inflection circle for a gear rolling inside a gear:

1. Draw arbitrary line PB from the contact point P.
2. Draw its parallel O_2A, and its perpendicular, PA. Locate A.
3. Draw line AO_1 through the center of the rolling gear. Locate W_1.
4. Draw a perpendicular through W_1. Obtain W. Line WP is the diameter of the inflection circle. Point W_1, which is an arbitrary point on the circle, will trace a curve of repeated almost-straight lines, as shown.

(continued next page)

DESIGNING FOR DWELLS

6. Center of curvature—gear rolling on gear

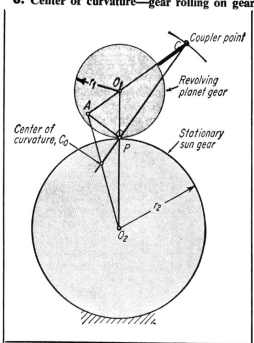

By locating the centers of curvature at various points, one can then determine the proper length of the rocking or reciprocating arm to provide long dwells.

1. Draw a line through points C and P.
2. Draw a line through points C and O_1.
3. Draw a perpendicular to CP at P. This locates point A.
4. Draw line AO_2, to locate C_o, the center of curvature.

7. Center of curvature—gear rolling on a rack

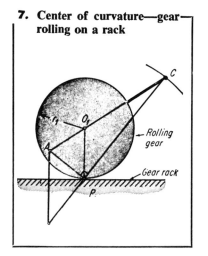

Construction is similar to that of the previous case.

1. Draw an extension of line CP.
2. Draw a perpendicular at P to locate A.
3. Draw a perpendicular from A to the straight suface to locate C_o.

8. Center of curvature—gear rolling inside a gear

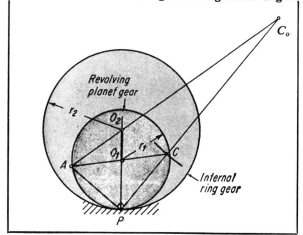

1. Draw extensions of CP and CO_1.
2. Draw a perpendicular of PC at P to locate A.
3. Draw AO_2 to locate C_o.

9. Analytical solutions

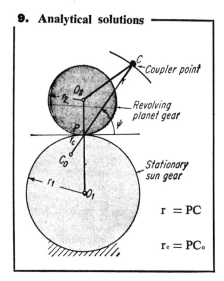

$$r = PC$$

$$r_c = PC_o$$

The centure of curvature of a gear rolling on a external gear can be computed directly from the Euler-Savary equation:

$$\left(\frac{1}{r} - \frac{1}{r_c}\right)\sin\psi = \text{constant} \quad (7)$$

where angle ψ and r locate the position of C.

By applying this equation twice, specifically to point O_1 and O_2 which have their own centers of rotation, the following equation is obtained:

$$\left(\frac{1}{r_2} + \frac{1}{r_1}\right)\sin 90° =$$

$$\left(\frac{1}{r} + \frac{1}{r_c}\right)\sin\psi$$

or

$$\frac{1}{r_2} + \frac{1}{r_1} = \left(\frac{1}{r} + \frac{1}{r_c}\right)\sin\psi$$

This is the final design equation. All factors except r_c are known; hence solving for r_c leads to the location of C_o.

For a gear rolling inside an internal gear, the Euler-Savary equation is

$$\left(\frac{1}{r} + \frac{1}{r_c}\right)\sin\psi = \text{constant}$$

which leads to

$$\frac{1}{r_2} - \frac{1}{r_1} = \left(\frac{1}{r} - \frac{1}{r_c}\right)\sin\psi$$

Designing crank-and-rocker links with optimum force transmission

Four-bar linkages can be designed with a minimum of trial and error by a combination of tabular and iteration techniques

The science of determining optimum crank-and-rocker linkages has always been computer-oriented because of the complexity of the equations and the iterations involved. Thanks to recent work at Columbia University's Department of Mechanical and Nuclear Engineering, all you need now is a pocket calculator and the accompanying computer-generated tables. The computations were made by Mr. Meng-Sang Chew, at the University.

A crank-and-rocker linkage, $ABCD$, is shown in the first figure. The two extreme positions of the rocker are shown schematically in the second figure, in which ψ denotes the rocker swing angle and ϕ the corresponding crank rotation, both measured counterclockwise from the extended dead-center position, AB_1C_1D.

The problem is to find the proportions of the crank-and-rocker linkage for a given rocker swing angle, ψ, a prescribed corresponding crank rotation, ϕ, and optimum force transmission. The latter is usually defined in terms of the transmission angle, μ, the angle between coupler BC extended and rocker CD.

Considering static forces only, the closer the transmission angle is to 90°, the greater is the ratio of the driving component of the force exerted on the rocker to the component exerting bearing pressure on the rocker. The control of transmission-angle variation becomes especially important at high speeds and for heavy duty.

How to find the optimum. The steps in the determination of crank-and-rocker proportions for a given rocker swing angle, corresponding crank rotation, and optimum transmission, are:

■ Select (ψ, ϕ) within the following range:

$$0° < \psi < 180°$$

$$(90° + \tfrac{1}{2}\psi) < \phi < (270° + \tfrac{1}{2}\psi)$$

■ Calculate: $t = \tan \tfrac{1}{2}\phi$

$$u = \tan \tfrac{1}{2}(\phi - \psi)$$

$$v = \tan \tfrac{1}{2}\psi$$

■ Using table, find ratio λ_{opt} of coupler to crank length, that minimizes the transmission-angle deviation from 90°. The most practical combinations of (ψ, ϕ)

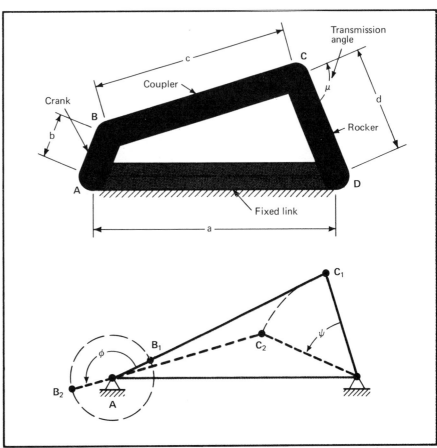

Four-bar crank-and-rocker is classic mechanism problem that now has optimum solution using only accompanying table and portable hand calculator

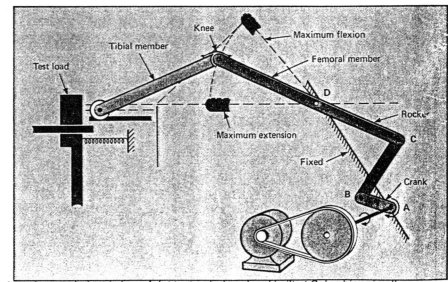

Actual example is this knee-joint tester designed and built at Columbia using the design and calculating procedures outlined in this article

are included in table. If the (ψ, ϕ) combination is not included, or if $\phi = 180°$, go to next steps (a,b,c):

■ (a) If $\phi \neq 180°$ and (ψ, ϕ) fall outside the range of table, determine arbitrary intermediate value Q from the equation:

$$Q^3 + 2Q^2 - t^2Q - (t^2/u^2)(1 + t^2) = 0$$

where $(1/u^2 < Q < t^2)$.

This is conveniently accomplished by numerical iteration:

Set $Q_1 = \frac{1}{2}(t^2 + \frac{1}{u^2})$

Calculate $Q_2, Q_3, \ldots$ from the recursion equation:

$$Q_{i+1} = \frac{2Q_i^2(Q_i + 1) + (t^2/u^2)(1+t^2)}{Q_i(3Q_i + 4) - t^2}$$

Iterate until the ratio $|(Q_{i+1} - Q_i)/Q_i|$ is sufficiently small, so that you obtain the desired number of significant figures. Then:

$$\lambda_{opt} = t^2/Q$$

(b) If $\phi \neq 180°$ and determination of λ_{opt} requires interpolation between two entries in table, let $Q_1 = t^2/\lambda^2$, where λ corresponds to the nearest entry in table, and continue as in (a) above to determine Q and λ_{opt}. Usually 1 or 2 iterations will suffice.

(c) $\phi = 180°$. This important special case is discussed in (PE—Mar6'61p47). In this case, $a^2+b^2=c^2+d^2$; $\psi = 2\sin^{-1}(b/d)$; and the max deviation, Δ, of the transmission angle from 90° is equal to $\sin^{-1}(ab/cd)$.

■ Determine linkage proportions as follows:

$$(a')^2 = \frac{u^2 + \lambda^2_{opt}}{1 + u^2}$$

$$(b')^2 = \frac{v^2}{1 + v^2}$$

$$(c')^2 = \frac{\lambda^2_{opt}\ v^2}{1 + v^2}$$

$$(d')^2 = \frac{t^2 + \lambda^2_{opt}}{1 + t^2}$$

Then: $a = ka'$; $\quad b = kb'$; $c = kc'$; $\quad d = kd'$

where k is a scale factor, such that the length of any one link, usually the crank, is equal to a design value. The max deviation, Δ, of the transmission angle from 90° is:

Optimum values of lambda ratio for given ø and ψ

ø deg	160	162	164	166	ψ, deg 168	170	172	174	176	178
10	2.3532	2.4743	2.6166	2.7873	2.9978	3.2669	3.6284	4.1517	5.0119	6.8642
12	2.3298	2.4491	2.5891	2.7570	2.9636	3.2272	3.5804	4.0899	4.9224	6.6967
14	2.3064	2.4239	2.5617	2.7266	2.9293	3.1874	3.5324	4.0283	4.8342	6.5367
16	2.2831	2.3988	2.5344	2.6964	2.8953	3.1479	3.4848	3.9675	4.7482	6.3853
18	2.2600	2.3740	2.5073	2.6664	2.8615	3.1089	3.4380	3.9080	4.6650	6.2427
20	2.2372	2.3494	2.4805	2.6368	2.8282	3.0704	3.3920	3.8499	4.5848	6.1087
22	2.2145	2.3250	2.4540	2.6076	2.7954	3.0327	3.3470	3.7935	4.5077	5.9826
24	2.1922	2.3010	2.4279	2.5789	2.7631	2.9956	3.3030	3.7388	4.4338	5.8641
26	2.1701	2.2773	2.4022	2.5505	2.7314	2.9594	3.2602	3.6857	4.3628	5.7524
28	2.1483	2.2539	2.3768	2.5227	2.7004	2.9239	3.2185	3.6344	4.2948	5.6469
30	2.1268	2.2309	2.3519	2.4954	2.6699	2.8893	3.1779	3.5847	4.2295	5.5472
32	2.1056	2.2082	2.3273	2.4685	2.6401	2.8554	3.1384	3.5367	4.1668	5.4526
34	2.0846	2.1858	2.3032	2.4421	2.6108	2.8223	3.0999	3.4901	4.1066	5.3628
36	2.0640	2.1637	2.2794	2.4162	2.5821	2.7899	3.0624	3.4449	4.0486	5.2773
38	2.0436	2.1420	2.2560	2.3908	2.5540	2.7583	3.0259	3.4012	3.9927	5.1957
40	2.0234	2.1205	2.2330	2.3657	2.5264	2.7274	2.9903	3.3587	3.9388	5.1177
42	2.0035	2.0994	2.2103	2.3411	2.4994	2.6971	2.9556	3.3175	3.8868	5.0430
44	1.9839	2.0785	2.1879	2.3169	2.4728	2.6675	2.9217	3.2773	3.8364	4.9712
46	1.9644	2.0579	2.1659	2.2931	2.4468	2.6384	2.8886	3.2383	3.7877	4.9023
48	1.9452	2.0375	2.1441	2.2696	2.4211	2.6100	2.8563	3.2003	3.7404	4.8358
50	1.9262	2.0174	2.1227	2.2465	2.3959	2.5820	2.8246	3.1632	3.6945	4.7717

$$\sin\Delta = \frac{|(a \pm b)^2 - c^2 - d^2|}{2cd}$$

$0° \le \Delta \le 90°$

$+$ sign if $\phi < 180°$

$-$ sign if $\phi > 180°$

An actual example. A simulator for testing artificial knee joints, built by the Department of Orthopaedic Surgery, Columbia University, under the direction of Dr. N. Eftekhar, is shown schematically. The drive includes an adjustable crank-and-rocker, ABCD. The rocker swing angle ranges from a maximum of about 48° to a minimum of about ⅓ of this value, the crank being 4 in. long and rotating at 150 rpm. The swing angle adjustment is obtained by changing the length of the crank. Find the proportions of the linkage,

assuming optimum-transmission proportions for the maximum rocker swing angle, as this represents the most severe condition. For smaller swing angles, the maximum transmission-angle deviation from 90° will be less.

Crank rotation corresponding to 48° rocker swing is selected at approximately 170°. Using the Table, we find λ_{opt} = 2.6100. This gives a' = 1.5382, b' = 0.40674, c' = 1.0616 and d' = 1.0218. For a 4 in. crank, k = 4/0.40674 = 9.8343 and a = 15.127 in., b = 4 in., c = 10.440 in. and d = 10.049 in., which is very close to the proportions used. The max deviation of the transmission angle from 90° is 47.98°. The simulator, which is a relatively low-power, low-speed device, has worked well for several years.

The above procedure can be used not

only for the transmission optimization of crank-and-rocker linkages, but also for other crank-and-rocker design purposes. For example, if only rocker swing angle and the corresponding crank rotation are prescribed, the ratio of coupler to crank length is arbitrary and the equations can be used with any value of λ^2 within the range (1, u^2t^2). The ratio, λ, can then be tailored to suit a variety of design requirements, involving size, bearing reactions, transmission-angle control, or combinations.

The method also was used to design dead-center linkages for landing-gear retraction systems, and can be applied to any design of four-bar linkages that meet the stated requirements here.

Professor Ferdinand Freudenstein,
Columbia University

Optimum values of lambda ratio for given ⌀ and ψ

⌀ deg	182	184	186	188	ψ, deg 190	192	194	196	198	200
10	7.2086	5.3403	4.4560	3.9112	3.5318	3.2478	3.0245	2.8428	2.6911	2.5616
12	7.0369	5.2692	4.4227	3.8969	3.5282	3.2507	3.0317	2.8528	2.7030	2.5748
14	6.8646	5.1881	4.3795	3.8739	3.5174	3.2478	3.0341	2.8589	2.7117	2.5855
16	6.6971	5.1013	4.3287	3.8435	3.5000	3.2392	3.0317	2.8610	2.7171	2.5934
18	6.5371	5.0121	4.2726	3.8071	3.4768	3.2252	3.0245	2.8589	2.7189	2.5982
20	6.3857	4.9226	4.2131	3.7663	3.4487	3.2065	3.0129	2.8528	2.7171	2.5998
22	6.2431	4.8344	4.1518	3.7221	3.4167	3.1837	2.9972	2.8428	2.7117	2.5982
24	6.1090	4.7484	4.0900	3.6759	3.3818	3.1575	2.9780	2.8293	2.7030	2.5934
26	5.9830	4.6652	4.0284	3.6284	3.3447	3.1286	2.9558	2.8127	2.6911	2.5855
28	5.8644	4.5849	3.9676	3.5804	3.3062	3.0976	2.9311	2.7833	2.6763	2.5748
30	5.7527	4.5079	3.9080	3.5324	3.2669	3.0652	2.9045	2.7718	2.6592	2.5616
32	5.6472	4.4339	3.8500	3.4849	3.2272	3.0318	2.8764	2.7484	2.6399	2.5461
34	5.5475	4.3630	3.7936	3.4380	3.1875	2.9979	2.8473	2.7236	2.6190	2.5287
36	5.4529	4.2949	3.7388	3.3920	3.1480	2.9636	2.8175	2.6977	2.5967	2.5097
38	5.3631	4.2296	3.6858	3.3470	3.1089	2.9294	2.7873	2.6711	2.5734	2.4894
40	5.2776	4.1669	3.6345	3.3031	3.0705	2.8953	2.7570	2.6440	2.5492	2.4680
42	5.1960	4.1067	3.5848	3.2602	3.0327	2.8615	2.7266	2.6166	2.5246	2.4459
44	5.1180	4.0487	3.5367	3.2185	2.9956	2.8282	2.6964	2.5891	2.4996	2.4232
46	5.0432	3.9928	3.4901	3.1779	2.9594	2.7954	2.6665	2.5617	2.4744	2.4001
48	4.9715	3.9389	3.4450	3.1384	2.9239	2.7631	2.6369	2.5344	2.4491	2.3767
50	4.9025	3.8869	3.4012	3.0999	2.8893	2.7314	2.6076	2.5073	2.4239	2.3533

New design curves and equations for

Gear-slider mechanisms

These little-known devices—there
are several new ones now—can produce
a wide variety of output motions

NICHOLAS P. CHIRONIS

WHAT is a gear-slider mechanism? In reality, it's little more than a crank-and-slider with two gears meshed in line with the crank (Fig 1). But because one of the gears (planet gear, 3) is prevented from rotating owing to its fixation to the connecting rod, the output is taken from the sun gear, not the slider. This produces a variety of cyclic output motions, depending on the proportions of the members.

In his investigation of the capabilities of the mechanism, Professor Preben Jensen of the University of Bridgeport, Conn nailed down the equations defining its motion and acceleration characteristics—then came up with some variations of his own (Fig 5 through 8). These he believes will outperform the parent type.

Speaking at the ASME Mechanism Conference held at Purdue University, Jensen illustrated how the output of one of the new devices, Fig 8, can come to dead stop during each cycle, or "progressively oscillate" to new positions around the clock. A machinery designer, therefore, can obtain a variety of intermittent motions out of the arrangement and, by combining two such units, can tailor the dwell period of the mechanism to fit the automatic feed requirements of a machine.

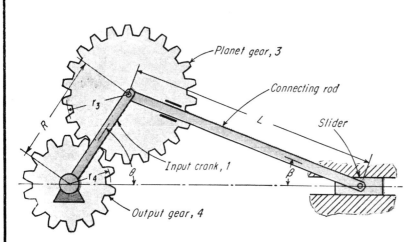

1. Basic gear slider mechanism. It differs from the better known three-gear drive in that a slider is employed to restrict the motion of the planet gear. Note that the output is taken from the gear which is concentric with the input shaft, and not from the slider.

Symbols

L = length of connecting rod, in.

r_3 = radius of gear fixed to connecting rod, in.

r_4 = radius of output gear, in.

R = length of crank, in.

α = angular acceleration of the input crank, rad/sec²

β = connecting rod displacement, deg

γ = output rotation, deg

θ = input rotation, deg

θ_o = crank angle rotation during which the output gear reverses its motion, deg

ϕ = angle through which the output gear rotates back

ω = angular velocity of input crank, rad/sec

Single prime mark denotes angular velocity, rad/sec; double prime marks denote angular acceleration, rad/sec².

The basic form

Input motion is to crank 1; output is from gear 4. As the crank rotates, say counterclockwise, it causes the planet gear 3 to oscillate while following a satellite path around gear 4. This imparts a varying output motion to gear 4 which rotates twice in the counterclockwise direction (when $r_3 = r_4$) for every revolution of the input.

Jensen's equations for angular displacement, velocity, and acceleration of gear 4, when driven at a speed of ω by crank 1, are as follows:

Angular displacement

$$\gamma = \theta + \frac{r_3}{r_4}(\theta + \beta) \qquad (1)$$

where β is computed from the following relationship (see also list of symbols at bottom of previous page):

$$\sin \beta = \frac{R}{L} \sin \theta \qquad (2)$$

Angular velocity

$$\gamma' = \omega + \frac{r_3}{r_4}(\omega + \beta') \qquad (3)$$

where

$$\frac{\beta'}{\omega} = \frac{R}{L} \frac{\cos \theta}{\left[1 - \left(\frac{R}{L}\right)^2 \sin^2 \theta\right]^{1/2}} \qquad (4)$$

Angular acceleration

$$\gamma'' = \alpha + \frac{r_3}{r_4}(\alpha + \beta'') \qquad (5)$$

where

$$\frac{\beta''}{\omega^2} = \frac{R}{L} \frac{\sin \theta \left[\left(\frac{R}{L}\right)^2 - 1\right]}{\left[1 - \left(\frac{R}{L}\right)^2 \sin^2 \theta\right]^{3/2}} \qquad (6)$$

For a constant angular velocity, Eq 5 becomes

$$\gamma'' = \frac{r_3}{r_4} \beta'' \qquad (7)$$

Design charts

The equations were then solved by Prof Jensen for various L/R ratios and positions of the crank angle θ to obtain the design charts in Fig 2, 3, and 4. Thus, for a mechanism with

$$L = 12 \text{ in.} \qquad r_3 = 2.5$$
$$R = 4 \text{ in.} \qquad r_4 = 1.5$$
$$\omega = 1000/\text{sec} = \text{rad/sec}$$

the output velocity at crank angle $\theta =$

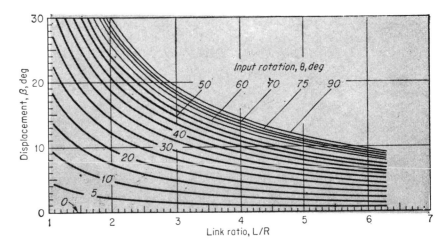

2. Angular displacement diagram for the connecting rod.

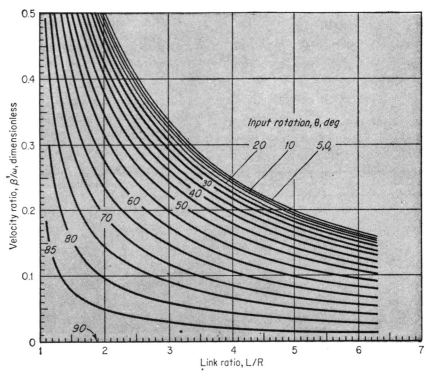

3. Angular velocity curves for various crank angles.

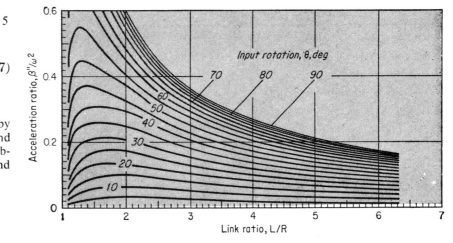

4. Angular acceleration curves for various crank angles.

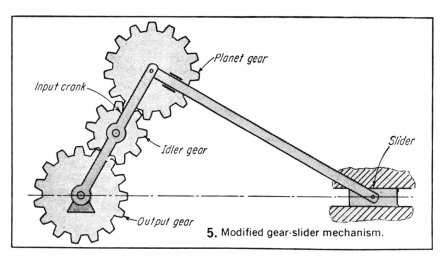

5. Modified gear-slider mechanism.

60 deg can be computed as follows:
$$L/R = 12/4 = 3$$
From Fig 3 $\beta'/\omega = 0.175$
$$\beta' = 0.175 (1000) = 175 \text{ rad/sec}$$
From Eq 3
$$\gamma' = 2960 \text{ rad/sec}$$

Three-gear variation

One interesting variation, Fig 5, is obtained by simply adding an idler, gear 5, to the drive. If gears 3 and 4 are then made equal, the output gear, 4, will then oscillate—and in exactly the same motion as the connecting rod 2.

One use for this linkage, Jensen says, is in machinery where a sleeve is to ride concentrically over an input shaft, and yet must oscillate to provide a reciprocating motion. The shaft can drive the sleeve with this mechanism by making the sleeve part of the output gear.

Internal-gear variations

By replacing one of the external gears of Fig 1 with an internal one, two mechanisms are obtained (Fig 6 and 7) which have wider variable output abilities. But it is the mechanism in Fig 7 that interested Jensen. This could be proportioned to give either a dwell or a progressive oscillation, that is, one in which the output rotates forward, say 360 deg, turns back for 30 deg, moves forward 30 deg, and then proceeds to repeat the cycle by moving forward again for 360 deg.

In this mechanism, the crank drives the large ring gear 3 which is fixed to the connecting rod 2. Output is from gear 4. Jensen derived the following equations:

Output motion

$$\omega_4 = -\left(\frac{L - R - r_4}{Lr_4}\right)R\,\omega_1 \quad (8)$$

When $r_4 = L - R$, then $\omega_4 = 0$ from Eq 8, and the mechanism is pro-

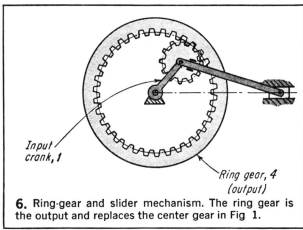

6. Ring-gear and slider mechanism. The ring gear is the output and replaces the center gear in Fig 1.

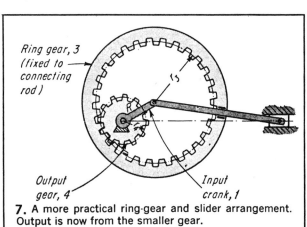

7. A more practical ring-gear and slider arrangement. Output is now from the smaller gear.

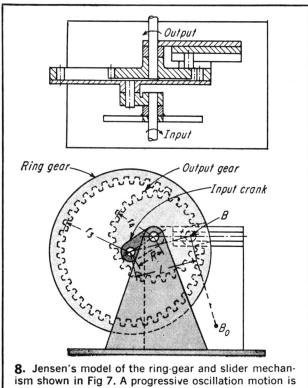

8. Jensen's model of the ring-gear and slider mechanism shown in Fig 7. A progressive oscillation motion is obtained by making r_4 greater than **L-R**.

portioned to give instantaneous dwell. To obtain a progressive oscillation r_1 must be greater than $L - R$, as shown in Jensen's model (Fig 8).

If gear 4 turns back and then starts moving forward again there must be two positions where the motion of gear 4 is zero. These two positions are symmetrical with respect to A_0B. Letting θ_0 = crank angle rotation (of input) during which the output gear reverses its motion, and ϕ = angle through which gear 4 rotates back, then

$$\cos \frac{\theta_o}{2} = \left[\frac{L^2 - R^2}{r_4(2R + r_4)} \right]^{1/2} \quad (9)$$

and

$$\gamma = \theta_o - \frac{r_3}{r_4} (\theta_o - \beta_o) \quad (10)$$

where

$$\sin \beta_o = \frac{R}{L} \sin \frac{\theta_\bullet}{2} \quad (11)$$

Chart for proportioning

The chart in Fig 9 helps proportion the mechanism of Fig 8 to provide a specific type of progressive oscillation.

It is set up for $R = 1$ in. For other values of R, you must convert the chart values for r_4 proportionally, as shown below.

For example, assume that you want the output gear, during each cycle, to rotate back 9.2 deg. Thus $\phi = 9.2$ deg. Also given is $R = 0.75$ in. and $L = 1.5$ in. Thus $L/R = 2$.

From the right side of the chart, go to the ϕ-curve for $L = 2$, then upward to the θ_0-curve for $L = 2$ in. Read $\theta_0 = 82$ deg at the left ordinate.

Now return to the second intersection point and proceed upward to read on the abscissa scale for $L = 2$, a value of $r_4 = 1.5$. Since $R = 0.75$ in., and the chart is for $R = 1$, convert r_4 as follows: $r_4 = 0.75 (1.5) = 1.13$ in.

Thus if the mechanism is built with output gear of radius $r_4 = 1.13$ in., then during 82-deg rotation of the crank, the output gear 4 will go back 9.2 deg. Of course, during the next 83 deg gear 4 will have reversed back to its initial position—and then will keep going forward for the remaining 194 deg of the crank rotation.

Future modifications

The mechanism in Fig 8 is designed

to permit easy changing of the output motion from progressive oscillation to instantaneous dwell or non-uniform CW or CCW rotation. This is accomplished by shifting the position of the pin which acts as the sliding piece of the centric slider crank. It is also possible to use an eccentric slider crank, a four-bar linkage, or a sliding-block linkage as the basic mechanism.

Two mechanisms in series will give an output with either a prolonged dwell or two separate dwells. The angle between the separated dwells can be adjusted during the time of operation by interposing a gear differential so that the position of the output shaft of the first mechanism can be changed relative to the positioin of the input shaft to the second mechanism.

The mechanism can also be improved by introducing an additional link, B-B_0, to guide pin B along a circular arc instead of a linear track. This would result in a slight change for the better in the performance of the mechanism.

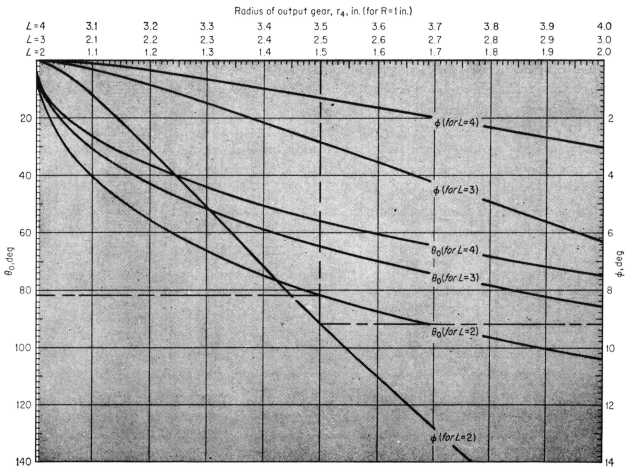

9. Chart for proportioning a ring-gear and slider mechanism.

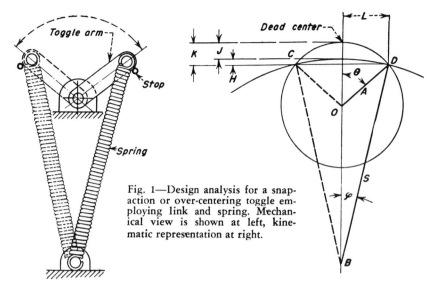

Fig. 1—Design analysis for a snap-action or over-centering toggle employing link and spring. Mechanical view is shown at left, kinematic representation at right.

SYMBOLS

A = length of toggle arm

S = free length of toggle spring in detented position

θ = angle swept by toggle arm moving from detented to dead center position

φ = angle swept by toggle spring in moving from detented to dead center position

CD = chordal distance between detent points

L = chordal distance between detent point and dead center

K = height of arc swept by toggle arm

0 = pivot point of toggle arm

B = pivot point of toggle spring

Designing Snap-Action

ERWIN F. C. SCHULZE

Over-centering toggle mechanisms, Fig. 1, are widely used in mechanical and electrical switches, latch mechanisms and mechanical overload controls. Such toggles also serve as: (1) detent (holds other parts in selected position); (2) overload device in a mechanical linkage (shifts to opposite position when sufficiently loaded); and (3) energy storage device.

Two applications shown in Fig. 2 illustrate the "snap-action" of a toggle. As the toggle passes dead center it is snapped ahead of the actuating force by the toggle spring. In most applications it is desirable to obtain maximum snap action.

Snap-action is a function of the elongation per length of the toggle spring as it moves over dead center. Elongation at dead center is equal to:

$$J = K - H \qquad (1)$$

The elongation e in per cent of length is equal to:

$$e = (100)J/S \qquad (2)$$

Since the resisting force of the spring increases with elongation but decreases with an increase in length, the ratio J/S should be as large as possible within the capacity of the spring for best snap-action performance.

The ratio J/S as a function of angle θ can be derived as follows:

$$H = S - S \cos \varphi \qquad (3)$$

and

$$K = A - A \cos \theta \qquad (4)$$

Substituting Eqs (3) and (4) into Eq (1),

$$J = A(1 - \cos \theta) - S(1 - \cos \varphi) \qquad (5)$$

or

$$J/S = (A/S)(1 - \cos \theta) - (1 - \cos \varphi) \qquad (6)$$

The relationship between θ and ϕ is:

$$L = A \sin \theta = S \sin \varphi$$

or

$$\sin \varphi = (A/S)(\sin \theta) \qquad (7)$$

By trigonometric identity,

$$\sin \theta = (1 - \cos^2 \theta)^{1/2} \qquad (8)$$

Substituting Eq (8) into Eq (7) and squaring both sides,

$$\sin \varphi^2 = (A/S)^2(1 - \cos^2 \theta) \qquad (9)$$

By trigonometric identity,

$$\cos \varphi = (1 - \sin^2 \varphi)^{1/2} \qquad (10)$$

Substituting Eq (9) into Eq (10),

$$\cos \varphi = [1 - (A/S)^2 + (A/S)^2 \cos^2 \theta]^{1/2} \qquad (11)$$

and Eq (11) into Eq (6),

$$J/S = (A/S)(1 - \cos \theta) - 1 + [1 - (A/S)^2 - (A/S)^2 \cos^2 \theta]^{1/2} \qquad (12)$$

Eq (12 can be considered as having only three variables: (1) the spring elongation ratio J/S; (2) the toggle arm to spring length ratio, A/S; and (3) the toggle arm angle θ.

A series of curves are plotted from Eq (12) showing the relationship between J/S and A/S for various angles of θ. The curves are illustrated in Fig. 3; for greater accuracy each chart uses a different vertical scale.

Maximum Snap-Action

Maximum snap-action for a particular angle occurs when J/S is a maximum. This can be determined by setting the first derivative of Eq (12) equal to zero and solving for A/S.

Differentiating Eq (12),

$$\frac{d(J/S)}{d(A/S)} = 1 - \cos \theta + \frac{[-2(A/S) + 2(A/S)(\cos^2 \theta)]}{2[1 - (A/S)^2 + (A/S)^2 \cos^2 \theta]^{1/2}} \qquad (13)$$

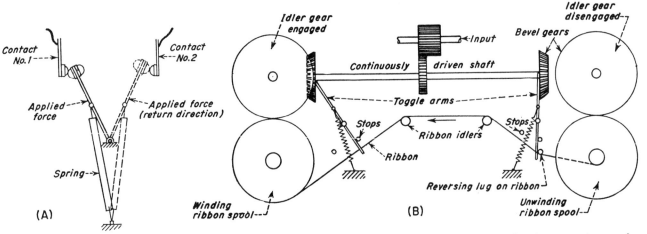

Fig. 2—Typical applications of toggles: (A) snap-action switches; (B) ribbon reversing mechanism for typewriters and calculators. The toggle in (B) is activated by a lug on the ribbon. As it passes dead center it is snapped ahead of the lug by the toggle spring, thus shifting the shaft and reversing the direction of the ribbon before next key is struck.

Toggles

Theory, formulas and design charts for quickly determining toggle dimensions to obtain maximum snap- action.

Setting Eq (13) equal to zero and rearranging terms,

$$\frac{\cos \theta - 1}{\cos^2 \theta - 1} =$$

$$\frac{A/S}{(A/S)[(S/A)^2 - 1 + \cos^2 \theta]^{1/2}} \quad (14)$$

Cross-multiplying, squaring and simplifying,

$$(S/A)^2 - 1 + \cos^2 \theta = \cos^2 \theta + 2 \cos \theta + 1$$

Reducing,

$$(S/A)^2 = 2 \cos \theta + 2$$

and finally simplifying to the following equation of A/S when J/S is a maximum:

$$A/S = [2(\cos \theta + 1)]^{-1/2} \quad (15)$$

The maximum value of J/S can be determined by substituting Eq (15) into Eq (12):

$$J/S_{max} = \frac{1 - \cos \theta}{[2(\cos \theta + 1)]^{1/2}} - 1 +$$

$$\left[1 - \frac{1}{2(\cos \theta + 1)} + \frac{\cos^2 \theta}{2(\cos \theta + 1)}\right]^{1/2} \quad (16)$$

which is simplified into the following expression:

$$J/S_{max} = \frac{2 - [2(\cos \theta + 1)]^{1/2}}{[2(\cos \theta + 1)]^{1/2}} \quad (17)$$

The locus of points of J/S_{max} is a straight line function as shown in Fig.

3. It can be seen from Eq (15) that the value of A/S at J/S_{max} varies from 0.500 when $\theta = 0$ to 0.707 when $\theta = 90$ deg. This relatively small range gives a quick rule-of-thumb to check if a mechanism has been designed close to the maximum snap-action point.

Elongation of the spring, Eq (2) is based on the assumption that the spring is installed in its free length S with no initial elongation. For a spring with a free length E smaller than S, the total elongation in per cent when extended to the dead center position is:

$$e = 100[(S/E)(1 + J/S) - 1] \quad (18)$$

Relationship between ϕ and θ at the point of maximum snap-action for any value of θ is:

$$\theta = 2\varphi \quad (19)$$

This can be proved by substituting Eqs (9) and (11) into the trigonometric identy:

$$\cos 2\varphi = \cos^2 \varphi - \sin^2 \varphi \quad (20)$$

and comparing the resulting equation with one obtained by solving for cos θ in Eq (15). This relationship between the angles is another means of quickly evaluating a toggle mechanism.

Design Procedure

A toggle is usually designed to operate within certain space limitations. When the dimensions X and W as shown in Fig. 4 are known, the angle θ resulting in maximum snap-action can be determined as follows:

$$A \sin \theta = S \sin \varphi = W/2 \quad (21)$$

Substituting Eq (19) into Eq (21),

$$A \sin \theta = S \sin (\theta/2) = W/2 \quad (22)$$

From Fig. 4:

$$X = S \cos (\theta/2) + A - A \cos \theta \quad (23)$$

Substituting Eq (22) into Eq (23),

$$X = \frac{W \cos (\theta/2)}{2 \sin (\theta/2)}$$

$$+ \frac{W}{2 \sin \theta} - \frac{W \cos \theta}{2 \sin \theta} \quad (24)$$

Converting to half-angle functions and simplifying,

$$X = W/[2 \sin (\theta/2) \cos (\theta/2)] \quad (25)$$

Using the trigonometric identity,

$$\sin \theta = 2 \sin (\theta/2) \cos (\theta/2) \quad (26)$$

Eq (25) becomes:

$$X = W/\sin \theta$$

or

$$\sin \theta = W/X \quad (27)$$

Solving for θ permits determination

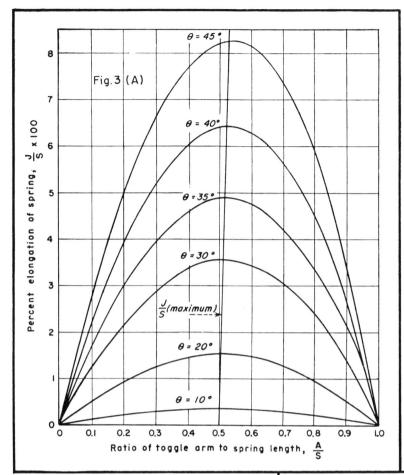

Fig. 3—Design charts for evaluating toggle arm and spring length for maximum spring elongation. Chart (B) is an extension of chart (A).

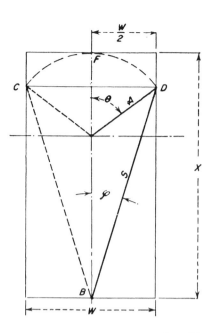

Fig. 4—Designing a toggle to lie within space boundaries W and X. It can be shown that for maximum snap-action, sin $\Theta = X/W$.

of ratios A/S and J/S from the charts in Fig. 3 when using the J/S(max) line. The values of S and A can then be obtained from Eq (22).

It can be seen from Fig. 3 that $\theta = 90$ deg results in maximum snap-action. Substitution of the sin of 90 deg in Eq (27) results in $W = X$; in other words, the most efficient space configuration for a toggle is a square.

EDITOR'S NOTE—In addition to the type discussed here, the term "toggle" is applied to a mechanism containing two links which line up in a straight line at one point of their motion giving a high mechanical advantage. A two-page mechanism spread, "Toggle Linkage Applications in Different Mechanisms" Thomas P. Goodman, appeared on p. 212, describing applications of this type to obtain: (1) high mechanical advantage, (2) high velocity ratio (3) variable mechanical advantage.

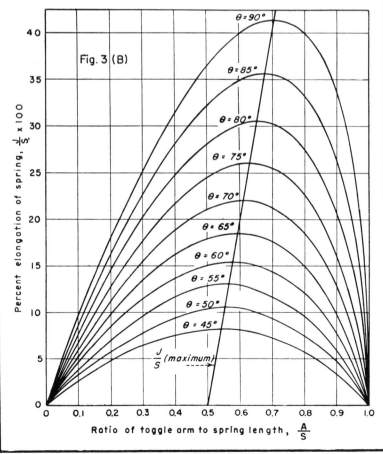

FEEDER MECHANISMS—
angular motions

*How to use four-bar linkages to
generate continuous or intermittant angular motions
required by feeder mechanisms.*

H SCHAEDLER AND G MARX

In feeder mechanisms it is often necessary to synchronize two sets of angular motions. A four-bar linkage offers one way. For example, in Fig 1 two angular motions, ϕ_{12} and ϕ_{13}, must synchronize with two others, ψ_{12} and ψ_{13}, about the given pivot points A_o and B_o and the given crank length A_oA. This means that crank length B_oB must be of such length that the resulting four-bar linkage will coordinate angular motions ϕ_{12} and ϕ_{13} with ψ_{12} and ψ_{13}. Procedure is:

1. Obtain point A'_2 by revolving A_2 about B_o through angle $-\psi_{12}$ but in the opposite direction.

2. Obtain point A'_3 similarly by revolving A_3 about B_o through angle $-\psi_{13}$.

3. Draw lines $A_1A'_2$ and $A_1A'_3$ and the perpendicular bisectors of the lines which intersect at desired point B_1.

4. The quadrilateral $A_oA_1B_1B_o$ represents the four-bar linkage that will produce the required relationship between the angles ϕ_{12}, ϕ_{13}, and ψ_{12}, ψ_{13}.

Three angles with four relative positions can be synchronized in a similar way. In Fig 2 it is desired to synchronize angles ϕ_{12}, ϕ_{13}, and ϕ_{14} with corresponding angles ψ_{12}, ψ_{13}, and ψ_{14}, using freely chosen pivot points A_o and B_o. In this case crank length A_oA as well as B_oB is to be determined, and the procedure is:

1. Locate pivot points A_o and B_o on a line that bisects angle $A_3A_oA_4$, the length A_oB_o being arbitrary.

2. Measure off $\frac{1}{2}$ of angle $B_3B_oB_4$ and with this angle draw B_oA_4 which establishes crank length A_oA at intersection of A_oA_4. This also establishes points A_3, A_2 and A_1.

3. With B_o as center and B_oA_4 as radius mark off angles $-\psi_{14}$, $-\psi_{13}$, $-\psi_{12}$, the negative sign indicating they are in opposite sense to ψ_{14}, ψ_{13} and ψ_{12}. This establishes points A'_2, A'_3 and A'_4, but here A'_3 and A'_4 coincide because of symmetry of A_3 and A_4 about A_oB_o.

4. Draw lines $A_1A'_2$ and $A_1A'_4$, and the perpendicular bisectors of these lines, which intersect at the desired point B_1.

5. The quadrilateral $A_oA_1B_oB_1$ represents the four-bar linkage that will produce the required relationship between the angles ϕ_{12}, ϕ_{13}, ϕ_{14} and ψ_{12}, ψ_{13}, ψ_{14}.

The illustrations show how these angles must be coordinated within the given space. In Fig 3(A) input angles of crank must be coordinated with output angles of forked escapement. In (B) input angles of crank are coordi-

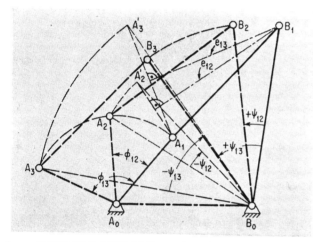

1 . . . FOUR-BAR LINKAGE synchronizes 2 angular movements, ϕ_{12} and ϕ_{13}, with ψ_{12} and ψ_{13}.

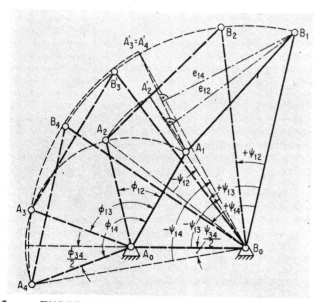

2 . . . THREE ANGULAR POSITIONS, ϕ_{12}, ϕ_{13}, ϕ_{14}, are synchronized by four-bar linkage here with ψ_{12}, ψ_{13}, and ψ_{14}.

nated with output angles of tilting hopper. In (C) input angles of crank are coordinated with output angles of segment. In (D) a box on a conveyor is tilted 90° by output crank, which is actuated by input crank

through coupler. Other mechanisms shown similarly coordinate input and output angles—some have dwell periods between the cycles, others give linear output with dwell periods.

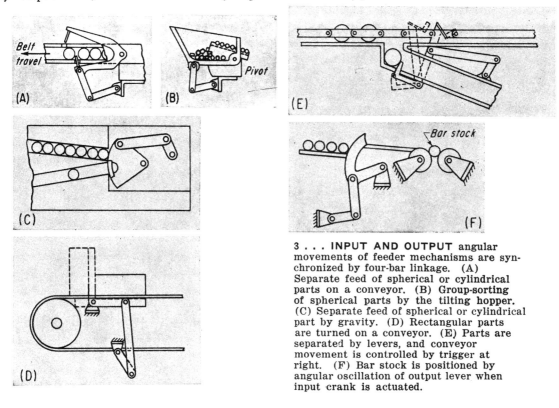

3 . . . INPUT AND OUTPUT angular movements of feeder mechanisms are synchronized by four-bar linkage. (A) Separate feed of spherical or cylindrical parts on a conveyor. (B) Group-sorting of spherical parts by the tilting hopper. (C) Separate feed of spherical or cylindrical part by gravity. (D) Rectangular parts are turned on a conveyor. (E) Parts are separated by levers, and conveyor movement is controlled by trigger at right. (F) Bar stock is positioned by angular oscillation of output lever when input crank is actuated.

FEEDER MECHANISMS—
curvilinear motions

Four-bar linkages can be combined into 6, 8 or more linkages for the feeder mechanisms in film cameras, automatic lathes, farm machinery, and torch cutters.

KURT HAIN

When feeder mechanisms require complex curvilinear motions, it may be necessary to use compound linkages of more than four links. However, four-bar linkages can be synthesized to produce curvilinear motions of various degrees of complexity, and all possibilities of four-bar linkages should be considered before complex types.

For example, a camera film-advancing mechanism, Fig 1, has a simple four-bar linkage with coupler point d which generates a curvilinear and straight-line motion a resembling a "D". Another more complex curvilinear motion, Fig 2, is also generated by a coupler point E of a four-bar linkage which controls an automatic profile cutter. Four-bar linkages can be used to generate a number of curvilinear motions as in Fig 3. Here the points of the coupler prongs, g_1, g_2, and g_3 on coupler b, and g_4 and g_5 on coupler e, are chosen in such a way that their motions result in the desired progressive feeding of straw into a press (manufactured by Gebr. Welger, Wolfenbuttel, Germany).

A similar feeding and elevating device (designed by Moertl in Germany) is shown in Fig 4. The rotating drive crank a moves coupler b and swinging lever c, which actuates the guiding arm f through the link e. The bar h carries the prone fingers g_1 through g_7. They generate coupler curves a_1 through a_7.

As another practical example, consider the torch-cutting machine in Fig 5(A) which is designed to cut sheetmetal along a curvilinear path a. Here the points A_0 and B_0 are fixed in the machine and the lever A_0A_1 is of adjustable length to suit different curvilinear paths a desired.

The length B_0B_1 is also fixed. The problem is to find the length of the levers A_1B_1 and E_1B_1 in the four-bar linkage to give the desired path a which is to be traced by the coupler point E on which the cutting torch is mounted.

The graphical solution for this problem as shown in (B) requires selection of the points A_1 and E_1 in such manner that the distances A_1E_1 to A_8E_8 are equal and the points

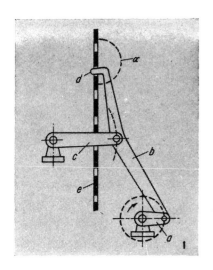

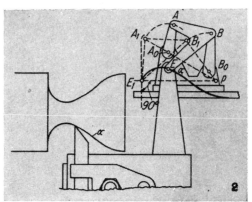

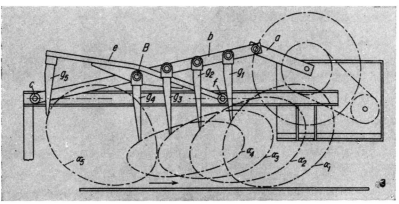

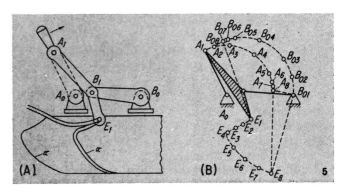

E_1 to E_8 lie on the desired coupler curved a. In this case only the points E_4 to E_8 represent the desired profile to be cut. The correct selection of points A_1 and E_1 depends upon making the following triangles congruent:

$$\Delta E_2 A_2 B_{01} = \Delta E_1 A_1 B_{02}$$
$$\Delta E_3 A_3 B_{01} = \Delta E_1 A_1 B_{03}$$
$$\Delta E_8 A_8 B_{01} = \Delta E_1 A_1 B_{08}$$

and so on until $E_8 A_8 B_{01} = E_1 A_1 B_{08}$. At the same time all points A_1 to A_8 must lie on the arc having A_0 as center, and all the points B_{01} to B_{08} must lie on the arc having B_1 as center.

Synthesis of an 8-bar linkage

Let us now design a linkage with 8 precision points, as shown in Fig 6. In this mechanism the curvilinear motion of one 4-bar linkage is coordinated with angular oscillation of a second four-bar linkage. The first four-bar linkage consists of AA_0BB_0 with coupler point E which

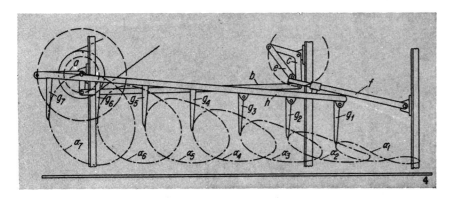

generates γ with 8 precision points E_1 through E_8 and drives a second four-bar linkage HH_0GG_0. Coupler point F generates curve δ with precision points F_1 through F_8. The coupler points F_2, F_4, F_6, F_8 are coincident, because straight links GG_0 and GH are in line with one another in these coupler positions. This is what permits HH_0 to oscillate despite the continuous motion of the coupler point F. The coupler points F_1 coincident with F_5, and F_3 coincident with F_7, have been chosen in such a way that F_1 is the center of a circle k_1 and F_3 is the center of a circle k_3. These circles are tangent to coupler curve γ at E_1, and E_5, E_3, and E_7; and they indicate the limiting positions of the second four-bar linkage HH_0GG_0.

The limiting angular oscillation of HH_0, which is one of the requirements of this mechanism, is represented by positions H_0H_1 and H_0H_3. It oscillates four times for each revolution of the input crank AA_0, and the positions H_1 to H_8 correspond to input crank positions A_1 to A_8.

Synthesis of a compound linkage with dwell periods and coordinated intermittent motion is shown in Fig 7. The four-bar linkage AA_0BB_0 generates an approximately triangular curve with coupler point E, with six precision points E_1 through E_6. A linkage to do this is not unusual and can be readily proportioned from known methods of four-bar linkage synthesis. However, the linkage incorporates dwell periods which are used to produce coordinated intermittent motion by means of a second four-bar linkage FF_0HB_0. Here the tangent arc k_{12}, k_{34} and k_{56} are drawn with EF as radius from centers F_{12}, F_{34} and F_{56}, and these centers establish the circle with F_0 as the center and pivot point for the second four-bar linkage. Each tangent arc causes a dwell of the link FF_0, while AA_0 rotates continuously. Thus the link FF_0, with three rest periods in one revolution, can produce intermittent curvilinear motion in the second four-bar linkage FF_0HB_0. In laying out the center F_0 it must be so selected that the angle EFF_0 deviates only slightly from 90° because this will minimize the required torque that is to be applied at E. The length of B_0H can be made to suit, and the rest periods at H_{34}, H_{12} and H_{56} will correspond to crank angles ϕ_{34}, ϕ_{12} and ϕ_{56}.

A compound linkage can also produce a 360° oscillating motion with a dwell period as in Fig 8. The two four-bar linkages are AA_0BB_0 and BB_0FF_0, and the output coupler curve γ is traversed only through segment E_1E_2. The oscillating motion is produced by lever HH_0 connected to the coupler point by EH. The fixed point H_0 is located within the loop of the coupler curve γ. The dwell occurs at point H_3 which is the center of circular arc k tangent to the coupler curve γ during the desired dwell period, and in this case the dwell is made to occur in the middle of the 360° oscillation. The coincident positions H_1 and H_2 indicate the limiting positions of the link HH_0, and correspond to the positions E_1 and E_2 of the coupler point.

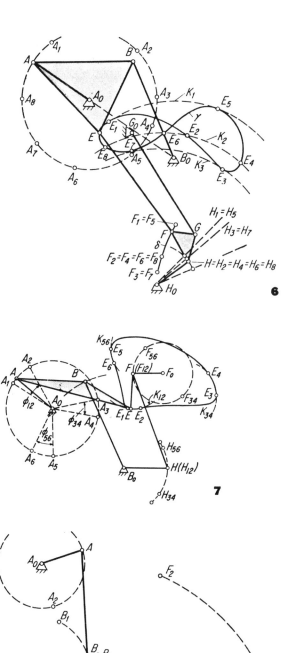

6

7

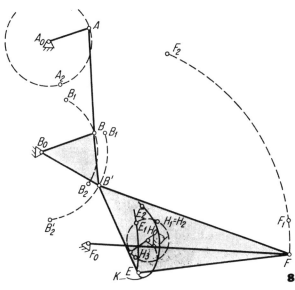

8

Roberts' law helps you find alternate four-bar linkages

R. T. HINKLE

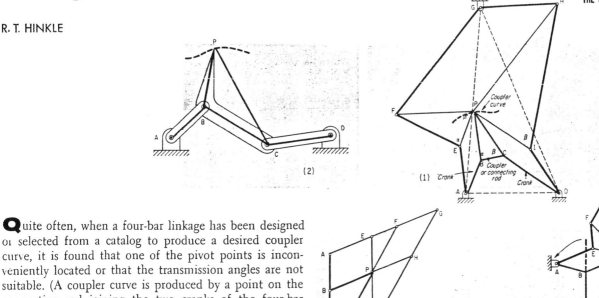

Quite often, when a four-bar linkage has been designed or selected from a catalog to produce a desired coupler curve, it is found that one of the pivot points is inconveniently located or that the transmission angles are not suitable. (A coupler curve is produced by a point on the connecting rod joining the two cranks of the four-bar linkage). According to Roberts' Law there are at least two other four-bar linkages that will generate the same coupler curve. One of these linkages may be more suitable for the application.

Roberts' Law is not widely known and does not seem to be mentioned in any English-language textbook. The law states that the two alternate linkages are related to the first by a series of similar triangles. This leads to graphical solutions of which three examples are shown below—the first involving similar triangles, the second being a more convenient method developed by the author, and the third illustrating solution of a special case where the coupler point lies along the connecting rod.

Method of Similar Triangles

Four-bar linkage ABCD in Fig. 1 uses point P, which is actually an extension of the connecting rod BC, to produce desired curve. Point E is found by constructing EP parallel to AB, and EA parallel to PB. Then triangle EFP is constructed similar to triangle BPC. This involves laying out angle α and β.

Point H is found similarly and point G is located by drawing GH parallel to FP and GF parallel to HP.

The two alternate linkages to ABCD are GFEA and GHID. All use point P, to produce the desired curve; and given any one of the three, the other two can be determined.

The Author's Method

With the similar-triangle method described above, slight errors in constructing the proper angles lead to large errors in link dimensions. The construction of angles can be avoided by laying off the link lengths along a straight line.

Thus, linkage ABCD in Fig. 2 is laid off as a straight line from A to D in Fig. 3. Included in the transfer is point P. Points EFGHI are quickly found by either ex-

Fig. 1—Method of similar triangles

Figs. 2, 3, 4—Author's method, step by step

Fig. 5—Special case shows simplicity of applying Robert's Law

tending the original lines or constructing parallel lines. Fig. 3, which now has all the correct dimensions of all the links, is placed under a sheet of tracing paper and, with the aid of a compass, links AB and CD are rotated (see Fig. 4) so that linkage ABCD is identical with that in Fig. 2. Links PEF and PHI are rotated parallel to AB and CD, respectively. Completion of the parallelogram gives the two alternate linkages AEFG and GHID.

Special Case

It is not uncommon for the coupler point P to lie on a line through BC, Fig. 5. Links EA, EP and ID are quickly found by constructing the appropriate parallel lines. Point G is located by using the proportion: $CB:BP = DA:AG$. Points H and F are then located by drawing lines parallel to AB and CD.

RATCHET LAYOUT ANALYZED

EMERY E. ROSSNER

The ratchet wheel is widely used in machinery, mainly to transmit intermittent motion or to allow shaft rotation in one direction only. Ratchet-wheel teeth can be either on the perimeter of a disc or on the inner edge of a ring.

The pawl, which engages the ratchet teeth, is a beam pivoted at one end; the other end is shaped to fit the ratchet-tooth flank. Usually a spring or counterweight maintains constant contact between wheel and pawl.

It is desirable in most designs to keep the spring force low. It should be just enough to overcome the separation forces—inertia, weight and pivot friction. Excess spring force should not be relied on to bring about and maintain pawl engagement against the load.

To insure that the pawl is automatically pulled in and kept in engagement independently of the spring, a properly layed out tooth flank is necessary.

The requirement for self-engagement is

$$Pc + M > \mu Pb + P \sqrt{(1 + \mu^2)} \, \mu_1 r_1$$

Neglecting weight and pivot friction

$$Pc > \mu Pb$$

but $c/b = r/a = \tan \phi$, and since $\tan \phi$ is approximately equal to $\sin \phi$

$$c/b = r/R$$

Substituting in term (1)

$$r R > \mu$$

For steel on steel, dry, $\mu = 0.15$. Therefore, using

$$r/R = 0.20 \text{ to } 0.25$$

the margin of safety is large; the pawl will slide into engagement easily. For internal teeth with ϕ of $30°$, c/b is $\tan 30°$ or 0.577 which is larger than μ, and the teeth are therefore self engaging.

When laying out the ratchet wheel and pawl, locate *points* O, A and O_1 on the same circle. AO and AO_1 will then be perpendicular to one another; this will

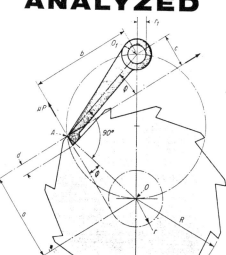

Pawl in compression . . .
has tooth pressure *P* and weight of pawl producing a moment that tends to engage pawl. Friction-force μP and pivot friction tend to oppose pawl engagement.

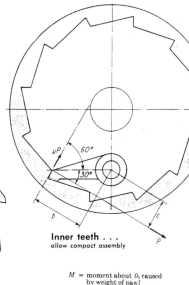

Inner teeth . . .
allow compact assembly

M = moment about O_1 caused by weight of pawl
$O, \, O_1$ = ratchet and pawl pivot centers respectively
P = tooth pressure = wheel torque/a
$P\sqrt{(1+\mu^2)}$ = load on pivot pin
μ, μ_1 = friction coefficients
Other symbols as defined in diagrams

insure that the smallest forces are acting on the system.

Ratchet and pawl dimensions are governed by design sizes and stress. If the tooth, and thus pitch, must be larger than required in order to be strong enough, a multiple pawl arrangement can be used. The pawls can be arranged so that one of them will engage the ratchet after a rotation of less than the pitch.

A fine feed can be obtained by placing a number of pawls side by side, with the corresponding ratchet wheels uniformly displaced and interconnected.

Slider-crank mechanism

MERL D. CREECH

THE slider crank—an efficient mechanism for changing reciprocating motion to rotary—is widely used in engines, pumps, automatic machinery, and machine tools.

The equations developed here for finding such factors are in a more streamlined form than is generally available.

SYMBOLS

L = length of connecting rod
R = crank length; radius of crank circle
x = distance from center of crankshaft, A, to wrist pin, C
x' = slider velocity (linear velocity of point C)
x'' = slider acceleration
θ = crank angle measured from dead center (when slider is fully extended)
ϕ = angular position of connecting rod; $\phi = 0$ when $\theta = 0$
ϕ' = connecting-rod angular velocity = $d\phi/dt$
ϕ'' = connecting-rod angular acceleration = $d^2\phi/dt^2$
ω = constant crank angle velocity

Displacement of slider
$x = L \cos \phi + R \cos \theta$

Also:

$$\cos \phi = \left[1 - \left(\frac{R}{L} \right)^2 \sin^2 \theta \right]^{1/2}$$

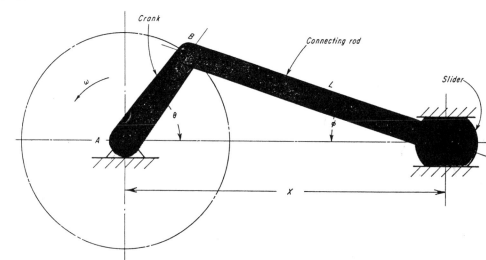

Angular velocity of the connecting rod

$$\phi' = \omega \left[\frac{(R/L) \cos \theta}{[1 - (R/L)^2 \sin^2 \theta]^{1/2}} \right]$$

Linear velocity of the piston

$$\frac{x'}{L} = -\omega \left[1 + \frac{\phi'}{\omega} \right] \left(\frac{R}{L} \right) \sin \theta$$

Angular acceleration of the connecting rod

$$\phi'' = \frac{\omega^2 (R/L) \sin \theta [(R/L)^2 - 1]}{[1 - (R/L)^2 \sin^2 \theta]^{3/2}}$$

Slider acceleration

$$\frac{x''}{L} = -\omega^2 \left(\frac{R}{L} \right) \left[\cos \theta + \frac{\phi''}{\omega^2} \sin \theta + \frac{\phi'}{\omega} \cos \theta \right]$$

INDEX

About the Author

Nicholas P. Chironis is a consulting engineer who practices in St. James, New York. He previously worked as a mechanical engineer with IBM, Mergenthaler Linotype, and Allied Processes. Mr. Chironis also headed the product design department at Grant Pully and Hardware, and taught courses at Cooper Union's School of Engineering. He was long associated with *Product Engineering* magazine as a specialist in mechanical components and design analysis. Mr. Chironis received both his bachelor's and master's degrees in mechanical engineering from the Polytechnic University located in New York City.